Contemporary Trends and Issues in Science Education

Volume 57

The book series Contemporary Trends and Issues in Science Education provides a forum for innovative trends and issues impacting science education. Scholarship that focuses on advancing new visions, understanding, and is at the forefront of the field is found in this series. Authoritative works based on empirical research and/or conceptual theory from disciplines including historical, philosophical, psychological and sociological traditions are represented here. Our goal is to advance the field of science education by testing and pushing the prevailing sociocultural norms about teaching, learning, research and policy. Book proposals for this series may be submitted to the Publishing Editor: Claudia Acuna E-mail: Claudia.Acuna@springer.com

Xiufeng Liu • William J. Boone
Editors

Advances in Applications of Rasch Measurement in Science Education

Springer

Editors
Xiufeng Liu
Department of Learning & Instruction
Graduate School of Education
University at Buffalo
Buffalo, NY, USA

William J. Boone
Department of Educational Psychology,
Program in Learning Sciences and Human Development
Miami University
Oxford, OH, USA

ISSN 1878-0482 ISSN 1878-0784 (electronic)
Contemporary Trends and Issues in Science Education
ISBN 978-3-031-28775-6 ISBN 978-3-031-28776-3 (eBook)
https://doi.org/10.1007/978-3-031-28776-3

This Springer imprint is published by the registered company Springer Nature Switzerland AG
The registered company address is: Gewerbestrasse 11, 6330 Cham, Switzerland

Foreword

At the end of the last century, I applied Rasch measurement for the first time with my research group. We were able to apply a TIMSS 1995 test in a study of physics teaching in 8th grades before and after an intervention. For us, it was fascinating to know the Rasch item difficulty for each item already when selecting the tasks, the standard error of the difficulty estimate, and an index of the goodness-of-fit of the item to the Rasch model. In addition, the fact that it was possible at all to select items from another test instrument and thus adapt them to our own needs and still locate our sample in the sample of 40 countries convinced us to go ahead with Rasch.

Until now, Rasch became an important method in science education research for the development of tasks in all sciences and for the clear and meaningful analysis of test data. An important reason for this is that Rasch, unlike classical statistical methods, is connectable to research methods in the natural sciences themselves. A fundamental affinity is that empirical research in the natural sciences and in measuring with Rasch require theoretical models that need empirical evidence, and that the models and the results in both cases are expected to be prescriptive.

As for example in physics, Rasch measurement assumes that the applied theoretical model is correct and that the measured data must fit the model. Therefore, for all tests or surveys, it must first be clarified which theoretical model or construct is to be surveyed. In physics education, for example, we may need a theoretical model of knowledge to measure knowledge of electrodynamics or a model of ability or competence to measure the ability/competence to apply scientific working methods. The test constructed according to these models should represent the models. The Rasch measurement then generates prescriptive measures of item-difficulties and person-abilities, which are invariant, and of equal distance.

Another important aspect of Rasch analysis is the possibility of a content oriented analysis by means of differential item functioning (DIF), combined with task difficulty and student ability. DIF allows for a substantive discussion of the test results before and after an intervention or when comparing samples of different educational systems with regard to their effect. The change in the score of difficulty of individual tasks provides information about the change in student ability in terms of content.

In addition, an essential feature of Rasch-scaled tests is the possibility to indicate both the difficulty of the tasks and the ability of the students on the same scale and to select tasks from a Rasch-scaled task collection, to combine them for a new test, e.g., about content just taught in an intervention, without the test losing its validity. Since we often deal with occasional samples in science education research, we can still classify the studied groups into larger samples and, vice versa, generalize the test results for the larger sample.

Furthermore, the measurements with Rasch are not limited to the measurement of subject content. Motivation, self-efficacy or other psychological models can also be empirically tested with Rasch. To measure those models, it is often necessary to distinguish between not only the degrees wrong/right and disagreement/agreement. As e.g. with Likert scales, even intermediate levels have to be considered, therefore there are more than two possible answers. For tests, research questions, exam questions, etc. with multi-level answer categories, which are additionally based on a certain order, the Partial Credit Model is provided for the analysis, which is also called ordinal Rasch model. Like the Rasch model, it assumes that a latent variable common to all items can be inferred from the answers given.

I am very pleased that this new book exists, in which all the essential topics of Rasch analysis are covered in great depth and care, and I sincerely hope that it will convince more young and also established researchers that the application of the Rasch model with all its facets can improve science education research.

Professor Emeritus, Faculty of Physics, University of Duisburg-Essen, Essen, Germany

Hans E. Fischer

Contents

Chapter 1
Introduction to Advances in Applications of Rasch Measurement in Science Education

Xiufeng Liu and William J. Boone

Abstract This chapter provides an overview of applications of Rasch measurement in science education. This chapter discusses issues and best practices in applications of Rasch measurement related to difference between one-parameter Item Response Theory (IRT) models and Rasch models, role of theory in Rasch measurement, sample size, fit statistics, local independence and unidimensionality, Wright maps, linking measures, and using Rasch scale measures for statistical analysis. Finally, the chapter introduces the other 17 chapters included in this book.

Rasch measurement is a domain of psychometrics that concerns the theories and applications of Rasch models. Originally conceptualized by Georg Rasch in the 1960s, Rasch measurement has been applied in numerous fields such as education, psychology, and medicine, to name a few. As the first comprehensive introduction of Rasch measurement to science education research, Liu and Boone (2006) edited a volume entitled *Applications of Rasch Measurement in Science Education*. In that volume, science education researchers from Australia, Germany, Hong Kong-China and the US introduced basic concepts of Rasch measurement and reported varied applications of Rasch measurement. Some of those applications were utilizing Rasch techniques to develop items and validate instruments for assessing science conceptual understanding and affective variables.

Since 2006 many advances have taken place not only in new Rasch models but also technologies (e.g., open-source software) for conducting Rasch analysis. As a result of these advances, there have been increased uses of Rasch measurement in science education research. For example, leading journals in science education,

X. Liu (✉)
Department of Learning & Instruction, Graduate School of Education, University at Buffalo, Buffalo, NY, USA
e-mail: xliu5@buffalo.edu

W. J. Boone
Department of Educational Psychology, Program in Learning Sciences and Human Development, Miami University, Oxford, OH, USA

X. Liu, W. J. Boone (eds.), *Advances in Applications of Rasch Measurement in Science Education*, Contemporary Trends and Issues in Science Education 57,
https://doi.org/10.1007/978-3-031-28776-3_1

Journal of Research in Science Teaching, *Science Education* and *International Journal of Science Education*, now regularly publish studies that involve applications of Rasch measurement. Despite increasing visibility, the application of Rasch measurement in science education research remains limited to a modest number of researchers. Further, published studies involving applications of Rasch measurement adopt various assumptions of Rasch measurement and use varied criteria and conventions in reporting and interpreting results, which results in inconsistencies and even contradictions. For example, published studies often use IRT and Rasch interchangeably disregarding differences in key assumptions; there is currently no agreement on which fit statistics (infit, outfit, standardized infit, and standardized outfit) and what acceptable fit criteria should be used and reported. Further, there is no consistency among published studies on if and when unidimensionality or local independence should be examined and how. Besides such technical issues as noted above, there also can be misconceptions/confusions regarding the relationship between theories and Rasch models. We believe that a lack of consensus and conventions may hinder further progress in applying and communicating applications of Rasch measurement in science education research.

Over the years, we have been writing about applications of Rasch measurement (e.g., Boone & Staver, 2020; Boone et al., 2014; Liu 2010/2020). We have also been teaching courses on applications of Rasch measurement to graduate students and reviewing for and editing journals that publish applications of Rasch measurement in science education. We believe there are a core set of issues to consider when applying Rasch measurement in the field of science education. Below we discuss these issues. These issues are of importance regardless of the sophistication of the Rasch analysis conducted.

1.1 Item Response Theory Models vs. Rasch Models

One statement often made in articles is that the Rasch model is a one-parameter item response theory (IRT) model. While this statement is correct mathematically (the formula expressing both models is identical), from a measurement perspective, there are major differences between the two models. For an IRT approach, which model is the best model depends on the data set. When there is not a good model-data-fit for one IRT model, a different IRT model is applied. In general, the more parameters (and dimensions) a model has, the better the model-data-fit. This is expected from a statistical point view because the more variables (in this case parameters) that are included in a model, the more variance in data can be explained. However, the Rasch perspective is that the Rasch model defines the construct (where items and respondents are located along the same single construct), and when there is not a good model-data-fit, the problem is not the Rasch model. The model is not altered, rather what needs to be investigated are issues such as the items used to define the construct. For example, for a science attitudinal survey consisting of a set of Likert type items, the rating scale Rasch model can be applied. If there is not optimal model-data fit, then the Rasch model is not altered. Rather an attempt for better model-data-fit is attempted through steps such as the revision of items, the removal of items, or the addition of new items. Such revisions may take multiple iterations. The two different approaches (one for IRT, one for Rasch) have been called two paradigms of measurement (Andrich, 2004). IRT models are altered to fit the data, the Rasch perspective is that the data should fit the model.

Why do those using Rasch measurement insist on data fitting the Rasch model, not choosing a model (e.g., the 2-parameter IRT model, the 3-parameter IRT model) to fit the data? It can be mathematically demonstrated that in Rasch models the difference in log odds (i.e., logits) of responding to an item in a positive way (e.g., correctly for a multiple-choice question or agreeing to an attitudinal survey statement) is only determined by the difference in difficulty measures between items for a given individual, or by the difference in ability measures between examinees for a given item (Liu, 2020, p. 36). Only when both item and ability measures are linear and for the same construct, they can be compared directly, and Wright maps are the direct result of this dual linearity of item and ability measures. This dual linearity is the foundation of many unique and innovative applications of Rasch measurement such as standard setting, learning progression research, etc. In IRT, item measures and ability measures are not directly comparable, because ability measures depend on more than one parameter (e.g., both item difficulty and discrimination in the 2-parameter IRT model).

Another important reason for the use of Rasch is the invariance properties of measures. As we know for any measurement to be objective, measures of items should not be dependent upon the sample used to calibrate them (person invariant item calibration), and measures of subjects should not dependent upon the items used to estimate them (item invariant person estimation). Engelhard and Wang (2021) show that only the Rasch model, not the two-parameter IRT model for example, can result in person and item invariant measures. The invariant properties are necessary conditions for objective measurement and Rasch measurement is objective measurement. Thus, if one is to conduct objective measurement, which is what one should always conduct, then it is the Rasch model and not the IRT models that should be used.

There are other reasons why fitting data to Rasch models is a better approach (Andrich & Marais, 2019) than choosing the model to fit a data set (the IRT approach). Because Rasch models represent idealized measurement scenarios about the interaction between items and examinees, Rasch models will more likely indicate more items for mis-fitting, which create more opportunities for improving items and producing invariant measures. As the result, Rasch models will help produce higher quality items and measures, which is desirable. Of course, having high quality items and measures means that one has high quality instruments.

Because of the above reasons, it is best not to refer Rasch models as one-parameter IRT models. This is not a mathematical distinction made, but rather a philosophical perspective. When applying Rasch models, we are committed to producing measures that are linear and invariant through constructing the best possible items and instruments. Rasch models are based upon the idea that the

Rasch model is a definition of what it means to measure. When a Rasch model is utilized, we see if the data fit the model. This makes for a more robust and rigorous measurement instrument.

1.2 Theory of Construct

One basic requirement of measurement in social sciences is that the construct to be measured is clearly defined. For Rasch measurement, this requirement goes even further; it requires that instrument developers conceptualize the construct as a linear variable defined by different levels of the attribute. Wilson (2005) calls this conceptualization a construct map. The most important features of a construct map include "(a) a coherent and substantive definition for the content of the construct; and (b) an idea that the construct is composed of an underlying continuum – this can be manifested in two ways – an ordering of the respondents and/or an ordering of item responses" (Wilson, 2005, p. 26). Essentially, a construct map is a hypothesized Wright map, and the process of applying Rasch measurement to develop a measurement instrument is to test this hypothesis, which means a construct is conceptualized and then a Rasch analysis is used to evaluate this conceptualization, for example does item ordering on a Wright Map match predictions?

The above requirement is one aspect that differentiates Rasch measurement and Classical Test Theory (CTT) and IRT. When using CTT and IRT, there is no such expectation for a hypothesized linear variable, although there may well be an implicit assumption that the construct is unidimensional. The advantage for such an explicit hypothesis for different levels of the variable is that items can be purposefully constructed for different levels of abilities. Thus, unlike CTT in which items are considered parallel (i.e., equivalent to each other), and IRT in which item difficulties are known after data analysis, Rasch item difficulties are hypothesized a priori, i.e., we predict the difficulty of items before an analysis based upon theory. This approach makes using Rasch measurement to develop measurement instruments a theory-driven process.

Purposefully constructing items with a range of difficulties based on a theory of the construct is critical for constructing a scale. Ben Wright when teaching his Rasch courses at the University of Chicago would draw a vertical line on the black board. He would stress that if one were creating a math test, one should have a theory as to the construct one was wishing to measure. What would be examples of tasks/topics that would be at the easy end of the variable, which would be example of tasks/topics at the middle part of the variable, and what tasks/topics would be at the difficult end? If one could not supply such information using theory, then it meant that one did not understand what one was measuring, and one should stop. One aim of applying Rasch measurement is that "we want to find out whether or not calibrated items spread out in a way that shows a coherent and meaningful direction." (p. 83, Wright & Stone, 1979).

1.3 Sample Size for Rasch Analysis

A common question about Rasch measurement is "what sample size do I need to do a Rasch analysis"? Related to this question is a view that Rasch analysis requires large sample sizes. There is no one hard and fast rule about the required sample size for Rasch analysis, because "sample size is influenced by a number of factors (e.g., the distribution of items along a trait and the distribution of persons along the trait)" (Boone et al., 2014, p. 365). Even if you collect data from many students, if your items are not distributed along the variable of interest, you could have a large amount of person measurement error. Likewise, if your respondents do not vary in their ability levels, more than likely there will be a limited number of items to define the measures of each respondent. And in this situation, you will have a large item measurement error. Thus, the spread and the relationship between the location of person measures on the trait and the location of item measures on the trait are important to consider when determining the required sample size for Rasch analysis. The more spread and aligned the item and person measures are, the smaller the required sample size is, and vice versa.

When conducting a Rasch analysis it is essential to consider the acceptable level of measurement errors. Wright (1977) shows that for a typical objective test (i.e., dichotomously scored items) with a raw score between 20% and 80% correct, i.e., a 5-logit difference range, the minimal sample size is

$$N = 6/SE^2$$

where SE is the standard error of Rasch measures. For example, if we want our SE to be smaller than 0.35, which is typically adequate for pilot-testing, then the required minimal sample size is 50; for SE to be smaller than 0.25, which is typically considered acceptable for low-stakes testing situations, the required minimal sample size is 96; for SE to be smaller than 0.15, which is typically considered excellent for most testing situations, the required minimal sample size is 267. In response to common misconceptions about the minimal sample size for Rasch modeling, Wright and Tennant (1996) state that the misconception of a large required minimal sample size being needed for Rasch modeling is based on a misconception that Rasch parameter estimation requires a normally distributed sample. In fact, Rasch modeling escapes from such a requirement by focusing on the separation between item parameters and person parameters. Another possible reason for this misconception is that Rasch analysis has been conducted on large scale assessment data. However, Rasch analysis also has been applied to small data sets.

Another issue impacting sample size has to do with the nature of data and types of Rasch models (Linacre, 1994). For survey data that are analyzed using the rating scale Rasch model, because there are many response options (such as Strongly Agree, Agree, Disagree, Strongly Disagree), i.e., test information, to be used to determine the rating scale structure of a scale, only a minimum of 10 observations per response category is required. However, when the partial credit model is used, because each item is viewed as having its own unique rating scale structure, 100 responses per item may still be too few, thus a larger sample would be needed to allow the confident computation of, for example, the location of the thresholds.

Finally, we should also take into consideration the statistical analysis which is often applied to the Rasch measures of a study. For example, when you compare different groups of students in terms of their Rasch ability estimates, you would need sufficient subjects for each group in order for such statistical analyses as ANOVA to be conducted. Thus, the consideration of sample size for specific statistical analyses is also of importance for your Rasch analysis.

1.4 Uses of Fit Statistics

There are two types of common fit statistics—Outfit and Infit, and each can be standardized, which gives four fit statistics, i.e., Infit MNSQ, Infit ZSTD, Outfit MNSQ and Outfit ZSTD. The development of ZSTD was to address inherit limitations of MNSQ, that is "critical values for detecting misfit with this mean square depend on the number of persons and W_{ni}, so they will vary from item to item and sample to sample" (Smith et al., 1998), where W_{ni} is the variance of probability of an examinee responding to an item correctly (P_{ni}), which equals to P_{ni} $(1 - P_{ni})$.

Common questions about these fit statistics are: which of them should be used and what criteria should be followed to decide on acceptable fit? Infit statistics (Infit MNSQs and Infit ZSTDs) are weighted means by assigning more weights for those persons' responses close to the probability of 50/50, while Outfit statistics (Outfit MNSQs and Outfit ZSTDs) are arithmetic means of MNSQs and standardized MNSQs (ZSTDs) over all persons. Thus, Outfit statistics are more sensitive to extreme responses –outliers. For person fit, those same four statistics are applicable. It has been recommended that good model-data-fit has Infit and Outfit MNSQs within the range of 0.7–1.3 for multiple-choice items and 0.6–1.4 for rating scale items, and Infit and Outfit ZSTDs within the range of −2 to +2 for both multiple-choice and rating scale items (Bond & Fox, 2015, p. 273).

Researchers have used various acceptable ranges of fit statistics to decide on item fit. One common misconception about the above fit statistics is that only ZSTDs are sensitive to sample size, i.e., the bigger the sample is, the more likely ZSTDs will be statistically significant. In fact, both MNSQs and ZSTDs are sensitive to sample sizes. Increasing the sample size will move MNSQs toward the expected value of 1, which may result in under-detecting mis-fitting items; similarly, increasing the sample size will increase ZSTDs, which may result in over-detecting mis-fitting items. This sensitivity to sample size is not so much an issue for person fit statistics, because the number of items for most measurement instruments is not large (e.g., rarely over 100 items). Through a simulation study, Smith et al. (1998) found that when sample size was over 500, the inflation to type I error, i.e., falsely rejecting null hypothesis of MNSQs to be within 0.7–1.3, becomes significant. Interestingly, overall, MNSQs is more sensitive to sample size than ZSTDs. Similarly, Linacre

(2022) reports a simulation study that show that ZSTDs "are insensitive to misfit with less than 30 observations and overly sensitive to misfit when there are more than 300 observations" (p. 673).

Given the above findings, one recommended practice is to report all four statistics and examine them by taking the sample size into consideration. Specifically, when sample size is large (e.g., >300) or too small (e.g., <30), we rely more on MNSQs and pay particular attention to items with MNSQ fit statistics outside the acceptable range. In order to make sure that the significant ZSTDs were due to large sample size, we may randomly select 300 students to conduct Rasch analysis to compare item fit between two sample sizes. When sample size is ordinary (e.g., between 30 and 300), we examine all items with both MNSQs and ZSTDs fit statistics outside the acceptable ranges.

When sample size is large (e.g., >300) a correction to MNSQ criterion values may be made (Smith et al., 1998). The formulae for adjusting Infit MNSQ and Outfit MNSQ criteria are:

$$\text{Adjusted Infit MNSQ} = 1 + 2/\text{sqrt}(x)$$
$$\text{Adjusted Outfit MNSQ} = 1 + 6/\text{sqrt}(x)$$

Where x = sample size. For example, for a sample size of 500, the adjusted Infit MNSQ would be 1.09, and adjusted Outfit MNSQ would be 1.27, which translates to a new acceptable Infit MNSQ range to be 0.91–1.09 and acceptable Outfit MNSQ range to be 0.73–1.27. For a sample size of 800, the adjusted Infit MNSQ would be 1.07, and the adjusted Outfit MNSQ would be 1.21, which translates to a new acceptable Infit MNSQ range to be 0.93–1.07 and acceptable Outfit MNSQ range to be 0.79–1.21.

Linacre (2022) recommends that before examining fit statistics, negative point-measure or point-biserial correlations should be examined first. Then, the following principle may be followed when examining fit statistics: (a) examine outfit before infit, (b) examine MNSQs before ZSTDs; (c) examine high values before low or negative values, and (d) consider high MNSQs (or positive ZSTDs) to be a much greater threat to validity than low MNSQs (or negative ZSTDs) (p. 671).

Finally, keep in mind that fit statistics provide a quality check (Bond & Fox, 2015). That is, fit statistics flag/signal issues with items and respondents as well. It just is that most the time, it is the misfitting items which can mess things up as we often have so few items defining a construct. In order to know exactly what the issues with the possible misfiting items are, it is necessary to examine the response patterns of items and the content of items themselves. For example, upon examination one could identify a few respondents (maybe 1, maybe 2, maybe a few more) who have answered an item unexpectedly. But all other respondents generally match our Rasch model predictions. Once those unexpected respondents are identified, one could remove those respondents for the computation of fit statistics and item difficulties while still computing person measures for all the respondents. Of course, item examination may uncover issues with the items themselves, such as ambiguity in wording, excessive length, double-negativity, etc. In these cases, items will need to be revised, re-tested and re-analyzed for fit. No misfitting items should be removed without a detailed examination. Sometimes misfitting items may still be retained if they are extremely difficult or easy because such items are easily flagged as misfitting, simply due to a few unpredictable responses from few respondents. Retaining such items would help define the construct in terms of the ranges of item difficulties and student abilities.

1.5 Local Independence and Dimensionality

For any measurement instrument, more than one item is needed. "Each item needs to provide related but independent information, or relevant but not redundant information" (Andrich & Marais, 2019, p. 173). Such a measurement requirement is called local independence. Local independence states that "all variation among responses to an item is accounted for by the person parameter β, and therefore that for the same value of β, there is no further relationship among responses" (Andrich & Marais, 2019, p. 174). That is, in Rasch measurement, only the person measure β is the source of dependence among responses to items. Local independence is a foundation of Rasch measurement; meeting this requirement must be explicitly evaluated because other indications such as point-measure correlation and item and person fit are not sufficient for determining local independence (Bond & Fox, 2015).

The requirement of local independence can be violated in two ways (Andrich & Marais, 2019). First there may be person parameters other than β that impact the responses, which is called multidimensionality. A second potential violation of local independence is the case in which the response to one item may still be dependent on the response to another item after controlling for β. The first violation is the violation of unidimensionality, and the second violation is the violation of response independence. Multidimensionality and response dependence are related but distinct; they should be examined separately.

There are procedures for examining multidimensionality. Conventional factor analysis is not adequate, because it is based on a sample dependent structure of variance and covariance (Bond & Fox, 2015). Principal Components Analysis of Rasch Residuals (PCAR) has been specifically developed for examining the degree to which a data set deviates from unidimensionality. Specifically, in Winsteps, a PCAR analysis provides a variety of metrics and produces a diagram to help identify the severity of departure from unidimensionality and items that potentially measure additional constructs. Linacre (2022) provides useful advice on how to use these metrics, such as the eigenvalue of the first contrast in the residual is <2, the variance explained by the Rasch measures is large (e.g., $>40\%$) and the unexplained variance by the first contrast is small (e.g., $<5\%$), and the ratio between eigenvalue of the first contrast in the residual over the eigenvalue explained by Rasch measures is small (e.g., <0.1). Bond and Fox (2015) also recommend examination of the contrasting content of items with high factor loadings (beyond ±0.4) (p. 288). The examination of contrasting content of items is particularly important because it could be that those items are just representing different extremes of one dimension. Boone and Staver (2020) also suggest using cross plotting to examine dimensionality; for example, if one has a 20-item test, and there seems to be evidence that items 1–9 are one dimension and items 10–20 are another dimension. In this scenario, one can conduct two separate analyses with the two sets of items and cross plot the computed person measures. If the ordering of respondents in the two analyses are not that different, then that suggests one dimension. If the ordering of respondents changes greatly, then that might suggest more than one dimension.

The RUMM2030 (https://www.rummlab.com.au/) Rasch analysis software provides statistics to evaluate the magnitudes of potential additional dimensions as correlation coefficients between the additional dimensions and the targeted dimension. Similarly, ConQuest (Adams et al., 2020), a software package for both unidimensional and multidimensional Rasch analyses, computes the correlations between different dimensions. The R package TAM (Robitzsch et al., 2021) computes the DETECT statistic for dichotomous item responses and the polyDETECT statistic for polytomous item responses. Two other statistics, ASSI and RATIO can also be computed. The DETECT, ASSI and RATIO statistics can be interpreted according to the following classification scheme (Jang & Roussos, 2007; Zhang, 2007):

Strong multidimensionality	DETECT > 1.00	
Moderate multidimensionality	0.40 < DETECT < 1.00	
Weak multidimensionality	0.20 < DETECT < 0.40	
Essential unidimensionality	DETECT < 0.20	
Maximum value under simple structure	ASSI = 1	RATIO = 1
Essential deviation from unidimensionality	ASSI > 0.25	RATIO > 0.36
Essential unidimensionality	ASSI < 0.25	RATIO < 0.36

It is important to emphasize that unidimensionality must be defined by theory, not by data. When we examine dimensionality following the above procedures, we should always keep in mind how the construct is defined and how items have been written according to the defined construct. Also, unidimensionality is not a dichotomous property, i.e., either unidimensional or multidimensional, it is a degree. As Linacre (2022) states, "Unidimensionality is never perfect. It is always approximate. The Rasch model constructs from the data parameter estimates along the unidimensional latent variable that best concurs with the data. But, though the Rasch measures are always unidimensional and additive, their concurrence with the data is never perfect. Imperfection results from multidimensionality in the data and other causes of misfit." (p. 638–639). Multidimensionality always exists to a lesser or greater extent. Also, Rasch models can tolerate a certain degree of a violation of local independence (e.g., multidimensionality, item dependence) without significant negative effect to model-data-fit and invariance properties of measures, which is called robustness (Liu, 1993). The vital question is: "Is the multi-dimensionality in the data big enough to merit deleting certain items, dividing the items into separate tests, or constructing new tests, one for each dimension?" (Linacre, 2022, p. 639).

Response dependence can occur in a number of situations, for example, when items share a common statement or scenario, e.g., item bundles, which is the case for PISA and assessment of Next Generation Science Standard performance expectations. Another situation is repeated testing in longitudinal studies. In these situations, a different scoring mechanism, such as treating each item bundle as one item or a same subject at each repeated testing as a new subject may be implemented to avoid item dependence. Item response dependence may be reflected in item fit, e.g., overfitting and over-discriminating. Item response dependence can be further assessed by examining correlation patterns among the standardized item residuals. Significant correlation coefficients between standardized item residuals indicate a violation of local independence. Rasch analysis software such as Winsteps allows easy export of the standardized item residuals for correlation analysis.

1.6 Wright Map

The Wright Map is one of the most significant innovations resulting from Rasch measurement. Wright Maps have changed the world of assessment, i.e., how instruments are developed, revised, and used. It has been noted that too little effort is taken by researchers to present their Wright Maps, and too little effort to interpret their Wright Maps, be it issues such as considering what the strengths and weaknesses of the instrument are with regard to item distribution, and/or interpreting the ordering and spacing of items, For example, what does the ordering reveal about student learning, how does the ordering and spacing of items match (or not match) what has been suggested from theory?

The Wright Map is named after Benjamin Wright and was previously named the person item map. The "map" includes persons and items plotted on the same linear scale. There are different formats of Wright Maps depending upon the software being used, but a commonality is that person abilities (or some summaries of person measures) are presented in one part of the map, and item difficulties are presented somewhere else on the same map. Perhaps the most common Wright Map presented is one that looks like a thermometer. Person abilities are plotted to the left side of the thermometer, and item difficulties are plotted on the right hand of the thermometer.

On the Wright Map one can review just the persons, one can review just the items, and one can look at the relationship of persons and items. For just the persons, one can review the distribution of respondents by ability. Are the respondents distributed as one might predict? Are there respondents who are at a ceiling or a floor? That might suggest a test that is too easy or difficult, or a survey where respondents are selecting the lowest rating scale category for all items, or selecting the highest rating scale category for all items.

Another aspect of a Wright Map is that persons and items are on the same scale. This enables one to explain the performance of a respondent in terms of a set of test items. Rather than simply stating that a student has a particular raw score (or Rasch measure), one can state which items a student with a particular measure will most likely answer correctly, and what items they will most likely answer incorrectly. This allows teachers and researchers to explain the meaning of a measure.

Of great importance is the way in which the items are distributed along the construct. If one considers that each item marks a part of the construct, then one can appreciate that the distribution of items will reveal how well a construct is marked by a set of items. Are there regions of the construct lacking items? That would suggest the need for items to fill the gap. Are there too many items in certain parts of the construct? If so, it might be advisable to remove some of these items.

Perhaps most important in a Wright Map is the ordering and spacing of items. What is the story that is revealed by the ordering of items? Does the content of items match theory for different parts of the construct? In the field of science education, there is currently a great deal of research being conducted with regard to learning progressions. A learning progression can be better understood, better explained, better investigated by reviewing the ordering and spacing on a Wright Map.

Given the rich information presented by the Wright map, all reports on application of Rasch measurement to develop instruments should include a Wright Map. Further, we believe that enhanced Wright Maps should be constructed for papers and presentations. Often researchers take off the shelf a computer output and use such an output directly in their papers. The problem is that often such an output is of poor quality because it is in text format and contains details that are not needed and distracting. Thus, computer output Wright maps should be edited to make them more readable and meaningful. Improvements to computer output Wright Maps may include making sure to note the units of the scale and detailing what it means to go up and down each side of the Wright Map. A lot will depend on the number of items being presented in a Wright Map. No matter the number of items, it is important to clearly identify each item with some sort of name. And if items can be grouped into logical categories, it is helpful to include such a grouping nomenclature in the item names. Be careful about the side-by-side presentation of different Wright Maps. That is, one Wright map for one measurement instrument is not directly comparable to another Wright map for an entirely different instrument, because the meaning of units of the two maps is different (unless there was some sort of item anchoring utilized to link scales).

Other possible improvements/edits to computer output Wright maps may also be considered. For example, Wright maps can be improved through scaling of the plot. That is, if one is only going to consider item difficulty, then it is helpful to scale a Wright map from the highest to lowest item difficulty. This will help provide more detail to your plot and make your Wright Map of more interest to readers. Sometimes researchers edit the Wright maps so that they present the location of an average difficulty of items. For example, in a learning progression measurement instrument, if five items measure level 1, five items measure level 2, and five items measure levels 3, then one might edit a Wright Map so that the location of the average difficulty of the five items for each level is marked on the map. By doing so it would be easier to see patterns in the data.

1.7 Linking Rasch Measures

Linking measures is commonly conducted in the application of Rasch measurement for a variety of reasons, e.g., different forms of a measurement instrument (e.g., pre- and post-tests), measurement of a learning progression across grades, item banking, score equating, etc. Two linking approaches are commonly used: common item linking and common person linking. Linking can also be achieved by anchoring, i.e., setting the item difficulty or thresholds for some items at particular values (Boone & Staver, 2020).

No matter what linking method is used, two important questions must be answered: (a) is the linkage strong? (b) is the linkage of high quality. The first question is about the number of linking or common items (or persons) to be used. It is expected that the more linking items are used, the stronger the linkage is. But the more linking items are used, the more duplication of items are between linked forms, thus the less unique each of the linked forms become. The current literature provides no uniform recommendation regarding the minimal number of linking or common items needed to link. Based on a literature review, Wolfe (2004) suggests the optimal number of linking items ranges from 15 to 25. This optimal number of items may be far too many for many measurement instruments in science education that are typically not too long (e.g., 20 items or 30 items are common). Some other studies suggest a smaller number of linking items, as few as 10 (Hills et al., 1988) or even 2 (Wingersky & Lord, 1984). Angoff (1971) recommends 20% of common items, or 20 items, whichever number is larger in each form. Since there is no absolute required number of linking items, other factors such as the distribution of difficulties of the linking items should also be considered. For example, difficulties of linking items should be widely dispersed, with a uniform distribution preferred. Two linking items of very similar difficulty provide in essence a single link.

The quality of linking items should also be evaluated. Fit statistics for those individual linking items as well as for linking items as a set (i.e., the overall fit) should be evaluated. Evaluating linking items individually follows the same rules as evaluation of fit for any items, which is based on MNSQs, ZSTDs, standard errors of measurement, point-measure correlations, category structure, and item characteristic curves. Evaluating the overall fit of linking items as a set can be based on two overall statistics: Item-Within-Link fit statistics and the Item-Between-Link statistics (Wright & Bell, 1984). In order to compute the above two statistics, each linking item needs to be coded as two different items within the two linked forms. After submitting all items to a simultaneous Rasch calibration, we will have two difficulty estimates and two sets of fit statistics for each linking item.

Item-Within-Link fit statistics IWL is defined as follows:

$$\mathrm{IWL} = \frac{\sum_{i=1}^{L} \left(\mathrm{INFIT}_{ij} + \mathrm{INFIT}_{ik}\right)}{2L}$$

where i represents a linking item, L is the total number of linking items, INFIT_{ij} is the weighted mean-square residual fit statistics (INFIT MNSQ) for item i within Form J, and INFIT_{ik} is the weighted mean-square residual fit statistics (INFIT MNSQ) for item i within Form K. IWL has an expected value of 1.

Item-Between-Link fit statistics is defined as follows:

$$X_{IBL}^2 = \sum_{i=1}^{L} \left(\frac{(dik - dij)}{\sqrt{SE_{dik}^2 + SE_{dij}^2}} \right)^2$$

where X_{IBL}^2 is a chi-square statistics for Item-Between-Link, L is the total number of linking items, d_{ik} is the item difficulty estimate of linking item i on Form K, and d_{ij} is the item difficulty estimate of linking item i on Form J, and SE_{dik} is the standard error of measurement for the difficulty estimate of linking item i on Form K, and SE_{dij} is the standard error of measurement for the difficulty estimate of linking item i on Form J. X_{IBL}^2 has a chi-square distribution with L-1 degrees of freedom. A value greater than the critical chi-square value indicates inadequate performance of linking items as a whole. A scatterplot graph between d_{ik} and d_{ij} may also be made to visually examine the overall stability of the estimates of linking items between the two forms.

1.8 Using Rasch Measures for Subsequent Analyses

Key advantages of the application of Rasch measurement to developing instruments are that: (a) measures are linear, thus interval, (b) there are error values computed for each person measure and for each item measure, and (c) measures are invariant with respect to items or persons (Liu & Boone, 2006). Since parametric inferential statistical analysis tests such as t-test, ANOVA, regression, etc. requires that measures of the variables are interval, Rasch scale measures, either in logits or in other forms of linear transformation of logits, should be used in subsequent analysis. Over the years, we have noted in some science education publications that researchers still use raw scores for conducting statistical analysis such as comparing the difference in abilities among different groups, despite that many efforts had been devoted to establishing validity and reliability of the measures based on Rasch measurement theories and principles. This is unfortunate. The purpose of developing a measurement instrument is to use the measures for answering important research questions. If Rasch scale measures are not used in subsequent statistical analysis, one is missing a key point of applying Rasch measurement.

Following the Standards for Educational and Psychological Testing (joint committee of the AERA, APA and NCME, 2014), it is highly desirable to produce a supporting document of the developed measurement instrument in terms of its content – appropriate use, content – test development, and content – test administration and scoring. Specifically, for any measurement instrument developed following Rasch measurement, we suggest a raw score to Rasch scale measure conversion table be provided so that users of the measurement instrument will not need to conduct Rasch analysis to obtain person measures. Similarly, for subsequent analyses of item difficulties, only Rasch difficulty measures should be used, not the conventional percentage correct difficulty indices.

This book includes 18 chapters written by scholars from China, Canada, Germany, Philippines, and the US. Chapter 1, the current chapter, provides an overview of current status, issues and best practices in applications in Rasch measurement in science education; it also provides an overview of the chapters included in this book.

Chapter 2 written by Lin Ding provides an evaluative review of relevant empirical studies, featuring the diverse applications of Rasch measurement in Physics Education Research (PER) that targets various constructs, instrument formats, scoring schemes and analytical techniques. It also highlights confusions and improper practices related to the theory-driven nature of Rasch measurement, its basic principles and operations, confirmatory bias in practice, and inconsistent benchmarks for data interpretation. To mitigate these issues, recommendations are made for stricter peer-review processes and more professional development opportunities.

Chapter 3 by Ki Cole and Insu Paek provides an overview of the open-source, freely available R software and introduces free Rasch item response modeling programs in R for unidimensional and multidimensional data that are dichotomously or polytomously scored. It provides instructions for installing the software, writing and executing syntax in the R console, and loading packages. The 'eRm' package is utilized for performing the simple Rasch analysis for unidimensional, dichotomous data. The 'TAM' package is used for analyzing the Partial Credit Model for unidimensional, polytomous data. The 'mirt' package is utilized for performing between-item multidimensional Rasch analysis for dichotomous data.

Chapter 4 by Xingyao Xiao, Mingfeng Xue and and Yihong Cheng introduces the Bayesian estimation procedure in R with a Stan package for Partial Credit Model Rasch analysis. It uses a Programme for International Student Assessment (PISA) dataset to show qualitatively meaningful relationships between explanatory variables and students' ability.

Chapter 5 by Yizhu Gao, Xiaoming Zhai, Ahra Bae, and Wenchao Ma introduces an application of integrating the Rasch model and cognitive diagnosis model (CDM), the Rasch-CDM, to measure learning progressions. The Rasch-CDM approach provides students' ability, the difficulty of individual attributes, as well as students' attribute mastery patterns. The information can be visualized in a map.

Chapter 6 by Martina Brandenburger and Martin Schwichow introduces latent class analysis (LCA) and how LCA can supplement Rasch analysis. It presents a concrete example involving measuring student experimental design errors.

Chapter 7 by Haider Ali Bhatti, Smriti Mehta, Rebecca McNeil, Shih-Ying Yao and Mark Wilson describes the BEAR Assessment System, a measurement framework built on Wilson's four "building blocks" (Wilson, 2005). It also presents an application of this framework to assess middle school students' proficiency with Arguing from Scientific Evidence (ARG).

Chapter 8 by Amanda A. Olsen, Silvia-Jessica Mostacedo-Marasovic and Cory T. Forbes reports a study to compare two Rasch parameter estimation methods, the joint maximum likelihood (JML) and the marginal maximum likelihood (MML) using data from a student epistemic understanding assessment. Overall, there is little difference in person and item estimates and in item fit statistics between the two estimation methods.

Chapter 9 by Jonathan M. Barcelo and Marlene B. Ferido describes how Rasch analysis was used to refine the reasoning progression of medical technology students in the chemistry-based health literacy test (CbHLT), an instrument that measures how chemistry concepts were linked to health promotion and disease prevention activities in the contexts of nutrition, diagnostics, and pharmacology. The differences in the types of explanations in these contexts are described, and the role of Rasch analysis in the revision of the chemistry reasoning progression is elucidated.

Chapter 10 by Cari F. Herrmann-Abell, Molly A.M. Stuhlsatz, Christopher D. Wilson, Jeffery Snowden, and Brian M. Donovan details the role that Rasch measurement played in the development of an assessment instrument that can be used to measure the complex science learning described in the Next Generation Science Standards. Four model-based reasoning (MBR) tasks were developed and tested along with content-focused (CF) items with high school biology students, high-school biology teachers and crowd-sourced adults. Rasch modeling was used to investigate the relative difficulties of the items within the tasks, explore the relationship between performance on the MBR tasks and the CF items, and compare the performance of students, teachers, and adults.

Chapter 11 by Shaohui Chi, Zuhao Wang, and Ya Zhu reports a study to develop a measurement instrument to assess students' learning progressions on the crosscutting concept of Stability and Change across middle school grades (from Grades 7 to 9). A partial credit Rasch model analysis was employed to inform instrument development and evaluation. Specifically, this study used step calibrations and item measures anchoring to express student performance across three grades on the same linear scale. Results provided evidence of reliability, content validity, construct validity, and predictive validity of measures of the instrument, suggesting the measurement instrument meets the quality benchmarks.

Chapter 12 by Amber Todd and William Romine describes the use of Rasch analysis on data from the Association of American Medical Colleges (AAMC) Medical School Graduation Questionnaire (GQ) for one allopathic undergraduate medical school to determine the impact of the clinical clerkship experience and participation in extracurricular activities on perception of preparedness for residency. It shows how to use Rasch analysis to elucidate the types of activities that could be beneficial to students given their clerkship experience and how other institutions can do the same to help inform curricular changes.

Chapter 13 by Peng He, Xiaoming Zhai, Namsoo Shin and Joseph Krajcik presents an application of many-facet Rasch measurement (MFRM) to assess students' knowledge-in-use in middle school physical science. It used three online knowledge-in-use classroom assessment tasks and then developed transformable scoring rubrics to score students' responses, including a task-generic holistic rubric (across multiple tasks), a task-specific holistic rubric (in a specific task), and a task-specific analytic rubric (in a specific task).

Chapter 14 by Dennis L. Danipog, Nona Marlene B. Ferido, Rachel Patricia B. Ramirez, Maria Michelle V. Junio, and Joel I. Ballesteros presents the results of a research project implemented in the Philippines that assesses senior high school STEM students' understanding of chemistry concepts and skills using Rasch measurement. This approach determined student readiness in engaging with the general college chemistry course.

Chapter 15 by Zoë Buck Bracey, Molly Stuhlsatz, Christopher Wilson, Tina Cheuk, Marisol M. Santiago, Jonathan Osborne, Kevin Haudek, and Brian Donovan reports a study using multi-facet Rasch modeling (MFRM) to examine the extent to which computer scoring models for assessing students' argumentation in science might be more or less severe when scoring students who have been designated as English Learner (EL) students than humans scoring the same data. It was found that while no one machine scoring approach produced significant bias, performance on certain items demonstrated that one machine model had significant potential to widen performance gaps.

Chapter 16 by William Romine, Amy Lannin, Maha Kareem, and Nancy Singer describes the application of the multi-faceted Rasch model to validate open-ended written scenario-based assessments of argumentation around socio-scientific issues which are subject to errors associated with the argumentation competency being assessed, the rater being assigned, and the particular socio-scientific issue given to the student. Through inspection of the hierarchy within each facet and misfit of particular elements, it was possible to tease out the strengths and limitations of particular scenarios and raters, and ultimately derive a better understanding of how students' observed argumentation changes as their skill in argumentation increases.

Chapter 17 by Ye Yuan and George Engelhard, Jr. presents an application of the linear logistic Rasch models (LLRMs) to explore item difficulty on formative assessments. LLRMs provide the opportunity to include item covariates in a measurement model. These covariates add to an understanding of the item characteristics that can be used to predict item difficulty. Data from a high school biology assessment is used to illustrate the model. Results indicated that word count, word concreteness, deep cohesion, and cognitive complexity are strong predictors of item difficulty.

Finally, Chapter 18 by Gavin W. Fulmer, William E. Hansen, Jihyun Hwang, Chenchen Ding, Andrea Malek Ash, Brian Hand, and Jee Kyung Suh reports on the development and pilot testing of a new instrument for teachers' knowledge of argument as an epistemic tool. The construct-based approach involved domain analysis, item writing, expert reviews, piloting, and applications of the Rasch rating scale model. The instrument is applicable for researchers working on argument-driven inquiry science to study potential growth in teachers' views of argumentation.

Together the above chapters provide a snapshot of current advances in applications of Rasch measurement in science education.

References

Adams, R. J., Wu, M. L., Cloney, D., & Wilson, M. R. (2020). *ACER ConQuest: Generalised item response modeling software [computer software]*. Version 5. Camberwell, Victoria: Australian Council for Educational Research.

Andrich, D. (2004). Controversy and the Rasch model: A characteristics of incompatible paradigms? In E. V. Smith & R. M. Smith (Eds.), *Introduction to Rasch measurement (chapter 7)* (pp. 143–166). JAM Press.

Andrich, D., & Marais, I. (2019). *A course in Rasch measurement theory: Measuring in the educational, social and human sciences*. Springer Nature.

Angoff, W. H. (1971). Scales, norming, and equivalent scores. In R. L. Thorndike (Ed.), *Educational measurement* (2nd ed., pp. 508–600). American Council on Education.

Bond, T. G., & Fox, C. M. (2015). *Applying the Rasch model: Fundamental measurement in the human sciences*. Routledge.

Boone, W. J., & Staver, J. (2020). *Advances in Rasch analysis in the human sciences*. Springer.

Boone, W. J., Staver, J., & Yale, M. S. (2014). *Rasch analysis in the human sciences*. Springer.

Engelhard, G., Jr., & Wang, J. (2021). *Rasch models for solving measurement problems*. Sage.

Hills, J. R., Subhiyah, R. G., & Hirsch, T. M. (1988). Equating minimum-competency tests: Comparison of methods. *Journal of Educational Measurement, 25*, 221–231.

Jang, E. E., & Roussos, L. (2007). An investigation into the dimensionality of TOEFL using conditional covariance-based nonparametric approach. *Journal of Educational Measurement, 44*(1), 1–21.

Joint Committee of AERA, APA, NCME. (2014). *Standards for educational and psychological testing*. the author.

Linacre, J. M. (1994). Sample size and item calibration stability. *RMT, 7*(4), 328.

Linacre, J. M. (2022). *Winsteps® Rasch measurement computer program User's Guide*. Winsteps.com.

Liu, X. (1993). *Robustness revisited: An exploratory study of the relationships among model assumption violation, model-data-fit and invariance properties*. Unpublished Doctoral Dissertation, University of British Columbia.

Liu, X., & Boone, W. (2006). Introduction. In X. Liu & W. Boone (Eds.), *Applications of Rasch measurement in science education (chapter 1)* (pp. 1–22). JAM Press.

Liu, X. (2010/2020). *Using and developing measurement instruments in science education: A Rasch Modeling approach* (1st/2nd ed.). Information Age Publishing.

Robitzsch, A., Kiefer, T., & Wu, M. (2021). *TAM: Test analysis modules*. R package [Computer software]. Version 3.7-16. https://CRAN.R-project.org/package=TAM

Smith, R. M., Schumacker, R. E., & Bush, M. J. (1998). Using item mean squares to evaluate fit to the Rasch model. *Journal of Outcome Measurement, 2*, 66–78.

Wilson, M. (2005). *Constructing measures: An item response modeling approach*. Lawrence Erlbaum.

Wingersky, M. S., & Lord, F. M. (1984). An investigation of methods for reducing sample error in certain IRT procedures. *Applied Psychological Measurement, 8*, 347–364.

Wolfe, E. W. (2004). Equating and item banking with Rasch model. In E. V. Smith Jr. & R. M. Smith (Eds.), *Introduction to Rasch measurement (chapter 16)* (pp. 366–390). JAP Press.

Wright, B. D. (1977). Misunderstanding the Rasch model. *Journal of Educational Measurement, 14*(3), 219–225.
Wright, B. D., & Bell, S. R. (1984). Item banks: What, why, and how. *Journal of Educational Measurement, 21*, 331–345.
Wright, B. D., & Stone, M. H. (1979). *Best test design*. MESA Press.
Wright, B. D., & Tennant, A. (1996). Sample size again. *Rasch Measurement Transactions, 9*(4), 468.
Zhang, J. (2007). Conditional covariance theory and DETECT for polytomous items. *Psychometrika, 72*(1), 69–91.

Xiufeng Liu is a Professor of Science Education at University at Buffalo, State University of New York, and a Fellow of the American Association for the Advancement of Science (AAAS). Liu conducts research on applications of Rasch measurement in STEM education such as student long-term development of crosscutting concepts on matter and energy.

William J. Boone is a Professor of Educational Psychology (Program in Learning Sciences and Human Development) at Miami University, Oxford, Ohio, USA. Boone received his Ph.D from the University of Chicago's Measurement, Evaluation and Statistical Analysis Program (Education) with Benjamin Wright as his thesis director. His research involves applications of Rasch measurement in the fields of science teacher education, biology education, physics education, chemistry education, and geology education.

Chapter 2
Rasch Measurement in Discipline-Based Physics Education Research

Lin Ding

Abstract This chapter reports the advancement of Rasch measurement in discipline-based physics education research and the challenges that scholars in the community are facing. It provides an evaluative review of relevant empirical studies, featuring the diverse applications of Rasch theory in PER that targets various constructs, instrument formats, scoring schemes and analytical techniques. It also offers a critical review of published studies, highlighting confusions and improper practices related to the theory-driven nature of Rasch measurement, its basic principles and operations, confirmatory bias in practice, and inconsistent benchmarks for data interpretation. To mitigate these issues, recommendations are made for stricter peer-review processes and more professional development opportunities.

2.1 Motivation and Introduction

Rasch theory, as a psychometric model that meets the fundamental requirements for objective measurement, is increasingly utilized in science education research (Bond & Fox, 2007; Boone & Staver, 2020; Liu, 2010). Its presence in many areas of study, particularly in the development and validation of educational assessment instruments, has recently reached an unprecedented level that is both prolific and fruitful. As a result, new insights into learning and cognition have emerged, which in turn have further advanced the development of science curriculum and instruction. As a sharp contrast, however, it is not until recently that Rasch theory began to receive much attention in discipline-based education research (DBER). Take the field of physics education research (PER) for example. It is the first DBER community that developed and studied concept inventories in the 1990s, but applications of these inventories almost entirely relied on classical test theory. Rasch measurement did not appear to be in many researchers' toolkits until the late 2000s and early 2010s when

L. Ding (✉)
Department of Teaching and Learning, The Ohio State University, Columbus, OH, USA
e-mail: Ding.65@osu.edu

X. Liu, W. J. Boone (eds.), *Advances in Applications of Rasch Measurement in Science Education*, Contemporary Trends and Issues in Science Education 57,
https://doi.org/10.1007/978-3-031-28776-3_2

a handful of scholars started to try out Rasch analysis with some widely used concept inventories for re-validation, such as the Force Concept Inventory (Hestenes et al., 1992) and Brief Electricity and Magnetism Assessment (Ding et al., 2006). Even so, broader adoptions of Rasch theory for empirical studies were still slow. In 2019, the debut of the special collection on *Quantitative Methods in PER: A Critical Examination* (Knaub et al., 2019) published by Physical Review Physics Education Research (PRPER) placed the topic of educational measurement back to the center of attention. It revitalized the community's interest in exploring new theories and applications of quantitative research. Rasch measurement was among those highlighted in the collection and since then has lent itself to bursts of published empirical studies in peer-reviewed venues.

These emerging studies boast a diversity of target constructs under investigation, ranging from the most familiar conceptual understandings of physics concepts to previously under-researched constructs such as student confidence and career aspiration (see, for example, Oon & Subramaniam, 2013; Planinic, 2006; Potgieter et al., 2010), all of which are now of increasing interest to the PER community. Similarly, measurement instruments being studied also show a broad variety of formats and scoring schemes, featuring not only the commonly used multiple-choice tests and Likert-scale surveys but also the newly popularized multi-tiered questionnaires and ordered or partial credit multiple-choice questions. As a result, applications of Rasch measurement have gone beyond the typical analysis of dichotomous data and rightly extended to more frequent invocations of Rasch polytomous models. Recently, due to the rapid rise of international PER, the needs are unprecedentedly high for bringing together assessment findings obtained under different educational contexts for valid and meaningful comparisons. This has quickly fueled interests in seeking new ways to design assessment and data analysis strategies, and therefore advanced techniques of Rasch measurement, such as item anchoring and test linking, also begin to garner much attention in PER.

While it is encouraging to witness the community's growing interest, it is imperative to point out that confusions and misunderstandings of Rasch theory abound. Many published studies contain various issues, which may seem trivial but in fact speak directly to the fundamental philosophies of Rasch measurement. Overall, researchers do not appear to have a clear view of what Rasch theory is, when it is used, why it is used or how it is used. As revealed in some PER journal publications, the theory-laden nature of Rasch measurement is often downplayed or misunderstood, leading to a reductionist view which mistakenly refers to Rasch theory as part of the Item Response Theory or casually treats it as yet-another-statistical tool for estimating psychometric properties of test items (see, for example, Xiao et al., 2019b). When using Rasch theory to establish validity evidence for assessment instruments, many practitioners have the propensity to focus on confirmatory evidence but overlook or misinterpret disconfirming data that in effect stands against their claims (see, for example, Oon & Subramaniam, 2011a, b; Testa et al., 2020). In carrying out Rasch analysis, researchers use various benchmarks to judge the acceptability of their analysis outputs, creating many inconsistencies and even confusions that make it difficult, if not at all impossible, to carry out comparisons across multiple studies (Marzoli et al., 2021; Uccio et al., 2020).

In light of these issues, it behooves us to shine a spotlight on them by conducting a careful review of these recently published studies in PER. This chapter is designed to serve such a purpose by providing a summative review regarding the status quo of Rasch measurement in PER. Specifically, it highlights the current trends and development on this topic and reveals the challenges faced by PER scholars in implementing Rasch analysis. The ultimate goal of the chapter is to facilitate more robust understandings of the theory among scholars and help the PER community institute proper practices of Rasch measurement for empirical investigations.

2.2 Scope and Structure of Review

Differing from Planinic et al.'s (2019) position paper, in which the researchers provided a general overview of important theoretical and methodological aspects of Rasch measurement, this chapter draws on published studies in PER to illustrate fundamental principles of Rasch theory, its applications in building argument-based validity evidence, and practices of data analysis and interpretation. To make the chapter more focused and relevant to scholars in the PER community, studies chosen for review are identified from journals with a readership primarily consisting of physics education researchers and/or physics teachers. These include Physical Review Physics Education Research (PRPER), American Journal of Physics (AJP), and European Journal of Physics (EJP). Major publications in broader science education, such as Journal of Research in Science Teaching (JRST) and International Journal of Science and Mathematics Education (IJSME) are also searched for pertinent studies. Because the majority of Rasch measurement studies were published in the recent years, a backtracking search of 15 years is appropriate and sufficient. This time frame also largely overlaps the life span of PRPER, a flagship journal of the American Physical Society.

Following the above boundary conditions, the author used the keyword "Rasch" to search for relevant studies in these journals. This yielded an initial body of 239 published manuscripts. A further reading of the abstracts and methods sections resulted in excluding a vast majority which were not discipline-based physics education studies or did not use Rasch theory for empirical data analysis and interpretation. The author then conducted a second round of reading, during which a few more studies were found irrelevant and hence excluded. At the same time, pertinent studies in book chapters and proceedings papers that were referenced by the initially identified manuscripts were added to the collection. As a result, a body of 37 published studies were finally retained for review in the chapter (see Appendix for a list of the reviewed studies of Rasch measurement in PER and some key features thereof). Of the 37 studies, the majority were published in PRPER but none in AJP, and a few in other discipline-general science education journals. This search outcome is expected as the aims of AJP are less research-heavy but more practitioner-oriented, and the general science education journals, although publishing a great deal of Rasch related works overall, contained far fewer discipline-specific studies in PER. While this identified body of literature may not be exhaustive of all Rasch measurement studies in PER, it does sufficiently represent the core efforts of the community on this matter. Therefore, a review of this body of literature can capture the major trends and challenges in applications of Rasch measurement in PER.

In what follows, I first review the trends of Rasch measurement practices in PER with a focus on the diverse applications of the theory in the reported empirical studies. This portion of the review highlights the recent development of this topic and potential research opportunities it brings to the community. I then turn to major issues and challenges in the practice of Rasch measurement within the PER community through a critical review of the relevant literature. To illustrate key points, where possible, I refer to specific studies as examples. These issues largely represent common misconceptions and improper applications of Rasch theory among the PER scholars and hence warrant a close inspection. In the wake of these challenges, I then offer recommendations for improvement in the rigor of future studies, including more vigilant peer-review processes and professional development opportunities for interested scholars in the community.

2.3 Diverse Use of Rasch Measurement in Physics Education Research

2.3.1 Assessment Revalidation and Assessment Development

The PER community has witnessed increasing applications of Rasch theory in the recent years. An important pattern of this trend is that the use of Rasch measurement has shifted from ad hoc applications for *assessment revalidation* to a concurrent or integrative use for *assessment development*. These two categories of application reflect different levels of sophistication in practice. Earlier studies invoked Rasch theory as an ex post facto technique to evaluate the functions of broadly used assessment instruments, often long after they were initially created. The Force Concept Inventory (FCI, Hestenes et al., 1992), for example, was originally designed and validated through classical test theory (CTT) in 1992, and since then it had been placed under repeated scrutiny for reevaluation almost exclusively within the realm of CTT. It was not until nearly two decades later that the FCI was re-validated through Rasch theory by Planinic and colleagues (Planinic et al., 2010). For the first time, both FCI item difficulty measures and person ability measures were placed on the same scale, and parametric comparisons were assuredly carried out on Rasch-generated interval measures. Akin to this effort is Ding's (2012, 2014) retrospective revalidation of Brief Electricity and Magnetism Assessment (BEMA), another broadly used concept inventory in physics that probes students' understandings of electromagnetism. Ding conducted a thorough Rasch analysis of BEMA by evaluating item Infit and Outfit statistics (including mean square residuals and Z scores), contrast loadings and principal component analysis of residuals (PCAR), and potential differential item functioning (DIF). Collectively, the results pointed toward a reasonably good fit of student response patterns to the model (Ding, 2012, 2014).

Besides the above ad hoc validation studies, recent work on assessment development has begun to take full advantage of Rasch measurement as a means for hypotheses testing on item functioning. In a study of university students' conceptual understandings of wave optics, Mešić et al. (2019) developed a test bank consisting of 65 multiple-choice items. They drew on a learning path model to lay out a set of increasingly more complex topics on wave optics, by which the researchers delineated the target construct and created corresponding items. They further considered possible DIF on genders and, more importantly, provided theoretical rationale to support such a concern. Through field testing and Rasch analysis of 35 items, they examined item dimensionality, DIF, and local independence of items and used the findings to inform item revision. Similar practices have been implemented to test the hierarchy of item difficulties. Planinic et al. (2013) postulated from the framework of knowledge transfer and prior empirical studies that student difficulty with graphs in physics was greater than in math, and that the greater difficulty was due to a lack of either physics knowledge or general skills in graph reading. To test their ideas, Planinic et al. (2013) developed a measurement instrument consisting of eight sets of 3 parallel questions, each framed in the math, kinematics and non-physics contexts respectively. They applied Rasch analysis to verify the hierarchy of item measures and concluded that added contexts, be they physics or non-physics, were the potential source of students' difficulty with graphs.

Perhaps, the best exemplars of integrating Rasch measurement into assessment development are studies of learning progressions. An often tacit, philosophically important principle underpinning these studies is that Rasch model serves as a standard, by which a particular learning progression theory can potentially be falsified through identification of misfitting evidence. This essentially is a post-positivist epistemology of science. For example, Neumann and colleagues (Neumann et al., 2012) created a learning progression for energy. They hypothesized four conceptions of energy, which from low to high levels of difficulty were energy forms and sources, energy transfer and transformation, energy dissipation, and energy conservation. Within each conception, they further hypothesized four complexity levels of reasoning in the increasing order of facts, mapping, relations and concepts. To operationalize these levels, Neumann et al. (2012) designed a bank of multiple-choice questions such that each question was uniquely matched to one of the conception levels and one of the reasoning complexity levels. After pilot testing the questions, Neumann et al. (2012) conducted a Rasch analysis to examine the alignment between the hypothesized levels and the obtained hierarchy of item measures, thereby offering empirical support for the learning progression. In a similar vein, Fulmer et al. (2014) mapped 7 FCI questions onto an *a priori* force-motion learning progression developed by Alonzo and Steedle (2009). They further coded the response options of the questions into 4 progression levels, representing progressively more complex ways of reasoning about force and motion. Based on the data collected from high school and university students, Fulmer et al. (2014) employed both Rasch partial credit model and latent class analysis to examine the validity of the hierarchical levels. By examining fit statistics, category probability curves and latent group performances, they found that student response patterns to the FCI items fit the partial credit model and that the 4 progression levels followed the presupposed hierarchical order. As a result, they made the most of what Rasch measurement could offer to fulfill the goal of validating the learning progression and its associated assessment tools.

2.3.2 Diverse Constructs, Assessment Formats and Scoring Schemes

Another important pattern in the relevant literature is the diversity in assessment instruments for which Rasch theory is invoked. Notably, *the target constructs* of these instruments have expanded to cover not only commonly researched content areas of classical mechanics and electromagnetism (Ivanjek et al., 2021; Küchemann et al., 2021; Planinic et al., 2006; Saglam & Millar, 2006), but also advanced topics including quantum mechanics, wave optics and special relativity (Aslanides & Savage, 2013; Glamočić et al., 2021; Mešić et al., 2019). Encouragingly, some PER-related interdisciplinary work utilizing Rasch theory to assess, for example, student understandings of astrophysics ideas, has also emerged (Testa et al., 2015). Additionally, constructs that are less domain-specific but more of general skills in physics, such as reading and interpreting graphs and vector diagrams, has started to garner attention and hence become new subjects of interest for Rasch measurement (Planinic et al., 2013; Susac et al., 2018).

Such rapid expansion further reaches other areas of cognition and affect. Potgieter and colleagues (Potgieter et al., 2010) investigated students' confidence in learning force and motion concepts. They hypothesized that an incorrect answer together with a high level of confidence indicated a strong misconception. Building on this hypothesis, Potgieter et al. assembled 20 items from the FCI and Mechanics Baseline Test to assess a cohort of 33 African university students. They asked the students to answer each question, provide written explanations and indicate confidence in their answers by choosing from 4 levels: certain, almost certain, almost guessing and totally guessing. Potgieter et al. applied Rasch dichotomous and polytomous models respectively to analyze students' performance and confidence, thereby identifying students' strongly held misconceptions. Worth noting here is that Potgieter et al. also repeated the analysis by using raw data and found that the results so obtained under-identified items on which student alternative conceptions were prevalent. To that end, they cautioned against using raw data to quantify student confidence levels. Another under-researched construct, physics metacognition, was examined through Rasch measurement by Taasoobshirazi and colleagues (Taasoobshirazi et al., 2015).

They created a 26-item Likert-scale questionnaire to probe student metacognition of physics problem solving. Each item was framed as a statement about awareness or a behavior of learning physics, and respondents were required to indicate a level of endorsement by selecting one of the following options: never, rarely, sometimes, usually and always true. Taasoobshirazi et al. (2015) conducted a Rasch analysis (unspecified, but likely the rating scale model) to seek validity evidence for the measurement and concluded that the instrument demonstrated reasonably good psychometric properties and conformed to the unidimensionality requirement. Yet another less frequently examined construct, now being placed under the lens of Rasch measurement for analysis, is student interest and career aspiration in physics. Oon and Subramaniam (2011a, b) developed a Likert-scale survey to draw out Singapore secondary and junior college physics teachers' views about student choice of physics as a major or a career. The final version of the survey contained 29 items, each soliciting respondents' agreement on a 6-point scale, ranging from strongly disagree to strongly agree with no neutral option. To validate the survey, Oon and Subramaniam used Rasch rating scale model to examine, among others, item fit, functioning of response categories, dimensionality and DIF. Roughly at the same time, they also utilized Rasch measurement to create a parallel survey for use with students on this target construct (Oon & Subramaniam, 2013).

Besides the diverse target constructs, the *format of instruments* also shows a great variety in the recent Rasch measurement studies. Ordered multiple-choice questions, multi-tier select-response surveys and partial-credit tests, along with the familiar dichotomous multiple-choice and true-false items, as well as the various combinations thereof, have all come to the central stage of quantitative PER as regular subjects of Rasch analysis. Among them, ordered multiple-choice questions are worth a spotlight in that such a format takes the typical item-level analysis to the finer level of answer choices. Simply put, the usual item-level analysis now becomes answer-choice-level analysis, or sub-item level analysis. This undoubtedly imposes higher demands on researchers. As one of the pioneers in this area, Fulmer (2015) created an abridged version of an ordered multiple-choice FCI. He selected 17 items from the original inventory and mapped each answer choice to one of the hierarchical levels of a combined learning progression for force, motion and Newton's third law. This instrument then was administered to 174 Singapore secondary school students. Through a multi-dimensional rating scale Rasch analysis, Fulmer examined the extent that student reasoning of force concepts could match the hypothesized progression levels, thereby building validity evidence for the learning progression and the ordered-multiple-choice FCI.

Due to the increased variation in instrument formats, different *scoring schemes* begin to be examined through Rasch analysis. Uccio et al. (2019) developed a 2-tier 10-item questionnaire to assess students' conceptual understandings of quantum mechanics. The first-tier of each item contained 3 true-false questions testing student knowledge of quantum mechanics concepts (knowing); the second-tier was a multiple-choice question targeting reasoning and application of the concepts (reasoning). Uccio et al. scored the first-tier (T1) dichotomously by assigning 1 point for correctly answering at least 2 out of 3 true-false questions and 0 point for otherwise.

They also scored the second-tier (T2) question dichotomously by assigning 1 point to a correct answer and 0 to an incorrect answer. Given the four different possible score pairs (00, 01, 10, and 11), Uccio et al. proposed 6 different methods to combine the two tiers (T1 and T2) into one dataset for Rasch analysis; they were T1 × T2, T1 + T2, 2 × T1 + T2, T1 + 2 × T2, T1 × (1 + T2), and T2 × (1 + T1). According to Uccio et al. (2019), each of the six scoring approaches represented a unique measurement perspective that emphasized knowing and reasoning differently. For instance, 2 × T1 + T2 assumed knowing to be more demanding than reasoning, whereas T1 + 2 × T2 was the opposite. The researchers then employed the partial credit model of Rasch analysis to examine the fit of the data to the model for each of the 6 scoring schemes. By inspecting the results, Uccio et al. found that the simple arithmetic sum of the two tiers (T1 + T2) was the only case that yielded no misting items.

2.3.3 Diverse Models and Analytical Techniques

As a result of these assessment formats and scoring schemes, applications of Rasch measurement in PER begin to feature a greater variety of analytical models and techniques for data analysis. The usual dichotomous model, rating scale model and partial credit model in Rasch theory have all been regularly employed by PER scholars to support validation of existing and newly developed measurements. More importantly, sophisticated applications, such as multi-dimensional analysis and item linking and equating techniques, also have surfaced from the recent studies. Vo and Csapo (2021), for example, created a 24-item multiple-choice test to measure secondary school students' reasoning skills in applying control-of-variables strategy (CVS) in physics experiments. The development of the test was built on theoretical specifications of the CVS construct, which included 3 subskills: identifying controlled experiments, interpreting experiments, and understanding the indeterminacy of confounded experiments. To verify the dimensions of the hypothesized 3-subskill construct, Vo and Csapo collected data from 470 Vietnamese secondary school students and conducted both unidimensional and multidimensional Rasch analyses. They compared, among other outputs, the information criterion statistics (Aikake and Bayes information criteria) from the two sets of Rasch analysis and found that the data, albeit conforming to both models, better fitted the 3-dimensional model. Similarly, Kirschner and colleagues (2016) conceptualized teachers' professional knowledge as a multi-dimensional construct involving three distinct aspects: pedagogical content knowledge (PCK), content knowledge (CK), and pedagogical knowledge (PK). To measure physics teachers' professional knowledge, they developed a written test containing 15 items. Most of the items were open-ended except for two where respondents were required to rank or select from a list of given options. Responses to each question were first qualitatively scored based on a 3-point system: 0 for no correct answer, 1 for one correct answer, and 2 for more than one correct answer. Using the data collected from 186 physics teachers,

Kirschner et al. performed a multidimensional Rasch analysis (unspecified, presumably Rasch partial credit analysis). They evaluated 4 different models by varying the number of dimensions and found that the 4-dimension model (PCK, CK, PK declarative, and PK procedural) yielded the best fit statistics.

With the rise of international collaborations in PER, the need for calibrating measurements beyond local contexts becomes a new initiative in the recent studies. Glamočić et al. (2021) led a group of international scholars to build a globally standardized item bank for measuring students' understandings of wave optics in introductory physics. They distinguished item bank from item pools in that the former would include, besides the usual questions and scoring schemes, psychometric information on each item indicating the ways that it would contribute to the operationalization of a variable (target construct). Glamočić et al. used Rasch item linking to illustrate the expansion of an item bank on wave optics. They first selected, from an existing small bank, 18 items that displayed satisfactory fit statistics, low DIF, representative content coverage, and proper spacing in difficulty distribution. Then, they added 12 newly created questions to form a 30-item test and used it with 106 students from four universities in Slovenia, Croatia, Bosnia and Herzegovina. To evaluate the linking precision, Glamočić et al. conducted a Rasch analysis and evaluated item fit, DIF, dimensionality, item local dependency, and the ratio of standard deviations of linking item difficulties on the new test and in the existing bank. Results indicated a reasonably good linking precision, hence allowing for new items to be added into the bank.

Worth noting is that item linking has not only been applied to multiple testlets (item banks or pools) containing identical items as seen in the previous study, but has also been used for virtual linking where no common items are shared. In an attempt to identify predictors for physics item difficulty, Mesic and Muratovic (2011) analyzed two large-scale assessments. One was from the 2007 Trends in International Mathematics and Science Study (TIMSS), and the other was administered by the Standards and Assessment Agency (SAA) in Bosnia and Herzegovina in 2006. They selected 123 physics items from the two tests and conducted a content analysis to categorize the content and cognitive complexity of each item into one of the hierarchical levels, similar to those in a learning progression, except that the identified content or cognitive variables were not topic-specific but general and applicable for broader physics domains. Because no common items existed across the two tests and no student took both tests, Mesic and Muratovic used a virtual equating method, in which they located pairs of items from the two tests with similar physics content and cognition as the linking items and rescaled their difficulty measures from one test (SAA) into the reference frame of the other (TIMSS) through a linear regression. Thereby, items on the two tests were linked together onto the same common scale.

As evident from the above review, applications of Rasch measurement now become ever more integrated into empirical PER, where a variety of assessment instruments and scoring schemes are utilized for diverse content areas. As a result, the analytical models and techniques employed for data analysis also begin to feature a greater extent of complexity and sophistication. This undoubtedly elevates the status of Rasch measurement in PER. That said, confusions and improper practices coexist with the advancement and, in some cases, are rather rampant. Below, I turn to the challenges that practitioners in PER are facing.

2.4 Confusions and Improper Practices of Rasch Measurement in Physics Education Research

Many challenges stand in the way toward proper applications of Rasch theory. They may appear subtle, so that they can easily remain unnoticed in published works, and yet are significant enough to speak directly to the fundamental principles of measurement. To a large extent, these issues concern the theory-driven nature of Rasch measurement, its basic principles and operations, confirmatory bias in practice, and inconsistent benchmarks for data interpretation.

2.4.1 Theory-Driven Nature of Rasch Measurement

It is foremost to understand that Rasch measurement differs fundamentally from item response theory (IRT) in that the former is a theory-laden approach to quantitative measurement and is operated by fitting empirical data to model, whereas the latter is centered around data and seeks to fit different models to data. This nuanced difference underscores the higher standards placed on Rasch theory, whose results should uniquely meet the requirements for objective measurement. In Linacre's (2002) words, "if you want to measure, you've got to use a Rasch model!" Unfortunately, this critical centerpiece is distorted by some practitioners. They often hold a reductionist view to refer to Rasch theory as part of IRT and uses it merely as a mathematical tool for data crunching. In some cases where Rasch analysis is employed, claims that are only appropriate for IRT are indiscriminately formulated for Rasch measurement, which raises questions concerning the philosophical intent of such studies. For example, Xiao et al. (2019b) attempted to divide the Conceptual Survey of Electricity and Magnetism (CSEM, Maloney et al. 2001) into two comparable short tests. They used dichotomous unidimensional Rasch model to examine the psychometric properties of the original CSEM and its two abridged versions. On that, Xiao et al. (2019) explicitly reported that "[t]*he Rasch model was separately fitted to data of the three CIs [concept inventories]*" (Xiao et al., 2019b, p. 010122–5) This is puzzling as it is unclear whether their use of Rasch analysis was simply by chance of acceptable psychometrics but in fact should have been IRT instead. Indeed, a review of another work by Xiao and colleagues further deepens this concern. In a highly similar study, Xiao et al. (2019a) divided BEMA into two shorter versions and used 2-parameter IRT to perform psychometric analysis. They offered no rationale for why this model was chosen, although such information would have been critical, particularly in light of their parallel study of CSEM for which Rasch analysis was used. In their report, Xiao et al. repeatedly claimed that the Rasch model under a unidimensional assumption was one of the IRT models (Xiao et al., 2019a). These methodological ambiguities and inconsistencies largely muddled the meaning of their findings, making it impossible to compare and interpret the two otherwise parallel studies. Similar confusions are also witnessed in a number of recent attempts of seeking best-fitting Rasch models. For examples, in the above mentioned studies by Vo and Csapo (2021) and by Kirschner et al. (2016), the researchers empirically subjected their collected data to different Rasch models and resorted to fit indices, such as AIC and BIC, as a criterion to select the best-fitting validity model. Effectively, this is a model-fitting-data practice, which deviates from the theory-driven spirit of Rasch measurement.

The lack of clear understandings of theory-driven Rasch measurement is further manifested in studies that merely employed Rasch analysis to generate item and person estimates but missed the opportunity of using it as a guiding principle to predict and test, for example, the hierarchy of items. In some cases where item measures were already obtained from Rasch analysis, researchers reverted back to raw data for parametric comparisons. This raises a question on the philosophical rationale for why Rasch theory was invoked in the first place. For example, Plummer and Maynard (2014) attempted to build a learning progression for celestial motion. They used a self-developed test for assessment which consisted of 13 items. Among them, 6 were multiple-choice questions scored dichotomously for either 0 or 3 points, and 7 were open-ended questions scored for partial credit, ranging from 0 to 3 integer points. Rightly, Plummer and Maynard chose Rasch partial credit model to examine item difficulties and determined the hierarchy of progression levels (although they did not begin with a hypothesized progression of these levels). However, after analyzing item measures, they turned to raw scores for analyzing student performances, as if person measures had never been part of the Rasch outputs from their analysis.

Similar treatment was observed in Marzoli and colleagues' (2021) study of Italian university physics students' views about remote instruction during the Covid pandemic. They adopted and adapted five Likert-scale surveys, each targeting students' perceptions of emergency remote instruction, subjective well-being, motivating to learn physics, physics academic orientation, and attitudes toward physics respectively. Marzoli et al. used a retrospective pre-post design to collect data from 362 students and performed Rasch analysis (unspecified, likely the rating scale model) for only one of the surveys on student perceptions about remote instruction. It is puzzling that they did not consistently employ Rasch analysis on the other surveys despite their recognition that "*the reason for using Rasch analysis... is that we cannot assume linearity in the rating scale*". Even for the survey where Rasch analysis was conducted, Marzoli et al. only examined item fit statistics but did not utilize person measures to examine student pre-post responses. Instead, they reverted back to raw scores for pre and post comparisons.

In another study of teaching wave optics, Mešić and colleagues (2016) compared three different approaches to visualizing light waves, which used sinusoidal representations, electric field vectors, and phasor diagrams respectively. They developed a 19-item multiple-choice test to assess understandings of wave optics across three student groups, each exposed to one of the visualizing approaches. To validate the construct, Mešić et al. chose Rasch analysis for building "interval measures of student understandings of wave optics". However, after examining item measures and fit statistics, they resorted back to raw scores rather than use Rasch person measures to perform between-group comparisons. Their rationale was that "*for a Rasch-compliant set of items, it makes perfect sense to calculate participants' test scores by summing their scores over individual items.*" (Mesic et al., 2016) Here, what Mešić et al. referred to as Rasch-compliant items were in effect the items on which student data fit Rasch model reasonably well. Following this premise, one would expect the Rasch-generated, interval-level person measures to be meaningful (or otherwise the data would not fit the model). However, a good model fit should not be mistaken as the raw data being at the interval level. Mešić et al. used raw scores anyway to compare students' performances between the three groups on clusters of items (subscales) and on the individual items. Perhaps, it is because Rasch analysis does not yield CTT-equivalent item-level person measures that Mešić et al. resorted to raw scores for help. In fact, this issue can be partly resolved by invoking multi-dimensional Rasch analysis, in which item clusters (subscales) are treated as separate dimensions and hence person measures on each item cluster can be generated. As for comparing item-level student performance, one can calculate group average odds ratio as an alternative. Put differently, one can imagine a statistically aggregate person whose ability is the group average of person measures $\overline{\theta}$, and then calculate this aggregate person's odds ratio on an item of difficulty δ_i as an estimate for the group performance $\left(\frac{\overline{P}}{1-\overline{P}} = e^{\overline{\theta}-\delta_i}\right)$. Or, one can further calculate the probability $\left(\overline{P} = \frac{e^{\overline{\theta}-\delta_i}}{1+e^{\overline{\theta}-\delta_i}}\right)$ to represent group performance. Either way, the results so obtained are closer in meaning to the stochastic nature of Rasch modeling than raw percentages.

2.4.2 *Principles and Operations of Rasch Measurement*

Related to the above issue are confusions regarding the fundamentals of Rasch measurement, particularly regarding the meaning of some key outputs from the analysis. Person ability measures, for example, are often misinterpreted as representing an amorphous construct, equivalent to general academic ability or intelligence. In a study, Aslanides and Savage (2013) developed a relativity concept inventory to measure university students' understandings of special relativity. They conducted a Rasch analysis (which they also referred to as item response theory) on data collected from 53 Australian students in the hopes of identifying pairs of items targeting the same concept, or in their own words "conceptually related questions". According to Aslanides and Savage (2013), "*it is reasonable to assume that a major determinant of whether a student answers a question correctly is their academic ability. Given a question pair, strong students will tend to get both right and weak students will tend to get both wrong, strengthening the overall correlations. If this assumption is correct, then removing that part of students' performance due to academic ability may increase the correlations due to conceptual relations.*" From this assumption, Aslanides and Savage conducted Rasch analysis and calculated pair-wise item correlations of residuals to identify questions targeting the same concepts. As seen, Aslanides and Savage misinterpreted Rasch person measures as general academic abilities, which they argued after being removed would give place to "conceptually related questions". In fact, Rasch-generated person measures are well defined and unique to the target construct under investigation, and they do not represent nebulous ability or intelligence in a general sense. More importantly, Rasch residuals represent an unwanted, additional construct (or sometimes called secondary construct) which effectively pose a threat to validity. In Aslanides and Savage's case, all the questions, in theory, ought to be "conceptually related" within a small area of content, and the residuals, if anything, would likely signal local dependency between items other than the desired target construct. Indeed, Aslanides and Savage only identified three positive pair-wise correlations significant at the level of 3 standard deviations with the greatest value of only 0.08.

Another misunderstood piece fundamental to Rasch measurement concerns response characteristic curves. Testa et al. (2015) used Rasch theory to test a learning progression for change of seasons, solar and lunar eclipse and moon phases. They created and administered a test consisting of 36 true-false items and 12 multiple-choice questions. For the latter, they plotted response characteristic curves, showing the percentages of students choosing each answer option at different ability measures. Next, they calculated the so-called item response curve integrals (IR integrals) by summing up all the percentages across the entire ability range for each answer choice, which effectively was just the total percentage of the students selecting a specific answer. Testa et al. assumed that the IR integrals would be greater for answers representing lower progression levels and smaller for those representing higher levels; thereby, they ordered the answer choices. This use of Rasch analysis is erroneous in at least two ways. First, the assumption of higher-level choices being selected by fewer students and lower-level choices selected by more students is fundamentally flawed. The key issue here is that the *total* numbers or percentages of students do not determine the hierarchy of answer choices. The actual determinant should be the *stratified* numbers or percentages of students. Simply put, higher-level choices should be selected by *fewer low-ability* students (or *more high-ability* students), and vice versa for lower-level answer choices. Second, because the so-called IR integrals represented the total percentages of students selecting each choice, the results were sample-dependent and hence revealed no insight into the actual hierarchy of the answer choices. A correct approach should be inspecting the peaks of the curves against the person ability scale to see if the higher-level answer choices peak at the high-ability end and lower-level answers peak at low-ability end. Unfortunately, Testa et al. mishandled the two fundamental issues of Rasch response characteristic curves, and therefore their conclusions warrant a closer re-evaluation.

Also frequently misunderstood is the invariance feature of Rasch measures. Technically, Rasch-generated item difficulty and person ability measures are at the interval level, hence their *relative* values (or distances) should remain constant to the extent allowed by measurement error. This is because the difference in logits for any person n (ability θ_n) responding to two items i and j (difficulty δ_i and δ_j respectively) is independent of the person (ability).

$$\ln \frac{P_{n,i}}{1 - P_{n,i}} - \ln \frac{P_{n,j}}{1 - P_{n,j}} = (\theta_n - \delta_i) - (\theta_n - \delta_j) = -(\delta_i - \delta_j)$$

Similarly, the difference in logits for any two persons n and m responding to the same item i is independent of the item (difficulty).

$$\ln \frac{P_{n,i}}{1 - P_{n,i}} - \ln \frac{P_{m,i}}{1 - P_{m,i}} = (\theta_n - \delta_i) - (\theta_m - \delta_i) = \theta_n - \theta_m$$

Alas, this invariance of *relative* distance between measures has been largely misconstrued as invariance of the *absolute* measures of item difficulty and person ability in Rasch analysis. For example, Oon and Subramaniam (2013) developed a 54-item Likert-scale questionnaire to survey Singapore students' choice of physics as a major in post-secondary education. They collected responses from 1076 high school and junior college students and separated them into two groups according to their indication of planning or not planning to select physics as a future major. Oon and Subramaniam conducted Rasch analysis separately for the two groups and directly compared the item measures from the two sets of analysis to check invariance. They stated that Rasch estimates were "sample-and item-independent" and that the scale would "show the property of measure invariance". Clearly, Oon and Subramaniam confused Rasch-generated interval data with ratio data whose absolute values remain invariant due to the existence of a zero point. Technically, before comparing the two sets of item measures, calibration should be made to equate their means. This is equivalent to aligning two interval scales to the same zero point, so that they can be comparable. While some computing programs for Rasch analysis automatically sets the mean item difficulty to be zero each time, others may not always be the case. Therefore, checking the mean values of item measures is crucial. In Oon and Subramaniam's (2013) study, the two means of item measures indeed were not preset to be zero (−0.4 and −0.2 respectively), therefore a direct comparison without calibration for means could overestimate the difference, if any.

Similarly, Ene and Ackerson (2018), in a study of validating a semiconductors concept inventory, mistook Rasch generated interval measures as ratio data and claimed that "*the person's estimated ability should not depend on the specific items chosen from a large calibrated item pool. As all the items in a Rasch calibrated test have equal discrimination, it does not matter what items are selected for estimating the ability of the respondents.*" In examining the range of person measures, Ene and Ackerson further commented that "*although we did not obtain a sharp discrimination between person abilities, the crucial benefit of the Rasch calibrated scale is that persons are not compared among themselves but with fixed located items.*" Here, Ene and Ackerson dismissed the significance of between-person comparisons altogether and mistakenly argued for the absolute invariance of Rasch measures. Ironically, as discussed above, it is the between-person and between-item distances, not the "fixed located" persons or items, that remain invariant in Rasch analysis.

2.4.3 Confirmatory Bias in Practice

As Kane (2006) pointed out, validity is an argument-based process. It requires that all available evidence, whether confirming or disconfirming, be examined, so that adequate and appropriate interpretations or decisions can be reached for assessment results. Rasch theory offers a unique measurement framework that can afford rich information to help establish validity evidence. Rooted in falsificationism, Rasch theory in principle represents a scientific approach where anomalies, in the form of misfit statistics, are explicitly sought to refute an *a priori* model. Admittedly, making sense of a large body of Rasch statistics by no means is an easy job. As with other practitioners of Rasch theory, scholars in PER often fall short in this regard. In empirical studies, the unidimensionality requirement for Rasch measurement often is assumed or only verified by checking fit statistics alone. Even when a few other metrics are examined, their use and interpretation appear to be limited and mechanically formulaic. In some rather extreme cases, researchers even overlook or trivialize disconfirming evidence in efforts to support the construct validity of their measurement. An example is the study by Testa et al. (2020), where the researchers examined students' overconfidence in understandings of introductory quantum mechanics. Testa et al. administered an 18-item multiple-choice test created by Uccio et al. (2020) to assess 408 Italian students' conceptual performance on this topic. Additionally, they added a second-tier Likert-scale questions to probe student confidence in their answers to each item. Instead of analyzing conceptual understanding and confidence separately, Testa et al. combined them into one single construct for Rasch analysis. As careful readers may have suspected, conceptual understanding and confidence are two distinct constructs, and forcing them into the same scale may not be theoretically warranted. Indeed, Testa et al. performed a principal component analysis of Rasch residuals and found the eigenvalues for the first two contrasts were 3.48 and 2.26 respectively. Further, they examined the contrast loading plot, which clearly displayed two clusters of items with all the concept questions in the positive loading region (half of them above the +0.4 region) and nearly all except one of the confidence questions in the negative loading region. However, such strong evidence for multi-dimensionality was dismissed by Testa et al. Instead, they only took into account supporting evidence on item fit statistics to argue for the unidimensionality of the combined construct, despite the fact that conceptual understanding and confidence did not lend themselves well onto the same scale either theoretically or empirically.

In another study of Singapore teachers' perceived reasons for the declining student interest in choosing physics as a major, Oon and Subramaniam (2011a, b) created a 29-item Likert-scale survey, addressing topics such as physics-related social effects, career advancement, discipline utility and difficulty, and teaching and student experience with the subject matter. To verify the assumed unidimensionality of the scale targeted by the survey, Oon and Subramaniam conducted a principal component analysis of the Rasch residuals (PCAR) on the data collected from 190 Singapore physics teachers. They found a significant contrast of an eigenvalue 4.2 (equivalent to approximately 4 items), which suggested a strong secondary dimension. They further identified the 4 items of high loadings and re-ran Rasch analysis and PCAR after tentatively removing these 4 items from the survey. However, the results showed no noticeable change and did not "*significantly enhance the undimensionality of the…scale*". In the face of these findings, Oon and Subramaniam selectively dismissed disconfirming evidence and proceeded to claim that it was "*reasonable to assume the…scale is essentially unidimensional*", and that "*the 29-iten scale can be used in place of the 25-item scale*".

Such confirmatory bias also exists in interpretation and handling of DIF information. Testa et al. (2019) designed 10 ordered-multiple-choice questions based on a learning profession for quantum mechanics to assess 287 Italian university students' understandings of this topic. They conducted, among others, PCAR in partial credit Rasch modeling and found that one item targeting Heisenberg's uncertainty principle exhibited a significant DIF for students who were in different years of undergraduate education and received different levels of instruction on QM. As they reported, such a DIF persisted across three groups of students: physics freshmen never receiving any informal instruction on QM, second-year physics students and third-year math and material science students having completed a conceptually-oriented introductory QM course, and third-year physics students having completed both a conceptually oriented QM course and an upper-level formalism-driven QM course. This finding should have raised concerns regarding what was actually targeted by the item. However, other than an acknowledgement of the need for further inspection of the DIF, Testa et al. turned to the similarity in results between the second-year physics students and the third-year math and material science students as a talking point to argue for the dismissal of the previously identified DIF, thereby circumventing the needed inspection of the item itself. As an extension on this issue, the puzzling motivation behind Testa et al.'s DIF analysis is also worth discussing (although it goes beyond the topic of confirmatory bias). Note that DIF and normal variations in person estimates are two fundamentally different concepts. The former is a measurement bias triggered by the design of items in question which "*occur[s] when equally able test takers differ in their probabilities of answering a test item correctly as a function of group membership*" (Joint Committee et al., 2014, p. 51). Whereas, the latter is an expected distribution of person abilities along the scale of a target construct due to, for example, different levels of instruction the test-takers have received. While it is not impossible for DIF to exist among students at different academic levels, the driving force for seeking DIF in Testa et al.'s study appears highly questionable. In their own words, Testa et al. stated "*DIF is a technique to analyse whether items' responses are biased with respect to a trait of the sample. In our case, differences could have been due to a higher degree of familiarity of one or more groups with the topics targeted in the questionnaire.*" (Testa et al., 2019, p. 397). Here, the researchers ostensibly confused DIF with the anticipated distribution of person estimates stemming from the students' different levels of familiarity with the tested topics.

Understandably, dealing with DIF is not an easy task. In a re-revaluation of the FCI through Rasch analysis, Planinic et al. (2010) found several items "*significantly change their difficulty (by more than three standard errors) from non-Newtonian to Newtonian sample[s]*" and appropriately concluded "*a difference in the FCI construct in these two populations*". That said, Planinic et al. also labelled the potential DIF as "*not so surprising*", "*quite common*" and "*informative of instruction efficiency*", thereby creating much ambiguity in the description of their findings. While pinpointing the exact causes for DIF might be difficult, it should not become the reason for downplaying any evidence of DIF-displaying items.

2.4.4 Inconsistent Benchmarks for Analysis

Also clear from the review is the striking inconsistency in benchmarks for judging the extent to which Rasch outputs are satisfactory. This issue perhaps looms large not only in PER but also in other fields. Researchers often choose various criteria to evaluate empirical Rasch results, almost invariably leading to affirming the validity of their measurement. Certainly, such practices are at the expense of introducing more confusions and inconsistencies. As a typical case in Rasch analysis, item fit statistics are almost always inspected for model fit. Just for the Rasch dichotomous model alone, a variety of benchmarks have been adopted to judge item fit. One popular practice is to examine both infit and outfit mean-square residuals (MNSQ) and set [0.7, 1.3] as the acceptable range for identifying misfitting items. A number of studies in PER adopted this benchmark (see, for example, Cvenic et al., 2022; Testa et al., 2019; Uccio et al., 2020). However, other researchers adopted a much more liberal benchmark of MNSQ∈[0.5, 1.5] to evaluate item fit. In a study of student understandings of light waves, Mesic et al. (2016) designed a 19-item multiple-choice test and used Rasch analysis to examine item fit. On the one hand, they explicitly stated that "*for multiple-choice tests reasonable mean-square (MNSQ) infit and outfit statistics are between 0.7, and 1.3*", but on the other hand they used [0.5, 1.5] as the cutoff range to judge item fit. Such practice was also observed in Susac et al.'s (2018) work, in which they conducted a Rasch analysis of 889 students' responses to a 20-item multiple-choice test on understanding of vectors. They first referred to [0.7, 1.3] as the acceptable range for MNSQ but quickly switched to claiming that "*items with MNSQ in the range 0.5–1.5 will be productive for measurement.*" Contrary to Susac et al.'s claim, Cvenic et al. (2022) stated that "*although items with infit and outfit MNSQ values in a broader range, between 0.5 and 1.5, can be acceptable and not degrading for measurement, such item still need to be inspected for the reasons of their larger misfit.*" Perhaps, one of the most stringent benchmarks was adopted in Neumann et al.'s (2012) study of a learning progression for energy. The researchers chose [0.8, 1.2] as the cutoff range for infit MNSQ in their Rasch analysis of 1856 students' responses to a bank of multiple-choice questions on energy. Interestingly, Neumann et al. only considered this criterion as "moderate cutoff levels". Worth noting here is that the sample size in Neumann et al.'s study is large. As Smith, Schumacker, and Bush (1998) pointed out, MNSQ values are sensitive to large samples and hence acceptable ranges need be adjusted. Following this recommendation, Ding et al. (2016) adopted a strict criterion of $\left[0.7, 1+\frac{2}{\sqrt{N}}\right]$ and $\left[0.7, 1+\frac{6}{\sqrt{N}}\right]$ for judging infit and outfit MNSQ statistics respectively in a study of 1637 students' responses to the Lawson's Classroom Test of Scientific Reasoning (Lawson, 1978).

Such variation in item fit benchmarks is also seen in survey studies where Rasch rating scale and partial credit models are used. Taasoobshirazi et al. (2015), for example, adopted the [0.7, 1.3] range for MNSQ in the analysis of a 26-item Likert-scale survey on student metacognition of physics problem solving. The range was relaxed to [0.6, 1.4] in Oon and Subramaniam's (2013) validation of two Likert-scale surveys on student choice of physics as a major in university. When it comes to partial credit assessments, Fulmer (2015) and Fulmer et al. (2014) used [0.7, 1.3] as the acceptable range to examine fit statistics for ordered-multiple-choice FCI questions. Of all the reviewed studies, the loosest criterion of [0.5, 1.5] was used by Kirschner et al. (2016) to examine only outfit MNSQ in their study of teachers' responses to questions on professional knowledge.

Besides item fit, reliability is another measure whose benchmarks appear to vary significantly in empirical studies. In general, reliability reveals the extent to which measurement results may be replicated. In Rasch analysis, item and person indices, indicating the replicabilities of item and person ordering respectively, are often inspected. The former is less of a concern, as person samples usually are much larger than item numbers in empirical studies and hence Rasch analysis can yield high item reliability values. As for person reliability, it can be interpreted as similar to Cronbach's alpha in CTT and typically is evaluated at the criterion of >0.7 for group measurement (Fulmer, 2015; Fulmer et al., 2014) or > 0.8 for individual measurement (Cvenic et al., 2022; Ding & Beichner, 2009). Recent studies, however, have become increasingly tolerant of lower person reliabilities. Researchers resort to Linacre's work for advocating the use of 0.5 as the minimum standard for person reliability (see, for example, Marzoli et al., 2021; Testa et al., 2019). That said, other researchers, while mentioning this lower limit, consciously added commentary cautions on this matter. For instance, Susac et al. (2018) in evaluating the psychometrics of a multiple-choice test on understanding of vectors found their Rasch person reliability was 0.6. On that, they commented that "*the person reliability was rather low (0.6), which indicates a not very reliable person order if a similar test was administered to the sample students.*" Similarly, Cvenic et al. (2022) acknowledged the 0.5 minimum standard for person reliability but emphasized the need for a higher cutoff in case of "*decision making about students (which would require person reliabilities of 0.8 or more).*"

A caveat worth mentioning is that the above discussion is not meant to promote draconian practices of Rasch measurement. As Bond and Fox (2007) recommended, it is reasonable to choose different criteria for fit statistics based on the types of assessment instruments, the specific models of Rasch analysis, and the purposes and use of assessment results. In some cases, fit statistics may not be inspected for retaining or removing items but solely for quality check or item revision, and hence strict benchmarks do not carry much significance for works in progress. Similarly, item reliability may be affected by person samples, particularly by the spread of the sampled participants along the scale of a measured construct. This perhaps is why a *minimum* reliability standard, as low as 0.6, recently has been proposed for internal consistency by What Works Clearinghouse Procedures and Standards Handbook (WWC, 2022, p. 27). However, these minimum standards by no means should be interpreted as a greenlight for arbitrarily loosening criteria to establish confirming evidence for measurement. Nor should Rasch practitioners be content with only aiming for the lowest possible standards. Instead, sufficient arguments need be made to articulate why or how the selected criteria may help fulfill prescribed research goals, be they creating a run-of-the-mill version of a survey or finalizing a high-stakes test.

A separate but related issue concerns the standards for parametric statistics using Rasch generated measures. It is true that person and item measures from Rasch analysis meet the requirements of interval measurement and therefore can be subject to parametric statistical analysis. That said, being at the interval level is not the only assumption one needs to consider for parametric statistics. Therefore, caution must be taken to ensure other assumptions are also met. In a study of student conceptual understandings of quantum mechanics, Uccio et al. (2019) developed a 10-item, 2-tiered test covering three content topics: wave behaviors of matter, measurement, and atomic and electronic behaviors. The two tiers consisted of, respectively, true-false concept questions and multiple-choice application questions, with both tiers targeting the same concepts and scored dichotomously. Uccio et al. obtained item difficulty measures from Rasch analysis by treating each tier as separate questions and then performed a parametric t-test to compare item difficulties between the two tiers (10 each). They further took into account the 3 content topics and conducted a 3×2 ANOVA (3 content topics by 2 tiers). Conceivably, each cell in the 3×2 matrix contained no more than 4 items, and parametric ANOVA certainly would not be appropriate. However, the researchers overlooked this issue. In another study of student understandings of quantum mechanics, Uccio et al. (2020) developed an 18-item multiple-choice test, with each item addressing one of the five hierarchical content levels: discrete energy levels, the uncertainty principle, atomic emission and absorption, wave function and probability, and molecules and condensed matter. Using the Rasch-generated item measures, Uccio et al. conducted a one-way ANOVA to verify the hierarchy of item difficulty between the 5 levels. Nevertheless, the numbers of items in these 5 levels were 7,6, 4, 3 and 2 respectively, too small to be suitable for parametric F test. Uccio et al. overlooked the required assumptions for parametric ANOVA and mistakenly took the interval level as the sole consideration as if it alone could have overridden other requirements.

2.5 Discussion and Implication

The above review highlights the recent advancement in applications of Rasch measurement in PER as well as major challenges that practitioners are facing. On the one hand, the PER community has witnessed an unprecedented interest in Rasch theory and produced an expansive body of empirical studies on this topic. These studies feature diverse use of Rasch theory not only as an ad hoc technique to revalidate existing concept inventories but also as an integrative tool to create and operationalize new constructs. Thanks to the versatility of Rasch measurement, a variety of constructs have been investigated through different assessment formats. Besides the usual multiple-choice, true-false and Likert-scale questions, other formats such as ordered multiple-choice tests and multi-tiered questionnaires also have been frequently utilized to measure a broad array of constructs, including but not limited to student understandings of introductory and advanced physics concepts, general skills in graphical-diagrammatic representations, physics metacognition, and confidence and interest in physics learning. Accordingly, different types of Rasch analysis—dichotomous, rating-scale, and partial credit— in both the unidimensional and multi-dimensional models, have come into play in the recent studies.

On the other hand, challenges in applications of Rasch measurement are evident. The theory-driven nature of Rasch theory is often misunderstood and conflated with the data-driven IRT. As a result, confusions of when and why to use Rasch analysis are frequently exposed through practices of, for example, resorting to raw data for parametric analysis after Rasch item and person measures are obtained. Also misunderstood are basic principles and operations of Rasch measurement. This has given rise to interpreting person ability as a globally general construct, likening response characteristic curves to raw-data-like histograms, and applying "relative" invariance to the absolute values of person and item measures. In the face of conflicting evidence, researchers tend to focus on confirmatory results that support their predetermined ideas but downplay or overlook disconfirming findings. A corollary of such practice is that interpretation of construct dimensionality and DIF becomes moot. Unfortunately, this situation is further worsened by the lack of consensus on psychometric benchmarks for Rasch output. The various standards and criteria that researchers cite in empirical studies seem always to support the validity of their measurement. While this, of itself, may not necessarily be a problem, inconsistent use of drastically different standards without reasonable explanations can create much confusion and leaves such an impression that the current practice of Rasch measurement lacks sufficient rigor.

In light of these concerns, it is important that we take actions to identify improper practices, to increase the awareness of the community on key principles, and to improve the applications of Rasch measurement in empirical studies. As an initial step, this chapter serves to cast light on some chief challenges revealed from the recent work, which to a larger extent point directly to deep issues regarding the fundamentals of Rasch theory. It is worth noting that the above cited studies are only meant to be illustrative and not exhaustive, and many identified issues herein are likely more prevalent than seen from the discussion. It is also important to note that all of the reviewed studies were published in renowned journals that represent scholarship of the highest quality. With these contexts in mind, it is reasonable to say that the above identified issues are quite troubling, particularly given that these studies had gone through highly stringent peer-review processes by experts in the field. Perhaps, Rasch measurement did not fall into the reviewers' areas of expertise, and hence some issues were left undetected. This is not uncommon and completely understandable, as Rasch measurement theory just started to gain traction in the science education and PER communities not long ago, and development of communal expertise takes time. To move forward, careful attentions should be given to studies involving Rasch analysis during the entire editorial process to ensure ample expertise exists among the reviewers and handling editors. In the meantime, it also is crucial that more professional development opportunities be created and offered to interested practitioners in the larger community of science education. In that respect, the focus of professional development should not be solely focused on operating computing software programs; instead, emphasis needs to be at least partly placed on grasping the underlying principles and techniques of Rasch theory. This is not to say that practitioners must become specialists in Rasch measurement to be able to use it, but proper applications do require basic training on its theory much in the same way as engineers need to know what scientists know of fundamental principles in order to best apply such principles to engineering practices. In that sense, practitioners of Rasch theory are very much like engineers working toward new constructions or optimizations of existing systems of measurement. Ultimately, such constructions and optimizations are to lead us toward increasingly more valid and reliable measurement results.

Appendix: A Summary of Reviewed Studies of Rasch Measurement in Discipline-Based Physics Education Research

Published studies	Rasch model	Target construct	Sample	Journal
Aslanides and Savage (2013)	Unidimensional dichotomous model	Conceptual understanding of special relativity	University physics students in Australia	*PRPER*
Cvenic et al. (2022)	Unidimensional dichotomous model for individual items; partial credit model for item groups	Conceptual understanding of wave optics	High school students in Croatia	PRPER

(continued)

Published studies	Rasch model	Target construct	Sample	Journal
Ding (2012)	Unidimensional dichotomous model	Conceptual understanding of introductory electricity and magnetism	University science and engineering majors in US	PERC proceedings
Ding (2014)	Unidimensional dichotomous model	Conceptual understanding of introductory electricity and magnetism	University science and engineering majors in US	PRPER
Ding et al. (2016)	Multidimensional dichotomous model	Scientific reasoning skills	University students in China	RISE
Ene and Ackerson (2018)	Unidimensional dichotomous model	Conceptual understanding of semiconductors	University graduate and undergraduate students in US	PRPER
Fulmer et al. (2014)	Unidimensional partial credit model	Conceptual understanding of force and motion (force and motion learning progression)	High school and university students in US	IJSE
Fulmer (2015)	Multidimensional (2-d) rating scale model	Conceptual understanding of force and motion, including Newton's third law	Secondary school students in Singapore	IJSME
Glamočić et al. (2021)	Unidimensional dichotomous model; item linking	Conceptual understanding of wave optics	University physics students in Slovenia, Croatia, and Bosnia and Herzegovina	PRPER
Ivanjek et al. (2021)	Unidimensional dichotomous model	Conceptual understanding of simple electric circuits	Middle school students in Austria and Germany	PRPER
Kirschner et al. (2016)	Multidimensional Rasch analysis; model unspecified, but likely the rating scale model	Physics teachers' pedagogical content knowledge	Physics teachers; presumably in Germany (unspecified)	IJSE
Küchemann et al. (2021)	Unidimensional dichotomous model	Competency in field lines and vector-field representations	University STEM students in Switzerland and Germany	PRPER
Marzoli et al. (2021)	Unidimensional Rasch analysis; model unspecified, but likely the rating scale model	Views of emergency remote instruction	University students in Italy	PRPER

(continued)

Published studies	Rasch model	Target construct	Sample	Journal
Mesic et al. (2016)	Unidimensional dichotomous model	Qualitative understanding of wave optics	Secondary school students in Bosnia and Herzegovina	PRPER
Mesic and Muratovic (2011)	Unidimensional dichotomous model; virtual test equating	Competency in physics	Secondary school students in Bosnia and Herzegovina	PRPER
Mešić et al. (2019)	Unidimensional dichotomous model	Understanding of wave optics	University students in Bosnia and Herzegovina, Croatia, and Slovenia	PRPER
Neumann et al. (2012)	Unidimensional dichotomous model	Understanding of energy	Grade 6–10 students in Germany	JRST
Oon and Subramaniam (2011a)	Unidimensional rating scale model	Teachers' views on student declining interest in physics	Secondary and junior college physics teachers in Singapore	IJSE
Oon and Subramaniam (2011b)	Unidimensional rating scale model	Teachers' views on student declining interest in physics	Secondary and junior college physics teachers in Singapore	Book chapter
Oon and Subramaniam (2013)	Unidimensional rating scale model	Views on and interest in physics	Secondary and junior college students in Singapore	IJSE
Planinic et al. (2006)	Unidimensional rating scale model	Confidence on conceptions of Newtonian dynamics and DC circuits	High school students in Croatia	JRST
Planinic et al. (2019)	Review of unidimensional Rasch models	General discussion of constructs in physics education	General discussion of assessment respondents	PRPER
Planinic et al. (2010)	Unidimensional dichotomous model	Conceptual understanding of force	High school students in Croatia	PRPER
Planinic et al. (2013)	Unidimensional Rasch analysis; model unspecified, but likely the partial credit model	Understanding of graph slope and undercurve area	First-year university students in Zagreb	PRPER
Plummer and Maynard (2014)	Unidimensional partial credit model	Understanding of celestial motion and seasons	Middle school students in US	JRST

(continued)

Published studies	Rasch model	Target construct	Sample	Journal
Potgieter et al. (2010)	Unidimensional dichotomous model for multiple-choice concept items; unidimensional rating scale model for Likert-scale confidence items	Confidence on conceptions of mechanics	University physics students in South Africa	IJSE
Saglam and Millar (2006)	Unidimensional dichotomous model	Understanding of electromagnetism	High school students in Turkey and England	IJSE
Susac et al. (2018)	Unidimensional dichotomous model	Understanding of vectors	First-year university students in Zagreb	PRPER
Taasoobashirazi et al. (2015)	Unidimensional Rasch analysis, model unspecified but likely the rating scale model	Metacognition of problem solving in physics	University physics students in US	IJSE
Testa et al. (2019)	Unidimensional partial credit model	Understanding of quantum mechanics	University students in Italy	IJSE
Testa et al. (2020)	Unidimensional dichotomous model for concept items and rating scale model for confidence items	Understanding of and confidence on conceptions of quantum mechanics	High school students in Italy	PRPER
Testa et al. (2015)	Unidimensional dichotomous model	Understanding of seasons, solar and lunar eclipses, and moon phases	Secondary school students in Italy	PRPER
Uccio et al. (2019)	Unidimensional dichotomous and partial credit models for various item scoring methods	Knowledge and reasoning of quantum mechanics	University physics students and high school physics teachers	PRPER
Uccio et al. (2020)	Unidimensional dichotomous model	Reasoning about quantum mechanics	High school students in Italy	PRPER
Vo and Csapo (2021)	Unidimensional and multidimensional dichotomous models	Reasoning in control of variables	High school students in Vietnam	IJSE
Xiao et al. (2019a)	Purportedly unidimensional dichotomous Rasch model, but likely 1-parameter IRT	Conceptual understanding of electromagnetism	University students; nation unspecified	PRPER

(continued)

Published studies	Rasch model	Target construct	Sample	Journal
Xiao et al. (2019b)	Purportedly unidimensional dichotomous Rasch model, but likely 1-parameter IRT	Conceptual understanding of electromagnetism	University students; nation unspecified	PRPER

References

Alonzo, A. C., & Steedle, J. T. (2009). Developing and assessing a force and motion learning progression. *Science Education, 98*(3), 389–421.

Aslanides, J. S., & Savage, C. M. (2013). Relativity concept inventory: Development, analysis, and results. *Physical Review Special Topics – Physics Education Research, 9*(1), 1–10.

Bond, T. G., & Fox, C. M. (2007). *Applying the Rasch model: Fundamental measurement in the human sciences* (2nd ed.). Routledge.Taylor & Francis.

Boone, W. J., & Staver, J. R. (2020). *Advances in Rasch analyses in the human sciences*. Springer Nature.

Cvenic, K. M., Planinic, M., Susac, A., Ivanjek, L., Jelicic, K., & Hopf, M. (2022). Development and validation of the conceptual survey on wave optics. *Physical Review Physics Education Research, 18*(1), 010103. https://doi.org/10.1103/PhysRevPhysEducRes.18.010103

Ding, L. (2012). Applying Rasch theory to evaluate the construct validity of brief electricity and magnetism assessment. In N. S. Rebello, P. Engelhardt, & C. Singh (Eds.), *2011 physics education research conference proceedings* (Vol. 1413, pp. 175–178). American Institute of Physics. https://doi.org/10.1063/1.3680023

Ding, L. (2014). Seeking missing pieces in science concept assessments: Reevaluating the brief electricity and magnetism assessment through Rasch analysis. *Physical Review Special Topics – Physics Education Research, 10*(1), 010105. https://doi.org/10.1103/PhysRevSTPER.10.010105

Ding, L., & Beichner, R. (2009). Approaches to data analysis of multiple-choice questions. *Physical Review Special Topics – Physics Education Research, 5*(2), 020103. https://doi.org/10.1103/PhysRevSTPER.5.020103

Ding, L., Chabay, R., Sherwood, B., & Beichner, R. (2006). Evaluating an electricity and magnetism assessment tool: Brief electricity and magnetism assessment. *Physical Review Special Topics - Physics Education Research, 2*(1), 010105. https://doi.org/10.1103/PhysRevSTPER.2.010105

Ding, L., Wei, X., & Liu, X. (2016). Variations in university students' scientific reasoning skills across majors, years, and types of institutions. *Research in Science Education, 46*(5), 613–632. https://doi.org/10.1007/s11165-015-9473-y

Ene, E., & Ackerson, B. (2018). Assessing learning in small sized physics courses. *Physical Review – Physics Education Research, 14*(1), 010102. https://doi.org/10.1103/PhysRevPhysEducRes.14.010102

Fulmer, G. (2015). Validating proposed learning progression on force and motion using the force concept inventory: Findings from Singapore secondary schools. *International Journal of Science and Mathematics Education, 13*(6), 1235–1254.

Fulmer, G., Liang, L., & Liu, X. (2014). Applying a force and motion learning progression over an extended time span using the force concept inventory. *International Journal of Science Education, 36*(17), 2918–2936.

Glamočić, D. S., Mešić, V., Neumann, K., Sušac, A., Boone, W. J., Aviani, I., et al. (2021). Maintaining item banks with the Rasch model: An example from wave optics. *Physical Review Physics Education Research, 17*(1), 010105. https://doi.org/10.1103/PhysRevPhysEducRes.17.010105

Hestenes, D., Wells, M., & Swackhamer, G. (1992). Force concept inventory. *The Physics Teacher, 30*(3), 141–158.

Ivanjek, L., Morris, L., Schubatzky, T., Hopf, M., Burde, J.-P., Haagen-Schützenhöfer, C., et al. (2021). Development of a two-tier instrument on simple electric circuits. *Physical Review Physics Education Research, 17*(2), 020123. https://doi.org/10.1103/PhysRevPhysEducRes.17.020123

Joint Committee of AERA, APA, & NCME. (2014). *Standards for educational and psychological testing*. American Educational Research Association.

Kane, M. T. (2006). Validation. In R. L. Brennan (Ed.), *Educational measurement* (pp. 17–64). Praeger Publishers.

Kirschner, S., Borowski, A., Fischer, H., Gess-Newsome, J., & Aufschnaiter, C. v. (2016). Developing and evaluating a paper-and-pencil test to assess components of physics teachers' pedagogical content knowledge. *International Journal of Science Education, 38*(8), 1343–1372.

Knaub, A. V., Aiken, J. M., & Caballero, M. (2019). Quantitative methods in PER: A critical examination. *Physical Review – Physics Education Research, 15*(2), 020001. https://doi.org/10.1103/PhysRevPhysEducRes.15.020001

Küchemann, S., Malone, S., Edelsbrunner, P., Lichtenberger, A., Stern, E., Schumacher, R., Brünken, R., Vaterlaus, A., & Kuhn, J. (2021). Inventory for the assessment of representational competence of vector fields. *Physical Review Physics Education Research, 17*(2), https://doi.org/10.1103/PhysRevPhysEducRes.17.020126

Lawson, A. E. (1978). The development and validation of a classroom test of formal reasoning. *Journal of Research in Science Teaching, 15*(1), 11–24.

Linacre, J. (2002). Review of reviews of Bond & Fox (2001). Rasch measurement: Transactions of the Rasch measurement. *SIG American Educational Research Association, 16*(2), 871–872.

Liu, X. (2010). *Using and developing measurement instruments in science education: A Rasch modeling approach.* Information Age Publishing.

Maloney, D. P., O'Kuma, T. L., Hieggelke, C. J., & Van Heuvelen, A. (2001). Surveying students' conceptual knowledge of electricity and magnetism. *American Journal of Physics, 69*(S1), S12–S23.

Marzoli, I., Colantonio, A., Fazio, C., Giliberti, M., & Uccio, U. S. d., & Testa, I. (2021). Effects of emergency remote instruction during the COVID-19 pandemic on university physics students in Italy. *Physical Review Physics Education Research, 17*(2), 020130. https://doi.org/10.1103/PhysRevPhysEducRes.17.020130

Mesic, V., & Muratovic, H. (2011). Identifying predictors of physics item difficulty: A linear regression approach. *Physical Review – Physics Education Research, 7*(1), 010110.

Mesic, V., Hajder, E., Neumann, K., & Erceg, N. (2016). Comparing different approaches to visualizing light waves: An experimental study on teaching wave optics. *Physical Review – Physics Education Research, 12*(1), 010135. https://doi.org/10.1103/PhysRevPhysEducRes.12.010135

Mešić, V., Neumann, K., Aviani, I., Hasović, E., Boone, W. J., Erceg, N., et al. (2019). Measuring students' conceptual understanding of wave optics: A Rasch modeling approach. *Physical Review Physics Education Research, 15*(1), 010115. https://doi.org/10.1103/PhysRevPhysEducRes.15.010115

Neumann, K., Viering, T., Boone, W. J., & Fischer, H. (2012). Towards a learning progression of energy. *Journal of Research in Science Teaching, 50*(2), 162–188.

Oon, P.-T., & Subramaniam, R. (2011a). On the declining interest in physics among students-- from the perspective of teachers. *International Journal of Science Education, 33*(5), 727–746.

Oon, P.-T., & Subramaniam, R. (2011b). Rasch modeling of a scale that explores the take-up of physics among school students from the perspective of teachers. In R. Cavanagh & R. Waugh

(Eds.), *Applications of Rasch measurement in learning environments research*. Sense Publishers.

Oon, P.-T., & Subramaniam, R. (2013). Factors influencing Singapore students' choice of physics as a tertiary field of study: A Rasch analysis. *International Journal of Science Education, 35*(1), 86–118.

Planinic, M. (2006). Assessment of difficulties of some conceptual areas from electricity and magnetism using the conceptual survey of electricity and magnetism. *American Journal of Physics, 74*(12), 1143–1148.

Planinic, M., Boone, W. J., Krsnik, R., & Beilfuss, M. (2006). Exploring alternative conceptions from Newtonian dynamics and simple DC circuits: Links between item difficulty and item confidence. *Journal of Research in Science Education, 43*(2), 150–171.

Planinic, M., Ivanjek, L., & Susac, A. (2010). Rasch model based analysis of the force concept inventory. *Physical Review Physics Education Research, 6*(1), 1–11. https://doi.org/10.1103/PhysRevSTPER.6.010103

Planinic, M., Ivanjek, L., & Susac, A. (2013). Comparison of university students' understanding of graphs in different contexts. *Physical Review - Physics Education Research, 9*(2), 020103. https://doi.org/10.1103/PhysRevSTPER.9.020103

Planinic, M., Boone, W. J., Susac, A., & Ivanjek, L. (2019). Rasch analysis in physics education research: Why measurement matters. *Physical Review - Physics Education Research, 15*(2), 020111. https://doi.org/10.1103/PhysRevPhysEducRes.15.020111

Plummer, J., & Maynard, L. (2014). Building a learning progression for celestial motion: An exploration of students' reasoning about the seasons. *Journal of Research in Science Teaching, 51*(7), 902–929.

Potgieter, M., Malatje, E., Gaigher, E., & Venter, E. (2010). Confidence versus performance as an indicator of the presence of alternative conceptions and inadequate problem-solving skills in mechanics. *International Journal of Science Education, 32*(11), 1407–1429.

Saglam, M., & Millar, R. (2006). Upper high school students' understanding of electromagnetism. *International Journal of Science Education, 28*(5), 543–566.

Smith, R., Schumacker, R., & Bush, M. J. (1998). Using item mean squares to evaluate fit to the Rasch model. *Journal of Outcome Measurement, 2*(1), 66–78.

Susac, A., Planinic, M., Klemencic, D., & Sipus, Z. M. (2018). Using the Rasch model to analyze the test of understanding of vectors. *Physical Review - Physics Education Research, 14*(2), 023101. https://doi.org/10.1103/PhysRevPhysEducRes.14.023101

Taasoobshirazi, G., Bailey, M., & Farley, J. (2015). Physics metacognition inventory part II: Confirmatory factor analysis and Rasch analysis. *International Journal of Science Education, 37*(17), 2769–2786.

Testa, I., Galano, S., Leccia, S., & Puddu, E. (2015). Development and validation of a learning progression for change of seasons, solar and lunar eclipses, and moon phases. *Physical Review – Physics Education Research, 11*(2), 020102. https://doi.org/10.1103/PhysRevSTPER.11.020102

Testa, I., Capasso, G., Collantonio, A., Galano, S., Marzoli, I., Uccio, U. S. D., et al. (2019). Development and validation of a university students' progression in learning quantum mechanics through exploratory factor analysis and Rasch analysis. *International Journal of Science Education, 41*(3), 388–417.

Testa, I., Colantonio, A., Galano, S., Marzoli, I., Trani, F., & Uccio, U. S. d. (2020). Effects of instruction on students' overconfidence in introductory quantum mechanics. *Physical Review Physics Education Research, 16*(1), 010143. https://doi.org/10.1103/PhysRevPhysEducRes.16.010143

Uccio, U. S. D., Colantonio, A., Galano, S., Marzoli, I., Trani, F., & Testa, I. (2019). Design and validation of a two-tier questionnaire on basic aspects in quantum mechanics. *Physical Review Physics Education Research, 15*(1), 010137. https://doi.org/10.1103/PhysRevPhysEducRes.15.010137

Uccio, U. S. D., Colantonio, A., Galano, S., Marzoli, I., Trani, F., & Testa, I. (2020). Development of a construct map to describe students' reasoning about introductory quantum mechanics.

Physical Review Physics Education Research, 16(1), 010144. https://doi.org/10.1103/PhysRevPhysEducRes.16.010144

Vo, D. V., & Csapo, B. (2021). Development of scientific reasoning test measuring control of variables strategy in physics for high school students: Evidence of validity and latent predictors of item difficulty. *International Journal of Science Education, 43*(13), 2185–2205.

WWC. (2022). *What works clearinghouse procedures and standards handbook, Version 5.0*. US Department of Education, Institute of Education Sciences, National Center for Education Evaluation and Regional Assistance (NCEE).

Xiao, Y., Fritchman, J., Bao, J. Y., Nie, Y., Han, J., Xiong, J., et al. (2019a). Linking and comparing short and full-length concept inventories of electricity and magnetism using item response theory. *Physical Review Physics Education Research, 15*(2), 020149. https://doi.org/10.1103/PhysRevPhysEducRes.15.020149

Xiao, Y., Koenig, K., Han, J., Liu, Q., Xiong, J., & Bao, L. (2019b). Test equity in developing short version conceptual inventories: A case study on the conceptual survey of electricity and magnetism. *Physical Review Physics Education Research, 15*(1), 010122. https://doi.org/10.1103/PhysRevPhysEducRes.15.010122

Lin Ding is a professor of science education in the Department of Teaching and Learning at The Ohio State University. He has extensive experience in discipline-based physics education research, including students' conceptual learning, problem solving, scientific reasoning, curriculum development and assessment design.

Chapter 3
Using R Software for Rasch Model Calibrations

Ki Cole and Insu Paek

Abstract This chapter provides an overview of the open-source, freely available R software and introduces Rasch item response modeling programs in R for unidimensional and multidimensional data which are dichotomously or polytomously scored. An introduction provides instructions for installing the software, writing and executing syntax in the R console, and loading packages. The "eRm" package is utilized for performing the simple Rasch analysis for unidimensional, dichotomous data. The "TAM" package is used for illustrating the Partial Credit Model (Masters, Psychometrika, 47, 149–174, 1982) for unidimensional, polytomous data. The "mirt" package is utilized for performing between-item multidimensional Rasch analysis for dichotomous data.

3.1 Introduction

Advances in Applications of Rasch Measurement in Science Education provides research studies of various Rasch models to science education data using various software. Commonly used software for performing Rasch analyses are WINSTEPS (Linacre, 2022) and ConQuest (Adams et al., 2020). However, the R software (R Core Team, 2020) is open-source and freely available to anyone. R may be used for many types of analyses, including Rasch analyses.

This chapter provides an overview of R features and introduces open-source free Rasch item response modeling programs in R for unidimensional and multidimensional data which are dichotomously or polytomously scored.

K. Cole (✉)
Oklahoma State University, Stillwater, OK, USA
e-mail: ki.cole@okstate.edu

I. Paek
Human Resources Research Organization (HumRRO), Alexandria, VA, USA
e-mail: ipaek@humrro.org

X. Liu, W. J. Boone (eds.), *Advances in Applications of Rasch Measurement in Science Education*, Contemporary Trends and Issues in Science Education 57,
https://doi.org/10.1007/978-3-031-28776-3_3

Rasch modeling has been a major latent measurement approach in practice and research. In the past, specialized commercial programs for the Rasch modeling (e.g, WINSTEPS, ConQuest or recent SAS PROC IRT procedures (Cole & Paek, 2017, 2022)) were dominant choices to conduct Rasch analysis. However, the availability of fast personal computers and free open-source computing environments, such as R, generated more options to users and practitioners of Rasch analysis. Those introduced R Rasch packages ("eRm", "TAM", and "mirt" are used in this study) are free open-source programs. Because these programs are syntax-based, the very beginners may take some time to get used to how to use them, but the benefits of getting used to and learning more about them outweigh the efforts to learn those programs. First of all, as mentioned, they are free and cost-effective, so they are readily accessible to anyone who wants to learn. In this sense, for teaching purposes in a course for Rasch modeling, they could be very attractive. Secondly, there exists some evidence that the performance of some of those programs are comparable to recent commercial item response theory software and the differences between those R Rasch programs are practically minimal (e.g., Paek et al., 2021; Xu & Paek, 2016). As long as popular conventional Rasch models (e.g., simple Rasch model for dichotomous data or PCM for polytomous models) are used, these R Rasch programs have little or no problem in research settings. Also, between different R programs, when the simple Rasch or PCM is considered, it is a matter of the user's choice regarding which R programs should be adopted. The use of these R Rasch programs may be considered even for more high-stakes operational purposes. In that case, we recommend some replication studies suitable for the operational context, though it is expected to be no or little concern for the use of the simple Rasch or PCM implementation with the R Rasch programs.

For those unfamiliar with Rasch Measurement, readers are encouraged to review the first chapter of the 1st Edition of Applications of Rasch Measurement in Science Education titled, Introduction to Rasch Measurement in Science Education (Liu & Boone, 2006). The chapter provides details on types of measures, dimensionality, dichotomous and polytomous Rasch models, assumptions of Rasch modeling, model estimations, and additional resources.

3.2 What Is R?

R is a free software which provides a statistical computing environment and graphics. All Rasch item response calibrations in R require additional R programs. The programs in R are also referred to as "packages." The installation of R and the installation of the packages for the Rasch modeling are described in this section. This chapter utilizes applications on a PC computer. R is also available and may be utilized on a Mac computer.

3.2.1 *Installation of the R Software and RStudio*

The R software may be freely downloaded from cran.r-project.org for Windows, macOS, and Linux. Users should follow on-screen instructions for the complete download. Once completed, the software may be opened.

Though the console version of R can be used without any problems, the use of the shell program "RStudio" could facilitate a more user-friendly experience. The RStudio application is also freely available for download at rstudio.com/products/rstudio/download. Users should download the free open source license RStudio Desktop and follow on-screen instructions to complete the installation.

Users may view the RStudio window with an inset RConsole, Environment, and Files windows. We recommend opening a new script upon first opening the application. To do this, click File, then New File, then R Script. Users can rearrange the tiles by clicking Tools and then Global Options. Within the Options window, click Pane Layout. Here, the desired window can be selected for each panel. Figure 3.1 shows the window displays.

The "Untitled1" window is the new script where code may be written and saved (we recommend saving by clicking File and Save, choosing the desired location and file name). Saving the script allows users to open the script and using and editing the

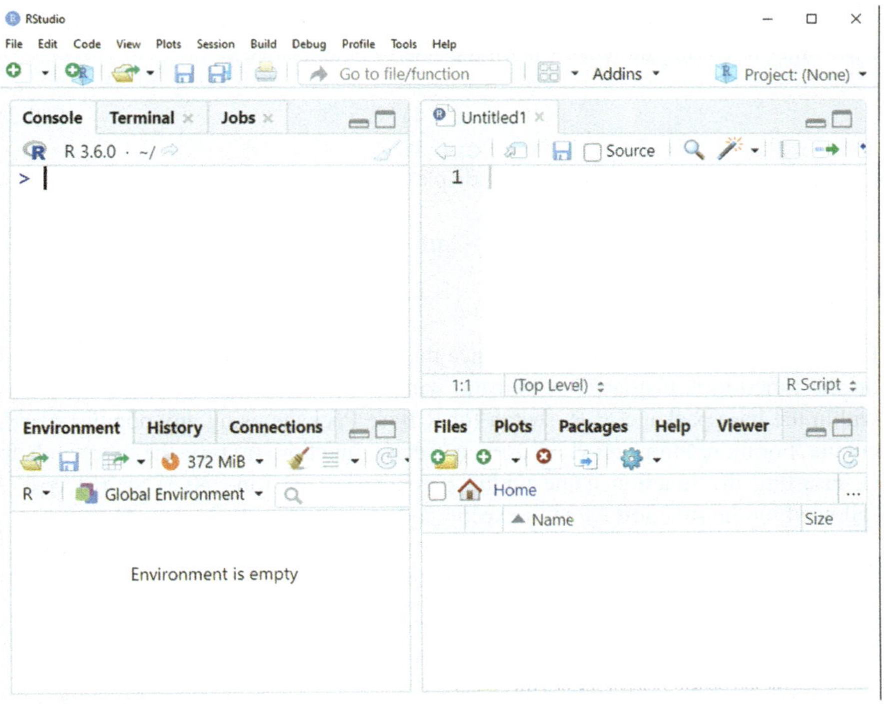

Fig. 3.1 RStudio Software windows

script as needed in future analyses. The Console is where the R script is executed and output is printed. The Environment displays all objects currently stored within the R session. And the fourth panel contains Files within the working directory, Plots displayed, Packages loaded within the session, and Help resources.

In the remaining chapter, lines of script are provided and numbered for reference. Note, within the script, the "#" can be used for comments. Any text after the "#" (within the same line) is not executed within the Console. The first line is provided below as Line #0.1. To get started, in the Script, users may type

$$1 + 1 \qquad (\#0.1)$$

Then, with the cursor still on the same line, push Ctrl + Enter to execute that line of script on the Console. On the Console panel, users should see a line with the executed script starting with ">" and a second line with the calculated solution, or output, next to "[1]":

```
> 1 + 1
[1] 2
```

Within R, functions are used to call pre-specified equations and sets of code. For example, the sqrt() function calculates the square root of a value provided within the parenthesis. Type the following line (#0.2) into the script and execute the line using Ctrl+Enter (ensuring the cursor is on the line to be executed).

sqrt(4) (#0.2)

On the console, the following should display:

```
> sqrt(4)
[1] 2
```

Users may notice that while typing, once the open parenthesis is typed, information on the function is displayed within the window. For a complete list of functions within the base package of R, search for R Base Package using any popular search engine. For more information on using a specific function, (a) type the function help (), inserting the function name within parenthesis (see Line #0.3), or (b) type ? followed by the function name to execute a documentation (see Line #0.4).

help(sqrt) (#0.3)

?sqrt (#0.4)

These open the R Documentation for the function of interest (in this example, sqrt()) in the Help panel. The R Documentations may be very helpful to users. Documentations provide a description of the function, how the functions may be used, the input arguments accepted, additional details, and examples of the function uses.

3.2.2 *Loading Data*

For uploading data, users should first set the working directory from which R operates. First, on the C: drive, create a folder and give it a name, e.g., workingdirectory. Now, in the R script, use the setwd() function to set the working directory to this new folder, shown in Line #0.5:

$$\text{setwd(“C : /workingdirectory”)} \qquad (\#0.5)$$

Next, in the Files pane, click “More” and select “Go to Working Directory”. This allows users to see files that are more easily accessible and where files may be written. In our C://workingdirectory file, we have the three datasets used in the examples of this chapter. These files are available in specific packages of R (instructions for accessing these are provided below) and are also available in the supplementary materials.

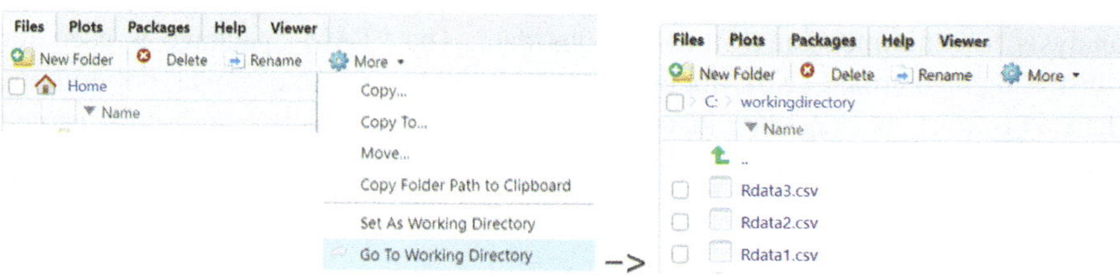

R may accept many file types. Most commonly, a .csv file is used with the function read.table(). In line #0.6 below, the read.table() function is used. The first argument within the parenthesis is the file name (i.e., “Rdata1.csv”; if the working directory has not been set, the file name “C:/workingdirectory/Rdata1.csv” may be used). The second argument indicates that the imported file is separated by a comma (e.g., sep=“,”) and the third argument indicates that the file has a header (i.e., header=TRUE). If a file does not have a header, the argument header=FALSE should be used.

$$\text{read.table(“Rdata1.csv”, sep = “, ”, header = TRUE)} \qquad (\#0.6)$$

3.3 Rasch Model Applications in R

3.3.1 R Programs/Packages for Rasch Modeling

One may think of the R software as a large library with thousands of books, or packages/programs, for which specific types of analysis may be performed. This chapter will introduce and focus on the use of three packages: "eRm" (Mair et al., 2021), "mirt" (Chalmers, 2012), and "TAM" (Robitzsch et al., 2021). For unidimensional dichotomous data modeling, we will use "eRm", which is equipped with producing an output like "Wright Map" and the conditional maximum likelihood (CML) estimation method, which fits the idea of "specificity" in the Rasch modeling best; the "eRm" also has capabilities of fitting a partial credit model (PCM) for unidimensional polytomous data. The "TAM" package provides the joint maximum likelihood estimation (JML) and the marginal maximum likelihood (MML) estimation. The "TAM" package can be used to conduct unidimensional modeling (with JML) and both unidimensional and multidimensional modeling (with MML) with both dichotomous and polytomous data. The "mirt" program which uses MML has a very wide scope of functionality for diverse models including the multidimensional modeling; it is used to illustrate the Rasch model for multidimensional dichotomous data. See Table 3.1 for details of each package, including dichotomous and polytomous models available, estimation procedure, and availability of multidimensional analysis and Wright Map.

In this chapter, the "eRm" package is utilized for performing the simple Rasch analysis for unidimensional, dichotomous data. The "TAM" package is used for illustrating the Partial Credit Model (Masters, 1982) for unidimensional, polytomous data. The "mirt" package is utilized for performing between-item multidimensional Rasch analysis for dichotomous data.

Table 3.1 Modeling and features for Rasch modeling R packages

			Packages	
		"eRm"	"TAM"	"mirt"
Dichotomous	Rasch	X	X	X
Polytomous	Partial Credit	X	X	X
	Rating Scale	X	X	X
Item Estimation	CML	X		
	JML		X (Uni)	
	MML		X (Uni, Multi)	X
Multidimensional			X	X
Wright Map		X		

Note. *Uni* unidimensional, *Multi* multidimensional

3.3.2 Package Installation

After installation of the R software and RStudio, users must install the three packages utilized in this chapter individually. Packages need to be installed on a computer system only once, and packages need to be uploaded within the R session upon each new execution of the R software. Use the install.packages() function, inserting the package name within quotation marks in the parenthesis. Lines #0.7, #0.8, and #0.9 present the script for installing the "eRm", "TAM", and "mirt" packages, respectively.

```
install.packages("eRm")     (#0.7)
install.packages("TAM")     (#0.8)
install.packages("mirt")    (#0.9)
```

Each line may be typed directly into the Console and executed individually. Or, the lines may be typed into a Script; to execute each line, place the cursor on the desired line to be executed and press Ctrl + Enter. Note, the R syntax is case sensitive, and upper- and lowercase letters need to be used as specified. After executing each install.packages() function, users may select a CRAN mirror from a pop-up window.

After installing the package, each package should be loaded into the R session using the function, library() and inserting the package name within parenthesis. It is not recommended to load all packages simultaneously, as some overlap in functions may cause a problem. This will be done individually in the following instructions.

3.3.3 Unidimensional Rasch Application for Dichotomous Data (Using the "eRm" package)

The "eRm" package is used to illustrate the Rasch Modeling of unidimensional dichotomous data. First, users should load the package (ensure that the package was previously installed using line #0.7). using the library() function, shown in #1.1

```
library(eRm)     (#1.1)
```

The package contains several artificial data sets for applying Rasch modeling. In this example, we will use the raschdat3 data set. This data contains dichotomous responses (i.e., 0s and 1s) to 6 items from 500 respondents. For application, readers may think of this as a 6-item science test, where items are answered correct (1) or incorrect (0), administered to 500 students. Following, load the raschdat3 dataset within the "eRm" package and assign it to the object named data1 using the following script:

data1 <- raschdat3 (#1.2)

In line #1.2, the "<-" operator is used to assign the contents of the object on the right-hand side to the object name on the left-hand side.

In RStudio, a second method can be used to load data that is in the working directory. The Rdata1.csv contains the data within the raschdat3 object. Users may download it and save it within the working directory folder. Within the Files pane, click the dataset to load, then click "Import Data". This opens a window that contains the file location, a preview of the data, and code to execute. Click "Import." The Rdata1 dataset is loaded and may be called using the Rdata1 name. To rename, type and execute, data1 <- Rdata1.

Now, contents and details of the data1 may be viewed using lines #1.3, #1.4, and #1.5.

head(data1) (#1.3)

dim(data1) (#1.4)

names(data1) (#1.5)

Line #1.3 prints the first six rows of the data. Line #1.4 prints the dimensions of the data, i.e., the number of rows and the number of columns. The raschdat3 data set has 500 rows (or person respondents) and 6 columns (or items). Line #1.5 prints the column names, or the item variable names. Figure 3.2 shows the RConsole output from lines #1.1 through #1.5.

Fitting the Rasch Model

Within the "eRm" package, the RM() function is used to fit the input data set to the unidimensional Rasch model for dichotomous data. In line #1.6, the calibrated result is saved in the object mod.rm.

```
> library(eRm)
> data1 <- raschdat3
> #data1 <- Rdata1  use if loaded data directly from the working directory.
> head(data1)
  I1 I2 I3 I4 I5 I6
1  1  1  1  1  1  0
2  1  1  0  1  0  1
3  1  1  1  1  1  1
4  1  1  0  1  0  1
5  1  1  0  0  0  0
6  0  0  1  1  1  0
> dim(data1)
[1] 500   6
> names(data1)
[1] "I1" "I2" "I3" "I4" "I5" "I6"
```

Fig. 3.2 R output of lines #1.1 to #1.5

mod.rm <- RM(data1) (#1.6)

After executing line #1.6, users may notice that no output is produced. If an error occurred, however, R would output an error script. Similar to lines #0.3 and #0.4, the help(RM) or ?RM may be used to provide the detailed documentation of the RM() function. A check of the model convergence should be done prior to reviewing the estimated item and person parameters. An attribute of the saved mod.rm object is a convergence check. This may be printed using the object name, followed by $conv shown in line #1.7:

mod.rm*conv* (#1.7)

An output of the value 1 indicates the calibration reached convergence within the default stopping criteria without problems.

Item Parameter Estimation

Estimated item parameters are also an attribute of the fitted mod.rm object using $betapar. The beta parameters are the *easiness* parameters. In a traditional Rasch model, a difficulty parameter is desired. This is printed by including a minus sign before the called object and argument (see Line #1.8). Additionally, the $se.beta attribute produces the standard errors of the estimated item parameter. Lines #1.8 and #1.9 print the item difficulty parameters and their standard errors, respectively.

-mod.rm*betapar* (#1.8)

mod.rmse.beta (#1.9)

Another option for printing the item easiness (or item difficulty) and the standard errors is to use the summary() function, which includes the easiness (beta) parameter estimate, standard error, and 95% confidence interval estimates (see Line #1.10):

summary(mod.rm) (#1.10)

Figure 3.3 shows the output of lines #1.8 through #1.10. Notice that the beta parameters from Line #1.8 are the Item Easiness Parameters printed from Line #1.10. From the output of Line #1.10, the first output is for the Item (Category) Difficulty Parameters (eta, η), which are the parameterizations under the $\sum b_i = 0$ constraint. The result of the first item is not shown, but can be obtained by the negative sum of all other item estimates. (The item difficulty of Item 1 is $-\eta_1 = 1.233$.) The Item Easiness Parameters (beta, β) are the easiness estimates under the same constraint.

In this set of 6 items, the first item is the easiest (having the lowest difficulty parameter) and the 6th item is the hardest (having the highest difficulty parameter).

```
> -mod.rm$betapar
   beta I1     beta I2     beta I3     beta I4     beta I5     beta I6
-1.2327218 -0.8460782 -0.2679352  0.2211616  0.8166714  1.3089022
> mod.rm$se.beta
[1] 0.10493664 0.09637705 0.08850867 0.08618090 0.08877572 0.09570038
> summary(mod.rm)

Results of RM estimation:

Call:  RM(X = data1)

Conditional log-likelihood: -1001.088
Number of iterations: 14
Number of parameters: 5

Item (Category) Difficulty Parameters (eta): with 0.95 CI:
   Estimate Std. Error lower CI upper CI
I2   -0.846      0.096   -1.035   -0.657
I3   -0.268      0.089   -0.441   -0.094
I4    0.221      0.086    0.052    0.390
I5    0.817      0.089    0.643    0.991
I6    1.309      0.096    1.121    1.496

Item Easiness Parameters (beta) with 0.95 CI:
         Estimate Std. Error lower CI upper CI
beta I1     1.233      0.105    1.027    1.438
beta I2     0.846      0.096    0.657    1.035
beta I3     0.268      0.089    0.094    0.441
beta I4    -0.221      0.086   -0.390   -0.052
beta I5    -0.817      0.089   -0.991   -0.643
beta I6    -1.309      0.096   -1.496   -1.121
```

Fig. 3.3 R output of lines #1.8 to #1.10

The difficulty parameter corresponds to the point on the Theta, or ability scale, where respondents at that ability level have a 50% probability of responding correctly to the item.

For Item 2 (η_2) is -0.846. Someone with an ability of -0.846 has a 50% probability of responding correctly to the item. Those with an ability level less than -0.846 are likely to answer the item incorrectly (less than 50% probability of correct), and those with an ability level greater than -0.846 are likely to answer the item correctly (greater than 50% probability of correct).

Item Characteristic Curve (ICC)

The item characteristic curve (ICC) plots the probability of answering each item correct on a continuum. An ICC has a lower asymptote of 0 and an upper asymptote of 1, and is monotonically increasing. The ICC is centered over $\theta = \eta$, where the curve at $\theta = \eta$ is .50.

The item characteristic curves that correspond to the estimated item parameter can be produced using the plotICC() function with arguments to specify which item(s) ICC to print. Lines #1.11, #1.12, and #1.13 are used to print ICCs of items or subsets of items.

```
plotICC(mod.rm)                                  (#1.11)
plotICC(mod.rm, 1 : 3)                           (#1.12)
plotICC(mod.rm, item.subset = c(1,3,5))          (#1.13)
```

Line #1.11 prints all ICCs in a sequence of 4x4 windows. When the RGraphics window displays, users click anywhere on the window to produce the first four plots. Clicking again produces the next four plots (until the maximum number of plots has been produced).

Line #1.12 prints the subset of items 1, 2, and 3 (starting at 1 and stopping at 3 as indicated by 1:3). Line #1.13 prints the subset of items 1, 3 and 5 using the argument item.subset=c(1,3,5).

If the Plot panel was used, it may need to be reset. To clear the Plot panel, use Line #1.14. If the RGraphics window was used, simply close the window.

```
dev.off(dev.list()["RStudioGD"])                 (#1.14)
```

Person Ability Parameter Estimation

A unidimensional Rasch score may be estimated for each unique raw sum score using the function person.parameter() and supplying the Rasch model object. In Line #1.15, the extracted person scores are assigned to the object p.rm.

```
p.rm < -person.parameter(mod.rm)                 (#1.15)
```

To print the Rasch person parameter estimates for each unique raw sum score, simply type the object name:

```
p.rm                                             (#1.16)
```

In this example containing 6 dichotomously scored items, the lowest score of 0 corresponds to answering all items incorrectly, and the highest score of 6 corresponds to answering all items correctly. Figure 3.4 shows the output for this dataset having six items, where the lowest raw score is 0 and the highest raw score is 6.

To view a plot of the raw sum score versus the person parameter estimate, use the function plot() with the person score argument; shown in Line #1.17 and printed in Fig. 3.5.

```
plot(p.rm)                                       (#1.17)
```

Fig. 3.4 R output of lines #1.15 and #1.16

```
> p.rm <- person.parameter(mod.rm)
> p.rm

Person Parameters:

 Raw Score    Estimate Std.Error
         0 -2.94837869        NA
         1 -1.85880537 1.1464002
         2 -0.82331722 0.9345989
         3 -0.00373927 0.8917881
         4  0.81854659 0.9374572
         5  1.86154722 1.1506001
         6  2.95961894        NA
```

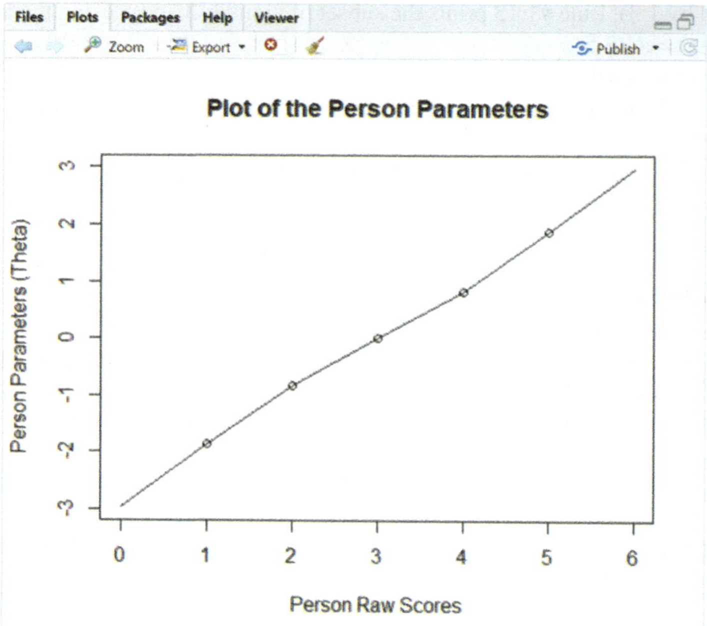

Fig. 3.5 Output of plot(p.rm), #1.17

Person-Item Map

In the "eRm" package, there is a function "plotPImap()" to create a Person-Item Map figure, which is similar to the Wright Map. Line #1.18 is used to print the Person-Item Map, supplying the mod.rm object.

plotPImap(mod.rm) (#1.18)

In the Person-Item Map, a bar graph of person abilities, or person parameters, is shown on the top panel. In the bottom panel, item difficulties are located using black dots. The Person-Item Map contains the same information as a Wright Map, which exhibits both person and item locations together on the latent continuum. This facilitates examining relative positions of person abilities and item difficulties. This Person-Item Map is part of the functionality in the "eRm" package. The other packages with Rasch modeling included in this article do not have this functionality. The output of the above command is shown below in Fig. 3.6.

There is also a stand-alone package which can create the Wright Map. The package is "WrightMap". One can download this package using the install.packages() function, as shown in Line #1.19, and load the packaging using the library() function, as shown in Line #1.20.

install.packages("WrightMap") (#1.19)

library(WrightMap) (#1.20)

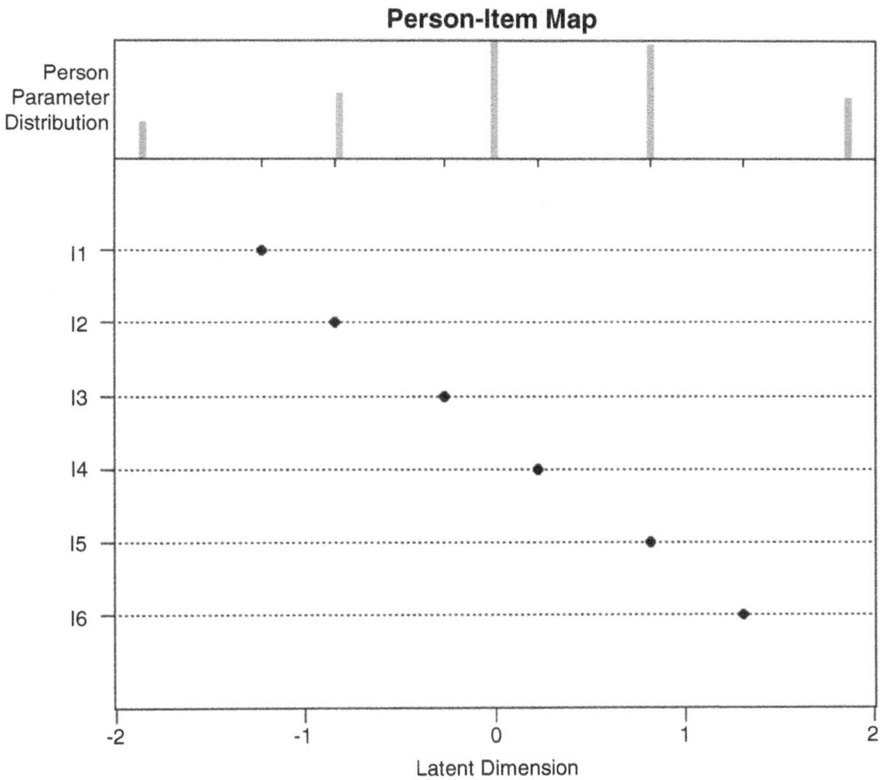

Fig. 3.6 Output of plotPImap(mod.rm), #1.18

Once installed and loaded, the required inputs are the person abilities and the item difficulties. The person abilities are obtained from the p.rm object (from Line #1.15) with the additional arguments $thetapar and $NAgroup1. Because the input of WrightMap must be a matrix, use the as.matrix() function. For the item difficulties, use the -mod.rm$betapar object (from Line #1.8). The full expression of the function is in Line #1.21.

$$\text{WrightMap(as.matrix(p.rm}thetapar\text{NAgroup1), -mod.rm}betapar) \qquad (\#1.21)$$

Execution of the line produces a plot similar to Fig. 3.6, though panels are side-by-side. Figure 3.7 is the WrightMap produced by Line #1.21. In this Wright Map, the histogram of person ability is shown on the left panel while the locations of item difficulties are shown on the right-hand side. Because the "WrightMap" is a stand-alone package, it can be used with any Rasch program estimates for person and items. The "WrightMap" package can be also used directly with the output files with ConQuest. More interested readers are referred to the package help file.

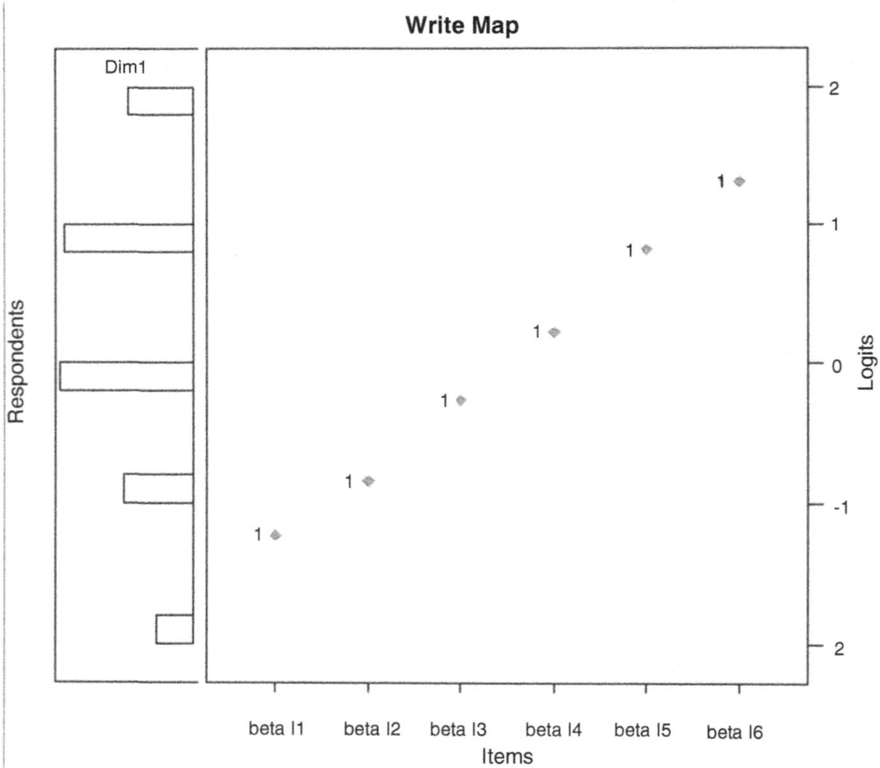

Fig. 3.7 Output of WrightMap(as.matrix(p.rm$thetapar$NAgroup1),-mod.rm$betapar), #1.21

Model Evaluation

In addition to ensuring that the Rasch calibration converged, the model-data fit should be evaluated. There exists numerous statistics for the model evaluation, and a single test should not be accepted exclusively for the absolute decision of a model-data fit. "eRm" provides a few global model-data fit statistics: Andersen's likelihood ratio (LR) test (Andersen, 1973) and the Wald test (Fischer & Molenaar, 1995).

Andersen's LR test (1973) splits the data into two groups based on the median raw score, comparing the calibration results of the two groups. If the data is a good fit to the model, the null hypothesis assumes that the calibration on each subset of the data is equal, or identical. Utilizing the LRtest() function is illustrated in Line #1.22.

$$\text{rm.lrt} < \text{-LRtest(mod.rm)} \qquad (\#1.22)$$

$$\text{rm.lrt} \qquad (\#1.23)$$

Line #1.22 executes the LR test using the LRtest() function and assigns the results to the object rm.lrt. Line #1.23 prints the results of the LR test, providing the statistics, degrees of freedom and p-value. As with all hypothesis tests, when a .05 level of significance is used, $p < .05$ indicates that the null hypothesis is rejected, and the model-data fit is not adequate. A result of $p > .05$ indicates that the null hypothesis is retained, the model-data fit is adequate.

For the data1 fit to the Rasch model, the output is shown in Fig. 3.8. The results indicate that the Rasch model fit is adequate for the data ($p > .05$).

A second test is the Wald Test, which is executed using the Waldtest() function, illustrated in Line #1.24. The Wald Test examines each item fit to the model, with a null hypothesis that the model-data fit is adequate for each item.

$$\text{Waldtest(mod.rm)} \qquad (\#1.24)$$

In order to control the family-wise Type I error rate to $\alpha = .05$, a Bonferroni adjustment may be considered with α divided by the number of items. For this case of 6 items, $\frac{.05}{6} = .008$ would be the level of significance. The p-value of all items is >.008 (shown in Fig. 3.9), therefore no items are flagged for misfitting the Rasch model.

In addition to the overall model-data fit above, fit indices at the item level are available. They are the infit and outfit measures which utilize a mean-square (MSQ) or a standardized statistic. Infit is an information-weighted fit statistic, whereas the

Fig. 3.8 R output of lines #1.22 and #1.23

```
> rm.lrt <- LRtest(mod.rm)
> rm.lrt
Andersen LR-test:
LR-value: 6.11
Chi-square df: 5
p-value:  0.296
```

Fig. 3.9 R output of Waldtest(mod.rm), line #1.24

```
> Waldtest(mod.rm)
Wald test on item level (z-values):
        z-statistic p-value
beta I1       0.827   0.408
beta I2      -0.123   0.902
beta I3       0.534   0.594
beta I4       1.067   0.286
beta I5      -2.162   0.031
beta I6      -0.224   0.823
```

```
> itemfit(p.rm)
Itemfit Statistics:
     Chisq  df p-value Outfit MSQ Infit MSQ Outfit t Infit t Discrim
I1 468.322 478   0.615      0.978     0.995   -0.177  -0.058  -0.181
I2 431.906 478   0.936      0.902     0.893   -1.193  -2.076   0.283
I3 451.794 478   0.800      0.943     0.949   -0.952  -1.235   0.174
I4 483.423 478   0.422      1.009     1.006    0.192   0.178  -0.052
I5 403.965 478   0.994      0.843     0.883   -2.550  -2.982   0.341
I6 418.494 478   0.977      0.874     0.909   -1.494  -1.887   0.012
```

Fig. 3.10 R output of itemfit(p.rm), line #1.25

outfit is an outlier-sensitive fit. The MSQ measures a size of randomness of the item response, where values close to 1.0 are desired, indicating the response is not too redundant and not too random. Values less than 1.0 correspond to a data overfit, or redundancy; values greater than 1.0 correspond to a data underfit, or unpredictable or noisy. The standardized statistic or *t*-statistic is a similar measure, where values less than 0.0 indicate an overfit, and values greater than 0.0 indicate an underfit. (The *t* statistic here is not the well-known student *t*-test in statistics.) The itemfit() function is utilized in Line #1.25, where the argument supplied is the person-fit object (see Line #1.15):

$$\text{itemfit(p.rm)} \qquad (\#1.25)$$

Figure 3.10 shows the item fit statistics. According to Linacre (2002), MSQ infit and outfit statistics between 0.5 and 1.5 indicate acceptable itemfit. The printed MSQ outfit and infit statistics would be treated as acceptable for all items. Items with a *t* statistic between -1.9 and 1.9 have an acceptable itemfit. The infit *t*-statistic for item 2 and the infit and outfit *t*-statistics for item 5 are less than the acceptable range.

In the closing of this section using the “eRm” package and preparing for the “TAM” package in the next section, the “eRm” package should be detached from the R session using the detach() function:

$$\text{detach(“package : eRm”)} \qquad (\#1.26)$$

3.3.4 Unidimensional Rasch Application for Polytomous Data (Using "TAM" for PCM)

The "TAM" package is used to illustrate the Rasch Modeling of unidimensional polytomous data, specifically using the partial credit model (PCM). First, users should load the package (ensure that the package was previously installed using line #1.26).

```
library(TAM)                                    (#2.1)
```

The data set provided in the TAM Package for this application is called data.gpcm. This data set contains responses from 392 respondents to 3 *Attitude to Science and Technology* items (Karlheinz & Melich, 1992): (1) Comfort: Science and technology are making our lives healthier, easier, and more comfortable; (2) Work: The application of science and new technology will make work more interesting; and (3) Benefit: The benefits of science are greater than any harmful effect it may have. Item responses were on a four-point scale where 1 = "strongly disagree", 2 = "disagree to some extent", 3 = "agree to some extent", and 4 = "strongly agree". Following, load the data.gpcm dataset within the "TAM" package (line #2.2) and assign it to the object named data2 (line #2.3) using the following script:

```
data(data.gpcm)                                 (#2.2)
data2 < -data.gpcm                              (#2.3)
```

The data within data.gpcm is provided within the supplementary materials as Rdata2.csv. Users may also download and save it in the working directory folder. Within the Files pane, click the dataset to load, then click "Import Data". This opens a window that contains the file location, a preview of the data, and code to execute. Click "Import." The Rdata2 dataset is loaded and may be called using the Rdata2 name. To rename, type and execute, data2 <- Rdata2.

Contents of the data2 dataset may be viewed using

```
head(data2)                                     (#2.4)
dim(data2)                                      (#2.5)
names(data2)                                    (#2.6)
```

The data contains responses from 392 respondents to 3 polytomous items, where responses were 0, 1, 2, or 3.

Fitting the Partial Credit Model

The "TAM" package can fit Rasch family models using the marginal maximum likelihood (MML) estimation or the joint maximum likelihood (JML) estimation.

The MML method is the preferred choice when ability distributions are assumed to follow a normal distribution. Within the "TAM" package, the tam() or tam.mml() function is used for the MML estimation, and tam.jml() is used for JML estimations. In tam.mml(), the mean of the normal ability distribution is fixed as zero for the model identification while the variance is freely estimated. Line #2.7 executes the MML estimation for the PCM.

$$\text{mod.pcm} < \text{-tam.mml}(\text{data2}, \text{irtmodel} = \text{``PCM''}) \qquad (\#2.7)$$

The first argument is the polytomous data set, and the second argument specifies the model. Additional model arguments are "PCM2" for the partial credit model parameterization used by ConQuest and "RSM" for the rating scale model. Characteristics of the object may be viewed by typing and executing the object name, mod.pcm.

The tam.mml() function also prints the estimation changes after each iteration of the calibration, final item parameters, variance-covariance matrix, EAP reliability, and time to execute.

Item Parameter Estimation

The partial credit model estimates a threshold parameter for the estimated intersection point between adjacent categories. The threshold parameter estimation attributes and corresponding standard errors are stored as an attribute to the mod.pcm object using $xsi, shown in Line #2.8.

$$\text{mod.pcmxsi} \qquad (\#2.8)$$

For the 4-category response data, three threshold parameters are estimated. These are titled using the item name and appending "_Cat1", "_Cat2", and "_Cat3". The round () function is often useful to print numbers rounded to a specified decimal point. In Line #2.9, values are rounded to the third decimal point:

$$\text{round}(\text{mod.pcmxsi}, 3) \qquad (\#2.9)$$

Figure 3.11 prints the three threshold estimates and their standard errors for the three items.

The three category threshold estimates are ordered within each item. On this *Attitude to Science and Technology* construct scale, lower values correspond to a more negative attitude, and higher values correspond to a more positive attitude. For item 1 (Comfort), those with an attitude score less than −1.302 are most likely to respond "strongly disagree" to the item, those with an attitude score between −1.302 and 2.456 are most likely to response "disagree to some extent", those with an attitude score between 2.456 and 2.727 are most likely to response "agree to some extent", and those with an attitude score greater than 2.727 are most likely to respond "strongly agree".

Fig. 3.11 R output of round (mod.pcm$xsi, 3), line #2.9

```
> round(mod.pcm$xsi,3)
                  xsi se.xsi
Comfort_Cat1 -1.302  0.128
Comfort_Cat2  2.456  0.181
Comfort_Cat3  2.727  0.464
Work_Cat1    -1.706  0.155
Work_Cat2     0.859  0.118
Work_Cat3     1.681  0.195
Benefit_Cat1 -1.233  0.137
Benefit_Cat2  0.829  0.120
Benefit_Cat3  2.210  0.236
```

Category Characteristic Curve

The category characteristic curves (CCCs) display the probability of responding in each category at any θ ability level (in this example, *attitude*). Due to the ordering of the threshold parameters, the CCC of the first category is monotonically increasing, the CCC of the last category is monotonically decreasing, and the CCCs of the inner categories are bell-shaped. When viewing the CCC, users can determine the probability of responding in any category for a specified θ. The CCCs are printed using the plot() function and specifying arguments: the TAM model, items, type (i.e., "expected" item response curves or "items" for fitted category curves), and export logical. The plot() function takes four arguments: the TAM object, the item number, type of plot requested, and an export request. Lines #2.10, #2.11, and #2.12 supply the fitted TAM model, followed by the requested item (1, 2, or 3, respectively); then, "items" type requests the item characteristic curve. Lastly, export=FALSE instructs R not to export the file. The execution of the commands below for the category characteristic curves require the "plyr" package, which should be installed prior to the implementation of the plot commands below. Users may need to first excute install.packages("plyr") and then load the package using library(plyr).

plot(mod.pcm, items = 1, type = "items", export = FALSE) (#2.10)

plot(mod.pcm, items = 2, type = "items", export = FALSE) (#2.11)

plot(mod.pcm, items = 3, type = "items", export = FALSE) (#2.12)

Executing the plot function loads a new R Graphics window printing the category characteristic curves for the specified item. Figure 3.12 shows the category characteristic curves for Item 1, Comfort. Notice that the adjacent category curves intersect at the estimated thresholds estimated in Line #2.8.

Review the estimated category thresholds from Line #2.10 of item 1 (Comfort) and align the values with the θ scale. Notice that the threshold values correspond to the intersection between adjacent category curves.

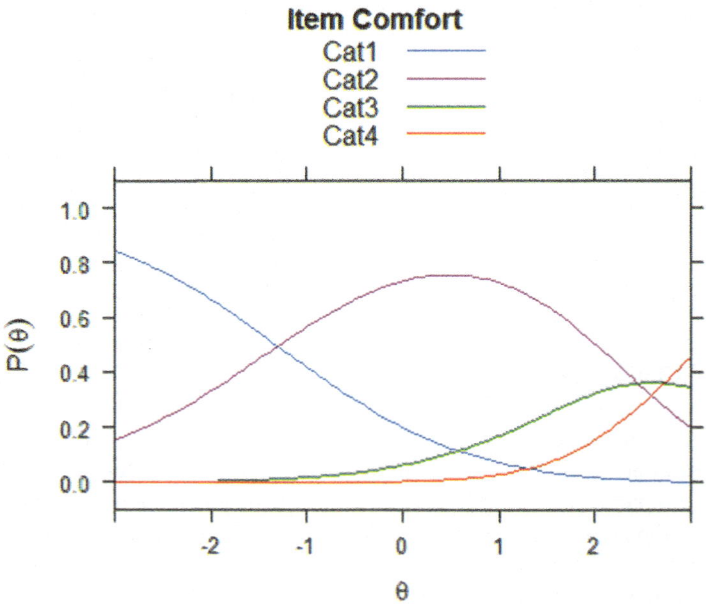

Fig. 3.12 R output of plot(mod.pcm, items=1, type="items", export=FALSE), line #2.10

Person Ability Parameter Estimation

When the PCM is fit using the tam.mml() function, the object contains the attribute $person, which contains the unidimensional EAP person scores. In line #2.13, the extracted EAP person scores are assigned to the object p.pcm.

$$\text{p.pcm} < \text{-mod.pcm}person \tag{\#2.13}$$

To print the maximum likelihood (ML) Rasch person parameter estimates for each sum score, simply type the object name:

$$\text{p.pcm} \tag{\#2.14}$$

Use head(p.rm) to print only the first six rows. Figure 3.13 shows the output of the EAP estimates (Lines #2.13 and #2.14) for this dataset having three polytomous items, where the columns of the person estimates object contains the raw sum score (titled, "score"), the EAP estimate, and the EAP standard deviation (i.e., the standard deviation of the posterior distribution).

In addition to EAP person estimates, TAM also provides Warm's weighted likelihood estimate (WLE) person estimates (Warm, 1989). To extract the WLE person estimates, first use the tam.wle() function to compute the estimated scores,

```
> p.pcm <- mod.pcm$person
> head(p.pcm)
  pid case pweight score max        EAP    SD.EAP
1   1    1       1     2   9 -0.4706645 0.6104115
2   2    2       1     3   9 -0.1036192 0.6012995
3   3    3       1     4   9  0.2525834 0.5923908
4   4    4       1     4   9  0.2525834 0.5923908
5   5    5       1     4   9  0.2525834 0.5923908
6   6    6       1     1   9 -0.8490869 0.6200517
```

Fig. 3.13 R output of lines #2.13 and #2.14

and then the $theta argument to extract the values. See #2.15 and #2.16 for demonstration with the mod.pcm object.

$$\text{p.pcm.wle} < \text{-tam.wle(mod.pcm)} \tag{\#2.15}$$

$$\text{p.pcm.theta} < \text{-p.pcm.wle}\textit{theta} \tag{\#2.16}$$

$$\text{p.pcm.theta} \tag{\#2.17}$$

Model Evaluation

Standardized infit and outfit statistics of the items are produced using the tam.fit() function, supplying the TAM model object and specifying the $itemfit attribute, as shown in Line #2.18.

$$\text{tam.fit(mod.pcm)}\textit{itemfit} \tag{\#2.18}$$

The t-statistic and corresponding p-values are printed. If applying the Bonferroni adjustment for the 9 parameter estimates, each statistic should be evaluated at a $\alpha = .006$ level of significance. The Holm procedure is another way of adjusting for controlling the familywise Type I error rate; the adjusted p-values according to Holm are also printed. None of the item parameters are flagged for misfit.

In addition to the item fit statistics, several descriptive model-data fit measures for relative model comparison are provided in the output using the summary() function, as shown in Line #2.19.

$$\text{summary(mod.pcm)} \tag{\#2.19}$$

The use of the Rasch modeling is typically positioned in the perspective of constructivism, rather than the conventional model-data fitting idea. Therefore, the use of AIC and BIC as presented here may be viewed as not consistent with the application of the Rasch modeling. One should be aware of this when applying to Rasch modeling. From #2.19, two notable indices are the two information criteria:

AIC and BIC. They are useful measures for relative model comparisons (e.g., comparing PCM vs. RSM or a constrained version of PCM). The model with a smaller value of AIC (and/or BIC) is preferred as a better model.

In the closing of this section using the "TAM" package and preparing for the "mirt" package in the next section, the "TAM" package should be detached from the R session using the detach() function:

$$\text{detach(“package : TAM”)} \tag{\#2.20}$$

3.3.5 *Multidimensional Rasch Application for Dichotomous Data (Using "mirt")*

The "mirt" package is used to illustrate the Rasch Modeling of multidimensional dichotomous data with a simple structure, i.e., subsets of items measuring a single dimension. First, users should load the package (ensure that the package was previously installed using line #2.20).

$$\text{library(mirt)} \tag{\#3.1}$$

The "mirt" package contains two unidimensional datasets, the LSAT6 and LSAT7 data. LSAT6 contains five dichotomously scored items from the LSAT, section 6 (Thissen, 1982). LSAT7 contains five dichotomously scored items from the LSAT, section 7 (Bock & Lieberman, 1970). The data sets will be combined into a single data set to illustrate multidimensional data.

Each of these datasets is loaded into the R session using #3.2 and #3.3. An arbitrary data set which combines these two LSAT data sets will be used to illustrate a multidimensional Rasch modeling application. The cbind() function in line #3.5 is to combine these two datasets, "L6" and "L7", by their columns randomly sampling $N = 300$ from each set. Line #3.5 combines these datasets and saves the larger data set as the object data3. The "set.seed(111)" must be executed to create the same data set for the illustration here, thereby yielding the same calibration result.

$$\text{L6} < \text{-expand.table(LSAT6)} \tag{\#3.2}$$

$$\text{L7} < \text{-expand.table(LSAT7)} \tag{\#3.3}$$

$$\text{set.seed(111)} \tag{\#3.4}$$

$$\text{data3} \\ < \text{-cbind(L6[sample(nrow(L6), 300),], L7[sample(nrow(L7), 300),])} \tag{\#3.5}$$

In RStudio, a second method can be used to load data that is in the working directory. The combined data presented in Line #3.5 is supplied in the supplementary material as Rdata3. Within the Files pane, click the dataset to load, then click "Import Data". This opens a window that contains the file location, a preview of the data, and code to execute. Click "Import." The Rdata3 dataset is loaded and may be called using the Rdata3 name. To rename, type and execute, data3 <- Rdata3.

Now, contents of the data3 may be viewed using

$$\text{head}(\text{data3}) \qquad (\#3.6)$$

$$\text{dim}(\text{data3}) \qquad (\#3.7)$$

$$\text{names}(\text{data3}) \qquad (\#3.8)$$

data3 is a dataset containing two subsets of five-item unidimensional data, where the two dimensions are practically uncorrelated, that is near zero.

Fitting the Two-Dimensional Rasch Model

Within the "mirt" package, the mirt() function is used to fit the input data set to a specified multidimensional model. First, the multidimensional model must be specified. Line #3.9 uses a quotation to encompass the defined multidimensional model where F1 is factor 1 and F2 is factor 2. The names F1 and F2 are chosen arbitrarily and could be named Factor1 and Factor2, or any other identifiers. Following the given factor names is an equal sign followed by the items that load on each factor; in this case, items 1–5 load on F1 and items 6–10 load on F2. Without specifying a covariance between the two factors in the specification (i.e., line 3 in #3.9), the factors are assumed to be uncorrelated. If users have correlated factors, a third line, as shown below in line #3.9, can be added to the specification which indicates COV = F1*F2, making the covariance a freely estimated parameter. The defined model is saved as the object spec.

$$\begin{aligned} \text{spec} < \text{-``F1} &= 1-5 \\ \text{F2} &= 6-10 \\ \text{COV} &= \text{F1}^*\text{F2''} \end{aligned} \qquad (\#3.9)$$

After the model is specified, the mirt() function is used in line #3.10 to calibrate the data according to the model specified and given item type. For the Rasch model, the itemtype="Rasch" is used. The last argument, SE=TRUE, estimates the item parameter standard errors during the calibration. The complete Line #3.10 fits the multidimensional data to the Rasch model and saves it as the object mod.md.

$$\begin{aligned} &\text{mod.md} < \\ &\quad - \text{mirt}(\text{data3}, \text{model} = \text{spec}, \text{itemtype} = \text{``Rasch''}, \text{SE} = \text{TRUE}) \end{aligned} \qquad (\#3.10)$$

After executing line #3.10, users will see the iterations printed with the loglikelihood and max-change values. For more details, the help(mirt) or ?mirt may be used to provide the detailed documentation of the mirt() function. Typing and executing the object name mod.md (line #3.11) will print the complete model statistics.

$$\text{mod.md} \qquad (\#3.11)$$

Item Parameter Estimation

Estimated item parameters may be printed in a simplified form using the coef() function, with the argument simplify=TRUE, shown in Line #3.12.

$$\text{coef(mod.md, simplify} = \text{TRUE)} \qquad (\#3.12)$$

Figure 3.14 shows the printed output of the coefficients:

The first attribute is $items which prints the two-dimensional item parameter estimates for the Rasch model. According to the model specification, Items 1–5 are loaded onto the first dimension (1 on $a1$ and 0 on $a2$); Item 6–10 are loaded onto the second dimension (0 on $a1$ and 1 on $a2$). The multidimensional Rasch modeling in the "mirt" package adopts notations for a more general model. The g is a lower-asymptote (fixed to 0 in the Rasch model) and u is an upper asymptote (fixed to 1 in the Rasch model). The d parameter is the estimated item intercept, or the *easiness* parameter, similar to the Rasch model estimates from the "eRm" package. The estimated item difficulty is equivalent to the negative value of the estimated d parameter in this between-item or simple structure Rasch model.

The $cov attribute contains the variance-covariance matrix of the factors (#3.12). The variances of factor 1 and 2 are printed on the diagonal, and the covariance

Fig. 3.14 R output of coef (mod.md, simplify=TRUE), line #3.12

```
> coef(mod.md, simplify=TRUE)
$items
       a1 a2     d g u
Item_1  1  0 2.758 0 1
Item_2  1  0 1.101 0 1
Item_3  1  0 0.379 0 1
Item_4  1  0 1.579 0 1
Item_5  1  0 2.174 0 1
Item.1  0  1 1.685 0 1
Item.2  0  1 0.675 0 1
Item.3  0  1 1.735 0 1
Item.4  0  1 0.497 0 1
Item.5  0  1 2.032 0 1

$means
F1 F2
 0  0

$cov
      F1    F2
F1 0.802 0.047
F2 0.047 1.380
```

between factors is printed on the off-diagonal. In this example, the covariance between the two dimensions (or factors) is 0.047.

For the standard errors of the estimated item intercepts, use the printSE=TRUE argument, as shown in Line #3.13. Note, the standard errors for the easiness and the difficulty parameters are equivalent.

$$\text{coef(mod.md, printSE} = \text{TRUE)} \qquad (\#3.13)$$

Item Characteristic Surface

An item characteristic surface (ICS) for multidimensional modeling that is an extension of an ICC and can be produced using the itemplot() function with arguments to specify which item(s) ICS to print. Line #3.14 prints the ICS of item 1.

$$\text{itemplot(mod.md, 1)} \qquad (\#3.14)$$

For the item information surface, the type="info" argument can be supplied, as shown in Line #3.15.

$$\text{itemplot(mod.md, 1, type} = \text{"info")} \qquad (\#3.15)$$

Person Ability Parameter Estimation

A two-dimensional Rasch (latent) scores can be requested by the commands below. The latent scores are printed for each unique item response string (or each user) using the function fscores() and supplying the Rasch model object. The "EAP" method is specified. Line #3.16 extracts person scores for each unique item response string, and Line #3.17 extracts scores for each person (or row), including the estimated standard errors.

$$\text{p.md} < \text{-fscores(mod.md, method} = \text{"EAP", full.scores} = \text{FALSE)} \qquad (\#3.16)$$

$$\begin{aligned}\text{p.md} < \text{-fscores(mod.md, method} = \text{"EAP", full.scores} = \text{TRUE,}\\ \text{full.scores.SE} = \text{TRUE)}\end{aligned} \qquad (\#3.17)$$

For either of these, head() can be used to view the first five lines of the output.

$$\text{head(p.md)} \qquad (\#3.18)$$

Figure 3.15 shows the output for Lines #3.17 and #3.18.

```
> p.md <- fscores(mod.md, method="EAP", full.scores=TRUE, full.scores.SE=TRUE)
> head(p.md)
              F1         F2     SE_F1     SE_F2
[1,]  0.60018831  0.9482920 0.7532337 0.9123923
[2,]  0.02674339 -0.4754519 0.7186316 0.7815422
[3,]  0.58180224  0.1885954 0.7520400 0.8342366
[4,]  0.55193226 -1.0507253 0.7501412 0.7571545
[5,] -0.45736698  0.1578039 0.6937350 0.8313097
[6,]  0.60018831  0.9482920 0.7532337 0.9123923
```

Fig. 3.15 R output of lines #3.17 and #3.18

Fig. 3.16 R output of line #3.20

```
> itemfit(mod.md, fit_stats = 'infit')
     item outfit z.outfit infit z.infit
1  Item_1  0.769   -1.013 1.088   0.504
2  Item_2  0.769   -3.227 0.864  -2.171
3  Item_3  0.772   -6.337 0.799  -6.047
4  Item_4  0.754   -2.373 0.918  -0.874
5  Item_5  0.808   -1.198 1.031   0.272
6  Item.1  0.737   -2.117 0.922  -0.826
7  Item.2  0.723   -4.553 0.797  -4.005
8  Item.3  0.663   -2.726 0.859  -1.505
9  Item.4  0.732   -4.927 0.795  -4.421
10 Item.5  0.733   -1.720 0.951  -0.401
```

Model Evaluation

An overall model-data fit test is available in "mirt", which is the M2 test. The syntax for conducting the M2 test is shown in Line #3.19. The value under "p" represents the p-value of the test. For this example, the p-value is > .05, retaining the currently fitted model.

$$\mathrm{M2(mod.md)} \tag{\#3.19}$$

For item fit indices, the infit and outfit measures, a mean-square (MSQ) or a standardized statistic t (under the name of "z.outfit" and "z.infit") can be requested by the following command line.

$$\mathrm{itemfit(mod.md, fit\_stats = “infit”)} \tag{\#3.20}$$

Figure 3.16 shows the item fit statistics for Line #3.20.

Epilogue

The purpose of this chapter is the introduction of R Rasch family modeling via "eRm", "TAM", and "mirt". As such it is not possible to cover wider ranges of functions in these packages. For those who want to pursue learning more about the R item response modeling, authors (Paek & Cole, 2020) have written a detailed book, *Using R for Item Response Theory Model Applications*, that describes further details of Rasch family modeling in R and multidimensional modeling using R. In addition to learning how to use R Rasch model programs, the study of the Rasch models through a course or article/book reading (e.g., Wilson, 2005) is also strongly recommended.

R Code

1+1	#0.1
sqrt(4)	#0.2
help(sqrt)	#0.3
?sqrt	#0.4
setwd("C:/workingdirectory")	#0.5
read.table("Rdata1.csv", sep=",", header=TRUE)	#0.6
install.packages("eRm")	#0.7
install.packages("TAM")	#0.8
install.packages("mirt")	#0.9
Unidimensional Rasch Application for Dichotomous Data (using the "eRm" package)	
library(eRm)	#1.1
data1 <- raschdat3	#1.2
head(data1)	#1.3
dim(data1)	#1.4
names(data1)	#1.5
mod.rm <- RM(data1)	#1.6
mod.rm$conv	#1.7
-mod.rm$betapar	#1.8
mod.rm$se.beta	#1.9
summary(mod.rm)	#1.10
plotICC(mod.rm)	#1.11
plotICC(mod.rm, 1:3)	#1.12
plotICC(mod.rm, item.subset=c(1,3,5))	#1.13
dev.off(dev.list()["RStudioGD"])	#1.14
p.rm <- person.parameter(mod.rm)	#1.15
p.rm	#1.16
plot(p.rm)	#1.17
plotPImap(mod.rm)	#1.18

(continued)

install.packages("WrightMap")	#1.19
library(WrightMap)	#1.20
WrightMap(as.matrix(p.rm$thetapar$NAgroup1),-mod.rm$betapar)	#1.21
rm.lrt <- LRtest(mod.rm)	#1.22
rm.lrt	#1.23
Waldtest(mod.rm)	#1.24
itemfit(p.rm)	#1.25
detach("package:eRm")	#1.26
Unidimensional Rasch Application for Polytomous Data (using "TAM" for PCM)	
library(TAM)	#2.1
data(data.gpcm)	#2.2
data2 <- data.gpcm	#2.3
head(data2)	#2.4
dim(data2)	#2.5
names(data2)	#2.6
mod.pcm <- tam.mml(data2, irtmodel="PCM")	#2.7
mod.pcm$xsi	#2.8
round(mod.pcm$xsi,3)	#2.9
plot(mod.pcm, items=1, type="items", export=FALSE)	#2.10
plot(mod.pcm, items=2, type="items", export=FALSE)	#2.11
plot(mod.pcm, items=3, type="items", export=FALSE)	#2.12
p.pcm <- mod.pcm$person	#2.13
head(p.pcm)	#2.14
p.pcm.wle <- tam.wle(mod.pcm)	#2.15
p.pcm.theta <- p.pcm.wle$theta	#2.16
head(p.pcm.theta)	#2.17
tam.fit(mod.pcm)$itemfit	#2.18
summary(mod.pcm)	#2.19
detach("package:TAM")	#2.20
Multidimensional Rasch Application for Dichotomous Data (using "mirt")	
library(mirt)	#3.1
L6 <- expand.table(LSAT6)	#3.2
L7 <- expand.table(LSAT7)	#3.3
set.seed(111)	#3.4
data3 <- cbind(L6[sample(nrow(L6),300),],L7[sample(nrow(L7),300),])	#3.5
head(data3)	#3.6
dim(data3)	#3.7
names(data3)	#3.8
spec <- "F1 = 1−5 F2 = 6−10 COV = F1*F2"	#3.9
mod.md <- mirt(data3, model=spec, itemtype="Rasch", SE=TRUE)	#3.10
mod.md	#3.11
coef(mod.md, simplify=TRUE)	#3.12
coef(mod.md, printSE=TRUE)	#3.13

(continued)

itemplot(mod.md, 1)	#3.14
itemplot(mod.md, 1, type="info")	#3.15
p.md <- fscores(mod.md, method="EAP", full.scores=FALSE)	#3.16
p.md <- fscores(mod.md, method="EAP", full.scores=TRUE, full.scores.SE=TRUE)	#3.17
head(p.md)	#3.18
M2(mod.md)	#3.19
itemfit(mod.md, fit_stats = "infit")	#3.20
detach("package:mirt")	#3.21

References

Adams, R. J, Wu, M. L, Cloney, D., & Wilson, M. R. (2020). *ACER ConQuest: Generalised Item Response Modelling Software* [Computer software]. Version 5. Australian Council for Educational Research.

Andersen, E. B. (1973). A goodness of fit test for the Rasch model. *Psychometrika, 38*(1), 123–140. https://doi.org/10.1007/bf02291180

Bock, R. D., & Lieberman, M. (1970). Fitting a response model for *n* dichotomously scored items. *Psychometrika, 35*(2), 179–197.

Chalmers, R. P. (2012). mirt: A multidimensional item response theory package for the R environment. *Journal of Statistical Software, 48*, 1–29.

Cole, K., & Paek, I. (2017). PROC IRT: A SAS procedure for item response theory. *Applied Psychological Measurement, 41*, 311–320.

Cole, K., & Paek, I. (2022). Using SAS PROC for multidimensional item response theory analysis. *Measurement: Interdisciplinary Research and Perspectives, 20*, 49–55.

Fischer, G. H., & Molenaar, I. (1995). *Rasch models – Foundations, recent developments, and applications*. Springer.

Karlheinz, R., & Melich, A. (1992). *Euro-Barometer 38.1: Consumer protection and perceptions of science and technology*. INRA (Europe).

Linacre, J. M. (2002). What do Infit and Outfit, mean-square and standardized mean? *Rasch Measurement Transactions*. https://www.rasch.org/rmt/rmt162f.htm

Linacre, J. M. (2022). *Winsteps® Rasch measurement computer program* [Computer software]. Version 5.2.2. Winsteps.com

Liu, X., & Boone, W. (2006). Introduction to Rasch measurement in science education. In X. Liu & W. J. Boone (Eds.), *Applications of Rasch measurement* (pp. 1–22). JAM Press.

Mair, P., Hatzinger, R., & Maier, M. J. (2021). *eRm: Extended Rasch modeling* [Computer software]. Version 1.0-2, https://cran.r-project.org/package=eRm

Masters, G. (1982). A rasch model for partial credit scoring. *Psychometrika, 47*, 149–174.

Paek, I., & Cole, K. (2020). *Using R for item response theory model applications*. Routledge.

Paek, I., Liang, X., & Lin, Z. (2021). Regarding item parameter invariance for the Rasch and the 2-paramaeter logistic models: An investigation under finite non-representative sample calibrations. *Measurement: Interdisciplinary Research and Perspectives, 19*, 39–54.

R Core Team. (2020). *R: A language and environment for statistical computing*. R Foundation for Statistical Computing. https://www.R-project.org/

Robitzsch, A., Kiefer, T., & Wu, M. (2021). *TAM: Test analysis modules. R package* [Computer software]. Version 3.7-16. https://CRAN.R-project.org/package=TAM

Thissen, D. (1982). Marginal maximum likelihood estimation for the one-parameter logistic models. *Psychometrika, 47*, 175–186.

Warm, T. A. (1989). Weighted likelihood estimation of ability in item response theory. *Psychometrika, 54*, 427–450.

Wilson, M. (2005). *Constructing measures: An item response modeling approach.* Psychology Press.

Xu, J., & Paek, I. (2016, November). *The 3-parameter logistic model parameter recovery with different prior distributions when using flexMIRT and the mirt package in R.* The 61st annual meeting of the Florida Educational Research Association (FERA).

Dr. Ki Cole is an Associate Professor of Research, Evaluation, Measurement, and Statistics at Oklahoma State University. Her major research interest is in item response theory (IRT), factor analysis, test and survey development and evaluation, and evaluating statistical software.

Dr. Insu Paek is a Senior Scientist at Human Resources Research Organization (HumRRO). His major research centers on item response theory (IRT), differential item functioning, test score equating, and practical issues in large scale assessment.

Chapter 4
Bayesian Partial Credit Model and Its Applications in Science Education

Xingyao Xiao, Mingfeng Xue, and Yihong Cheng

Abstract Partial Credit Model (PCM), belonging to the Rasch family model, is widely used in analyzing ordinal polytomous response data in education and psychology. This chapter aims to introduce science education researchers unfamiliar with it to a user-friendly and flexible tool for conducting Rasch analyses. We first introduce what the PCM is and demonstrate the Bayesian estimation procedure in R with a Stan package using a simulated dataset. We then show how to apply the PCM in an explanatory way using the Programme for International Student Assessment (PISA) dataset and illustrate how to obtain qualitatively meaningful relationships between explanatory variables and students' scientific literacy. Bayesian Partial Credit Model and Its Applications in Science Education.

4.1 Introduction

4.1.1 Rasch Measurement in Science Education

Rasch models have been widely applied in science education to validate and refine instruments. For example, Wei, Liu, and Jia (2013) used Rasch measurement to validate the instrument of Students' Understanding of Models in Science (SUMS), which was originally developed under Classical Test Theory (CTT) and identified aspects for further improvement. In addition, Boone and Scantlebury (2006) used a science achievement test from a state level systemic reform project as an example to

---

X. Xiao (✉) · M. Xue
Berkeley School of Education, University of California, Berkeley, CA, USA
e-mail: xiaoxg@berkeley.edu; mingfengxue@berkeley.edu

Y. Cheng
Lynch School of Education and Human Development, Boston College, Chestnut Hill, MA, USA
e-mail: chengyz@bc.edu

X. Liu, W. J. Boone (eds.), *Advances in Applications of Rasch Measurement in Science Education*, Contemporary Trends and Issues in Science Education 57,
https://doi.org/10.1007/978-3-031-28776-3_4

discuss the strength of Rasch models in measuring and documenting changes in science education.

4.1.2 Different Estimation Approaches to Rasch Analyses in Science Education

Although Rach measurement models and Item Response Theory (IRT) models are different in their philosophical goals, in this chapter, we take a mathematical lens in that we view the Rasch Model as a one-parameter IRT model. Specifically, we use the terms Rach model and one-parameter IRT model interchangeably to describe a model that has an item difficulty parameter, an item discrimination parameter fixed to be equal across items, and no pseudo-guessing parameter.

The estimation algorithms for IRT models can be divided into the frequentist approach and the Bayesian approach (Lord, 1986). The former has a more extended history and relies on the maximum likelihood estimation (MLE) and its variants, including joint ML, conditional ML, expectation maximization, etc. They are available in a hold of commercial IRT software programs, such as BIGSTEPS, WINSTEPS, and ConQuest.

On the other hand, compared to the frequentist approach, some education researchers choose the Bayesian approach because Bayesian languages (Stan, JAGs, and BUGs) provide convenient functions that enable researchers to customize their models. Besides, researchers can incorporate additional information from previous data and expert knowledge into estimating person locations through Bayes' Theorem. The critical part is that researchers can create a probabilistic description of the belief about person location information before collecting data, which is known as a *prior*. They can then use collected data and a model to update their belief, a process known as forming the *posterior distribution*. After that, they can base their inference about person locations on the full posterior distribution. The Bayesian approach can estimate individuals with extreme scores (i.e., zero and perfect scores), while various MLE-based programs cannot obtain or provide different kludges for handling these scores (Wyse, 2010). Moreover, the mean squared error from Bayesian estimation with proper setting is less than that obtained using MLE. Researchers normally use the mean of the posterior distribution as the location estimates, in that the posterior mean minimizes the expected loss (if the loss function is the mean squared error).

A middle ground between the frequentist and Bayesian approaches is known as *empirical Bayes* (EB). Instead of basing inference completely on the point estimates (i.e., MLE) or assigning a full distribution for a prior, the EB approach uses the data to estimate the prior parameters and then combines the prior with likelihood to obtain posterior distribution via Bayes' theorem. In *maximum a posteriori* (MAP), researchers use the mode of the posterior distribution as the estimation of person location, and in *expected a posteriori* (EAP), researchers use the mean of the posterior distribution as person's location. The commonality between frequentist's MLE and EB's MAP and EAP is that they all treat the item parameters' estimates as known and thus ignore the real uncertainty about person locations. However, unlike MLE, the EB approach can obtain location estimates for extreme scores.

Thus, using the fully Bayesian approach, we can estimate person locations for all individuals and incorporate additional information into the person location estimates as a way to take into account all the parameters' uncertainty fully.

4.1.3 The Bayesian Approach

The Bayesian approach takes in prior information, data, and likelihood and its general equation is given below:

$$P(H|E) = \frac{P(E|H)P(H)}{P(E)} \tag{4.1}$$

where P(E|H) stands for the likelihood of observing the evidence or data given the hypotheses or parameters, P(H) denotes the prior distribution of the hypotheses or parameters, P(E) is the marginal distribution of the data, and P(H|E) is the posterior distribution of the parameters given the data observed and what we are interested in. In other words, the posterior distribution is the compromise between prior information about the parameters and the likelihood of observing the data conditional on the parameters. Wagenmakers et al. (2018) pointed out the benefits of Bayesian parameters estimation, including incorporating prior knowledge, quantifying confidence of the estimated parameters as a natural product, coherent estimation, and flexibility in extending to complicated models. However, the estimation of posterior distribution requires integration, which will be tricky, especially when there are many parameters typically seen in the setting of IRT. A feasible approach to addressing this problem is Markov chain Monte Carlo (MCMC), which relies on the Markov chain property (i.e., the probability of each event depends only on the most recent event) and Monte Carlo integration. This chapter will not elaborate on explaining the MCMC algorithm, and interested researchers can refer to Chapter 3 in The Bug's Book (Lunn, Jackson, Best, Thomas, & Spiegelhalter, 2012).

Thanks to advances in MCMC implementation software, we have the luxury of using various programs to implement this Bayesian approach. One of the most promising ones is Stan, an open-source C++ program that performs Bayesian inference (Stan Development Team, 2017). There are two major reasons to choose Stan. First, Stan is open source such that researchers can run directly and freely using a variety of interfaces, including R (i.e., RStan). Second, Stan implements Bayesian inference with a superior sampling algorithm, an improved version of Hamiltonian Monte Carlo (HMC) known as the No-U-Turn Sampler (NUTS), to estimate parameters, which is more efficient than other MCMC software such as BUGS (Bayesian inference Using Gibbs Sampling) and JAGS (Another Gibbs Sampler) (Hoffman & Gelman, 2014). Among the few published Bayesian IRT studies, many show that NUTS performs better than other fully Bayesian or classical frequentist methods in estimating Rasch measurement models and various IRT models (Chang, 2017; Grant et al., 2017; Martin-Fernandez & Revuelta, 2017). Besides, some studies have also taken advantage of NUTS to fit complex IRT models (Caughey & Warshaw, 2014; Copelovitch et al., 2015) in political science and economic applications.

After fitting model using MCMC software, we have access to the posterior distribution of parameters instead of point estimates obtained from frequentist methods. Nonetheless, before extracting information from the posterior distribution, we have to ensure the convergence and effectiveness of MCMC results. Firstly, the effective sample size (ESS) calculates the equivalent number of independent samples that would contain the same posterior accuracy as the correlated samples from MCMC, and we can obtain the MCMC efficiency by dividing ESS by the computation time (in seconds). We expect our ESS to be large. Secondly, to diagnose MCMC convergence, R-hat statistic (Rhat) and traceplot graphs of the parameters' posterior values are the two commonly used criterion. Rhat measures the ratio of the average variance of the MCMC iterations within each chain to the variance of the pooled iterations across all chains (after discarding the warm-up iterations). If all chains are at equilibrium (meaning that the same stationary distribution has been reached), these variances will be the same so Rhat will equal one. At convergence, we expect all Rhat to be smaller than 1.05.

Moreover, traceplot displays the sampled values per chain and iteration (after discarding the warm-up iterations). It allows us to evaluate the convergence and mixing of the chains visually. We examine three signs of a healthy trace plot: (1) stationarity, (2) good mixing, and (3) convergence. Specifically, stationarity refers to the mean value of each chain being quite stable from the beginning to the end, good mixing means that the chain rapidly zig-zags around the full region, and convergence means that multiple chains stick around the sample region of high probability.

4.1.4 Programme for International Student Assessment

To illustrate how to apply Bayesian PCM to science education, we use the well-known Programme for International Student Assessment (PISA) dataset in the sample empirical study. PISA is an internationally recognized study initiated by the Organisation for Economic Co-operation and Development (OECD) and held every three years since 2000, which targets 15-year-old students (OECD, 2019). Each round of PISA encompasses two major components – cognitive assessments and background questionnaires – and focuses on students' literacy in reading, mathematics, and science. Specifically, in terms of science education, OECD (2017a) argued that an understanding of science was so important that it should be a feature of every young person's education and defined scientific literacy by three competencies, that is, explaining phenomena scientifically, evaluating and design scientific enquiry, and interpreting data and evidence scientifically.

In this chapter, we employ PISA 2015 data, where scientific literacy is the major domain (OECD, 2017a), and specifically choose the U.S. data. Note also that the majority of participating countries of PISA 2015 (including the U.S.) have switched to computerized testing, so a flexible way of testing is present (OECD, 2017b). In total, there were 184 items in the domain of scientific literacy, among which 99 items were newly developed for PISA 2015. The formats of the 184 items fell into two broad categories – multiple-choice items ($n = 120$) and open-ended items ($n = 64$). In terms of scaling students' scientific proficiency, OECD (2017b) adopted unidimensional IRT models, which is also the reason why we used unidimensional Bayesian PCM in this chapter. For the U.S. data, there were 5712 students in the cognitive assessments with a nearly even gender distribution. For explanatory variables on the person side, this chapter chooses gender and socioeconomic status ($M =$ 0.06, $S.D. = 1.01$) as examples, given their widely recognized impact on education. Socioeconomic status is composed of parental education, highest parental occupation, and household possessions via principal component analysis (Avvisati, 2020).

4.2 Objectives

This chapter aims to achieve three objectives. First, we show how to rewrite PCM in a Bayesian procedure using a probability interpretation and how to transcribe it into the Stan code. Second, we provide intuitive explanations of the Stan code and the workflow of Bayesian inference and use a simulated dataset to investigate software efficiency to validate our Stan code. Third, we use a PISA dataset as an example of how to apply the Bayesian PCM in science education, where the PCM is extended in an *explanatory* way on the student side (i.e., the latent regression PCM) and shows how student properties can explain the differences in their scientific literacy. Eventually, we show a comparison in analysis results between Bayesian PCM and frequentist PCM estimated in ConQuest on the PISA data. We hope science education researchers will find this tutorial a helpful reference as they work on applying Bayesian Rasch measurement models in instrument validation and refinement in more flexible and rigorous ways on the theoretical side and in more explanatory ways on the empirical side. Dataset and R codes can be found from shorturl.at/hxQR5.

4.3 Methods

PCM is a generalized Rasch model that can apply to various types of items ranging from dichotomous items (i.e., items with two possible response categories) to ordinal, polytomous items. And the items can have different numbers of response categories. A standard PCM specifies that the conditional probability that a student with latent ability θ obtains a category score of y, which is presented in Eq. 4.2, and θ is typically assumed to follow a normal distribution with a mean of zero. An explanatory version is presented in Eq. 4.3 by including a latent regression on the person side in the mean of the normal distribution.

$$\Pr\left(Y_{ij}=y, y>0 \middle| \theta_j, \beta_i\right) = \frac{exp \sum_{s=1}^{y} \left(\theta_j - \beta_{is}\right)}{1 + \sum_{k=1}^{m_i} \exp \sum_{s=1}^{y} \left(\theta_j - \beta_{is}\right)} \tag{4.2}$$

and

$$\theta_j \sim \mathrm{N}\left(\omega_j' \lambda, \sigma^2\right) \tag{4.3}$$

- where observed variables are:
  - $Y_{ij} \in \{0, \ldots, m_i\}$ is the response of person $j \in \{1, \ldots, J\}$ to item $i \in \{1, \ldots, I\}$,
  - m_i is the number of "steps" required to correctly answer an item i.
  - w_j is the vector of covariates for person j, in which the first element is 1 for the model intercept. For an explanatory model with $K - 1$ person covariates and J persons, w_j can be generalized to a $J \times K$ covariate matrix W.
- where parameters are:
  - β_{is} the $s -$ step difficulty for item i, indicating the relative difficulty in endorsing category s over category $s - 1$. Across all items, the average difficulty parameter $\frac{\sum_{i=1}^{I} \beta_i}{I}$ is constrained to be zero, as the last step difficulty parameter is constrained to be the negative sum of the other difficulties, $\beta_i = -\sum_{i=1}^{I-1} \beta_i$ (the constraint is removed from the item difficulties in the Stan model for ease of demonstration),
  - θ_j is the ability for person j,
  - λ is a vector of latent regression parameters of length K, where the first element is the intercept,
  - σ^2 is the variance of the distribution of θ_j.

Equation 4.2 is used to formulate the likelihood function. To specify the model in a Bayesian framework, we also have to specify a prior distribution for each parameter of interest.

- Below are the weakly informative priors used in this chapter:
  - the person standard deviation as $\sigma \sim$ Exponential (.1),
  - the step difficulty parameter as $\beta_{is} \sim$ N(0, 3),
  - the latent regression parameters as $\lambda \sim$ StudentT(3, 0, 1) given degrees of freedom $\nu = 3$, location $\mu = 0$, and scale $\sigma = 1$.

4.4 Formulation of the PCM in Stan

This section introduces how to write the Stan code that corresponds to the PCM defined in the Methods section. The complete model code is presented in Appendix A, including a standard PCM with three required blocks: data (lines 10–20), parameters (line 25–30), and model (lines 31–39) and one optional block generated quantities (lines 41–50), where:

1. the **`data`** block is where you define the data and the dimensions of the data;
2. the **`parameters`** block is where you describe the unknown parameters that you want to estimate;
3. the **`model`** block is where you list the priors and define the likelihood function for the model; and
4. the **`generated quantities`** block will generate quantities of interest based on previous blocks. Here we generate the log likelihoods, which then can be used to compute model fits, such as widely applicable information criterion (WAIC).

In the **`data`** block, we declare the type and constraint of the data. For example, we define the number of items, the number of persons, and the number of responses as **`int<lower=1> I`**, **`int<lower=1> J`**, and **`int<lower=1> N`**, respectively, where **`int`** represents integers and **`<lower=1>`** puts a constraint on the minimum as one. The corresponding variables, including indexing items, person, and binary outcome variables, are denoted as **`ii[N]`**, **`jj[N]`**, and **`y[N]`**, respectively.

In the **`parameters`** block, we define parameters **`beta`** as a $I \times m$ array, with each row representing the step difficulties for item **`I`**. Parameters **`theta`** and **`lambda`** are defined in the same way. The standard deviation parameter **`Sigma`** is constrained to have a lower bound of zero.

Finally, the **`model`** block is the core that calculates the log-probability function of the observed (data) and unobserved (parameters) variables. This is the numerator in Bayes' rule, and Stan uses it to run the Hamiltonian simulation. We assign the prior for each parameter. In this chapter, we use the weakly informative priors as specified in the Methods section. Researchers can also specify their own priors based on their theoretical framework or prior knowledge. Note that here we use a **`for`** loop for specifying the prior for an **`array`**-type parameter, **`beta`**, but we do not need to do that for a **`vector`** type. For the likelihood, we use **`target +=`** , which is equivalent to **`target = target +`** , to increment the log-probability function.

4.5 Data Simulation for Parameter Recovery

To illustrate the estimation accuracy of the Bayesian PCM using Rstan, we simulate responses through the explanatory PCM and then estimate the values of parameters using the Rstan code introduced above. The **`R`** code that we use to simulate the dataset is shown in Appendix B (line 18–23).

The simulated dataset contains 250 respondents who respond to 7 items with 4 response categories. We generate two person-level covariates: (1) a continuous variable **V1** following a standard normal distribution (with mean zero and standard deviation one), and (2) a dichotomous variable **V2** with values of zero and one. The generated coefficient values *lambda* for latent regression are 0.5 and 0.6 for **V1** and **V2**, respectively. Then, we draw σ from a normal distribution with mean zero and standard deviation 0.9. Thus, we can generate **theta** for person j. After simulating data, we define a list of objects (see lines 25–32 in Appendix B) and pass it to the data option to the Rstan sampling (see line 34 in Appendix B). We sample the PCM model with four chains and 2000 iterations and fit it to the simulated data by using the **sampling()** function (see lines 34–35).

First, we check the model convergence by printing Rhat using the **print()** function (see line 36 in Appendix B) and displaying Rhat in graph (see line 38 in Appendix B). Figure 4.1 shows the convergence statistics (Rhat) are all around one and below the rule of thumb, 1.05, which is desired. Moreover, we can display the traceplot graphs for the parameter of interest (see lines 40–41 in Appendix B). Figure 4.2 shows the other criterion we use to judge the model convergence. For the purpose of demonstration, we choose some examples of key parameters of interest to show an example of good mixing. Researchers can choose the parameters of their interest by including parameters in **pars= c()**. For the chosen parameters of interest, the displayed traceplot graphs show that chains are intertwined together around the mean, which can be considered as evidence of convergence.

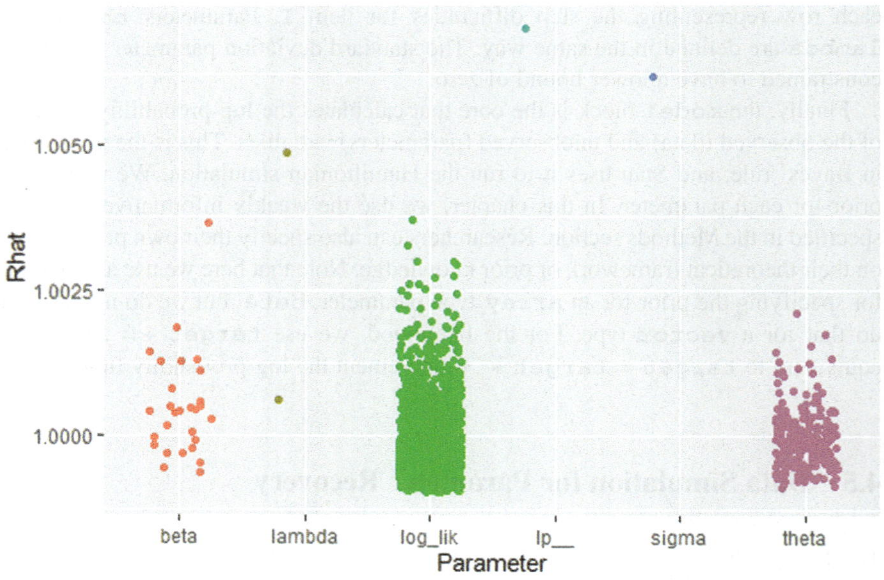

Fig. 4.1 Convergence statistics (Rhat) by parameters of interest, all values are less than 1.05 to indicate convergence

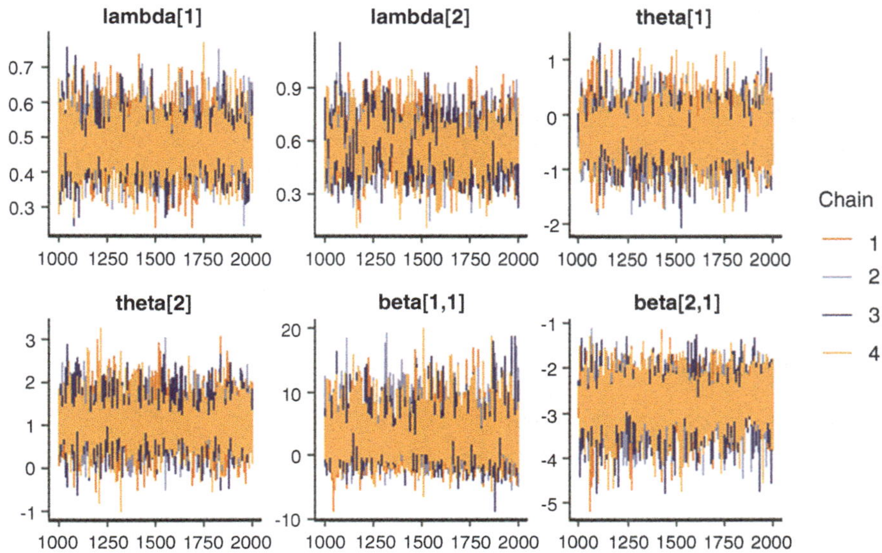

Fig. 4.2 Traceplot graphs of the parameter values of interest

Next, we assess the estimation accuracy of the Bayesian PCM using Rstan by checking the model's ability to recover parameters. Figure 4.3 shows the differences between estimated and simulated parameter values (see lines 42–18 in Appendix B for the code to generate Fig. 4.3). Note that while we merely choose parameter vector λ to demonstrate, researchers can adapt the code to check other parameters. Specifically, points show the differences between the posterior means and simulated parameter values for each element of the parameter vector, λ, and horizontal lines indicate 95% posterior intervals for the differences. We find that all differences are around zero, indicating that the Bayesian PCM using Rstan successfully recovers the parameter values specified in the data simulation.

Finally, for explanatory purposes, Fig. 4.4 shows the estimated ability means as a function of a continuous variable and a dichotomous variable (see lines 72–81 in Appendix B for the code to generate Fig. 4.4). The black line is for the focus group (when the dichotomous variable equals one), and the red line is for the reference group (when the dichotomous variable equals zero). Since there is no interaction, these two lines are parallel. In an empirical study, the y-axis, for example, can be the estimates for science ability means for a classroom, the x-axis can be students' math scores, and the dichotomous variable can be gender (with female = 1 and male = 0). Thus, Fig. 4.4 can show the relationship between students' science abilities and their math scores and how this relationship is different for different gender identities. Furthermore, we can generalize the model by including more covariates and their interaction terms.

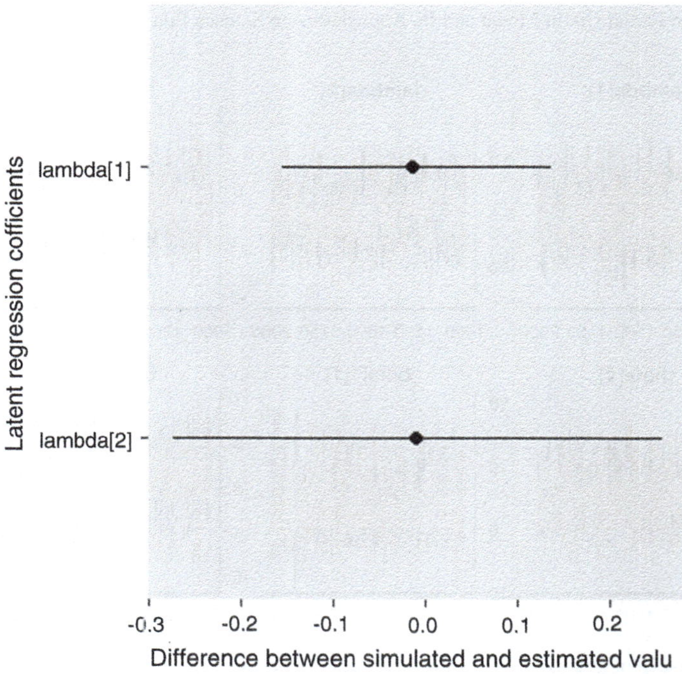

Fig. 4.3 Differences between estimated and simulated parameter values

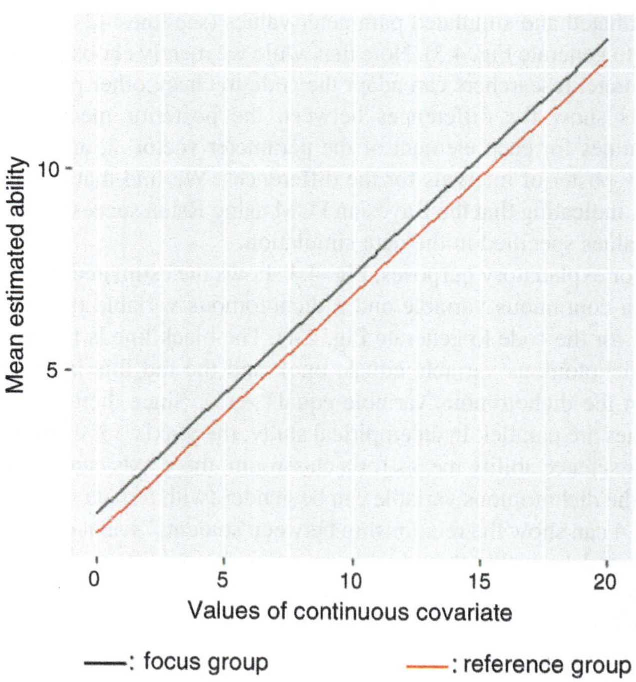

Fig. 4.4 Mean of estimated abilities as a function of a continuous covariate and a dichotomous covariate

4.6 Empirical Application Results and Model Checking

This section illustrates our approach to analyzing PISA 2015 scientific literacy data using three chains and 3000 iterations for each chain, with first half of each chain as warm-up (also known as burn-in), which means that 1500 out of 3000 iterations for each chain were intentionally removed in order to remove the influence of the starting values. Since we start each chain with some arbitrary starting values for the parameters, we need a long warm-up phase to reach the stationary distribution. In the following sections, we will demonstrate the convergence of Rstan results, interpret parameter estimates, and display graphical results to facilitate the understanding (see Appendix C for R code). For comparison between frequentist and Bayesian methods, we used ConQuest to run the same explanatory PCM analysis on the same PISA 2015 scientific literacy data and extracted weighted likelihood estimates (WLEs), as a representation of frequentist estimates in one-parameter IRT (i.e., Rach) model.

4.7 Convergence and Efficiency Diagnostics

Before demonstrating and interpreting the results, it is necessary to examine the convergence of the MCMC. Similar to the simulation study, three techniques are employed, namely Rhats, effective sample size, and traceplots. Results show that Rhats range from 0.9993 to 1.0186. According to the rule of thumb (Rhat < 1.05), all parameters obtained from MCMC are convergent. The average effective sample size is 6778.32. Finally, regarding the traceplot graphs and given the limited space, Fig. 4.5 only shows the traceplot graphs of three example parameters, including one step difficulty, one ability estimate, and the one coefficients of the explanatory variables (i.e., socioeconomic status). The three trace plots look like caterpillars, which is an indication of convergence. In sum, the estimates of the parameters from MCMC are suitable for the following interpretation.

4.8 Estimated Parameters

Figure 4.6 illustrates the posterior distributions of scientific literacy of five randomly selected students obtained from MCMC with vertical lines indicating the means. The distributions of students 1, 6, 15, and 20 overlap to a large extent, and the means of the distributions are close to one another, which implies that the three students have matching scientific literacy. Nevertheless, the distribution of student 15 is more spread out than those of students 1 and 6, and 20, so there is more uncertainty around the estimates of student 15's scientific literacy than around students 1 and 20. Third,

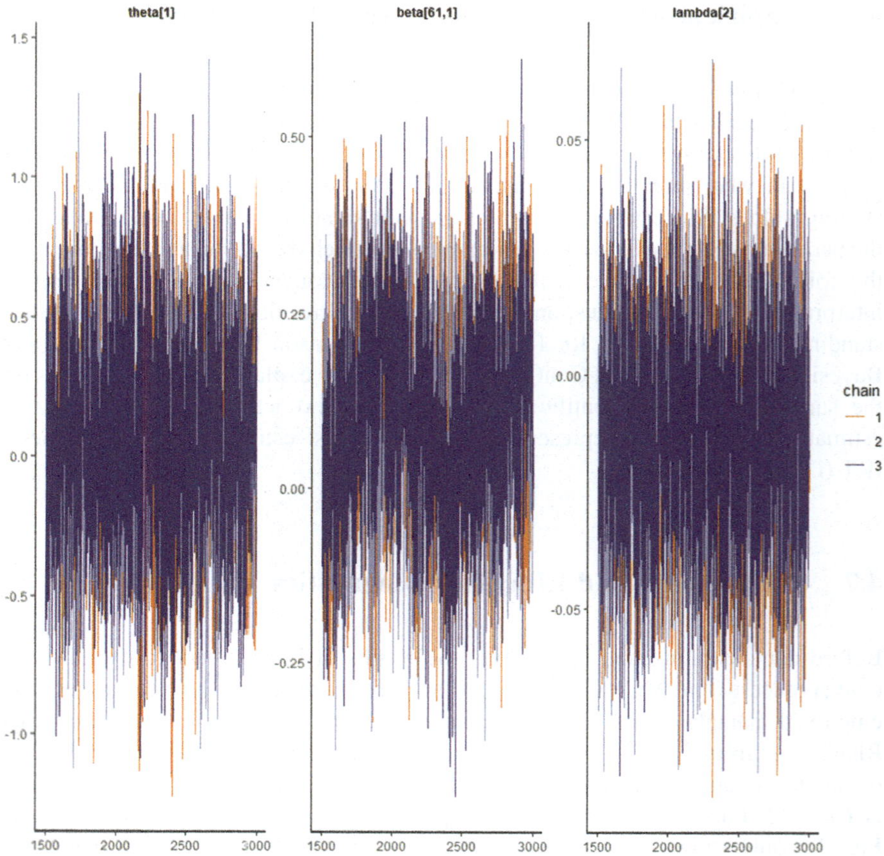

Fig. 4.5 Trace plots of four parameters in MCMC

the 95% credible interval of student 71's distribution does not overlap with that of student 20, which means that student 71 outperforms student 20 significantly.

Likewise, Fig. 4.7 demonstrates the posterior distributions of the step difficulties of items 52, 103, and 161 from top to bottom. For item 52, the two step difficulties did not differ from each other as their posterior distribution overlapped to a large extent. However, for items 103 and 161, the step difficulties from zero to partial credit were significantly higher than that from partial credit to full credit as their posterior distributions were largely separate.

Figure 4.8 shows the posterior distributions of the regression coefficients of socioeconomic status and gender on students' scientific literacy. For the regression coefficient of socioeconomic status, its 95% credible intervals included zero and its mean was very close to zero, which means that the regression coefficients were not significantly different from zero. For the regression coefficient of gender, the upper bound of its posterior distribution barely touched zero, and the mean of posterior

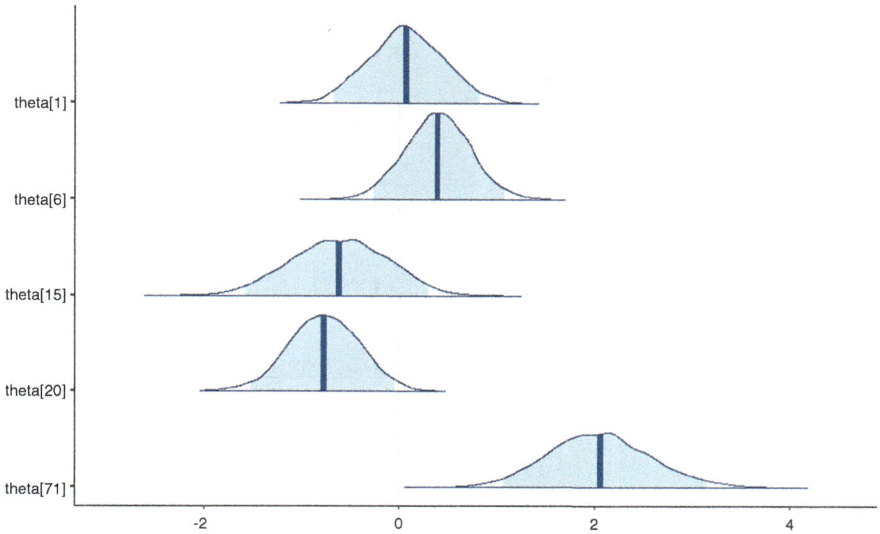

Fig. 4.6 Posterior distribution of student 1, 6, 15, 20,and 71's scientific literacy. Blue shades indicate 95% credible intervals; Dark blue bars indicate means of the distributions

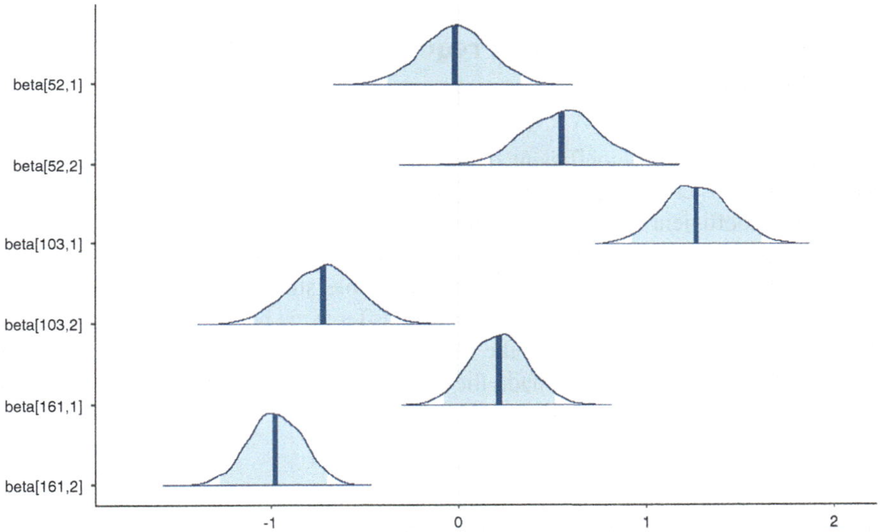

Fig. 4.7 Posterior distribution of the step difficulties of item 61, 63, and 86. Blue shades indicate 95% credible intervals; Dark blue bars indicate means of the distributions

distribution was −0.08. That is, girls' scientific literacy was estimated to be 0.08 units higher than boys' scientific literacy on average, after controlling for socioeconomic status.

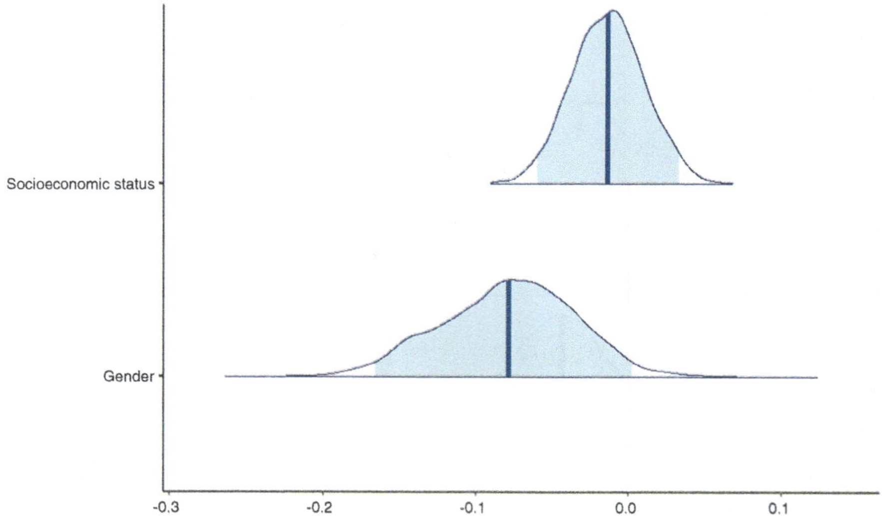

Fig. 4.8 Posterior distribution of the regression coefficients of socioeconomic status and gender on students' scientific literacy. Blue shades indicate 95% credible intervals. Dark blue bars indicate means of the distributions; For gender, girl is treated as the reference group

4.9 Comparisons Between Frequentist and Bayesian Methods

Overall, the correlation coefficient of scientific literacy obtained from WLEs and means of the posterior distributions was 0.997 ($p < 0.001$), and the corresponding correlation coefficient for step difficulties was 0.998 ($p < 0.001$). Therefore, in our empirical study, these two methods show perfect agreement.

Table 4.1 presents a closer look at the example students, example items, and regression coefficients mentioned above. The differences between the two methods were tiny, and this might be due to the vague prior distribution we used and the large volume of data we got, which made the estimation more data-driven and closer to frequentist methods based on MLE.

4.10 Flexibility in Extending the Model

RStan allows for great flexibility in extending modeling by code modifications. For example, in Rasch models, we could reparameterize the latent traits with covariates of interest to explain between-person variability and formulate item parameters in terms of various item properties to explore the source of item difficulties. We can

Table 4.1 Comparison table of example students, example items, and regression coefficients between frequentist and Bayesian method

	Frequentist method	Bayesian method
Students'scientific literacy		
θ_1	0.14 [−0.65, 0.93]	0.07 [−0.68, 0.82]
θ_6	0.47 [−0.21, 1.15]	0.39 [−0.26, 1.08]
θ_{15}	−0.61 [−1.64, 0.42]	−0.61 [−1.57, 0.31]
θ_{20}	−0.75 [−1.50, 0]	−0.77 [−1.52, −0.04]
Item step difficulty		
β[52, 1]	0.078	0 [−0.38, 0.33]
β[52, 2]	0.648	0.55 [0.17, 0.94]
β[103, 1]	1.296	1.27 [0.93, 1.62]
β[103, 2]	−0.69	−0.72 [−1.09, −0.36]
β[162, 1]	0.27	0.22 [−0.08, 0.52]
β[162, 2]	−0.92	−0.98 [−1.27, −0.69]
Regression coefficient		
socioeconomic status	0.04 [−0.06, 0.13]	−0.01 [−0.06, 0.03]
gender	−0.03 [−0.08, 0.02]	−0.08 [−0.17, 0]

[1]Confidence and credibility intervals axe presented in []
[2]Note. Conquest dosen't provide standard errors for threshold parameters

even design our own Rasch models that accommodate our theories and needs using Rsatn.

Another flexible feature of the Bayesian PCM is its ability to incorporate different prior information through setting the prior distribution. When there is no or insufficient information, we can adopt a vague or non-informative prior distribution, like distributions with large variance, and then the Bayesian PCM will be more data-driven. On the other hand, when we have rich information about the parameters, we can set a narrow prior distribution, thus making the Bayesian approach more theory-driven. For example, if based on literature, we are strongly confident that socioeconomic status is positively and largely related to students' scientific literacy, we can adopt a log normal distribution with a large positive mean and small variance. Besides setting the prior distribution directly, we can even create hyperprior distributions for parameters of the prior distributions (i.e., hyperparameters) if needed.

On the other hand, we have to admit the flexibilities of Bayesian PCM come with prices. First, it will be challenging for researchers, who are new to Stan, to write the codes that address their research questions. Therefore, one of the purposes of this study is to familiarize those researchers with Stan, and we suggest they can start by running our codes, then modify the codes step by step, and finally achieve their research purpose. It is also worth mentioning that Stan has an active community and various resources online that help starters overcome the challenges. Secondly, MCMC is 'notorious' for its time-consuming iterations. To speed up the Stan execution, researchers can consider vectoring the manipulation of data and parameters and running Stan codes in parallel, like performing three chains in three different processors of their computers. However, we would argue that though Bayesian PCM might be challenging with respect to coding and running time, the promising features of Bayesian PCM, mentioned in the introduction section, make it worthwhile.

4.11 Discussion

This chapter introduces how to perform Bayesian PCM with latent regression in Rstan by a simulation study and an empirical study on PISA scientific literacy data and demonstrates the advantages of Bayesian Rasch modeling, such as incorporating various prior information and providing a posterior distribution for each estimated parameter.

In the simulation study, we show the Bayesian PCM using Rstan successfully recovers the true parameter values. Specifically, we demonstrate the whole process and the Rstan code to perform data simulation, model convergence checking, estimation accuracy checking, and the plot generated for explanatory purposes. In the empirical study, we explore the relationship between students' gender, socioeconomic status, and scientific literacy, where we only find a significant gender difference in scientific literacy, which is in line with a lot of research on science education (Baker, 2016). Moreover, the non-significant relation between socioeconomic status index and scientific literacy could be due to the limitation of OECD's socioeconomic status index in the American setting as it only involves parental education, highest parental occupation, and household possessions, where income and other important factors are not present.

4.12 Conclusion

This chapter aims to help science education researchers apply Bayesian Rasch measurement analyses using a user-friendly tool called RStan. We hope that by following the demonstrated procedure of (1) using a graphical display to facilitate communication, (2) extending the model to add more explanatory capability, and (3) performing model checking to assess if the written code works correctly and if the model is adequate, science education researchers can, when necessary, estimate Bayesian Rasch measurement models and interpret their parameters of interest more rigorously and flexibly.

Appendices

Appendix A: Stan Code for a Simple PCM

```
functions {
  real pcm(int y, real theta, vector beta) {
    vector[rows(beta) + 1] unsummed;
    vector[rows(beta) + 1] probs;
    unsummed = append_row(rep_vector(0.0, 1), theta - beta);
    probs = softmax(cumulative_sum(unsummed));
    return categorical_lpmf(y + 1 | probs);
  }
}
data {
  int<lower=1> I;                  // # items
  int<lower=1> J;                  // # persons
  int<lower=1> N;                  // # responses
  int<lower=1,upper=I> ii[N];      // i for n
  int<lower=1,upper=J> jj[N];      // j for n
  int<lower=0> y[N];               // response for n; y = 0, 1 ... m_i
  int<lower=0> max_rsp[I];         // highest response of each item
  int<lower=1> C;                  // # of covariates
  matrix[J,C] covariate;              // covariates
}
transformed data {
  int m;                           // # parameters per item (same for all items)
  m = max(y);
}
parameters {
  vector[m] beta[I];
  vector[J] theta;
  real<lower=0> sigma;
  vector[C] lambda;
}
model {
  for(i in 1:I)
  beta[i] ~ normal(0, 10);
  lambda ~ student_t(3, 0, 1);
  theta ~ normal(covariate*lambda, sigma);
  sigma ~ exponential(.1);
  for (n in 1:N)
    target += pcm(y[n], theta[jj[n]], segment(beta[ii[n]], 1, max_rsp[ii[n]]));
}

generated quantities {
  vector[N] log_lik;
  for (n in 1: N){
        vector[max_rsp[ii[n]]+1] unsummed;
        vector[max_rsp[ii[n]]+1] probs;
        unsummed = append_row(rep_vector(0.0, 1), theta[jj[n]] - segment(beta[ii[n]], 1, max_rsp[ii[n]]));
        probs = softmax(cumulative_sum(unsummed));
        log_lik[n] = categorical_logit_lpmf(y[n]+1| probs);
      }
}
```

Appendix B: R Code for Data Simulation

```r
##############################
###### LOAD R PACKAGES######
##############################
library(haven)
library(stringr)
library(reshape2)
library(rstan)
library(catR) # for the genPattern() function
library(ggplot2)
rstan_options(auto_write = TRUE)
options(mc.cores = parallel::detectCores())


##############################
### DATA-GENERATING MODEL###
##############################

simcov <- data.frame(v1=rnorm(250),v2=rbinom(250,1,c(0.7,0.3)))
theta <- 1+0.5*simcov$v1+0.6*simcov$v2+rnorm(250,mean=0,sd=0.9)
item <- matrix(rnorm(28,0,3),ncol = 4)
simData <- as.data.frame(genPattern(th = theta, item,model='PCM'))
simData$id <- 1:nrow(simData)
simData <- melt(simData,id.vars = 'id',value.name='response')

sim_data_reg <- list(I=length(unique(simData$variable)),J=length(unique(simData$id)),
                     N=nrow(simData),
                     ii=as.integer(factor(x=simData$variable)),
                     jj=as.integer(factor(x=simData$id)),
                     y=simData$response,
                     max_rsp=rep(4,7),
                     covariate=simcov,
                     C=2)
model_pcm_sim_reg <- stan_model(file='pcm_s_latent_regression.stan')
fit_pcm_sim_reg <- sampling(model_pcm_sim_reg, data=sim_data_reg,chain=4, iter = 2000, cores = 8 )
# print result of the latent regression coefficients
print(fit_pcm_sim_reg,pars ='lambda',probs=c(0.05,0.95))
# inspect Rhat of all parameters
stan_columns_plot(fit_pcm_sim_reg)
# display traceplots for some parameter example
traceplot(fit_pcm_sim_reg, pars = c("lambda", "theta[1]", "theta[2]",
                                    "beta[1,1]", "beta[2,1]"))
generated_lambda<-c(0.5, 0.6)


estimated_lambda<- summary(fit_pcm_sim_reg, pars = "lambda",
                           probs = c(.025, .975))
estimated_lambda<- estimated_lambda[["summary"]]

pcm_diff <- data.frame(par = rownames(estimated_lambda),
                       mean = estimated_lambda[, "mean"],
                       p025 = estimated_lambda[, "2.5%"],
                       p975 = estimated_lambda[, "97.5%"],
                       gen = generated_lambda)
pcm_diff$par <- with(pcm_diff, factor(par, rev(par)))
pcm_diff$lower <- with(pcm_diff, p025 - gen)
pcm_diff$middle <- with(pcm_diff, mean - gen)
pcm_diff$upper <- with(pcm_diff, p975 - gen)

# plot the discrepancy

ggplot(pcm_diff) +
  aes(x = par, y = middle, ymin = lower, ymax = upper) +
  scale_x_discrete() +
  labs(y = "Difference between simulated and estimated values",
       x = "Latent regression coefficients") +
  geom_linerange() +
  geom_point(size = 2) +
  theme(panel.grid = element_blank()) +
  coord_flip()

# explanatory plot

f0 <- function(v1) 1 + v1*estimated_lambda["lambda[2]", "mean"]
f1 <- function(v1) 1 + estimated_lambda["lambda[1]", "mean"] + v1*estimated_lambda["lambda[2]", "mean"]


ggplot(data.frame(v1 = c(0, 20))) +
  aes(x = v1) +
  stat_function(fun = f0, color = "red") +
  stat_function(fun = f1) +
  ylab("Mean estimated ability") +
  xlab("Values of continuous covariate")
```

Appendix C: R Code for the PISA Example

```
library(haven)
library(readxl)
library(reshape2)
library(rstan)
library(bayesplot)

USA <- read_sav("USA.sav")
USA_questionnaire <- read_sav("USA questionnaire.sav")

## please find PISA 2015 Main Survey Items.xlsx on the website of PISA

### Find out science items in pisa data set
item_info0 <- read_excel("PISA 2015 Main Survey Items.xlsx",
sheet = "Table C2.1", skip = 7)
item_info1 <- read_excel("PISA 2015 Main Survey Items.xlsx",
sheet = "Table C2.2", skip = 7)
item_info <- na.omit(c(item_info0$`CBA Item ID in MS Analysis Output`,item_info1$`Item ID in Analysis Output`))

USA_CS <- USA[,c('CNTSTUID',item_info)]
USA_CS <- merge(USA_CS,USA_questionnaire[,c('CNTSTUID','ST004D01T','ESCS')],by='CNTSTUID')
USA_CS <- USA_CS[!is.na(USA_CS$ST004D01T)&!is.na(USA_CS$ESCS),]
USA_CS$ESCS <- round(USA_CS$ESCS,5)
USA_CS$DS131Q02C[USA_CS$DS131Q02C<10]=0
USA_CS$DS131Q02C[USA_CS$DS131Q02C>10]=1
USA_CS$DS131Q04C[USA_CS$DS131Q04C<10]=0
USA_CS$DS131Q04C[USA_CS$DS131Q04C>10]=1
USA_CS$DS269Q03C[USA_CS$DS269Q03C<10]=0
USA_CS$DS269Q03C[USA_CS$DS269Q03C>10]=1
USA_CS$DS438Q03C[USA_CS$DS438Q03C<10]=0
USA_CS$DS438Q03C[USA_CS$DS438Q03C>10]=1
USA_CS$DS510Q04C[USA_CS$DS510Q04C<10]=0
USA_CS$DS510Q04C[USA_CS$DS510Q04C>10]=1
USA_CS$DS657Q04C[USA_CS$DS657Q04C<10]=0
USA_CS$DS657Q04C[USA_CS$DS657Q04C>10]=1


covariate <- USA_CS[,c('CNTSTUID','ST004D01T','ESCS')]


USA_CS <- melt(USA_CS[,1:185],id.vars = 'CNTSTUID',value.name='response')
USA_CS <- na.omit(USA_CS)

CNTSTUID <- unique(USA_CS$CNTSTUID)

## remove students who have no reponses to cognitive assessments

covariate <- covariate[match(CNTSTUID,covariate$CNTSTUID),]

max_rsp <- aggregate(response~variable,data=USA_CS,max)[,2]
##################
### SIMPLE PCM ###
##################
data_pisa <- list(I=length(unique(USA_CS$variable)),J=length(unique(USA_CS$CNTSTUID)),
                  N=nrow(USA_CS),
                  ii=as.integer(factor(x=USA_CS$variable)),
                  jj=as.integer(factor(x=USA_CS$CNTSTUID)),
                  y=USA_CS$response,
                  max_rsp=max_rsp)
fit_pcm <- stan(file='pcm_s.stan',data=data_pisa,chain=3, iter = 3000)


#################################
### SIMPLE PCM with REGRESSION###
#################################

data_pisa_reg <- list(I=length(unique(USA_CS$variable)),J=length(unique(USA_CS$CNTSTUID)),
                  N=nrow(USA_CS),
                  ii=as.integer(factor(x=USA_CS$variable)),
                  jj=as.integer(factor(x=USA_CS$CNTSTUID)),
                  y=USA_CS$response,
                  max_rsp=max_rsp,
                  covariate=covariate[,-1],
                  C=ncol(covariate[,-1]))
fit_pcm_reg <- stan(file='pcm_s_latent_regression.stan',
                 data=data_pisa_reg,chain=3,cores=3, iter = 3000)

save(fit_pcm_reg,file = 'fit_pcm_reg.Rdata')
print(fit_pcm_reg,probs = c(0.025,0.975))
fit_summary <- summary(fit_pcm_reg)

summary(fit_summary$summary[,'Rhat'])
summary(fit_summary$summary[,'n_eff'])
fit_summary$summary['lambda[1]',]

traceplot(fit_pcm_reg,pars=c('theta[1]','beta[61,1]','lambda[2]'))
mcmc_areas(x=fit_pcm_reg,pars=c('lambda[1]','lambda[2]'),point_est ='mean',prob=0.95)+scale_y_discrete(
  labels = c("lambda[1]" = "Gender","lambda[2]" = "Socioeconomic status"))

mcmc_areas(x=fit_pcm_reg,pars=c('theta[1]','theta[6]','theta[15]','theta[20]','theta[71]'),prob=0.95,point_est ='mean')
mcmc_areas(x=fit_pcm_reg,pars=c('beta[52,1]','beta[52,2]','beta[103,1]','beta[103,2]','beta[161,1]','beta[161,2]'),prob=0.95,point_est ='mean')
```

References

Avvisati, F. (2020). The measure of socio-economic status in PISA: A review and some suggested improvements. *Large-scale Assessments in Education, 8*(1), 1–37.

Baker, D. R. (2016). Equity issues in science education. In *Understanding Girls* (pp. 127–160). Springer.

Boone, W. J., & Scantlebury, K. (2006). The role of Rasch analysis when conducting science education research utilizing multiple-choice tests. *Science Education*, 253–269.

Caughey, D., & Warshaw, C. (2014). Dynamic representation in the American states, 1960–2012. In *Proceedings of American Political Science Association 2014 annual meeting.*

Chang, M. (2017). *A comparison of two MCMC algorithms for estimating the 2pl IRT models* (Unpublished doctoral dissertation). ProQuest Dissertations Theses AI.

Copelovitch, M., Gandrud, C., & Hallerberg, M. (2015). Financial regulatory transparency, international institutions, and borrowing costs. Retrieved from https://doi.org/10.2139/ssrn.2701852

Grant, R. L., Furr, D. C., Carpenter, B., & Gelman, A. (2017). Fitting Bayesian item response models in Stata and Stan. *The Stata Journal*, 343–357.

Hoffman, M. D., & Gelman, A. (2014). The no-U-turn sampler: Adaptively setting path lengths in Hamiltonian Monte Carlo. *Journal of Machine Learning Research, 15*, 1593–1623.

Lord, F. M. (1986). Maximum likelihood and Bayesian parameter estimation in item response theory. *Journal of Educational Measurement*, 157–162.

Lunn, D., Jackson, C., Best, N., Thomas, A., & Spiegelhalter, D. (2012). *The bugs book: A practical introduction to Bayesian analysis*. https://doi.org/10.1201/b13613

Martin-Fernandez, M., & Revuelta, J. (2017). Bayesian estimation of multidimensional item response models. *Psicológica, 38*(1), 25–55.

OECD. (2017a). *Pisa 2015 science framework.*

OECD. (2017b). *Pisa 2015 technical report.*

OECD. (2019). *Pisa 2018 assessment and analytical framework*. OECD publishing.

Stan Development Team. (2017). *Stan modeling language users guide and reference manual, version 2.17.0.* Retrieved from http://mc-stan.org

Wagenmakers, E.-J., Marsman, M., Jamil, T., Ly, A., Verhagen, J., & Love, J. (2018). Bayesian inference for psychology. Part i: Theoretical advantages and practical ramifications. *Psychonomic Bulletin & Review, 25*(1), 35–57.

Wei, S., Liu, X., & Jia, Y. (2013). *Using Rasch measurement to validate the instrument of students' understanding of models in science (SUMS).* SpringerLink.

Wyse, A. (2010). R. J. de Ayala (2009) the theory and practice of item response theory. *Psychometrika, 75*, 778–779. https://doi.org/10.1007/s11336-010-9179-z

Xingyao Xiao is a Ph.D. student studying Social Research Methodologies in education at the University of California, Berkeley. Her research interests include statistical modeling development and applications in education.

Mingfeng Xue is a Ph.D. student in the School of Education at the University of California, Berkeley. His research interests focus on psychometrics and educational measurement.

Yihong Cheng is a Ph.D. student studying Curriculum and Instruction at Boston College. His research interests include K-12 STEM education and teacher preparation and professional development.

Chapter 5
Rasch-CDM: A Combination of Rasch and Cognitive Diagnosis Models to Assess a Learning Progression

Yizhu Gao, Xiaoming Zhai, Ahra Bae, and Wenchao Ma

Abstract Learning progressions (LPs) have been highlighted in science education over the past ten years. LPs have the potential to guide curriculum development, instruction, and assessments. An LP has to be validated using empirical evidence, and Rasch model is one of the most broadly used measurement models to connect the empirical evidence with the hypothesized LP. However, Rasch model has limitations as it only provides students' overall ability measures and item difficulties, offering limited information about the specific cognitive skills (i.e., attributes) needed to complete specific tasks. This study thus develops an approach that combines Rasch model and cognitive diagnosis models (CDMs) to assess students' ability and fine-grained attributes. Students' ability, the difficulty of individual attributes, as well as students' attribute mastery patterns estimated from the Rasch-CDM approach can be used to validate LPs. Additionally, attribute mastery patterns of LP levels, ability distribution, and attribute difficulty can be visualized on a map, named as MGZA, showing how students progress towards more sophisticated thinking about a topic. We demonstrate this approach by validating the LP of buoyancy.

5.1 Introduction

Learning progressions (LPs), which depict children's "successively more sophisticated ways of thinking about a topic that can follow one another as children learn about and investigate the topic" (NRC, 2007, p. 219), have drawn increasing attention in science education in the past decade. Research on LPs primarily

---

Y. Gao · X. Zhai (✉) · A. Bae
Department of Mathematics, Science, and Social Studies Education, University of Georgia, Athens, GA, USA
e-mail: Xiaoming.Zhai@uga.edu

W. Ma
College of Education, University of Alabama, Tuscaloosa, AL, USA

X. Liu, W. J. Boone (eds.), *Advances in Applications of Rasch Measurement in Science Education*, Contemporary Trends and Issues in Science Education 57,
https://doi.org/10.1007/978-3-031-28776-3_5

examines students' cognitive development of thinking during scientific practices, aiming to provide evidence for meaningful curriculum development, instruction, and assessments. However, examining students' LP faces challenges. One challenge is due to the nature of children's cognitive development, which takes years and progresses diversely depending on the instruction (Gao et al., 2021). Such a continuous progression challenges researchers in regard to the efforts placed to track the progression, particularly if students' progression trajectories are diverse. It would be ideal if researchers could continually observe students' science performances, collect evidence, and infer students' cognitive development in a considerable time span. However, such work would be time and cost-intensive, and exemplary longitudinal research does not yet exist (Jin et al., 2019a, b).

Instead, many researchers in science education tend to rely on theoretical hypotheses and statistical approaches such as the Rasch model to assess students' LP. For example, Neumann et al. (2013) applied the Rasch model to calibrate students' understanding of energy and validated a four-level progression hypothesis. Researchers prefer the Rasch model because of its many advantages, such as transforming non-interval raw scores into interval and linear student abilities (Boone, 2016; Liu, 2010). As Rasch modeling is broadly used in LP research, researchers realized that this approach overly relies on the overall construct of measurement instrument (Liu, 2020, p. 1), and may lack the power to detect the nuance of students' diverse cognitive abilities required for completing the instrument. For example, the item below may require students to develop multiple cognitive skills (Gao et al., 2020) to respond to the item, such as (a) being able to identify buoyancy (and gravity), (b) knowing the direction of buoyancy (upward, which is opposite to gravity and normal force from the bottom of the container), and (c) knowing that objects stay at rest only if the net force is zero (Gao et al., 2020) (Fig. 5.1).

Rasch modeling could calibrate students' overall ability based on their performance on a test including items like the example, but it is difficult to assess students' specific cognitive skills required to accomplish an individual item. More importantly, measurement evidence is needed for assessment design and development; otherwise, we might not be able to develop assessment items that assess cognitive

Objects 1, 2, 3, and 4 stay still in water. They have different shapes but the same mass. According to their states in water (see the picture below), which of these objects has the smallest buoyancy?

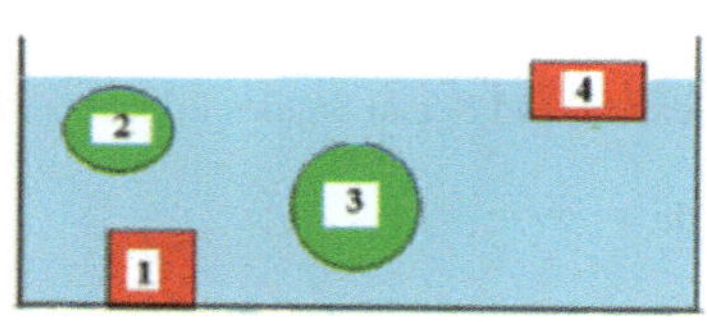

Fig. 5.1 Different statuses of objects in water

skills as intended. It is possible that additional cognitive skills are also required but may be ignored by assessment designers. To overcome these challenges, this study introduces an approach (i.e., Rasch-CDM) incorporating the Rasch model and cognitive diagnosis model (CDM) that can assess fine-grained cognitive skills to validate students' LPs. CDM could examine the individual cognitive skills required for one assessment item (Zhai et al., 2022), and thus could supplement the Rasch model that can calibrate students' overall ability based on performance on a test.

In our prior studies (e.g., Gao et al., 2020, 2021), we applied a similar framework in examining middle school students' LP of buoyancy, but the methodologies were not the focus of those publications. This posed challenges for science education researchers to fully understand the novel approach and the findings.

In this study, we introduce the Rasch-CDM approach for assessing LPs. We aim to provide science education researchers access to the Rasch-CDM approach using plain language that articulates its rationale and procedures. We also introduce the *GDINA* R package (Ma & de la Torre, 2020) that was developed for CDM. Finally, we present an example to demonstrate the approach and discuss the difference between the Rasch-CDM and other statistical approaches in assessing students' LP and other cognitive developments.

5.2 Learning Progressions and Rasch Model

LPs hold many promises to inform curriculum development and instruction and have attracted tremendous attention from researchers in science education. LPs are usually developed based on "big ideas" in science, such as matter, force and interaction, and energy, and highlight students' use of content knowledge in scientific practices (e.g., modeling, argumentation). Though the field has not developed a consensus on LPs, most researchers regard LPs as a tool that depicts students' cognitive development when they learn big ideas. LPs are cognitive models developed based on theories and are verified using empirical data. Conventionally, a hypothesized LP acts as a starting point; empirical assessment data are then collected to validate and refine the assumed LP via psychometric analyses. An LP can be considered as a set of discrete levels or a continuum of cognitive development, both of which are powerful in developing cognitive models. As argued by Jin et al. (2019a, b), neither of them alone provides an all-inclusive description of science learning. LPs are different from construct maps in that they are considered to be more complex (Wilson, 2009). Construct maps specify content being assessed so that they can be relevant to the curriculum (Wilson, 2009). LPs consist of "a set of construct maps, each comprising a 'dimension' of the learning progression," though the complexity of LPs will vary depending on the purpose of the assessment (Wilson, 2009, p. 723).

Researchers have developed different approaches to validating LPs, among which Rasch modeling has been one of the most popular ones (Liu, 2010, p. 212). Rasch model assumes that there is a linear hierarchy in student performances of a construct. Constructs denote the latent traits or attributes of examinees that tests intend to assess, such as students' conceptual understanding of buoyancy (Cronbach & Meehl, 1955; Zhai et al., 2021). Instruments and measurement models are needed to assess constructs. Rasch modeling provides information regarding a person's ability and item difficulty (Johnson & Tymms, 2011; Morell et al., 2017). Furthermore, Rasch modeling transforms raw data into a scale that places "item locations and individual ability estimates on one scale using equal intervals of relative difference" (Osborne et al., 2018, p. 830), allowing for data from different tests on the same metric using common items (Neumann et al., 2013). Rasch model can be used to plot the distribution of person ability and item difficulty on a Wright map on the same scale, with lower item difficulty and person ability towards the bottom and higher item difficulty and person ability towards the top (Boone, 2016; Neumann et al., 2013; Herrmann-Abell & DeBoer, 2018). In order for students to correctly answer more difficult items, the logit value of the person's ability and location on the map must be higher than the item difficulty. If the item's logit value is greater than the person's ability, the likelihood for this person to answer the question correctly would be less than 50%. In contrast, if the item's logit value is smaller than the person's ability, the likelihood of answering the question correctly would be greater than 50%. If an item's logit value is equal to the person's ability, then the chances of answering the question correctly would be 50% (Boone, 2016; Morell et al., 2017; Neumann et al., 2013).

Although the Rasch model is applied to validate LPs varying on topics, grade levels, and theoretical foundations, the general procedures are identical. LP research usually starts by identifying hypothesized levels. Wilson (2009) termed such levels as a construct map, which includes the highest level, the lowest level, and the messy middle. Generally, researchers rely on one or a combination of three guidelines to identify the hypothesized progression levels:

Curriculum standards as guidelines. Curriculum Standards are usually developed by scientists, education researchers, and expert science teachers, representing a consensus on science learning expectations. Therefore, some researchers analyzed the performance expectations of Curriculum Standards to identify the hypothesized levels of an LP, especially for the upper anchors that correspond to the performance expectations in the Standards. For example, Kaldaras et al. (2021) employed the performance expectation of the Next Generation Science Standards (NGSS; NRC, 2012) to identify the upper anchor of the LP of electrical interactions for high school students. Zhai et al. (2018) employed the *Physics Curriculum Standards in China* to revise a LP of energy validated by Wei (2014) by removing the two highest levels of the LP in order to adapt the LP to a high school unit.

Empirical findings as guidelines. The most challenging part of an LP is to identify the middle levels that students progress through before reaching the highest level of expectation. Gotwals and Songer (2010) suggest, "while it may be relatively clear to illustrate an endpoint of what we want students to be able to do (e.g., create a scientific explanation about disturbances to a food web), and it is clear when students do not have any knowledge of the content, defining the pathways or middle knowledge that students may have as they move from a beginning point to the endpoint is much trickier given that the middle is often messy and may not be the same for all students" (p. 277). They termed this middle knowledge as a "messy middle," which is most challenging to identify and assess. In a LP study, researchers can use empirical findings to help organize the "messy middle" to develop the middle levels of a hypothesized LP. For example, Alonzo and Steedle (2009) reviewed literature about students' misunderstanding of force and motion and analyzed the relations of these misconceptions to develop the middle and lower levels of a hypothesized LP of force and motion.

Learning theories as guidelines. LPs are derived from the research of developmental science, which suggests that learning progresses as students engage in various learning activities. Learning theories are used to explain students' cognitive development (Lehrer & Schauble, 2012; Zhai & Li, 2021). Some researchers thus explicitly refer to learning theories to help identify the hypothesized levels. For example, Neumann et al. (2013) referred to the theory of the complexity of knowledge (Kauertz & Fischer, 2006) to identify four levels of conceptual understanding: facts (i.e., a fragmental knowledge base indicating that students have pieces of knowledge unconnected to each other), mappings (i.e., pieces of knowledge are simply connected), relations (i.e., qualified connections exist between pieces of knowledge), and concepts (i.e., knowledge pieces are connected, structured, and intertwined). Based on the theory, they identified four levels of energy understanding for their LP: the types of energy, transformations of energy, energy degradation, and energy conservation. Likewise, Osborne et al. (2018) developed their hypothesized LP levels of argumentation based on cognitive load theory (Chandler & Sweller, 1991). They assume that argumentation would demand varied working memory based on the complexity of arguments because the working memory is accountable for students' LP of argumentation. They thus developed three levels that increase depending on the number of connections students made with claims, evidence, and warrants in their arguments.

Rasch model. Once the hypothesized LP levels are identified, researchers can develop assessment items linked to the levels and then employ the Rasch model to validate the hypothesized LP. In most cases, an LP is centered on one big idea or scientific practice, thus researchers usually assume that the latent trait (i.e., construct) is unidimensional. Thus, a unidimensional Rasch model may be employed to analyze students' performance on the items. For example, Yao et al. (2017) developed an instrument of Energy Concept Progression Assessment to validate and revise a four-level LP with two variables: key ideas and conceptual development. They found the unidimensional Rasch model best fit their data and assumption. Unidimensional Rasch model allows for the measurement of a single trait (Planinic et al., 2019). Moreover, unidimensional Rasch model involves fewer parameters and thus demands a smaller sample size.

In some instances, however, unidimensional Rasch is insufficient to examine the complexity of the assessment construct; researchers thus have to employ the multidimensional Rasch model. A multidimensional Rasch model assumes that the assessment construct includes subdimensions. That is, an assessment item may address one or multiple subdimensions of the construct. The multidimensional Rasch model is appropriate for assessments with robust theoretical foundations that support the dimensionality assumption. For example, Hadenfeldt et al. (2016) used a multidimensional Rasch to examine grades 6 to 13 students' LP on four big ideas of matter (i.e., structure and composition, physical properties and change, chemical properties reactions, and conservation). They compared a four-dimensional Rasch model with two-dimensional and unidimensional Rasch models and found the four-dimensional Rasch model yielded the most robust psychometric properties, which supported the assumption that the four big ideas are distinct. An in-depth investigation through the four-dimensional Rasch model suggests that students' understanding of each of the four big ideas showed five progression levels, though three of them developed in parallel. Similar research can be referred to by Morell et al. (2017) in which the authors investigated students' LP of the structure of matter using multidimensional analysis. Though multidimensional Rasch may better reflect the authentic construct in assessment, it involves more parameters and requires larger samples than unidimensional models. Therefore, researchers need to determine which Rasch model to adopt based on theoretical assumptions and data.

Given the complexity of LPs, other complementary statistical approaches have also been used alongside the Rasch model. For example, latent class analysis (LCA) has been applied together with the Rasch model in validating LPs. LCA can be used to examine the latent class of the samples based on the similarity of students' responses, which is appropriate to identify or validate students at different levels of the LP (e.g., Gao et al., 2020, 2021). Fulmer et al. (2014) used the Rasch model to analyze high school and university students' LP of force and motion. In this analysis, student responses were transformed into ability measures, which provided evidence with regards to the consistency of student performances on the five-level LP. They then applied LCA to examine the latent classes of student performance, which suggested four classes. Although not consistent with the five levels LP, the LCA result demonstrated that students' average ability progressed along with the four classes.

ANOVA is another analysis that is frequently employed to examine the difference in student performance at different levels of LPs (e.g., Gotwals & Songer, 2013; Todd et al., 2016). Lee and Liu (2009) employed ANOVA and Tukey's post hoc to compare data across sixth-, seventh-, and eighth-grade students taking earth, life, and physical science, respectively. The analysis revealed that the mean knowledge integration increased from sixth to eighth grade, with the mean of eight graders being significantly higher. Testa et al. (2019) also used Rasch, ANOVA, Tukey's post hoc, and exploratory factor analysis to revise their hypothesized LP on quantum mechanics. Through the analyses, they revised their factors to two big ideas, atomic description and measurement and wave function, and its properties in the measurement process.

The development of an LP is usually an iterative process during which researchers revise their hypothesized LP using empirical evidence. In the validation process, measurement models such as the Rasch model play a significant role in linking the observation with the construct. However, Rasch model has limitations, so other methods (e.g., LCA) are usually accompanied. Below, we provide another approach, CDM, that can provide more nuanced evidence to examine students' LP.

5.3 Learning Progressions and Cognitive Diagnosis Modeling

CDMs are statistical models that provide diagnostic information regarding examinees' mastery patterns of skill sets in problem-solving. Unlike the Rasch model, in which estimating an individual's general ability is the primary goal, CDMs are restricted latent class models that aim to assess an individual's profile on a set of skills used to solve problems (de la Torre & Douglas, 2004). These skills are usually referred to as attributes and assumed to be binary with two possible statuses—mastery and non-mastery. That is, by analyzing students' responses, CDMs provide information about whether the examinee has mastered a given attribute or a combination of several attributes. Individuals' attribute profiles reveal their strengths and weaknesses in problem-solving (i.e., which attributes they have and have not mastered), making them potentially useful for remediation and classroom instruction. For example, by analyzing the buoyancy item referred above, along with other items in the test, CDMs can estimate whether a student is able to (a) identify buoyancy (and gravity), (b) know the direction of buoyancy, and (c) understand that objects stay at rest only if the net force is zero, etc. Students with the same total score or ability value generated through Rasch modeling can have different attribute mastery profiles. Therefore, CDMs could be another complement to Rasch to provide additional nuances of students' cognitive abilities on an LP (Chen et al., 2017; Gao et al., 2020).

The advantages of CDMs in constructing LPs have been recognized in previous studies. For example, Chen et al. (2017) used the Rule Space Method (RSM; Tatsuoka, 1983, 1995, 1996) to construct the LP of thermochemistry for senior high school students. Their four-level LP gradually describes how senior high school students master seven attributes of thermochemistry (e.g., exothermic and endothermic reactions, chemical bonds, and heat quantity change). The RSM is one of the first psychometrical methods for diagnosing cognitive attributes (Gierl, 2007). It is designed to yield specific information about examinees' mastery of specified cognitive skills (Tatsuoka, 1995). With the RSM, we validated and revised the LP of buoyancy for grade eight students (Gao et al., 2020; Gao et al., 2021). We found that students' mastery statuses of seven attributes featured a four-level LP of buoyancy. For example, students who only grasped the density attribute but failed to master the rest of the attributes were located at the basic LP level. In these studies, students at different LP levels are featured with specific attribute mastery patterns. That is, LP levels are distinguished from each other at the attribute scale.

However, the RSM in analyzing LPs has limitations. First, it may fail to obtain the correct attribute hierarchical relationships (Yu et al., 2011; Yuan et al., 2022), which consequently affects the validation of LPs. Second, there is no publicly available software programs to perform the RSM analysis, which essentially limits its application in practice. Third, the RSM does not formally define a measurement model, which makes it difficult to accommodate different response functions. To overcome these shortcomings, we put forward the Rasch-CDM approach, which involves both the Rasch model and CDMs. The Rasch-CDM approach does not utilize hierarchical structures among attributes when fitted to assessment data. Comparisons between attributes' difficulties generated by the Rasch-CDM approach can be used to verify theory-driven attribute hierarchical relationships that correspond to a hypothesized LP. In addition, the Rasch-CDM analysis can be easily performed using an existing R package. In the following, we present the application of the Rasch-CDM in LPs by taking the buoyancy LP as an example.

5.4 Rasch-CDM: Integration of Rasch Model and CDM for Examining Student Learning Progressions

The Rasch-CDM approach, which was originally proposed in de la Torre and Douglas (2004), is a two-layer model, where the lower layer is a CDM, which assumes that students' item responses are affected by their attribute mastery profiles (see Fig. 5.2). The upper layer is the Rasch model where the overall ability is assumed to affect students' attribute mastery. A distinguishing feature of this approach is that researchers can estimate students' ability from the Rasch component and students' attribute profiles from the CDM component simultaneously.

The CDM component defines the relationship between attributes measured by an item and the item responses. The CDM we adopted is the generalized deterministic input noisy and gate (G-DINA; de la Torre, 2011) model. Technical details of the G-DINA model can be found in de la Torre (2011). To simplify the presentation, we take item 5 from the buoyancy test as an example. Item 5 measures two attributes, α_3 (i.e., A3) and α_6 (i.e., A6). Based on the G-DINA model, the probability of answering item 5 correctly given α_3 and α_6, each of which can take 0 or 1, is written by

$$P(X_5 = 1|\alpha_3, \alpha_6) = \delta_0 + \delta_1\alpha_3 + \delta_2\alpha_6 + \delta_{12}\alpha_3\alpha_6,$$

where δ_0 is the baseline, representing the probability of success for students who do not master either required attributes, δ_1 and δ_2 are the increase in probability of success due to the mastery of attributes 3 and 6, respectively, and δ_{12} is the interaction effect due to the mastery of both attributes.

The Rasch component specifies the relationship between overall ability and the mastery of attributes. Let α_{ik} be the mastery status (i.e., 0 stands for non-mastery and

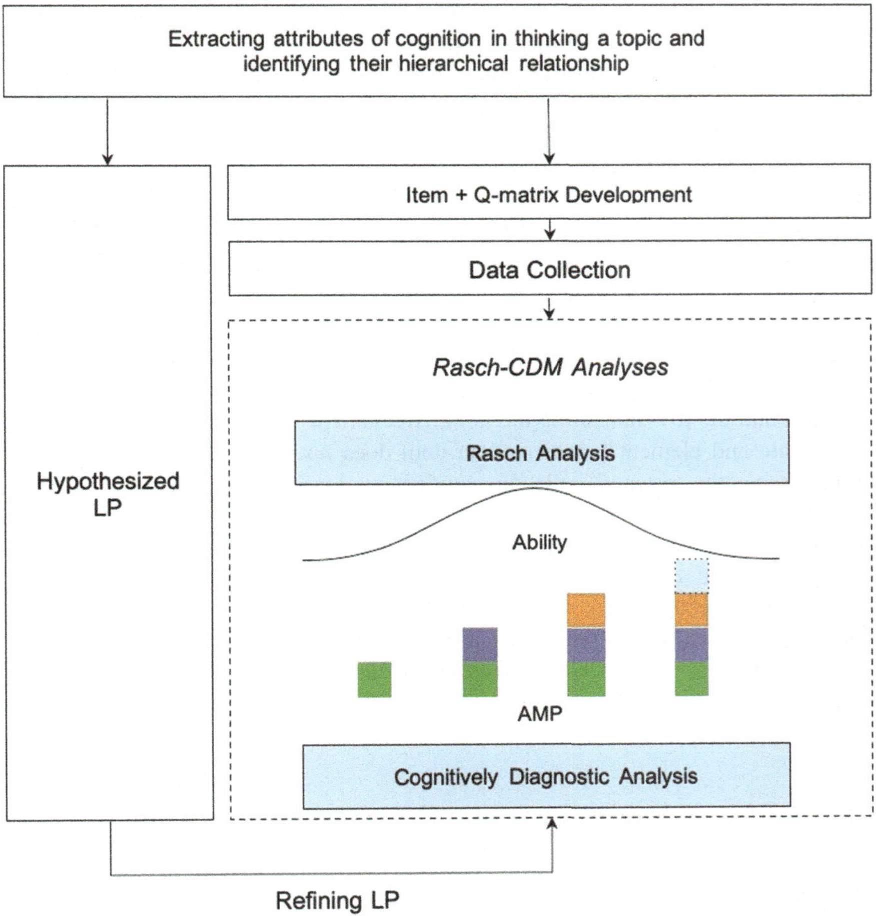

Fig. 5.2 Flowchart of using the Rasch-CDM to refine an LP
Note. AMP: Attribute Mastery Pattern. Q-matrix demonstrates the relationship between items and attributes

1 for mastery) of student i on attribute k. The probability of mastering attribute k for student i is

$$P(\alpha_{ik} = 1|\theta_i) = \frac{1}{1 + \exp[-(\theta_i - b_k)]}$$

where b_k indicates the difficulty of attribute k and θ_i is the overall ability of student i. Note that this is not the traditional Rasch model, which connects the ability with observed item responses; however, we view this as an extension of Rasch model to establish the relationship between attribute mastery and overall ability. It should also be noted that the presented model above is not a multilevel model; rather, it was originally labelled as a higher-order model (de la Torre & Douglas, 2004), similar to the higher-order confirmatory factor analysis.

To validate LPs using this integrated framework, researchers need to define the attributes of interest and then derive hypothesized LPs based on curriculum standards, empirical evidence, and learning theories. The hypothesized LPs may consist of hierarchical relationships among these attributes. A hierarchy between two attributes means that the mastery of one attribute is a prerequisite for the mastery of the other one. The hierarchical relationship can be developed by subject matter experts or by reviewing the literature on the topic. To validate the hypothesized LPs empirically, researchers need to construct test items and develop a Q-matrix that shows the relationship between each item and the attributes. The Q-matrix includes a given number of attributes. To respond to the items, students need at least one of the cognitive skills. The Q-matrix is a binary matrix, with the column indicating the attributes and the row indicating the item. An element 1 indicates an item measures an attribute and element 0 indicates an item does not measure an attribute. After administering the test and collecting student responses, researchers can evaluate whether the data fits the model adequately. In addition to estimating an individual student's ability and attribute profile, the difficulties of all attributes can also be estimated. Based on overall ability, students can be divided into groups with different proficiency levels. Further analysis can be performed to examine the attribute profiles in each proficiency group. Altogether, the attribute mastery proportion of each group, which is derived by calculating the average attribute mastery probability of group students, will be used to refine the hypothesized LPs.

5.5 A Sample Study: Learning Progression of Buoyancy

In this section, we present a sample study with real data (i.e., the Buoyancy test data from Gao et al., 2020, 2021) as a practical tutorial for the Rasch-CDM approach. In accordance with the procedure shown in Fig. 5.2, we first present the attributes and a hypothesized LP about buoyancy. We then articulate the item development and Q-matrix construction to assemble a buoyancy test. We employed a data set of 1,127 eighth-grade students to validate the hypothesized LP of buoyancy.

5.5.1 Step 1: Identification of Attributes

Buoyancy is an upward force exerted by a fluid or gas due to the pressure difference on different parts of an object, which explains the phenomenon that objects sink and float in liquid or gas (Şahin & Çepni, 2012). Buoyancy is an important concept in everyday life, but it is challenging for students to understand because it includes multiple sub-concepts. Moreover, more than 20 misconceptions may impede students' understanding of buoyancy (Gao et al., 2021). Additionally, advanced reasoning skills are needed for students to use buoyancy to explain phenomena (Inhelder & Piaget, 1958). Thus, it is important for educators and researchers to understand how students progress toward a holistic understanding of buoyancy. The current study employed the Rasch-CDM approach to validate the progression levels from the lower anchor to the upper anchor.

First, attributes relevant to buoyancy were identified. Two important factors were recognized when we abstracted attributes of buoyancy: misconceptions and hierarchical property (Gao et al., 2021). Misconceptions help identify critical issues that influence students' understanding of buoyancy whereas hierarchical relationships indicate the premises of understanding the big idea of buoyancy. In Gao et al. (2021), we have presented the rationale and approach to abstracting seven attributes of buoyancy. These attributes and their operational definitions are shown in Table 5.1. For example, the first attribute *Direction* indicates the direction of buoyancy is vertically upward regardless of object status (e.g., float or sink). Each of the seven attributes is essential for students to fully explain and predict the buoyancy phenomena.

Alongside the attributes, hierarchical relationships among the seven attributes were also identified according to literature (see Gao et al., 2020 for details). As shown in Fig. 5.3, *Direction* (A1) and *Identification* (A2) form the basis of *Relation with motion* (A5), *Density* (A3) and *Submerged volume* (A4) are the basis of *Archimedes' theory* (A7), and *Density* (A3) is the basis of *Relative density* (A6) and *Archimedes' theory* (A7). That is, students are supposed to master attributes 1 and 2 before they master attribute 5, understand attribute 3 before grasping attribute 6, and know attributes 3 and 4 before they master attribute 7.

After identifying attributes and hierarchical relationships, we developed a hypothesized LP of buoyancy. Two premiers were first identified for constructing hypothesized LP levels, which corresponded to divergent attribute mastery patterns. First, students are more likely to master attribute patterns at lower hierarchical levels

Table 5.1 The seven attributes of buoyancy and their operational definitions

Attribute	Operational definition
Direction (A1)	Know that buoyancy is vertically upward
Identification (A2)	Can identify not only gravity but also buoyancy that is exerted on an object when it is floating, hanging, or sinking in liquid
Density (A3)	Know that density is a property of an object, which is equal to mass divided by volume and will not change even though the mass or the volume changes
Submerged volume (A4)	Understand the relationship between submerged volume in liquid and object volume
Relation with motion (A5)	Can develop the relationship between the magnitude of buoyancy and the state of motion
Relative density (A6)	Can judge floating and sinking by comparing the densities of the object and of the liquid
Archimedes' theory (A7)	Can employ Archimedes' theory to calculate buoyancy

Note. Attributes 1–7 (e.g., *Direction*, *Identification*) are simplified as A1–A7

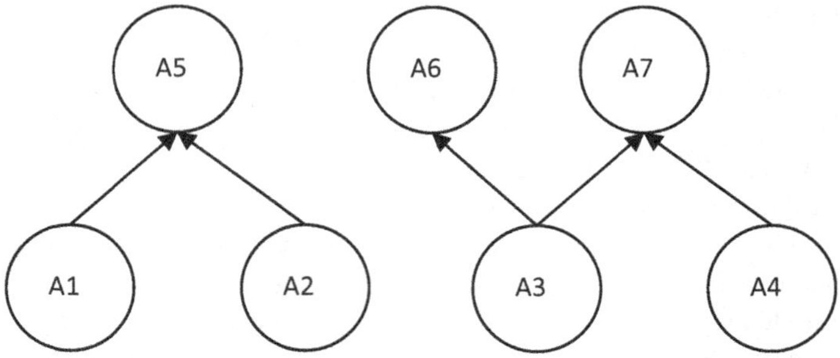

Fig. 5.3 The hierarchical relationships of the seven attributes

rather than those at higher hierarchical levels. For example, students are unlikely to master attribute 5 before knowing attributes 1 and 2 accurately. In other words, the mastery probability of attribute 5 would be enhanced after students acquire attributes 1 and 2. Thus, attribute mastery patterns with a mastery status of attribute 5 and non-mastery statuses of attributes 1 and/or 2 are unexpected in the current study. Second, students at higher LP levels are assumed to display more complex attribute mastery patterns as compared to students at lower LP levels. That is, the higher LP levels are students at, the more attributes they would master.

Based on the two premiers, we adopted a four-level modeling theory to construct our hypothetical LP (Clement, 2008). Science education engages students to use scientific knowledge to construct models to explain phenomena (Zhai, 2022). However, scientific knowledge is not essential for students to construct their models. Students with intuitive knowledge can also develop models to explain phenomena. For example, in Chinese buoyancy indicates "floating" which explicitly suggests an upward force. Therefore, students have no difficulty understanding the direction of buoyancy (A1). We thus assigned A1 to level 1. As students get mature and knowledgeable through learning, they may revise their models using scientific knowledge. In this case, the students first learned density and it would be reasonable for them to use density to explain buoyancy phenomena. We thus assigned A3 (density) and A6 (relative density) to level 2. As more knowledge is accumulated, students would be able to quantitatively explain phenomena. In level 3, the quantitative level, we assume that the students are able to employ Archimedes' theory to calculate the magnitude of buoyancy. Clement (2008) argued that quantitative modeling was usually more complex than qualitative modeling. Finally, we asserted that the integrative level should indicate the relationship between buoyancy and motion. The topic of force and motion is a big idea that requires students to understand not only buoyancy but also other forces such as gravity and also Newton's laws. Considering this complexity, the panel split students' understanding of force and motion into two levels. The attribute A2 is added to level 4. This attribute indicates that students should be able to identify buoyancy by observing the motion of an object. Although this level requires a higher ability, it is qualitative.

Table 5.2 The attributes corresponding to the hypothesized LP and descriptions

Level	A1	A2	A3	A4	A5	A6	A7	Description
1	X							Students use intuitive knowledge and understand the direction of buoyancy.
2	X		X			X		Students are able to use density and relative density to explain sinking and floating.
3	X		X	X		X	X	Students are able to use Archimedes' theory to quantitatively explain sinking and floating.
4	X	X	X	X		X	X	Students are able to identify the forces that exert on the objects results in sinking or floating.
5	X	X	X	X	X	X	X	Students are able to use the relations between (net) force and motion to explain sinking and floating.

Whereas in the higher quantitative level—level 5, students should be able to quantitatively establish the relationship between buoyancy and motion.

The hypothesized LP of buoyancy is presented in Table 5.2, with five levels in total, each of which is associated with a specific attribute mastery pattern. The symbol "X" indicates a mastery status of an attribute; otherwise, the attribute has not yet been mastered at the LP level. For example, students at Level 1 only master attribute 1, and students at Level 2 are expected to master attributes 1 and 2. That is, students at Level 1 only know that buoyancy is a vertically upward force. When students progress to Level 2, they understand that density is a property of an object as well as the direction of buoyancy. The hypothesized LP of buoyancy indicated that students gradually master more attributes when progressing to higher LP levels (i.e., Levels 3, 4, and 5).

5.5.2 *Step 2: Item Development and Q-Matrix Construction*

A panel including physics researchers, physics teachers, and measurement experts collaborated to design and revise items to assess students at different LP levels. After conducting a pilot test examining item quality (e.g., item difficulty, item discrimination), one constructed response and thirteen multiple-choice items were included in the final buoyancy test (see Gao et al., 2020 for the test). Table 5.3 shows attributes measured by each item (i.e., Q-matrix). Most items involved more than one attribute, whereas four items measured only one attribute. For example, item 1 only involved attribute 1, while item 14 measured all the seven attributes. At least four items measure a specific attribute in the buoyancy test.

We used two example items to illustrate item formats and item scoring rubrics. As shown in Fig. 5.4, item 2 is the constructed response item of the buoyancy test, which requires students to identify what forces are exerted on three objects which are floating, hanging, and sinking in liquids (i.e., A2). In contrast, item 14 is one of the multiple-choice items, which measures students' mastery of all the seven attributes.

Table 5.3 Attributes that correspond to test items

Item	Attributes	Number of attributes
1	A1	1
2	A2	1
3	A3	1
4	A4	1
5	A3, A6	2
6	A3, A6	2
7	A1, A2, A5	3
8	A1, A2, A5	3
9	A3, A4	2
10	A3, A4, A7	3
11	A1, A2, A3, A5, A6	5
12	A1, A3, A4, A6, A7	5
13	A1, A2, A3, A4, A6, A7	6
14	A1, A2, A3, A4, A5, A6, A7	7

2. Draw the force diagrams for a ball in each of the following states and compare the forces of gravity and buoyancy.

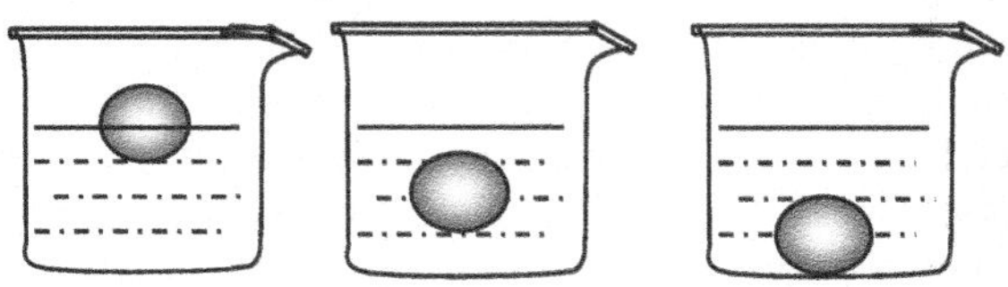

14. Two identical blocks which are made of paraffin are placed into sufficient amounts of water and alcohol separately. What is the ratio of the two blocks' submerged volumes in these liquids? (Note. $\rho_{water} = 1.0 \times 10^3\ kg/m^3$, $\rho_{paraffin} = 0.9 \times 10^3\ kg/m^3$, $\rho_{alcohol} = 0.8 \times 10^3\ kg/m^3$*)*

A. 9:8 *B. 1:1*

C. 8:9 *D. It cannot be determined*

Fig. 5.4 Two example items from the buoyancy test

That is, to answer item 14 correctly, students need to understand gravity, buoyancy, and how they relate to objects' motion in fluid (i.e., A1, A2, and A5). In addition, students are able to identify statuses of the two objects in the two liquids (i.e., A3 and A6) when they retrieve densities of water, paraffin, and alcohol from the item stem. Also, students can complete calculations of submerged volumes using Archimedes' theory (i.e., A3, A4, and A7). To score students' answer to item 2, we examined whether they identified both gravity and buoyancy for all the three objects. Students who failed to point out gravity or/and buoyancy for either of the three objects were scored as "0"; otherwise, students were scored as "1". In contrast, the scoring rubric of item 14 is straightforward: students' response was dichotomously scored as "1" for correct and as "0" for incorrect in comparison with the right choice.

5.5.3 Step 3: Data Collection and Data Analysis

We used measurement data from 1127 eighth-grade students who took the buoyancy test. These students come from ten middle schools located in five east coast cities in China. All these students used identical physics textbooks, and all had formally learned buoyancy in their science classes by the time the buoyancy test was taken. The test lasted 40 minutes to ensure enough time for students to complete all the test items. Tests in different schools were not conducted at the same time because of a restricted number of test administrators. However, the schools did not communicate about the test, ensuring the validity of the test results. We performed the Rasch-CDM approach using the *GDINA* package (Ma & de la Torre, 2020) within the R computing environment (R Core Team, 2022). We first examined model-data fit statistics to evaluate whether the Rasch-CDM could fit measurement data adequately. Then, we compared the difficulties of the seven attributes, especially for those within hierarchical structures. Finally, we gathered estimations of students' overall abilities and their attribute mastery proportions generated from the Rasch-CDM analysis. Abilities and attribute mastery proportions provided evidence for the LP of buoyancy, and the hypothesized LP developed in Step 1 was revised and refined. The R codes with annotations have been attached to Appendix.

5.5.4 Step 4: Rasch-CDM Analyses and Refinement of the Hypothesized LP

Model-Data Fit and Q-Matrix Validation

To determine whether the Rasch-CDM fits our measurement data appropriately, we evaluated absolute fit statistics. One criterion of absolute model fit is the standardized root mean square residual (SRMSR; Maydeu-Olivares, 2013). Maydeu-Olivares suggests that SRMSR values smaller than 0.05 indicate well-fitting models. Results showed that the SRMSR value for the Rasch-CDM is 0.044, which is smaller than the benchmark (Maydeu-Olivares, 2013), 0.05. Thus, the absolute fit statistic suggested a robust model fit for the Rasch-CDM.

We also investigated the adequacy of the hypothesized Q-matrix. The *Hull* method (Nájera et al., 2021) was used to validate the Q-matrix specifications by identifying possible misspecified entries of each item (de la Torre & Chiu, 2016). Results of Q-matrix validation suggest attributes of items 5 and 9 may require modifications. Specifically, item 5 was suggested to measure three attributes (i.e., A3, A4, and A6) rather than A3 and A6, and item 9 was recommended to measure A4 instead of both A3 and A4. Domain experts further examined the two items (see Fig. 5.5) and disagreed with suggested modifications. First, students can answer item 5 correctly without mastering the relationship between submerged volume and object volume (i.e., A4). That is, students can use relative density and the block's status to conclude the block's density. Second, to answer item 9 correctly, students

5. Block A sinks in a container with water. As for the density of Block A, which of the following choices is most likely?

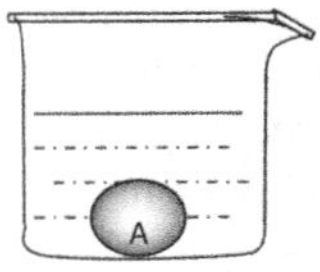

A. 0.3×10^3 kg/m^3
B. 0.5×10^3 kg/m^3
C. 0.8×10^3 kg/m^3
D. 1.1×10^3 kg/m^3

9. Blocks 1, 2, and 3 have the same mass, but they are made of different materials: copper, iron, and aluminum (Note. $\rho_{copper} > \rho_{iron} > \rho_{aluminum}$). If we put these blocks into three same containers full of water, they would sink, and the water would flow out. Which of these three circumstances has the most overflow water?

A. Copper *B. Aluminum*
C. Iron *D. All are same*

Fig. 5.5 Items 5 and 9 which attributes are suggested to be modified

need to understand object density (i.e., A3). Otherwise, students will not be able to compare the volumes of the three blocks, which is critical to compare their submerged volumes. We thus continued to use the original Q-matrix in the following analyses.

Attribute Difficulty

In addition to model-data fit statistics, the Rasch-CDM analysis estimated the difficulty of each attribute. As shown in Table 5.4, attribute 3 has the lowest difficulty value (i.e., -1.569), while the difficulty value of attribute 7 is the highest (i.e., 0.024). We also noted that the difficulty of attribute 5 is higher than attributes 1 and 2, attribute 6 is more challenging than attribute 3, and attribute 7 is more difficult in comparison with attributes 3 and 4. These results confirmed the hierarchical attribute relationships established in Step 1. That is, attributes at higher hierarchical levels have higher difficulty values than those at lower hierarchical levels.

Ability and Attribute Mastery Pattern

To verify the hypothesized LP of buoyancy, we used estimations of overall ability and attribute mastery proportion of the seven attributes, respectively. Given that the total number of attributes of buoyancy is seven, we assume the highest-level students mastered all seven attributes while the lowest-level students did not master any attribute. Between the lowest and the highest levels, we further segmented students into six groups with equal ability intervals. That is, the range of student abilities was evenly divided into eight groups. For each group, we then calculated students' average attribute mastery proportion per attribute. As shown in Table 5.5, average

Table 5.4 Difficulties of the seven attributes

Attribute	Difficulty
A1	−0.569
A2	−0.238
A3	−1.569
A4	−0.577
A5	0.013
A6	−0.385
A7	0.024

Table 5.5 The average abilities and attribute mastery proportions of eight groups

Group	Average ability	A1	A2	A3	A4	A5	A6	A7
1	−1.390	0.000	0.000	0.000	0.000	0.000	0.000	0.000
2	−1.080	0.015	0.029	0.471	0.096	0.081	0.110	0.029
3	−0.744	0.200	0.224	0.744	0.176	0.280	0.168	0.176
4	−0.398	0.390	0.314	0.790	0.448	0.343	0.419	0.276
5	−0.055	0.425	0.485	0.916	0.713	0.335	0.760	0.689
6	0.356	0.832	0.734	0.958	0.832	0.455	0.867	0.713
7	0.626	0.829	0.884	0.973	0.925	0.774	0.890	0.877
8	1.020	1.000	1.000	1.000	1.000	0.996	1.000	1.000

attribute mastery proportions were different across the eight groups. The higher average ability one group had, the higher average attribute mastery proportion the group displayed. For example, the average attribute mastery proportions of the first group were all zeros for the seven attributes, whereas the final group mastered all the seven attributes clearly.

Average attribute mastery proportions shown in Table 5.5 were used to derive students' attribute mastery patterns. Consistent with our previous studies (Gao et al., 2020, 2021), we employed 0.50 (50% chance to master an attribute) as the cut-off value to determine whether students in a group mastered an attribute or not. That is, if students at a given LP level had an average opportunity higher than 50% to acquire a given attribute, the attribute was featured at the specific LP level. According to the cut-off value, the first two groups did not master any attribute; the third and fourth groups only mastered attribute 3; the fifth group mastered three more attributes (i.e., A4, A6, and A7); the sixth group almost mastered all the attributes except attribute 5; the last two groups mastered all the seven attributes. Given that some groups shared identical mastery patterns, we combined them into one group with a unique attribute mastery pattern. We thus merged groups 1 with 2 to form Level 1, combined groups 3 with 4 to constitute Level 2, and integrated groups 7 and 8 as Level 5. Groups 5 and 6 were labeled as Levels 3 and 4, separately. Consequently, students were classified into five groups with increasing ability measures and distinct attribute mastery patterns (see Table 5.6). Generally, students at higher LP levels had higher abilities and were more likely to master the seven attributes. Students at Level 1 did not

Table 5.6 The average ability and attribute mastery proportion of five groups

Level	Average ability	A1	A2	A3	A4	A5	A6	A7
1	−1.180	0.010	0.020	0.315	0.064	0.054	0.074	0.020
2	−0.586	0.378	0.265	0.765	0.300	0.309	0.283	0.222
3	−0.055	0.425	0.485	0.916	0.713	0.335	0.760	0.689
4	0.356	0.832	0.734	0.958	0.832	0.455	0.867	0.713
5	0.867	0.935	0.956	0.990	0.971	0.911	0.958	0.953

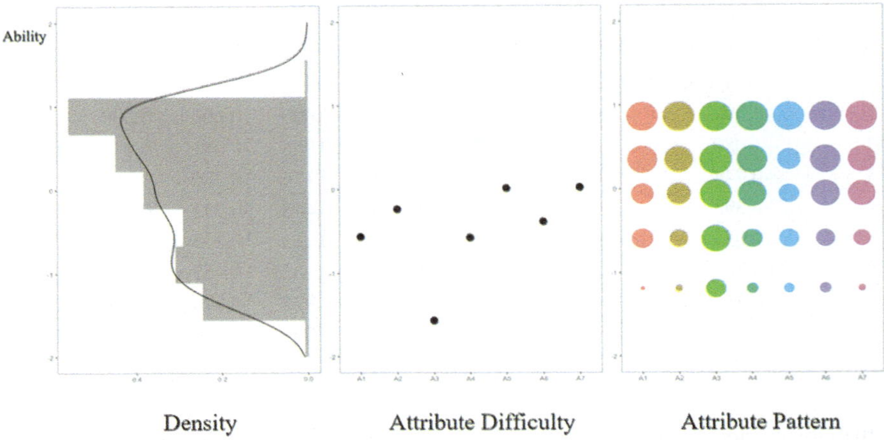

Fig. 5.6 MGZA map for student ability, attribute difficulty and attribute mastery patterns generated from the Rasch-CDM approach
Note. The larger the circle is, the higher the attribute mastery probability is

master any attribute. Average attribute mastery proportions were comparatively low. Students at Level 2 only mastered attribute 3. When students progressed to Level 3, they mastered three more attributes (i.e., A4, A6, and A7) together with attribute 3. Students at Level 4 mastered all attributes except attribute 5. Students at Level 4 mastered all the seven attributes at relatively high mastery proportions (i.e., all proportions were larger than 0.90).

Figure 5.6 shows students' overall ability distribution in the left panel, the attribute difficulty in the middle panel, and the attribute mastery pattern of each LP level in the right panel, which was referred to as the MGZA plot (using the authors' names). The y-axis denotes student ability and attribute difficulty estimates estimated from the Rasch-CDM approach. We noted that abilities calculated from the Rasch-CDM approach had a high correlation with abilities estimated by the unidimensional Rasch model ($r = 0.959$) and the abilities of most students ranged primarily from -1.5 to 1.0. The MGZA plot also shows that attribute 3 is the least challenging attribute while attributes 5 and 6 are the most demanding attributes for students. The right panel of the MGZA map displays the attribute patterns and the

Table 5.7 Comparisons between hypothesized LP and modified LP of buoyancy

	Hypothesized LP of Buoyancy								Modified LP of Buoyancy						
	A1	A2	A3	A4	A5	A6	A7		A1	A2	A3	A4	A5	A6	A7
1	X							1							
2	X		X			X		2			X				
3	X		X	X		X	X	3			X	X		X	X
4	X	X	X	X		X	X	4	X	X	X	X		X	X
5	X	X	X	X	X	X	X	5	X	X	X	X	X	X	X

average abilities of the five groups. The size of the dots indicates the probability of students in the group mastering each attribute. Results suggest that higher LP levels correspond to higher chance of mastery on most attributes.

Attribute mastery patterns inform revisions to the hypothesized LP. As shown in Table 5.7, the modified LP of buoyancy is different from the hypothesized LP of buoyancy in levels 1, 2, and 3. In conclusion, this study derived a verified LP of buoyancy by combining students' overall ability estimation and their attribute mastery patterns of the seven attributes.

5.6 Discussion of Differences Between Rasch-CDM and Other Methods

This study presented how to use the Rasch-CDM approach to validate LPs in science education and offers a sample study of buoyancy. Compared with the Rasch modeling approach to validating LPs, the proposed Rasch-CDM approach provides more fine-grained information regarding students' cognitive development of scientific thinking. The Rasch-CDM approach measures students' ability with Rasch modeling while simultaneously evaluating students' mastery or non-mastery of different attributes. The combined information can better inform instruction and assessment design because derived LP levels are distinguished from each other at an attribute scale.

In comparison with the RSM used in previous studies in science education (e.g., Chen et al., 2017; Gao et al., 2020), this study employed the existing *GDINA* R package. In addition to attribute mastery proportions, the Rasch-CDM approach estimates attribute difficulty, supplying empirical evidence for a theory-driven attribute hierarchical relationship.

However, additional work is needed to further explore the application of the Rasch-CDM approach in LPs. For example, approaches to dividing students into groups affect LP levels. Our previous studies used latent class analysis of students' test data to cluster students into different LP levels (Gao et al., 2020, 2021). Some researchers ordered attributes by the difficulty measure, and attributes with similar difficulties were assigned to the same LP level (e.g., Yuan et al., 2022). The current sample study divided students into eight groups based on the number of attributes.

The final LP levels were generated after combining groups with the same attribute mastery patterns. The LP levels are slightly inconsistent with findings from prior studies using different methods (i.e., use Rasch modeling and Rule space modeling separately), though with justifiable evidence. In the future, researchers have to select appropriate methods and justify them based on their assumptions and data.

5.7 Conclusions

The study contributes to the field by introducing a Rasch-CDM approach to validating LPs. This approach extends the Rasch model and incorporates CDMs, which provides an alternative way to verify hypothesized LPs or refine LPs with additional information. Because the attributes are finer grain-size than assessment items in measuring students' LP, we argue that this approach demonstrates superiority compared to traditional Rasch approach. The example study provides a practical tutorial to demonstrate the usability of this approach in LP research. The findings in this study support the development of effective instruction and diagnostic assessments to promote students' LP in science education.

Even though this approach provides additional nuance in probing students' LP, we noted one limitation. Different from the Rasch model which uses item difficulty to validate hypothesized LP, this approach starts by estimating students' probability of mastering attributes. Because students' mastery probability may vary by the sample, it is critical to recruit a sufficiently large sample size to validate LPs when using the Rasch-CDM approach.

Acknowledgment This study is supported by the National Science Foundation (Grant Number 2150601). Any opinions, findings, and conclusions or recommendations expressed in this material are those of the authors and do not necessarily reflect the views of the National Science Foundation.

Appendix

This appendix shows how to conduct the Rasch-CDM analyses in refining the learning progression of buoyancy. R codes are shown below.

Step 1: Import Buoyancy Data and Q-Matrix

```
library(readxl)
mydata <- read_excel("/Users/Desktop/Buoyancy Data.xls")
head(mydata)
Q <- read_excel("/Users/Desktop/Q.xls")
head(Q)
```

Step 2: Perform Rasch-CDM Analysis

```
library(GDINA)
est <- GDINA(dat = mydata, Q = Q, model = "GDINA", att.dist = "higher.order",
        higher.order = list(model = "Rasch"))
```

Step 3: Model-Data Fit

Standardized Root Mean Square Residual (SRMSR)

```
modlfit <- modelfit(est)
modlfit
```

Q-Matrix Validation

```
sug.Q <- cdmTools::valQ(est, verbose = TRUE, index = "PVAF")
sug.Q
```

Step 4: Attribute Difficulty

```
attribute.parameter <- coef(est,"lambda")
attribute.parameter <- as.data.frame(attribute.parameter)
attribute.parameter$difficulty <- -attribute.parameter$intercept
attribute.parameter
```

Step 5: Ability and Attribute Mastery Proportion

```
### EAP estimates of attribute patterns
person.amp <- personparm(est, what = "EAP")
### HO estimates of ability
person.ability <- personparm(est, what = "HO")
### Combine ability and attribute mastery pattern
person.profile <- cbind(person.amp,person.ability)
head(person.profile)
```

Step 6: Learning Progression Levels

```
library(dplyr)
min.eap <- min(person.profile$EAP)
max.eap <- max(person.profile$EAP)
unit.eap <- (max.eap-min.eap)/(ncol(Q)+1)
cut_off1 <- min.eap + unit.eap; cut_off2 <- min.eap + unit.eap*2
cut_off3 <- min.eap + unit.eap*3; cut_off4 <- min.eap + unit.eap*4
cut_off5 <- min.eap + unit.eap*5; cut_off6 <- min.eap + unit.eap*6
cut_off7 <- min.eap + unit.eap*7
person.profile$group <- ifelse(person.profile$EAP <= cut_off1, 1, ifelse(person.profile$EAP
<= cut_off2 & person.profile$EAP > cut_off1, 2, ifelse(person.profile$EAP <= cut_off3 & p
erson.profile$EAP > cut_off2, 3, ifelse(person.profile$EAP <= cut_off4 & person.profile$EA
P > cut_off3, 4, ifelse(person.profile$EAP <= cut_off5 & person.profile$EAP > cut_off4, 5, i
felse(person.profile$EAP <= cut_off6 & person.profile$EAP > cut_off5, 6, ifelse(person.prof
ile$EAP <= cut_off7 & person.profile$EAP > cut_off6, 7, 8)))))))

### Sample Size in Each Group
table(person.profile$group)

### Mean Attribute Mastery Proportion
attribute.proportion <- person.profile %>%
  group_by(group) %>%
  summarise_at(vars(A1,A2,A3,A4,A5,A6,A7),list(mean),na.rm=TRUE)
attribute.proportion

### Mean Ability
ability.group <- person.profile %>%
  group_by(group) %>%
  summarise_at(vars(EAP),list(mean))
ability.group

person.profile$group <- ifelse(person.profile$EAP <= cut_off1, 1, ifelse(person.profile$EAP
<= cut_off2 & person.profile$EAP > cut_off1, 1, ifelse(person.profile$EAP <= cut_off3 & p
erson.profile$EAP > cut_off2, 2, ifelse(person.profile$EAP <= cut_off4 & person.profile$EA
P > cut_off3, 2, ifelse(person.profile$EAP <= cut_off5 & person.profile$EAP > cut_off4, 3, i
felse(person.profile$EAP <= cut_off6 & person.profile$EAP > cut_off5, 4, ifelse(person.prof
ile$EAP <= cut_off7 & person.profile$EAP > cut_off6, 5, 5)))))))

### Sample Size in Each Group
table(person.profile$group)

### Mean Attribute Mastery Proportion
attribute.proportion <- person.profile %>%
  group_by(group) %>%
  summarise_at(vars(A1,A2,A3,A4,A5,A6,A7),list(mean),na.rm=TRUE)
attribute.proportion

### Mean Ability
ability.group <- person.profile %>%
  group_by(group) %>%
  summarise_at(vars(EAP),list(mean))
ability.group
```

Step 7: Plot: Ability, attribute difficulty, and attribute mastery proportion

```
library(grid)
library(gtable)
library(ggplot2)
library(patchwork)
library(tidyr)
theme_set(theme_bw())

theta <- data.frame(d1 = person.profile$EAP)
attribute <- c("A1", "A2", "A3", "A4", "A5", "A6", "A7")
d <- data.frame(attribute = factor(1:7), difficulty = attribute.parameter$difficulty)
d <- data.frame(attribute, difficulty = attribute.parameter$difficulty)
lp.wide <- merge(data.frame(ability.group), data.frame(attribute.proportion), by="group")
lp.long <- gather(lp.wide, attribute, proportion, A1:A7, factor_key=TRUE)

dp <- ggplot(d, aes(x = attribute, y = difficulty)) +
  geom_point(size=4) +
  ylim(c(-2,2)) +
  theme(legend.position = "none", panel.grid = element_blank(), axis.title = element_blank())

tp <- ggplot(theta,aes(x=d1)) +
  geom_histogram(aes(y=..density..), alpha= 0.5, position= "identity", bins = 10) +
  geom_density(alpha = .2, adjust = 1.5) + coord_flip() + xlim(c(-2,2)) +
  theme(legend.position = "none", panel.grid = element_blank(), axis.title = element_blank())
+
  scale_y_reverse()

mid <- 0.5; sizeRange <- c(2,18)
ap <- ggplot(lp.long, aes(x = attribute, y = EAP, size = proportion)) +
  geom_point(alpha = 0.8, aes(colour = factor(lp.long$attribute))) +
  scale_size(range = sizeRange, name="index") + ylim(c(-2,2)) +
  scale_alpha("proportion") +
  theme(legend.position = "none",
        panel.grid = element_blank(),
        axis.title = element_blank())

tp + wrap_ggplot_grob(ggplotGrob(dp)) + wrap_ggplot_grob(ggplotGrob(ap))
```

References

Alonzo, A. C., & Steedle, J. T. (2009). Developing and assessing a force and motion learning progression. *Science Education, 93*(3), 389–421.

Boone, W. J. (2016). Rasch analysis for instrument development: Why, when, and how? *CBE—Life Sciences Education, 15*(4), 1–7. https://doi.org/10.1187/cbe.16-04-0148

Chandler, P., & Sweller, J. (1991). Cognitive load theory and the format of instruction. *Cognition and Instruction, 8*(4), 293–332.

Chen, F., Zhang, S., Guo, Y., & Xin, T. (2017). Applying the rule space model to develop a learning progression for thermochemistry. *Research in Science Education, 47*, 1357–1378. https://doi.org/10.1007/s11165-016-9553-7

Clement, J. (2008). The role of explanatory models in teaching for conceptual change. In S. Vosniadou (Ed.), *International handbook of research on conceptual change* (pp. 417–452). Routledge.

Cronbach, L. J., & Meehl, P. E. (1955). Construct validity in psychological tests. *Psychological Bulletin, 52*(4), 281–302. https://doi.org/10.1037/h0040957

de la Torre, J. (2011). The generalized DINA model framework. *Psychometrika, 76*(2), 179–199. https://doi.org/10.1007/s11336-011-9207-7

de la Torre, J., & Chiu, C. Y. (2016). A general method of empirical Q-matrix validation. *Psychometrika, 81*(2), 253–273. https://doi.org/10.1007/s11336-015-9467-8

de la Torre, J., & Douglas, J. A. (2004). Higher-order latent trait models for cognitive diagnosis. *Psychometrika, 69*(3), 333–353. https://doi.org/10.1007/BF02295640

Fulmer, G. W., Liang, L. L., & Liu, X. (2014). Applying a force and motion learning progression over an extended time span using the force concept inventory. *International Journal of Science Education, 36*(17), 2918–2936. https://doi.org/10.1080/09500693.2014.939120

Gao, Y., Zhai, X., Andersson, B., Zeng, P., & Xin, T. (2020). Developing a learning progression of buoyancy to model conceptual change: A latent class and rule space model analysis. *Research in Science Education, 50*(4), 1369–1388. https://doi.org/10.1007/s11165-018-9736-5

Gao, Y., Zhai, X., Cui, Y., Xin, T., & Bulut, O. (2021). Re-validating a learning progression of buoyancy for middle school students: A longitudinal study. *Research in Science Education*. https://doi.org/10.1007/s11165-021-10021-x

Gierl, M. J. (2007). Making diagnostic inferences about cognitive attributes using the rule-space model and attribute hierarchy method. *Journal of Educational Measurement, 44*(4), 325–340.

Gotwals, A. W., & Songer, N. B. (2010). Reasoning up and down a food chain: Using an assessment framework to investigate student' middle knowledge. *Science Education, 94*(2), 259–281.

Gotwals, A. W., & Songer, N. B. (2013). Validity evidence for learning progression-based assessment items that fuse core disciplinary ideas and science practices. *Journal of Research in Science Teaching, 50*(5), 597–626. https://doi.org/10.1002/tea.21083

Hadenfeldt, J. C., Neumann, K., Bernholt, S., Liu, X., & Parchmann, I. (2016). Students' progression in understanding the matter concept. *Journal of Research in Science Teaching, 53*(5), 683–708. https://doi.org/10.1002/tea.21312

Herrmann-Abell, C. F., & DeBoer, G. E. (2018). Investigating a learning progression for energy ideas from upper elementary through high school. *Journal of Research in Science Teaching, 55*(1), 68–93. https://doi.org/10.1002/tea.21411

Inhelder, B., & Piaget, J. (1958). *The growth of logical thinking from childhood to adolescence*. Basic Books.

Jin, H., Mikeska, J. N., Hokayem, H., & Mavronikolas, E. (2019a). Toward coherence in curriculum, instruction, and assessment: A review of learning progression literature. *Science Education, 103*(5), 1206–1234.

Jin, H., van Rijn, P., Moore, J. C., Bauer, M. I., Pressler, Y., & Yestness, N. (2019b). A validation framework for science learning progression research. *International Journal of Science Education, 41*(10), 1324–1346. https://doi.org/10.1080/09500693.2019.1606471

Johnson, P., & Tymms, P. (2011). The emergence of a learning progression in middle school chemistry. *Journal of Research in Science Teaching, 48*(8), 849–877.

Kaldaras, L., Akaeze, H., & Krajcik, J. (2021). Developing and validating next generation science standards-aligned learning progression to track three-dimensional learning of electrical interactions in high school physical science. *Journal of Research in Science Teaching, 58*(4), 589–618. https://doi.org/10.1002/tea.21672

Kauertz, A., & Fischer, H. E. (2006). Assessing students' level of knowledge and analysing the reasons for learning difficulties in physics by Rasch analysis. In X. Liu & W. J. Boone (Eds.), *Applications of Rasch measurement in science education* (pp. 212–246). JAM Press.

Lee, H., & Liu, O. L. (2009). Assessing learning progression of energy concepts across middle school grades: The knowledge integration perspective. *Science Education, 94*(4), 665–688. https://doi.org/10.1002/sce.20382

Lehrer, R., & Schauble, L. (2012). Seeding evolutionary thinking by engaging children in modeling its foundations. *Science Education, 96*(4), 701–724. https://doi.org/10.1002/sce.20475

Liu, X. (2010). Using and developing instruments for measuring learning progression. In *Using and developing measurement instruments in science education: A Rasch modeling approach* (1st ed.). Information Age Publishing, Inc.

Liu, X. (2020). Using and developing instruments for measuring learning progression. In *Using and developing measurement instruments in science education: A Rasch modeling approach* (2nd ed.). Information Age Publishing, Inc.

Ma, W., & de la Torre, J. (2020). GDINA: An R package for cognitive diagnosis modeling. *Journal of Statistical Software, 93*(14), 1–26. https://doi.org/10.18637/jss.v093.i14

Maydeu-Olivares, A. (2013). Goodness-of-fit assessment of item response theory models. *Measurement: Interdisciplinary Research and Perspectives, 11*, 71–137.

Morell, L., Collier, T., Black, P., & Wilson, M. (2017). A construct-modeling approach to develop a learning progression of how students understand the structure of matter. *Journal of Research in Science Teaching, 54*(8), 1024–1048.

Nájera, P., Abad, F. J., & Sorrel, M. A. (2021). Determining the number of attributes in cognitive diagnosis modeling. *Frontiers in Psychology, 12*, 614470.

National Research Council. (2007). *Taking science to school: Learning and teaching science in grade K-8*. The National Academic Press.

National Research Council. (2012). *A framework for K-12 science education: Practices, crosscutting concepts, and core ideas*. National Academies Press.

Neumann, K., Viering, T., Boone, W. J., & Fischer, H. E. (2013). Towards a learning progression of energy. *Journal of Research in Science Teaching, 50*(2), 162–188.

Osborne, J. F., Henderson, J. B., MacPherson, A., Szu, E., Wild, A., & Yao, S. (2018). The development and validation of a learning progression for argumentation in science. *Journal of Research in Science Teaching, 53*(6), 821–846. https://doi.org/10.1002/tea.21316

Planinic, M., Boone, W. J., Susac, A., & Ivanjek, L. (2019). Rasch analysis in physics education research: Why measurement matters. *Physical Review Physics Education Research, 15*(2), 020111. https://doi.org/10.1103/PhysRevPhysEducRes.15.020111

R Core Team. (2022). *R: A language and environment for statistical computing*. R Foundation for Statistical Computing. https://www.R-project.org/

Şahin, Ç., & Çepni, S. (2012). Effect of different teaching methods and techniques embedded in the 5E instructional model on students' learning about buoyancy force. *Eurasian Journal of Physics and Chemistry Education, 4*(2), 97–127.

Tatsuoka, K. K. (1983). Rule space: An approach for dealing with misconceptions based on item response theory. *Journal of Educational Measurement, 20*(4), 345–354. https://doi-org.login.ezproxy.library.ualberta.ca/10.1111/j.1745-3984.1983.tb00212.x.

Tatsuoka, K. K. (1995). Architecture of knowledge structures and cognitive diagnosis: A statistical pattern recognition and classification approach. In P. D. Nichols, S. F. Chipman, & R. L. Brennan (Eds.), *Cognitively diagnostic assessment* (pp. 327–359). Erlbaum.

Tatsuoka, K. K. (1996). Use of generalized person-fit indexes, Zetas for statistical pattern classification. *Applied Measurement in Education, 9*(1), 65–75. https://doi-org.login.ezproxy.library.ualberta.ca/10.1207/s15324818ame0901_6

Testa, I., Capasso, G., Colantonio, A., Galano, S., Marzoli, I., di Uccio, U. S., et al. (2019). Development and validation of a university students' progression in learning quantum mechanics through exploratory factor analysis and Rasch analysis. *International Journal of Science Education, 41*(3), 388–417. https://doi.org/10.1080/09500693.2018.1556414

Todd, A., Romine, W. L., & Whitt, K. C. (2016). Development and validation of the learning progression-based assessment of modern genetics in a high school context. *Science Education, 101*(1), 32–65. https://doi.org/10.1002/sce.21252

Wei, X. (2014). *Developing a learning progression framework for energy in physics across the 8th–12th grades* (Unpublished doctoral dissertation). Beijing Normal University.

Wilson, M. (2009). Measuring progressions: Assessment structures underlying a learning progression. *Journal of Research in Science Teaching, 46*(6), 716–730.

Yao, J., Guo, Y., & Neumann, K. (2017). Refining a learning progression of energy. *International Journal of Science Education, 39*(17), 2361–2381. https://doi.org/10.1080/09500693.2017.1381356

Yu, X., Ding, S., Qin, C., & Lu, Y. (2011). Application of Bayesian networks to identify hierarchical relation among attributes in cognitive diagnosis (in Chinese). *Acta Psychologica Sinica, 43*(3), 338–346.

Yuan, L., Liu, Y., Chen, P., & Xin, T. (2022). Development of a new learning progression verification method based on the hierarchical diagnostic classification model: Taking grade 5 students' fractional operations as an example. *Educational Measurement: Issues and Practice*. https://doi-org.login.ezproxy.library.ualberta.ca/10.1111/emip.12488

Zhai, X. (2022). Assessing high-school students' modeling performance on Newtonian mechanics. *Journal of Research in Science Teaching*. https://doi.org/10.1002/tea.21758

Zhai, X., & Li, M. (2021). Validating a partial-credit scoring approach for multiple-choice science items: An application of fundamental ideas in science. *International Journal of Science Education, 43*(10), 1640–1666.

Zhai, X., Li, M., & Guo, Y. (2018). Teachers' use of learning progression-based formative assessment to inform teachers' instructional adjustment: A case study of two physics teachers' instruction. *International Journal of Science Education, 40*(15), 1832–1856.

Zhai, X., Haudek, K., Wilson, C., & Stuhlsatz, M. (2021). A framework of construct-irrelevant variance for contextualized constructed response assessment. *Frontiers in Education*. https://doi.org/10.3389/feduc.2021.751283

Zhai, X., Haudek, K., & Ma, W. (2022). Assessing argumentation using machine learning and cognitive diagnostic modeling. *Research in Science Education*. https://doi.org/10.1007/s11165-022-10062-w

Yizhu Gao, Assistant Professor of science education at the University of Georgia. Her research has focused primarily on developing innovative assessments and technology-enhanced learning environments (i.e., adaptive learning systems) for science education.

Xiaoming Zhai, Assistant Professor of Science Education and Affiliated Assistant Professor of Artificial Intelligence, Director of the AI4STEM Education Center at the University of Georgia. He is interested in developing innovative assessments and technologies to support science teaching and learning.

Ahra Bae is a graduate student at the University of Georgia in Science Education and a middle school teacher. She is interested in using assessments to support teachers and students in science education.

Wenchao Ma, Assistant Professor of educational research at the University of Alabama. He is interested in item response theory, cognitive diagnostic models, and their applications in education, psychology, and other social sciences.

Chapter 6
Utilizing Latent Class Analysis (LCA) to Analyze Response Patterns in Categorical Data

Martina Brandenburger and Martin Schwichow

Abstract Models based on Rasch's (1960) measurement theory build quantitative, continuous latent variables based on persons' categorical responses. In contrast, within latent class analysis (LCA), persons are represented by qualitative, categorical latent variables. LCA can be used to identify patterns within categorical responses. In this chapter, we present the general idea of the LCA, including conventions for interpreting results, compare the LCA with Rasch analysis and combine the LCA's results with a unidimensional Rasch model. Our presentation is based on a concrete empirical example that investigates experimental design errors and includes the used data set and R scripts as supplemental materials.

6.1 Introduction

One aim of educational assessment is to find out the strengths and difficulties of students regarding a particular concept or skill (Pelligrino et al., 2001). To achieve this, teachers and researcher construct items, score them as correct or incorrect and analyze which items a student solved and which not. However, if the number of students and items is increasing, an itemwise interpretation for every student is complex and time consuming. Rasch analysis offers a more efficient method to link students' scores (number of solved items) to the difficulty of items. A Rasch analysis represents students' abilities and item difficulties on the same continuous scale, so that both can be compared directly. If the student's ability fits the item difficulty, a student typically has

Supplementary Information The online version of this chapter (https://doi.org/10.1007/978-3-031-28776-3_6) contains supplementary material, which is available to authorized users.

M. Brandenburger (✉) · M. Schwichow
Department of Physics, University of Education Freiburg, Freiburg, Germany
e-mail: martina.brandenburger@ph-freiburg.de; martin.schwichow@ph-freiburg.de

X. Liu, W. J. Boone (eds.), *Advances in Applications of Rasch Measurement in Science Education*, Contemporary Trends and Issues in Science Education 57,
https://doi.org/10.1007/978-3-031-28776-3_6

a 50% chance to solve the item. For items with a larger difficulty, the probability for students to solve decreases, while for items with a lower difficulty it increases.

However, teachers and researchers are often interested in students' preconceptions, i.e., their ideas about natural phenomena which are based on their experiences, use of language, or inappropriate instruction. Preconceptions are often quite coherent ideas that have explanatory power with respect to multiple phenomena but are to some extent incompatible with scientific concepts, and thus cause erroneous responses to items (Vosniadou, 2019). If we only look at dichotomously coded data, we will not get insights into the students' preconceptions that influence a particular response. Instead, we have to code which preconceptions might have caused erroneous student responses. Since multiple ideas about a task can exist, such an analysis will not result in dichotomous data but in multiple categories representing the diverse ideas of students regarding the task. However, if we use multiple tasks, the data can be complex and hard to interpret as many combinations of tasks and categories can exist (e.g., 10 tasks with 3 categories each can result in up to $3^{10} = 59049$ combinations). Latent class analysis (LCA) is a method to analyze such complex data by grouping students with similar patterns of responses (combination of preconception and item) together into one class. A basic LCA requires only categorical data and no information about correct / incorrect responses or a hierarchical order of categories. By describing and interpreting the students' patterns of responses grouped together into the same class, or group, researchers can investigate whether and which coherent conceptions the students share and quantify the influence of each concept on a student's response. Thus, LCA offers qualitatively different information about students' responses than methods based on dichotomous coding, while additionally reducing the complexity of categorical data.

In this section, we describe the general idea of the LCA within the Rasch measurement framework based on a concrete example of a study by Schwichow et al. (2022), which utilizes an LCA to analyze patterns of students' responses on items asking to design and interpret controlled experiments. We describe the assumptions and mathematical model underlying LCA. We explain how missing data can be handled in LCA, as well as criteria and conventions for reporting and interpreting the results of LCA. Finally, we combine the results of the LCA with findings of a unidimensional Rasch model based on the same data. In an online appendix, we present a step-by-step guide to run the LCA with the open source R package poLCA (Linzer & Lewis, 2011; Linzer & Jeffrey, 2022). The R script to run the analysis as well as the data set of the example are available as an appendix, too.

6.2 Commonalities and Differences Between LCA and Rasch Analysis

Before discussing the details of the LCA, we will summarize general commonalities and differences between the LCA (details in Sect. 6.4) and the Rasch analysis (details e.g., Rost, 2004; Sick, 2010; Boone et al., 2014; Boone & Staver, 2020) (overview: Fig. 6.1).

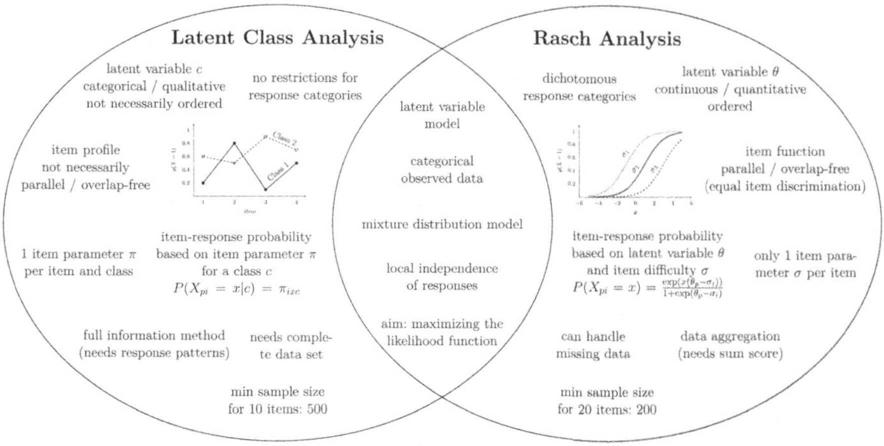

Fig. 6.1 Venn diagram of the commonalities and differences between LCA and Rasch analysis

Overall, LCA in its structure is completely parallel to the Rasch analysis – the only difference is the scaling level of the latent variable (Rost, 2004). Rasch models result in continuous ordered ability θ, while LCA models lead to unordered categorical class membership c, unless further specified. To describe the item-response probability based on a person's class membership class c, we need to take into account that c is a discrete number. We do not calculate (parallel) item functions, like in Rasch analysis, but item profiles.

LCA models are, like Rasch models, mixture distribution models, meaning the latent variable (the continuous θ or the categorical c) is derived in a way that the local independence of observed variables is given – dependencies between items will disappear if only persons from one class c or with one ability θ are considered (Rost, 2004). This assumption is a strong simplification of the complex reasons behind concrete response behavior and might only seldom be true. However, as "all models are wrong but some are useful" (Box, 1979, p. 202) we should still use LCA to reduce the complexity of categorical data but be aware that latent classes are one but not the only one way to analyze these data.

Beside these commonalities of LCA and Rasch analysis, there are some crucial differences. In Rasch models, there is only one item parameter, the difficulty σ, per item, while LCA models have one item parameter, the item-response probability π, for each item for every class c (assuming dichotomous data).

Additionally, when using a Rasch model, only the sum score is important for the person's ability θ, it doesn't matter which items were solved correctly (Rost, 2004). In contrast, LCA does not use such data aggregation. The class c depends on the frequency of the response patterns of all students and therefore LCA is a full information method (Rost, 2004).

Consequently, LCA needs a larger sample size than Rasch analysis – four to five items need 60–70 persons, ten items already need around 500 persons (Rost, 2004), whereas a Rasch model for 20 items needs around 200 persons (Yen & Fitzpatrick, 2006). Furthermore, LCA requires complete data samples without missing values while Rasch analysis can handle missings. As an advantage of LCA however, we get a more detailed picture of the response pattern within a population but as downside, we need complete data sets that is the detailed response category of each person for every item.

At last, it is to note that measuring a quantitative latent trait (Rasch analysis) is special case of measuring a qualitative latent trait (LCA) when persons with response patterns with an identical sum score are treated equally (Rost, 2004).

6.3 Empirical Example

We describe the assumptions of LCA and how it works based on a data set from a study that investigates response patterns on items asking students to plan, identify, and interpret controlled experiments. The skills associated with the design and interpretation of controlled experiments are summarized under the term control-of-variables-strategy (CVS). Research on elementary school students' experimentation skills shows that they have basic conceptions of what a controlled experiment is, as they can correctly identify controlled experiments to test given hypotheses and interpret the results of controlled experiments. Students perform poorly, however, when they have to plan experiments and interpret confounded experiments. Furthermore, Siler and Klahr (2012) have identified eight misleading preconceptions regarding designing experiments that cause invalid experimental designs, called design errors (Table 6.1). Examples are designing confounded experiments (beside the investigated variable, one or more additional variables differ between conditions) or non-contrastive experiments (contrasting identical conditions). Based on students' oral and written responses, they furthermore conducted interviews to identify underlying preconceptions that produced the design errors (Table 6.1).

The overview in Table 6.1 shows that the preconceptions numbered two to five cause the same design error resulting in a contrast of confounded or multiple confounded conditions (more than one variable differs between conditions). With respect to the underlying conceptual idea, these preconceptions are quite different from each other, but nonetheless all lead to the design of confounded experiments as a visible result. Reasons for designing confounded experiments are that students (1) believe that the additional varying variables have no causal effects, (2) just ignore them, (3) want to test multiple effects in the same experiment or (4) want to produce extreme differences on the dependent variable. The remaining four preconceptions cause unique design errors. Students who understand the logic of controlling variables but who have problems identifying the correct independent variable will compare controlled conditions but focus on the wrong independent variable (according to the given hypothesis). If students understand the importance of controlling variables, they might over-interpret this idea and thus build "totally fair conditions" that differ in no variable and thus are non-informative. Students who do not understand why conditions are compared will eventually design an "experiment" consisting of only one condition and neglect comparing at least two different conditions. Tschirgi (1980) describes a further design error that she calls "hold-one-thing-at-a-time" (hotat). In this case, students

Table 6.1 Overview of CVS preconceptions, visible errors and misconceived aspects of the CVS according to Siler and Klahr (2012) complemented by the hotat strategy described by Tschirgi (1980)

		Mistakes und rule violations when planning controlled experiments						
Preconception	Design error	Focus on variable level	Test one variable	Identify correct target variable[a] (R1)	Compare across conditions (R4)	Vary what testing (R2)	Control variables not Testing (R3)	Understand setup/outcome causal relationship
(1) *CVS for wrong variable*	Design controlled experiments for wrong independent variable (cwv)			X				
(2) *CVS-vary-"noncausal"*	Designing (multiple) confounded experiments (ce1, ce2)						X (noncausal belief)	
(3) *Test multiple variables*	Designing (multiple) confounded experiments (ce1, ce2)		X	X			X (no CVS logic)	X
(4) *Ignore confound(s)*	Designing (multiple) confounded experiments (ce1, ce2)						X (no CVS logic)	X
(5) *Whole condition comparison*	Designing (multiple) confounded experiments (ce1, ce2)	X	X	X			X (no CVS logic)	X
(6) *Noncontrastive target variable*	Non-contrastive experiments (nce)					X	X	X
(7) *Single-condition experiment*	Designing only one experimental condition (ooe)				X	X	X	X
(8) *Hold one constant and vary all additional variables*	Hold-one-thing-at-a-time (hotat)					X	X (no CVS logic)	X

Table from Schwichow et al. (2022)
[a]Failing to identify the correct target variable may occur in each of the approaches, but is only a necessary criterion in three

design experiments in which the independent variable is not varying between conditions, whereas all other variables differ. This design error is thus a special case of a confounded experiment. It is based on the idea that one has to find the one and only variable that has to be constant in order to get identical results under varying conditions.

6.3.1 Sample and Assessment Instrument

Students from 17 elementary schools (grades 2–4; ages 7–10) in southern Germany (in the state of Baden-Württemberg) participated in this study. The final sample of our analyses is $N = 496$ (48% female, 51% male). To assess students' CVS skills, we adopted and expanded the CVS Inventory (CSVI) by Schwichow et al. (2016) to an elementary school level. For each context (dropping of parachutes, floating of boats, scaling of lemonade, crawling of snails), we created four items: two items for the identification subskill (students identify one controlled experiment out of a selection of four experiments) and one each for the interpretation (students interpret the outcome of controlled experiments) and understanding subskill (students reject to draw causal inferences from confounded experiments). We did not create items for the CVS subskill planning because this subskill is very challenging for primary school students (Bullock & Ziegler, 1999; Peteranderl & Edelsbrunner, 2020) and we wanted to keep a reasonable test length. Every item presents a short story and a hypothesis about a causal relationship. The adapted version of the CVSI includes 16 multiple choice/multiple select items out of four contexts (questionnaire available at Schwichow et al., 2022).

6.3.2 Data Scaling and Descriptive Statistics

We assigned every response option to the represented design error. For missing responses, we used hot deck imputation (Sect. 6.4.2). We combined items of the same item type based on different contexts so that every student can be characterized by his/her two responses within one of the four item types. In the following sections, we will cut this original dataset in half and only use one item for each item type for a less complex and therefore easier to understand presentation of the general idea of the LCA in Sect. 6.4. Based on this categorization, we calculated the observed frequencies of design errors for every item type (Table 6.2).

6.3.3 Research Questions

Table 6.3 shows the categorical responses of four students for the four items X_i. The value 1 represents a correct understanding of CVS; values 2–7 represent a wrong

Table 6.2 Observed frequency of students' design errors

	Students'design errors						
	1	2	3	4	5	6	7
Item	cvs	cwv	ce1	ce2	nce	hotat	ooe
Identification 1 X_1	384	67	–	–	19	–	26
Identification 2 X_2	368	–	48	42	38	–	–
Interpretation X_3	173	39	168	65	27	24	–
Understanding X_4	179	57	104	27	83	46	–

Note. $N = 496$. cvs: correct application of the control-of-variables strategy; cwv: CVS for wrong variable; nce: non-contrastive target variable; ooe: single-condition experiment; ce1 & ce2: confounded experiments; hotat: hold-one-thing-at-a-time

Table 6.3 Response pattern of design errors for four students. p represents the person ID of the data set in the supplemental material

	Items X_i			
Student (person ID p)	X_1	X_2	X_3	X_4
Theo (22)	1	3	4	3
Julia (8)	5	1	4	3
Aylin (25)	1	4	5	5
Tim (20)	1	1	3	1

response with its associated design error (as described in Table 6.2). From a quantitative point of view, we can count the correct responses and design errors (as shown for all students in Table 6.2) but we cannot determine if there are qualitative differences between the response patterns. We presume that different pattern of design errors are based on different underlying CVS misconception. However, we do not have strong theoretical assumptions for the number of sub-groups within the students and how the misconceptions are manifested in the responses.

Thus, our first research question is: Based on an LCA: Which patterns of design errors regarding the CVS subskills can be identified within elementary school students? (Examine a categorical/discrete latent variable.)

Moreover, we expect that items representing different CVS subskills differ in their item difficulty because each subskills require specific cognitive tasks. Items of the identification subskill require only the selection of a controlled experiment and thus might be easier than items that require justified conclusions (subskills: interpretation and understanding).

This assumption is further investigated based on a Rasch analysis in research question 2: How difficult are the items of different CVS subskills for elementary school students? (Examine a continuous latent variable.)

At last, we expect to get different perspectives on students' CVS skills by LCA and Rasch analysis. Thus, our third research questions is: What additional insight can be gained by combining the results of the LCA and the Rasch analysis?

We assume that the reader of this chapter is familiar with the basic principles of Rasch analysis (for an introduction: e.g., Boone et al., 2014; Boone & Staver, 2020). However, we explain LCA as a different type of latent variable model further and by describing the theoretical background of LCA based on the empirical example.

6.4 Latent Class Analysis

6.4.1 Overview & Introduction

In the 1950s, Paul Lazarsfeld developed a conceptual foundation of latent structure models (Andersen, 1982). The concept and methods were refined over time and Lazarsfeld and Henry (1968) published the first detailed work (conceptual and mathematical) covering LCA (Collins & Lanza, 2010). By now, LCA is widely used in social sciences (e.g., Collins & Lanza, 2010; Davier & Lee, 2019; Fulmer et al., 2014; Vincent-Ruz & Schunn, 2021) along with other latent variable models.

Latent variable models aim to describe an underlying latent construct (i.e., a construct that is not directly measurable) based on observed data. The observed data as well as the latent variable can be continuous or categorical/discrete. As a result, there are four possible combinations that form the different latent variable models (Table 6.4). Models based on Rasch's (1960) measurement theory utilize discrete data (often dichotomous) to construct continuous latent variables. In contrast, the observed data in the LCA as well as the latent variable are discrete categories, while neither have to be necessarily ordinal (Rost, 1988). Thus, the aim of the LCA is not to arrange persons on a latent continuum but to group persons together into "classes" that show similar response patterns (Rost, 2004).

In our empirical example, the observed data are the categorical design errors, as shown in Table 6.2. The latent variable is the underlying CVS misconception, also a categorical variable. Our aim is to identify classes, or groups, of students that show qualitatively similar response patterns in their design errors.

Assumptions of the LCA and the General Model

LCA uses probabilistic procedures to identify classes of qualitatively different response patterns. Thereby, LCA makes no assumptions about the response patterns or the number of classes. Furthermore, it does not require that all items have the same number of categories (Collins & Lanza, 2010; Davier & Lee, 2019; Rost, 2004). In our empirical example, items 1 and 2 have four categories, items 3 and 4 have six categories (Table 6.2). However, as with all probabilistic methods, LCA has assumptions that we will explain next, together with the theoretical background and the general (mathematical) model of the LCA based on our empirical example. We will use the four students from Table 6.3 and an already estimated model with three classes. In section "Choosing a model", we will explain why we have chosen this model.

Table 6.4 Overview of latent variable models

Observed data	Latent variable	
	Continuous	Categorical/discrete
Continuous	Factor analysis	Latent profile analysis
Categorical/discrete	Rasch analysis	Latent class analysis

According to for example Collins and Lanza (2010, p. 7)

Annotation to Mathematical Notations

- Class: $c \in \{1,\ldots,C\}$
- Person: $p \in \{1,\ldots,N\}$
- Item: $i \in \{1,\ldots,j\}$
- Response: $x \in \{1,\ldots,m\}$
- Person's response to an item: $X_{pi} = x$; for example $X_{14} = 2$: the response of person 1 to item 4 is 2.
- Person's response pattern: $\underline{x}_p$; for example the response pattern for Theo is $\underline{x}_{Theo} = 1, 3, 4, 3$.
- Person's conditional probability: *P*(*response*| *condition*); for example $P(X_{p3} = 4 | c = 1)$: the probability of a student within class 1 responding to item 3 with 4.
- Unconditional probability: *P*(*response*); for example $P(X_{p3} = 4)$: the probability over all students responding to item 3 with 4.

Assumption 1: Constant Response Probability for all Persons Within a Class

Within a class, every person has the same probability of responding to a specific item X_i (Collins & Lanza, 2010; Rost, 2004). In the 3-class model of the LCA, we predict that Theo and Julia are both assigned to class 2. Therefore, both Theo and Julia have the same probability of 74% (Table 6.5) to respond to the item 1 with correct understanding of CVS: $P(X_{Theo,\,1} = 1 | c = 2) = P(X_{Julia,\,1} = 1 | c = 2) = .74$. Students in other classes, like Tim (class 1) and Aylin (class 3), may have different probabilities of responding to item 1 with correct understanding of CVS. In a model that fits well, the estimated expected frequencies of responses should closely match the observed frequencies (Collins & Lanza, 2010). Therefore, we can verify the probability estimated by the model by counting the observed responses in the class. In class 2, 79 out of 107 students responded with correct understanding of CVS which is, based on the law of large numbers, equivalent to a probability of 79/107 = 74%, matching the item-response probability estimated by the latent class model.

In general, we can equate the conditional item-response probability to response x for all students:

$$P(X_{1i} = x|c) = P(X_{2i} = x|c) = \ldots = P(X_{pi} = x|c) = \pi_{ixc} \qquad (6.1)$$

Table 6.5 Estimated model parameters π_{ixc} for item X_1 in the 3-class model

	Item-response probability for response x at item X_1 (π_{1xc})			
Class c	1	2	5	7
1	.89	.09	.00	.02
2	.74	.18	.06	.03
3	.54	.19	.11	.16

We use the probability parameter π_{ixc} to describe the item-response probability for the response x to item i in class c (Table 6.5 for item 1, Appendix Table 6.1 for all other parameters). Class wise, π_{ixc} of each item i across all responses x needs to sum up to 1 (e.g., each row in Table 6.5; within rounding errors). To have a good latent class separation, different classes should have different item-response probabilities for (at least some) items, so the classes are conceptually distinct and can be labeled according to these differences (e.g., correct CVS understanding vs. change of too many variables) (Collins & Lanza, 2010).

In the previous example, 74% might be a high probability for responding with correct CVS. Nevertheless, a student assigned to class 2 can still respond differently, like Julia (Table 6.3). The observed responses of a student are determined by their latent trait (class membership) and a random error (Collins & Lanza, 2010), so two students with the same latent trait (Theo and Julia) can respond differently, represented by the item-response probability. The closer the item-response probabilities within a class are to 0 or 1 (0 to .2 and .8 to 1), the higher is the homogeneity of the latent class – the members of the latent class are more likely to provide the same observed response patterns (Collins & Lanza, 2010). Furthermore, persons have a particular probability to be members of any class (section "Assigning the class membership").

Assumption 2: Disjunctive and Exhaustive Latent Classes With Unknown Prevalence

Every person is a member of one class (exhaustive) and one class only (disjunctive) which represents the person's latent trait (Collins & Lanza, 2010; Rost, 2004). We use the probability parameter π_c to describe the probability of a random student of the data set to be member of a specific class c – the expected class prevalence (Table 6.6). Moreover, because the class is a latent variable, the class membership of individual persons estimated by the model is probabilistic. Every person has a particular probability of membership of all classes (section "Assigning the class membership").

Additionally, we know, based on the second assumption (disjunctive and exhaustive classes), that the expected class prevalences across all classes need to sum up to 1 (Eq. 6.2).

$$\sum_{c=1}^{C} \pi_c = 1 \tag{6.2}$$

The number of the classes C is not a parameter of the model. Though it is unknown, it will not be estimated within our model (Rost, 2004). To decide the number of classes, we either have a strong theoretical assumption (confirmatory approach) or we need to estimate several models ($C = 1$, $C = 2$, $C = 3$, ...) and compare the goodness of fit of the models (Sect. 6.4.4, explorative approach).

Assumption 1 gives us π_{ixc} – the conditional item-response probability of every student within a particular class c responding to an item i with response x. With π_c

Table 6.6 Expected class prevalence π_c for the 3-class model

Class c	π_c
1	.53
2	.24
3	.23

from assumption 2, we know the expected class prevalence. Multiplying the expected class prevalence with the conditional item-response probability will give us, summed up over all classes, the unconditional item-response probability $P(X_{pi} = x)$ of all students (Rost, 2004; Eq. 6.3).

$$P(X_{pi} = x) = \sum_{c=1}^{C} \pi_c \cdot \pi_{ixc} \tag{6.3}$$

We can follow the plausibility of Eq. 6.3 with an example. We use the conditional item-response probability π_{11c} for responding to item 1 with correct CVS (Table 6.5) and the expected class prevalence estimated for the 3-class model (Table 6.6). We calculate the unconditional item-response probability following Eq. 6.3: $P(X_{p1} = 1) = .53 \cdot .89 + .24 \cdot .74 + .23 \cdot .54 = .77$. In a well-fitting model, the expected frequencies should again closely match the observed frequencies. In our sample, 384 students out of 496 students showed correct CVS at item 1 (Table 6.2). Based on the law of large numbers, the observed probability over all students for item 1 to respond with correct CVS is 384/496 = .77 – the same probability as calculated based on Eq. 6.3.

Assumption 3: Homogeneity of Items

Within the latent class model, we assume that all items measure the same latent variable (Collins & Lanza, 2010; Rost, 2004), the categorical underlying CVS misconception. Therefore, we can calculate an unconditional pattern probability $P(\underline{x}_p)$ (Eq. 6.4) in the same fashion as Eq. 6.3 by multiplying the expected class prevalence with the conditional pattern probability $P\left(\underline{x}_p|c\right)$.

$$P\left(\underline{x}_p\right) = \sum_{c=1}^{C} \pi_c \cdot P\left(\underline{x}_p|c\right) \tag{6.4}$$

Assumption 4: Local Independence of Observed Responses

Only a person's membership of a specific latent class c explains the different observed responses to an item (Collins & Lanza, 2010; Rost, 2004). Therefore, the probability of a student within a latent class c responding in a particular pattern

$P(1, 3, 4, 3|c)$, like Theo, is the same as the product of the probability of every item: $P(1, 3, 4, 3|c) = P(X_1 = 1|c) \cdot P(X_2 = 3|c) \cdot P(X_3 = 4|c) \cdot P(X_4 = 3|c)$.

We can formulate the general conditional probability for a response pattern $\underline{x}_p$:

$$P\left(\underline{x}_p|c\right) = P(X_{p1} = x|c) \cdot \ldots \cdot P(X_{pj} = x|c) = \prod_{i=1}^{j} P(X_{pi} = x|c) = \prod_{i=1}^{j} \pi_{ixc} \quad (6.5)$$

General Model

If we combine the assumptions and Eqs. 6.4 and 6.5, we can set up the general model for the LCA (Collins & Lanza, 2010; Rost, 2004). Eq. 6.6 describes the unconditional pattern probability $P\left(\underline{x}_p\right)$ dependent on the expected class prevalence π_c and the conditional item-response probability π_{ixc}. $P\left(\underline{x}_p\right)$ describes the estimated probability of a specific observed response pattern $\underline{x}_p$ being present in a data set, under the assumption that the LCA model is valid (Rost, 2004). For example, Theo's observed pattern $\underline{x}_{\text{Theo}} = 1, 3, 4, 3$ has an estimated probability of $4.66 \cdot 10^{-3}$ based on the parameters of our 3-class model (Table 6.6 and Appendix Table 6.1). For a good fitting model, we want to maximize the estimated probability of all observed pattern frequencies (maximum likelihood approach, Sect. 6.4.3).

$$P\left(\underline{x}_p\right) = \sum_{c=1}^{C} \pi_c \cdot \prod_{i=1}^{j} \pi_{ixc} \quad (6.6)$$

Assigning the Class Membership

In the previous sections, we calculated the probability for a response pattern conditional on a distinct class membership: $P\left(\underline{x}_p|c\right)$ (Eq. 6.5). However, to estimate how precisely the observed variables measure the underlying latent variable, we have to turn things around (Collins & Lanza, 2010). When we calculate the probability of the class membership conditional on a distinct response pattern $P\left(c|\underline{x}_p\right)$, we are able to figure out the probability of each person's latent class membership (based on the response pattern) and therefore assign a class membership.

We calculate the conditional probability of the class membership c of a person with the response pattern $\underline{x}_p$ in Eq. 6.7 based on Bayes' theorem (Rost, 2004) using Eq. 6.5. This probability is also called classification probability or posterior probability (Collins & Lanza, 2010).

$$P(c|\underline{x}_p) = \frac{\pi_c \cdot P(\underline{x}_p|c)}{\sum_{c=1}^{C} \pi_c \cdot P(\underline{x}_p|c)} \tag{6.7}$$

To assign a person to a class, we need to estimate the person's classification probability for all classes and assign the person to the class with the highest classification probability. We can check the assignment by calculating the classification probabilities based on Eq. 6.7 for the 3-class model by using the probability of Theo's response pattern in different classes (Eq. 6.5 with parameters π_{ixc} from Appendix Table 6.1) and the expected class prevalence π_c (Table 6.6). The classification probability is clearly the highest for class 2 (71%) – the known class membership of Theo.

$$P(1|1,3,4,3) = \frac{\pi_1 \cdot P(1,3,4,3|1)}{\sum_{c=1}^{C} \pi_c \cdot P(1,3,4,3|c)} = \frac{.53 \cdot 0 \cdot 10^{-3}}{4.66 \cdot 10^{-3}} = .00$$

$$P(2|1,3,4,3) = \frac{\pi_2 \cdot P(1,3,4,3|2)}{\sum_{c=1}^{C} \pi_c \cdot P(1,3,4,3|c)} = \frac{.24 \cdot 13.86 \cdot 10^{-3}}{4.66 \cdot 10^{-3}} = .71$$

$$P(3|1,3,4,3) = \frac{\pi_3 \cdot P(1,3,4,3|3)}{\sum_{c=1}^{C} \pi_c \cdot P(1,3,4,3|c)} = \frac{.23 \cdot 5.81 \cdot 10^{-3}}{4.66 \cdot 10^{-3}} = .29$$

Every person that is an assigned member of a class still has particular classification probabilities for alternative classes, like Theo has a probability of 29% to have the latent trait of class 3. Moreover, in all data sets, there always will be individuals with very ambiguous classification probabilities (Collins & Lanza, 2010). However, a good model with high homogeneity of latent classes and good class separation will – over all students – produce classes where the classification probability for one class is high and for the other classes low (Collins & Lanza, 2010). By the mean classification probability for all students (Table 6.7), we calculate the "hit rate" – a

Table 6.7 Mean classification probability; hit rate in the diagonal

	Mean classification probability (*SD*)		
Students' class membership	1	2	3
1	.85 (.14)	.08 (.10)	.08 (.10)
2	.05 (.09)	.82 (.16)	.13 (.14)
3	.09 (.15)	.08 (.14)	.82 (.20)

measure similar to reliability (Rost, 2004). For example, a person with class membership 2 has a hit rate of 82% to be in class 2, and only a probability of 5% (class 1) / 13% (class 3) to be a member of the other two classes. Our model produces a relatively good hit rate (>.85 is recommended; Rost, 2004) for all classes (diagonal in Table 6.7), thus we conclude a reasonably good reliability of the model based on the cut-in-half original data set.

When estimating a model several times using different starting values (Sect. 6.4.3), the same latent classes will be produced, but the order might vary: class 1 in solution 1 might be labeled class 3 in solution 2. This effect is called "label switching". Label switching is no statistical problem; however, it can be confusing (Collins & Lanza, 2010). It is helpful to re-name the classes in descending order of their expected prevalence, so you will always have the same labeling.

Still, the assignment of a class based on the classification probability is not recommended to be the base of subsequent analyses (except as rough exploratory or heuristic device) (Collins & Lanza, 2010), because of the potential of underestimating the uncertainty / error that lies in the class probability of every student.

6.4.2 *Missing Data*

So far, we have assumed that our sample is complete and we do not have to deal with missing data. However, while collecting data, especially in an educational context, missing values are frequent. We can distinguish two types of missing data: Either by students neglecting to choose a response for an item (missing at random) or by using test booklets (missing by design).

As we describe in Sect. 6.4.3, the LCA uses the complete response patterns to estimate the model parameters. Therefore, a complete sample is necessary for running an LCA. Nevertheless, imputation methods replace missing values with reliable, estimated values.

In performance tests, replacing missing values as "wrong response" is widely used. Even though the responses of our CVS test include a "correct CVS understanding," there is no general category for "wrong response" because we collected qualitative data about design errors – even though it is possible to use the category "missing" to gain further insight into reasons for missing data (Rost, 2004). Therefore, if we want to replace missing responses, we have to impute responses the student presumably would have given. In our original data sample, less than 3% of the data was missing. We utilized "hot deck imputation" to impute values for missing responses (Rubin, 1987). This technique selects a (weighted) random value out of the distribution of given responses. For example, if overall 77% of the responses in an item were correct CVS, an imputed response has a probability of 77% to be correct CVS. With the hot deck imputation, we generated a similar variation in the missing data as in our complete data. This non-parametric method has several advantages (i.e., any type of variable can be replaced and no strong distributional assumptions are required) (Pérez et al., 2002). If we have a distribution of the chosen responses for each item and the number of missing cases, we can just calculate how often each response should appear in the missing data and randomly allocate these responses to the missing cases.

6.4.3 Parameter Estimation With Maximum Likelihood Approach

As seen in section "General model", the model for the LCA only includes the expected class prevalence π_c and the conditional item-response probability π_{ixc}. Because these probability parameters sum up to 1 (Collins & Lanza, 2010), the number of model parameters t that has to be estimated for a model with C classes and j items with m categories each is:

$$t = C \cdot (j \cdot (m - 1) + 1) - 1 \tag{6.8}$$

For our example, a $C = 3$ class model with 4 items (two items ($j = 2$) with $m = 4$ categories and two items ($j = 2$) with $m = 6$ categories), we need to estimate $t = 3 \cdot (2 \cdot (4 - 1) + 2 \cdot (6 - 1) + 1) - 1 = 50$ parameters.

We use LCA in an explorative way and therefore we have no strong theoretical assumptions for restricting the parameters. However, it is possible to fix parameters (e.g., fixing the expected class prevalence π_c to equal sized classes) or constrain parameters (e.g., constraining that two item-response probability parameters π_{ixc} should be equal between the classes). With restricted parameters, LCA makes it possible to test different a priori hypotheses (Rost, 2004).

We will use the maximum of the likelihood function (Rost, 2004; Eq. 6.9) to estimate the parameters (MLE). That way, the found parameters π_c and π_{ixc} will represent a model where the probability of all observed response patterns is maximal. We find the likelihood function L in a similar way as in the Rasch analysis (Rost, 2004) as the product of the unconditional pattern probabilities $P\left(\underline{x}_p\right)$ out of the general model (section "General model").

$$L = \prod_{p=1}^{N} P\left(\underline{x}_p\right) \tag{6.9}$$

To find the maximum of the likelihood function, a basic idea is to derivate the likelihood function and set it equal to zero (Rost, 2004). As stated before, because of the latent variable being categorical, we cannot use this method for LCA. LCA uses the EM algorithm (E for "expectation" and M for "maximization"; Dempster et al., 1977) to estimate the model parameters for the maximum of the likelihood function (MLE). The EM algorithm is an iterative process that will estimate with every step a higher value of the likelihood function (Rost, 2004).

Overall, there are several aspects determining if a latent class model can be found with the MLE approach. A large sample size N, a large ratio between observed response patterns/sample size and possible response patterns (low sparseness), a good latent class homogeneity, and good class separation benefit the identification of a latent class model (Collins & Lanza, 2010).

One noteworthy problem of MLE in the LCA is that the likelihood function might have several local maxima, so we might only find a local, but no global, maximum (Collins & Lanza, 2010). This will happen more frequently with five or more classes and 12 or more items (Rost, 2004) or with complex categories. To minimize the probability of only finding a local maximum, we can use different (random) start values and compare the maximum likelihoods for the estimated model parameters. If we had found a global maximum, other start values should find the same maximum likelihood. Collins & Lanza (2010) recommend to estimate at least 10 models.

6.4.4 Goodness of Fit / Model Selection

Even after estimating a model, we do not know if the estimated model fits the data. We need this information to decide for example on the number of classes by comparing the fit of several models ($C = 1$, $C = 2$, $C = 3$, . . .) because the number of classes is no model parameter. However, there is no general "yes/no" to the model fit – every estimated model fits the data more or less well and we need to find criteria to set boundaries for fitting (Rost, 2004). Beside the model fit, we need to consider the complexity of the model that might result in a good fit. In LCA, the best fitting model, called saturated model, will be a model that has as many classes as there are response patterns (161 classes in our example), resulting in a high number of model parameters. In general, the more classes we choose the better the fit might be, although under a higher complexity of parameters. Lastly, we need to consider the state of research and the potential for theoretical insight. We need to keep in mind that model selection always requires statistical and theoretical considerations, and a good fit alone cannot be used as "proof" for a psychological structure (Edelsbrunner & Dablander, 2019). Therefore, model selection is, within certain statistical boundaries, a judgement call (Collins & Lanza, 2010).

Simplicity and Usefulness

Rost (2004) suggests that the aim of building a theory (in our example: explaining design errors by underlying misconceptions) is not solely rooted in finding a good as possible statistical fit but rather on being based on few and simple assumptions. Models should be empirically valid as well as simple as possible. This philosophical principal, quite similar to Ockham's Razor, is called "parsimony" (Collins & Lanza, 2010). Therefore, a simpler model should be preferred to a more complex one whenever the more complex model does not supply significant higher theoretical insight and has therefore a higher usefulness or model interpretability (Collins & Lanza, 2010).

Likelihood Ratio Test G^2 and Observed Pattern Frequency Test χ^2

The likelihood of a model gives insight into how well the model fits the data – the higher the likelihood, the better is the model fit. In Sect. 6.4.3, we calculated the likelihood of a latent class model by multiplying the unconditional pattern probabilities. The likelihood of one estimated model can be compared to the likelihood of other (nested) estimated models or the saturated model (Collins & Lanza, 2010; Rost, 2004). Of course, we can only compare the likelihoods of models that are estimated based on the same data.

The empirical comparison of the likelihood is based on a likelihood ratio test: $\mathrm{LR} = \frac{L_0}{L_1}$. The two compared models have to be nested models. Models are statistically nested when one model L_1 is the restricted version of another model L_0 (Collins & Lanza, 2010). Furthermore, the restrictions must not be achieved by restricting model parameters to 0. In other words, all compared models need to have the same number of classes and response pattern but can differ regarding the allocation of response pattern to specific classes based on fixed or constrained model parameters. If these requirements are met, we can calculate a χ^2 distributed test statistic G^2 (Eq. 6.10) with the difference in the number of model parameters t as degrees of freedom $df = t(L_1) - t(L_0)$ (Rost, 2004).

$$G^2 = -2 \cdot \log\left(\frac{L_0}{L_1}\right) \tag{6.10}$$

An alternative to the likelihood ratio test, we can compare the observed frequency of patterns ($o_{\underline{x}}$) to the expected frequency of patterns ($e_{\underline{x}}$) based on the model parameters, Pearson's χ^2 test statistic (Rost, 2004; Eq. 6.11). A lower value of χ^2 is associated with a better model fit (Collins & Lanza, 2010; Linzer & Lewis, 2011). Again, we will get a χ^2 distributed test statistic with the same degrees of freedom as in the likelihood ratio test against the saturated model. The likelihood ratio test and the χ^2-test usually lead to the same results (Rost, 2004).

$$\chi^2 = \sum_{\underline{x}} \frac{\left(o_{\underline{x}} - e_{\underline{x}}\right)^2}{e_{\underline{x}}} \tag{6.11}$$

Unfortunately, we cannot compare models with a different number of classes using G^2 because of an unknown distribution rather than χ^2 (Collins & Lanza, 2010; Rost, 2004). The model parameters of the LCA are the expected class prevalence π_c and the dependent item-response probabilities π_{icx}, but not the number of classes (section "General model"). Therefore, a lower class model is restricted by setting a model parameter of the higher class model to 0 (e.g., for three classes $C = 3$, the expected class prevalence of a fourth class is zero: $\pi_4 = 0$). Furthermore, comparisons to the saturated model with G^2 as well as the χ^2 test for the frequency of patterns are not allowed in data sets with a large sparseness, because then the test statistic will also have an unknown distribution (Collins & Lanza, 2010; Linzer & Lewis, 2011; Rost,

2004). To make sure data is not sparse, 80% of the possible patterns should have an expected frequency greater than 5 and no possible pattern should have an expected pattern frequency of less than 1 – therefore, as a rule of thumb, the sample size should equal at least the number of possible patterns multiplied by 5 when using a χ^2-test (McHugh, 2013). Sparseness happens quickly in more complex latent class models with many latent classes (Collins & Lanza, 2010; McHugh, 2013). Our data is sparse. The number of observed patterns (161) is a lot smaller than the number of possible patterns (576).

Bootstrapping

To solve the problem of an unknown distribution of the test statistic, Aitkin et al. (1981) (also see Davier, 1997; Rost, 2004) suggest performing a parametric bootstrap. During parametric bootstrapping, random data sets are resampled repeatedly out of a distribution based on the original model's estimated parameters. As we now have multiple data sets, we can estimate the distribution of the test statistics from these resampled data.

After resampling the data, we compare test statistics of the original model with the now known empirical distribution of the test statistics. As test statistic, we can use G^2 or χ^2 (section "Likelihood ratio test G^2 and observed pattern frequency test χ^2"). However, with sparse data, only the use of χ^2 is recommended (Davier, 1997).

We can localize the test statistic of the original model within the empirical distribution of the resampled models. Similar to a normal χ^2-test with a χ^2 distribution, we can conclude the original model fits the data if the p-value is bigger than .05 because then there is no significant difference between the original model and most of the resampled models (Rost, 2004).

Information Criteria

We can use information criteria to determine which model is better in relative terms (Collins & Lanza, 2010; Rost, 2004). The only restriction to the use of information criteria is that the models must be estimated based on the same data. On the downside, we do not know how much lower the information criteria of a model needs to be to underline a better fit.

Following the simplicity criterion, models should be empirically valid as well as simple as possible. Information criteria consider both requirements because they use the log-likelihood $\log(L)$ (measurement for empirical validity) and the number of model parameters t (measurement for complexity). Additionally, the sample size N is used. There are several information criteria with differences in weighting the number of model parameters t. We can calculate four information criteria: the well-known AIC (Akaike's information criterion; Akaike, 1974) and BIC (Bayes' information criterion; Schwartz, 1978) as well as CAIC (consistent AIC; Bozdogan, 1987) and aBIC (adjusted BIC; Sclove, 1987). Lower criteria are associated with a better fit.

$$\mathrm{AIC} = -2\log(L) + 2 \cdot t$$

$$\mathrm{BIC} = -2\log(L) + \log(N) \cdot t$$

$$\mathrm{CAIC} = -2\log(L) + [\log(N) + 1] \cdot t$$

$$\mathrm{aBIC} = -2\log(L) + \log\left(\frac{N+2}{24}\right) \cdot t$$

The AIC takes an equal weighting of the log-likelihood and the number of parameters. The BIC weights the number of model parameters more, based on the sample size. The CAIC is a correction of the AIC for bigger sample sizes. The aBIC further adjusts the weight of the sample size.

What information criteria to use depends on the sample size and aim of the analysis. Dziak et al. (2020) discuss several information criteria, including their sensitivity and specificity. Their simulations for latent class models (Dziak et al., 2020, pp. 13–14) show, for a 3-class model with N around 500 cases, like our empirical example, that the AIC has significant and aBIC marginal overfitting rates (choosing a model with too many classes) whereas BIC and CAIC have marginal underfitting rates (choosing a model with too few classes). Furthermore, BIC and aBIC have a similarly correct fit rate close to 100%, whereas AIC and CAIC have a lower fit rate. Based on their simulations, Dziak et al. (2020) provide heuristics for choosing a useful information criterion based on the aim of the analysis. For a sufficiently rich model with more theoretical insight where describing heterogeneity within the population is more important than the simplicity of the model, or when expecting similar classes that still have distinct differences, AIC or aBIC are preferable than BIC or CAIC. For obtaining fewer larger classes, BIC is more fitting. If the AIC favors a solution with a high, difficult to interpret class number, BIC is the better choice. In general, information criteria often are not used to identify one single, best fitting model, but to reduce the number of viable models (Collins & Lanza, 2010). To find an appropriate model, researchers should also consider the usefulness of the found model (Rost, 2004).

Hit Rate and Assigned Class Membership

As described in section "Assigning the class membership", we can calculate a hit rate as the mean probability of class membership for all students. The hit rate is a measure similar to reliability, and good models should have a high hit rate (high >.85; Rost, 2004). Additionally, we added the percentage of the students assigned to the found classes to provide a better understanding of the distribution of students over the classes.

Choosing a Model

Until now, we worked with the 3-class model without giving reason why we are using it. In this section, we will explain how we exploratively chose a latent class

Table 6.8 Goodness of fit statistics for the empirical example ($N = 496$)

Model fit	1-class model	2-class model	3-class model	4-class model
Model properties				
Parameters	16	33	50	67
Log-likelihood[a]	−2345.53	−2280.53	−2257.13	−2239.35
Hit rate	–	.83/.91	.84/.82/.82	.84/.78/.81/.95
Class membership		58%/42%	60%/22%/19%	57%/22%/19%/2%
Bootstrap				
χ^{2b}	1181.67	594.04	535.46	425.70
χ^2 emp. p	.000	.268	.146	.512
Information criteria				
AIC	4723	4627	4614	4613
BIC	4790	4766	4825	4895
CAIC	4806	4799	4875	4962
aBIC	4740	4661	4666	4682

Note. a: log-likelihood is the logarithmized likelihood $\log(L)$ that is more frequently used to describe models. Likelihoods are small numbers $0 < L < 1$, therefore $\log(L) < 0$; a lower absolute log-likelihood shows a better fit. b: we only use χ^2 as test statistic for the bootstrap because our data is sparse

model for our empirical example based on goodness of fit statistics (Table 6.8) discussed in Sect. 6.4.4.

As expected, the log-likelihood of the data shows a better fit for higher class models. The hit rate of the models decreases with more classes but is for the 2- and 3-class models still within reason. Looking at the empirical p-values of the bootstrap, we clearly need to reject the 1-class model ($p < .05$) because the empirical data differs significantly from the resampled data based on the model parameters. The 2-, 3- and 4-class models show adequate fit for the empirical χ^2 test statistic ($p > .05$). Following the simplicity criterion, we reject the 4-class model and focus further on comparing the 2-class and 3-class models.

Comparing the information criteria, the 2-class model has a slightly worse AIC than the 3-class model (13), but a better BIC (59). The CAIC is in favor of the 2-class model (76), whereas the aBIC of the 2-class model is only slightly better (5). As discussed in section "Information criteria", AIC and aBIC might favor overfitting models, whereas BIC and CAIC might lead to underfitting.

However, for choosing a model, we cannot rely solely on statistical considerations but need to include theoretical aspects like potential insight and usefulness. If we take a closer look at the actual class distribution, we notice that the potential theoretical insight of a 2-class model is only small. In the 2-class model, class 1 ($\pi_1 = .52$, 58% of the students) includes students with high CVS skill (overall item-response probability for correct CVS: .73) and class 2 ($\pi_2 = .47$, 42% of the students) students with low CVS skill (overall item-response probability for correct CVS: .36). A split in half of the sample into "good" and "bad" students will give us only small additional insight into patterns of design errors and, therefore, underlying CVS misconception. A 3-class model will split one class found in the 2-class model, therefore we will have similar classes with distinct differences, which will lead to a more differentiated model, favoring AIC for the model selection.

In conclusion, we will choose the 3-class model, which shows overall a good model fit (bootstrapping χ^2), potentially offers additional theoretical insight into distinct patterns of design errors (Sect. 6.5.1) and is still simple enough (AIC).

6.5 Findings

6.5.1 Finding for Research Question 1: Identifying Patterns of Design Errors With LCA

In this section, we will analyze and interpret the item-response probabilities in the classes and find common characteristics to label the three classes. Figure 6.2 shows the item-response probabilities (Appendix Table 6.1) for the four items.

In our example, we will describe the distinct underlying preconception in every class based on the probabilities of the design errors in the four item types:

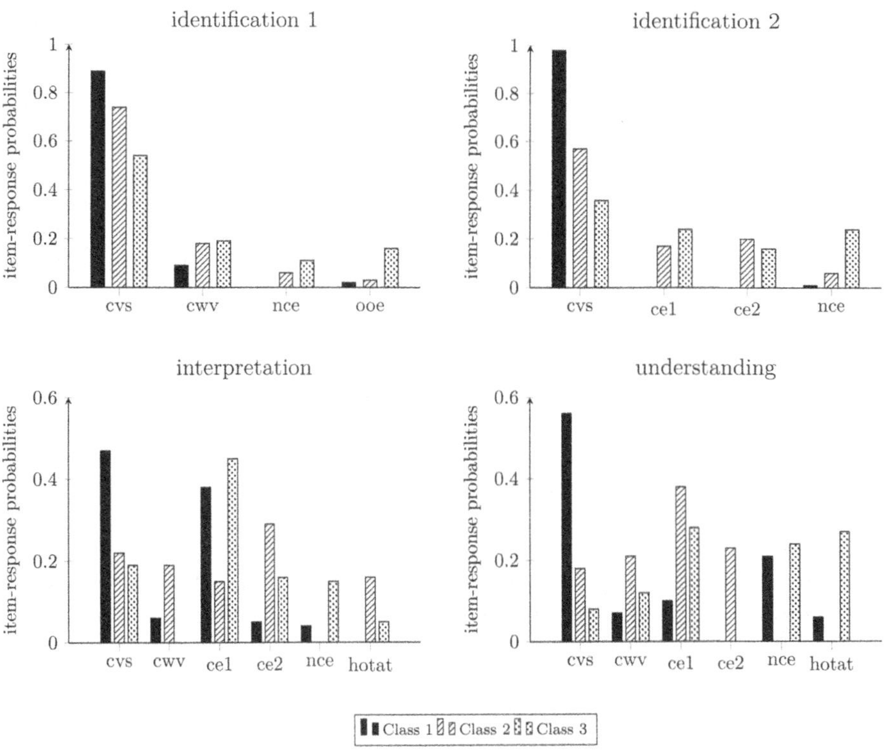

Fig. 6.2 Item-response probabilities. Note. cvs: correct application of the control-of-variables strategy; cwv: CVS for wrong variable, nce: noncontrastive target variable, ooe: single-condition experiment, ce1 & ce2: Confounded experiments, hotat: hold-one-thing-at-a-time

identification 1 + 2 (ID1, ID2), interpretation (IN), and understanding (UN) based on the CVS subskills. The first step in describing the latent classes is to find a label that describes well what the response patterns that are in the same class have in common. To do so, we can refer to dominant response categories (e. g. correct response) and the shape of the response patterns like "high ce1 and low correct" (Marsh et al., 2009). In the following, we will describe the three classes, label them, and discuss the extent of class homogeneity. We labeled the three classes only by dominant response pattern and not by the shape because the categories differ characteristically in the dominant design errors. Overall, our model shows a good class separation, because the item-response probabilities differ between the three classes for almost all response options. In general, a high class homogeneity and good class separation help to interpret the found model.

Class 1 (correct CVS understanding, $\pi_1 = .53$, 60% of the students): Students in class 1 are very likely to respond with correct CVS in items regarding all subskills. The item-response probabilities for correct CVS were close to 1 for ID1 (.89) and ID2 (.98). In the subskill 'interpretation' they still have the highest item-response probability of all classes (.47). However, the probability for the response ce1 is likely high (.38). We conclude, if students in class 1 make errors, it is mostly the choice of a confounded experiment (two variables change). Nevertheless, class 1 students understand the problem of confounded experiments as they have a quite high probability of .56 to respond that the experiments are poorly designed because too many variables differ between conditions. Overall, class 1 shows a relative high class homogeneity with item-response probabilities for correct CVS be close to 1 (cvs: ID1 .89, ID2 .98) or stand out in relation to the other response options (cvs: IN .47, UN .56).

Class 2 (change of too many variables, $\pi_2 = .24$, 22% of the students): Students in class 2 basically understand that variables need to differ between contrasted conditions. Nonetheless, they do not comprehend why all non-investigated variables should be equal. Accordingly, their item-response probabilities for responses representing design errors associated with the choice or justification of confounded experiments (ce1, ce2) are higher compared to class 1. In particular, such design errors appear in interpretation and understanding items. In interpretation items, students are likely to state that in a controlled experiment more variables have to be changed (.15 ce1, .29 ce2 within IN items). In understanding items, they are more likely to respond that a purely planned experiment has too few changed variables (ce1 .38, ce2 .23 within UN items). Overall, class 2 shows a lower class homogeneity than class 1, because there are no single item-response probabilities close to 1. However, item-response probabilities for changing too many variables (ce1/ce2) stand out in relation to the other response options different from correct CVS.

Class 3 (non-contrastive experiments, $\pi_3 = .23$, 19% of the students): Students in class 3 basically understand that it is part of CSV to keep variables unchanged. However, they tend to overgeneralize this concept. In the subskill 'identification', they are more likely to show the error 'only one experiment' than any other class (.16 within ID1) and stick to this concept even if the non-contrastive experiment is the only choice with unchanging variables (.24 within ID2). Furthermore, they are more likely to show these design errors in understanding items when they respond that the experiment is flawed because all variables are changed (nce .24, hotat .27). However, in interpretation items, the most likely design errors of class 3 students are associated with confounded experiments (ce1 .45). Overall, class 3 also shows a lower class homogeneity than class 1 but the item response probabilities for changing too few variables (nce/ooe/hotat) are higher than in the other classes.

6.5.2 Finding for Research Question 2: Difficulty of Items With Rasch Analysis

To get insight into the difficulty of the CVS items, we conducted a unidimensional Rasch analysis. To do this, we transformed the categorical data including the students' errors into dichotomous (correct CVS response / not correct) data. We did use the complete data set with 8 items, because a split-half data set with 4 items is not suitable for a Rasch analysis. Therefore, every item type is present twice, marked with .1 and .2. We utilized the R package TAM to fit our data to a unidimensional Rasch model (MML 1PL, Robitzsch et al., 2022). We set the average latent ability θ of the students to 0 because we want to make statements about the distribution within the students, not the items. To estimate the student latent ability, we draw five plausible values (PVs). The items' wMNSQ fits between .96 and 1.08, the EAP/PV reliability is .60 is acceptable. In summary, our data fit the Rasch model. We use PVs because, compared to EAP or WLE, the use of PVs allows measurement error adjusted estimations for the sample (Lüdtke & Robitzsch, 2017). It should be noted, however, that PVs are not suitable as individual estimators (Rost, 2004). For all following computations of statistics for the PVs, we followed the recommendations of OECD (2009). The final mean estimate is equal to the average of the five mean estimates. The final error variance has been calculated with the final sampling variance and the imputation variance (for details: OECD (2009): PISA Data Analysis Manual SPSS, p. 118).

The student's mean latent ability over all drawn PVs is $\overline{\theta} = 0.00$ ($SD = 1.04$). Table 6.9 lists the item difficulties and standard errors. The mean difficulty of the items is $-.24$, meaning the CVS test is overall a bit too easy for the students.

In the Wright Map, we see a clear distribution in the item difficulty (Fig. 6.3). ID1 and ID2 items ($M = -1.24$, $SD = .20$) are easier than IN and UN items ($M = .77$, $SD = .38$) with a large effect ($t(6) = -9.36$, $p < .001$, $d = -5.29$). We can conclude that identifying controlled experiments (ID1, ID2) is easier than interpreting controlled experiments (IN) or explaining why a confounded experiment can't be meaningfully interpreted (UN). The different difficulties are consistent with the different requirements of the subskills. For identification, only one controlled experiment had to be selected, whereas for interpretation and understanding, conclusions had to be drawn from (un-)controlled experiments and justified.

To answer research question 2, we used a unidimensional Rasch model. However, as discussed in Brandenburger et al. (2022), the data on students' CVS skills

Table 6.9 Item difficulties of the Rasch model

Item name	Item difficulty σ	Standard error
ID1.1	−1.19	.11
ID1.2	−1.49	.12
ID2.1	−1.00	.11
ID2.2	−1.28	.11
IN.1	1.27	.11
IN.2	.76	.10
UN.1	.70	.10
UN.2	.34	.10
	$M = -.24$, $SD = 1.11$	$M = .11$, $SD = .01$

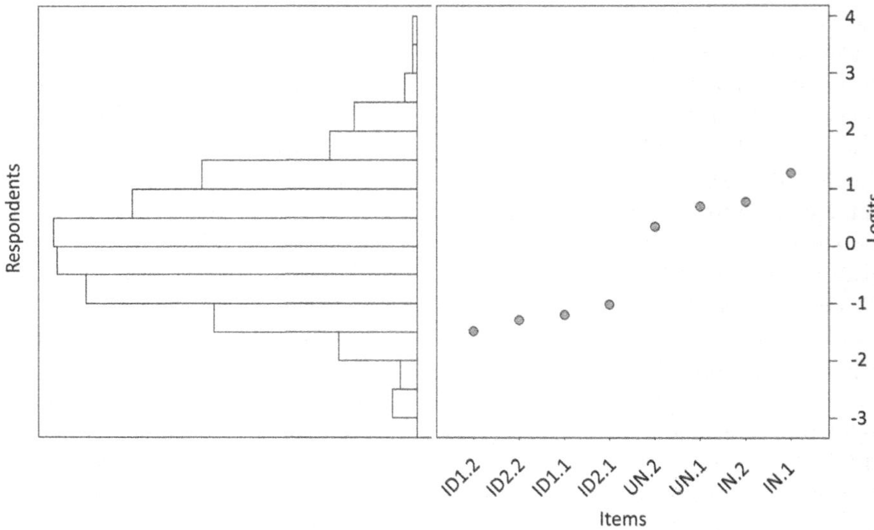

Fig. 6.3 Wright Map of the unidimensional Rasch model

better fit a multidimensional model based on the subskills. Nevertheless, the fit to the unidimensional model is acceptable for answering research question 2 and the planned combination with an LCA for research question 3.

6.5.3 Finding for Research Question 3: Combination of LCA and Rasch Analysis

To combine the found classes of the LCA with the student score of the unidimensional Rasch model, we calculated the mean student score within the unidimensional Rasch model within the 3 classes of the LCA, aware that the class membership holds

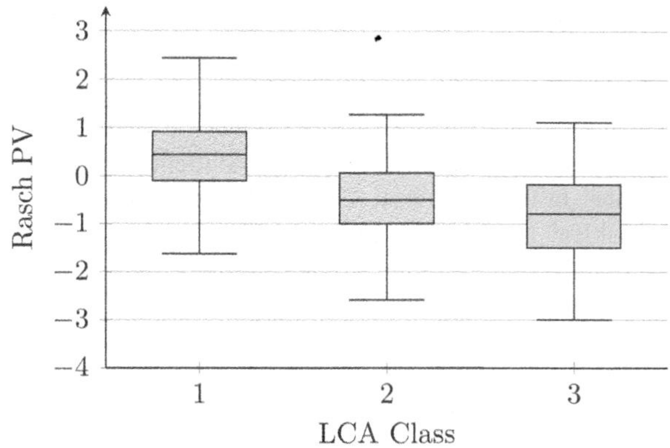

Fig. 6.4 Boxplot for Rasch PVs of every student per LCA class

unavoidable uncertainty (section "Assigning the class membership"). Overall, the mean student score of class 1 (correct CVS understanding) is above the average of 0 (M = .43, SD = .90, N = 296). The mean student score of class 2 (change of too many variables) and class 3 (non-contrastive experiments) are lower than 0. (class 2: M = −.50, SD = .86, N = 107; class 3: M = −.81, SD = .96, N = 93). An ANOVA confirms there are differences in the student score between the classes ($F(2, 493) = 87.35$, $p < .001$, $\eta^2 = .26$, large effect). A Tukey post-hoc analysis confirmed that the differences between all classes are significant ($p < .050$). We can conclude that students who follow the dominant design error of class 3 (non-contrastive experiments) have overall a lower CVS skill than students out of class 2 (change of too many variables) – a conclusion we couldn't make only with the LCA. However, we can see in the boxplot in Fig. 6.4 that there is quite an overlap between the student scores associated with different classes.

6.6 Discussion

By utilizing LCA, we identified three classes of patterns within students' responses to different items regarding the design and interpretation of experiments (research question 1, Sect. 6.5.1). The three identified classes are easy to interpret because the number of classes is low, and students within each class choose characteristic distractors that are related to common design errors. This was possible because the response options of our multiple-choice items correspond to typical experimental design errors. However, LCA is not restricted to data from multiple-choice tests but can be used to analyze all kinds of categorical data, including a categorical coding of open-response items or student drawings (e.g., Zarkadis et al., 2021). Researchers can utilize LCA to analyze data from all kinds of tasks that encourage students to respond according to their preconceptions. Additionally, to improve the performance of the LCA, covariates (like gender or reading skill) can be used to predict the latent class membership based on logistic regression (Collins & Lanza, 2010).

Our findings of the Rasch analysis (research question 2, Sect. 6.5.2) show that items of the interpretation and understanding subskills are more difficult than items of the identification subskill. We were able to identify a progression in students understanding of CVS by combining the findings of the LCA and Rasch analysis (research question 3, Sect. 6.5.3). Students of the class 3 "non-contrastive experiments" have the lowest CVS abilities. Students in class 2 "change of too many variables" have higher abilities while the students in class 1 "correct understanding of CVS" have the highest CVS abilities and can solve even the challenging interpretation and understanding items. Generally, LCA groups students together that have similar response patterns while Rasch analysis groups items together that have similar item difficulty.

LCA and Rasch analysis have their distinct areas of application. Based on a similar data set, the LCA allows to find classes of underlying leading design errors (research question1; Schwichow et al., 2022), whereas using a Rasch analysis gives insight into the difficulty of items (research question 2) or dimensionality of the CVS subskills (Brandenburger et al., 2022). By combining both analysis approaches (research question 3), researchers can get a more detailed picture of students' response behavior and the item characteristics (see also Fulmer et al., 2014).

However, LCA also have some disadvantages compared to a Rasch analysis. LCA needs a more complex handling of missing data because it is a full information method. Rasch analysis allows an easy handling of missing data because it uses aggregated data. Thus, researchers utilizing a Rasch analysis can divide the items of a bigger item pool into different test booklets and thus reduce the number of items each student has to work on without reducing the total number of items. A further disadvantage of LCA is that LCA models do not always have clear and easy to interpret outcomes. If the number of classes is high and if the patterns within classes are diverse, then the meaning of the classes is hard to understand.

If LCA delivers clear results, they offer detailed information that can help identify students' conceptions that underlie their incorrect responses. LCA seems to be an ideal tool to analyze student preconceptions because the methods align with theories of preconceptions (Vosniadou, 2019). These theories describe preconceptions as rather coherent ideas that can be applied by students to different tasks. Accordingly, students' responses to different tasks should cluster together so that specific preconceptions result in specific response patterns that are based on students' preconceptions. Indeed, as many cognitive models are analogies from statistical models (Gigerenzer, 1991), the mathematical model behind LCA can serve as a blueprint to understand and describe how specific preconceptions (represented as classes) result in probabilities for specific responses to tasks (represented as conditional item-response probability). This probabilistic view on the relation between preconception and item response matches with empirical finding that although students have prevalent preconceptions they do not respond consistently to these conceptions in all items (Vosniadou, 2019).

Appendix

Appendix Table 6.1 Estimated item-response probabilities π_{ixc} in the 3-class model

Item	Class	Category *k*						
		1	2	3	4	5	6	7
Identification 1 X_1								
	1	.89	.09	–	–	.00	–	.02
	2	.74	.18	–	–	.06	–	.03
	3	.54	.19	–	–	.11	–	.16
Identification 2 X_2								
	1	.98	–	.00	.00	.01	–	–
	2	.57	–	.17	.20	.06	–	–
	3	.36	–	.24	.16	.24	–	–
Interpretation X_3								
	1	.47	.06	.38	.05	.04	.00	–
	2	.22	.19	.15	.29	.00	.16	–
	3	.19	.00	.45	.16	.15	.05	–
Understanding X_4								
	1	.56	.07	.10	.00	.21	.06	–
	2	.18	.21	.38	.23	.00	.00	–
	3	.08	.12	.28	.00	.24	.27	–

Appendix Table 6.2 R script LCA

```
# install.packages("poLCA")
library(poLCA) # loading poLCA

setwd("")

file <- "" # name of the data file

nclasses <- 3 # number of classes
f1 <- as.formula(cbind(ID1, ID2, IN, UN)~1) # selecting the variables for the LCA

##########################################################################

mydata <- read.csv2(file) # reading the data file

# Running the analysis
# max number of iterations: 1000, convergence criteria: .001
# number of times to estimate the model, using different start values: 30
LCA <- poLCA(f1, data=mydata, nclass=nclasses, maxiter=1000, tol=.001, nrep=30)

##########################################################################
```

(continued)

Appendix Table 6.2 (continued)

```
# start bootstrap
bootstrapN <- 40 # max number of bootstrap samples

bootstrapCHIsq <- LCA$Chisq

for (i in 1:bootstrapN){
  # simulating a data set based on the calculated model parameters
  simdata <- poLCA.simdata(N=nrow(mydata),probs=LCA$probs,P=LCA$P)
  # running the LCA for the simulated data set
  simLCA <- poLCA(
    as.formula(paste0("cbind(",paste0("Y",1:ncol(LCA$y),collapse=","),")~1")),
    data=simdata$dat, nclass=nclasses, maxiter=1000, tol=.001, nrep=1)
  # save the simulated CHI^2 value in a list
  bootstrapCHIsq <- c(bootstrapCHIsq, simLCA$Chisq)
}

# rank the original CHI^2 within the simulated distribution
bootstrapCHIsq <- sort(bootstrapCHIsq)
empP <- 1-(match(LCA$Chisq, bootstrapCHIsq)/(bootstrapN+1))


# creating a histogram of the simulated CHI^2 distribution
hist(bootstrapCHIsq)
abline(v=LCA$Chisq, col="red")
abline(v=bootstrapCHIsq[floor((bootstrapN+1)-((bootstrapN+1)*0.05))],lty=3)

# End bootstrap
##########################################################################

# classification probability and assigned class membership
classMembership <- data.frame(LCA$posterior, class=LCA$predclass)
write.csv2(classMembership,paste0(format(Sys.time(),"%y%m%d-%H%M%S-
"),nclasses,"-classMembership.csv", collapse=""))

# hit rate / mean classification probability
hitrate <- aggregate(classMembership, list(classMembership$class), FUN=mean)

# we can also calculate the SD of the hit rate
hitrateSD <- aggregate(classMembership, list(classMembership$class), FUN=sd)

# Printing a model summary into a text file named with model and current time
sink(paste0(format(Sys.time(),"%y%m%d-%H%M%S-"),nclasses,"-
classSummaryLCA.txt", collapse=""))
cat(c("Date of analysis:", format(Sys.time(),"%x %X")))
```

(continued)

Appendix Table 6.2 (continued)

```
cat(c(
  "\n \n", paste0(rep("-", times=36), collapse=""), "\n Model summary for the", nclasses,
"class model \n ", paste0(rep("-", times=36), collapse=""),
  "\n number of iterations: ", LCA$numiter,
  "\n number of observations: ", LCA$Nobs,
  "\n number of estimated parameters: ", LCA$npar,
  "\n maximum log-likelihood: ", round(LCA$llik, digits=2),
  "\n \n", paste0(rep("-", times=23), collapse=""), "\n Goodness of fit \n", paste0(rep("-",
times=23), collapse=""),
  "\n AIC:", round(LCA$aic, digits=0),
  "\n BIC:", round(LCA$bic, digits=0),
  "\n CAIC:", round(-2*LCA$llik+(log(LCA$Nobs)+1)*LCA$npar, digits=0),
  "\n aBIC:", round(-2*LCA$llik+log((LCA$Nobs +2)/24)*LCA$npar, digits=0),
  "\n G^2:", round(LCA$Gsq, digits=2), "(Likelihood ratio/deviance statistic)",
  "\n CHI^2:", round(LCA$Chisq, digits=2),"(Chi-square goodness of fit)",
  "\n CHI^2 emp. p-value:", if(exists("empP")) round(empP, digits=3) else "no
bootstrap performed", "\n"
  ))

cat(c(
  "\n \n", paste0(rep("-", times=42), collapse=""), "\n Expected class prevalence for each
class \n", paste0(rep("-", times=42), collapse=""), "\n"
))

class_prevalence <- data.frame(lapply(LCA$P, round, digits=3))
colnames(class_prevalence) <- c(paste("class", 1:nclasses))
print(class_prevalence, row.names=FALSE)

cat(c(
  "\n", paste0(rep("-", times=51), collapse=""), "\n Conditional item-response (column)
probabilities,",
  "\n by outcome variable, for each class (row) \n", paste0(rep("-", times=51),
collapse=""), "\n \n"
  ))

lapply(LCA$probs, round, digits=3)

cat(c(
  "\n", paste0(rep("-", times=23), collapse=""), "\n Hit Rate \n", paste0(rep("-",
times=23), collapse=""), "\n"
))

print(round(hitrate, digits=3), row.names=FALSE)

cat(c(
  "\n \n", paste0(rep("-", times=23), collapse=""), "\n Hit Rate SD \n", paste0(rep("-",
times=23), collapse=""), "\n"
```

(continued)

Appendix Table 6.2 (continued)

```
))
print(round(hitrateSD, digits=3), row.names=FALSE)
sink()
```

Appendix Table 6.3 R script Rasch analysis

```
setwd("")

file <- "" # name of the data file

mydata <- read.csv2(file) # reading the data file

#install.packages("TAM")
library(TAM)

mod_1D <- tam.mml(mydat, irtmodel="1PL", constraint="cases",
control=list(convD=0.00010, conv=0.00010, Msteps=10))

# save model summary
sink(paste0(format(Sys.time(),"%y%m%d-%H%M%S-"),"Summary.txt", collapse=""))
summary(mod_1D)
tam.fit(mod_1D)
summary(tam.modelfit(mod_1D))
sink()

# save item difficulty
difficulty <- mod_1D$xsi
sink(paste0(format(Sys.time(),"%y%m%d-%H%M%S-"),"ItemDifficulty.txt",
collapse=""))
summary(difficulty)
difficulty
sink()

# save 5 plausible values
pvmod_1D <- TAM::tam.pv(mod_1D, nplausible=5)
write.csv2(cbind(mydata[grep("casenum",names(mydata))],pvmod_1D$pv),paste0(for
mat(Sys.time(),"%y%m%d-%H%M%S-"),"PV.csv", collapse=""))

# create Wright map
#install.packages("WrightMap")
library(WrightMap)
library(RColorBrewer)

pvmod_1D <- TAM::tam.pv(mod_1D, nplausible=5)
thetas <- pvmod_1D$pv$PV1

df <- mod_1D$item[order(mod_1D$item$xsi.item),]
thresholds <- df$xsi.item
wrightMap(thetas,thresholds
        , min.logit.pad = .5, max.logit.pad = 1.5
```

(continued)

Appendix Table 6.3 (continued)

```
, item.prop=0.5
, label.items=df$item
, label.items.ticks=TRUE
    , label.items.cex=.75
, dim.names=c(" ")
, thr.sym.col.fg=c("Black")
, thr.sym.cex =1.5
, thr.sym.pch = 21
, show.thr.lab=FALSE
, label.items.srt=45
, main.title=""
, axis.persons="Respondents"
, axis.items="Items"
, return.thresholds=FALSE)
```

References

Aitkin, M., Anderson, D., & Hinde, J. (1981). Statistical modelling of data on teaching styles. *Journal of the Royal Statistical Society. Series a (General), 144*(4), 419. https://doi.org/10.2307/2981826

Akaike, H. (1974). A new look at the statistical model identification. *IEEE Transactions on Automatic Control, 19*(6), 716–723. https://doi.org/10.1109/TAC.1974.1100705

Andersen, E. B. (1982). Latent structure analysis. A survey. *Scandinavian Journal of Statistics, 9*, 1–12. https://www.jstor.org/stable/4615848

Boone, W. J., & Staver, J. R. (2020). *Advances in Rasch analyses in the human sciences*. Springer. https://doi.org/10.1007/978-3-030-43420-5

Boone, W. J., Staver, J. R., & Yale, M. S. (2014). *Rasch analysis in the human sciences*. Springer. https://doi.org/10.1007/978-94-007-6857-4

Box, G. E. P. (1979). Robustness in the strategy of scientific model building. In R. L. Launer & G. N. Wilkinson (Eds.), *Robustness in statistics* (pp. 201–236). Academic Press. https://doi.org/10.1016/B978-0-12-438150-6.50018-2

Bozdogan, H. (1987). Model selection and Akaike's information criterion (AIC). The general theory and its analytical extensions. *Psychometrika, 52*(3), 345–370. https://doi.org/10.1007/BF02294361

Brandenburger, M., Salim, C. A., Schwichow, M., Wilbers, J., & Mikelskis-Seifert, S. (2022). Modellierung der Struktur der Variablenkontrollstrategie und Abbildung von Veränderungen in der Grundschule [modelling the structure of the control of variables strategy (CVS) and mapping changes in CVS through elementary school]. *Zeitschrift für Didaktik der Naturwissenschaften, 28*(1), 651. https://doi.org/10.1007/s40573-022-00140-x

Bullock, M., & Ziegler, A. (1999). Scientifc reasoning: Developmental and individual differences. In F. E. Weinert & W. Schneider (Eds.), *Individual development from 3 to 12: Findings from the Munich longitudinal study* (pp. 38–54). Cambridge University Press.

Collins, L. M., & Lanza, S. T. (2010). *Latent class and latent transition analysis. With applications in the social, behavioral, and health sciences*. Wiley. https://doi.org/10.1002/9780470567333

Davier, M. V. (1997). Bootstrapping goodness-of-fit statistics for sparse categorical data: Results of a Monte Carlo study. *Methods of Psychological Research Online, 2*(2), 29–48.

Davier, M. V., & Lee, Y.-S. (2019). *Handbook of diagnostic classification models*. Springer International Publishing. https://doi.org/10.1007/978-3-030-05584-4

Dempster, A. P., Laird, N. M., & Rubin, D. B. (1977). Maximum likelihood from incomplete data via the EM algorithm. *Journal of the Royal Statistical Society / B*, 1–38. https://www.jstor.org/stable/2984875

Dziak, J. J., Coffman, D. L., Lanza, S. T., Li, R., & Jermiin, L. S. (2020). Sensitivity and specificity of information criteria. *Briefings in Bioinformatics, 21*(2), 553–565. https://doi.org/10.1093/bib/bbz016

Edelsbrunner, P. A., & Dablander, F. (2019). The psychometric modeling of scientific reasoning: A review and recommendations for future avenues. *Educational Psychology Review, 31*(1), 1–34. https://doi.org/10.1007/s10648-018-9455-5

Fulmer, G. W., Liang, L., & Liu, X. (2014). Applying a force and motion learning progression over an extended time span using the force concept inventory. *International Journal of Science Education, 36*(17), 2918–2936. https://doi.org/10.1080/09500693.2014.939120

Gigerenzer, G. (1991). From tools to theories: A heuristic of discovery in cognitive psychology. *Psychological Review, 98*(2), 254–267. https://doi.org/10.1037/0033-295X.98.2.254

Lazarsfeld, P. F., & Henry, N. W. (1968). *Latent structure analysis*. Houghton Mifflin.

Linzer, D. A., & Jeffrey, L. (2022). *poLCA: Polytomous variable latent class analysis*. R package version 1.6. https://github.com/dlinzer/poLCA. Accessed: 25/05/2022.

Linzer, D. A., & Lewis, J. B. (2011). poLCA: An R package for Polytomous variable latent class analysis. *Journal of Statistical Software, 42*(10), 1–29. https://doi.org/10.18637/jss.v042.i10

Lüdtke, O., & Robitzsch, A. (2017). Eine Einführung in die Plausible-Values-Technik für die psychologische Forschung [An introduction to the plausible value technique for psychological research]. *Diagnostica, 63*(3), 193–205. https://doi.org/10.1026/0012-1924/a000175

Marsh, H. W., Lüdtke, O., Trautwein, U., & Morin, A. J. (2009). Classical latent profile analysis of academic self-concept dimensions: Synergy of person-and variable-centered approaches to theoretical models of self-concept. *Structural Equation Modeling, 16*, 191–225. https://doi.org/10.1080/10705510902751010

McHugh, M. L. (2013). The chi-square test of independence. *Biochem Med, 23*(2), 143–149. https://doi.org/10.11613/bm.2013.018

OECD, Publishing. (2009). *PISA Data Analysis Manual SPSS: SPSS* (2nd ed.). Organization for Economic Cooperation & Development.

Pellegrino, J., Chudowsky, N., & Glaser, R. (2001). *Knowing what students know: The science and design of educational assessment*. National Academies Press.

Pérez, A., Dennis, R. J., Gil, J. F. A., Rondón, M. A., & López, A. (2002). Use of the mean, hot deck and multiple imputation techniques to predict outcome in intensive care unit patients in Colombia. *Statistics in Medicine, 21*(24), 3885–3896. https://doi.org/10.1002/sim.1391

Peteranderl, S., & Edelsbrunner, P. A. (2020). The predictive value of the understanding of inconclusiveness and confounding for later mastery of the control-of-variables strategy. *Frontiers in Psychology, 11*, 531565. https://doi.org/10.3389/fpsyg.2020.531565

Rasch, G. (1960). *Probabilistic models for some intelligence and attainment tests* (p. 1980). Nielsen & Lydiche (Expanded Edition, Chicago, University of Chicago Press.

Robitzsch, A., Kiefer, T., & Wu, M. (2022). *TAM: Test Analysis Modules*. R package version 4.0–16, https://CRAN.R-project.org/package=TAM. Accessed: 06/07/2022.

Rost, J. (1988). Rating scale analysis with latent class models. *Psychometrika, 53*(3), 327–348. https://doi.org/10.1007/BF02294216

Rost, J. (2004). *Lehrbuch Testtheorie – Testkonstruktion. [Textbook test theory and test construction]*. Huber.

Rubin, D. B. (1987). *Multiple imputation for nonresponse in surveys. Wiley series in probability and mathematical statistics Applied probability and statistics*. Wiley. https://doi.org/10.1002/9780470316696

Schwartz, G. (1978). Estimating the dimension of a model. *The Annals of Statistics, 6*(2), 461–464. http://www.jstor.org/stable/2958889

Schwichow, M., Christoph, S., Boone, W. J., & Härtig, H. (2016). The impact of sub-skills and item content on students' skills with regard to the control-of-variables-strategy. *International Journal of Science Education, 38*(2), 216–237. https://doi.org/10.1080/09500693.2015.1137651

Schwichow, M., Brandenburger, M., & Wilbers, J. (2022). Analysis of experimental design errors in elementary school: How do students identify, interpret, and justify controlled and confounded experiments? *International Journal of Science Education*. https://doi.org/10.1080/09500693.2021.2015544

Sclove, S. L. (1987). Application of model-selection criteria to some problems in multivariate analysis. *Psychometrika, 52*(3), 333–343. https://doi.org/10.1007/BF02294360

Sick, J. (2010). Rasch measurement in language education Part 5: Assumptions and requirements of Rasch measurement. In *Shiken: JALT Testing & Evaluation SIG Newsletter, 14*(2) October 2010. 23–29. https://hosted.jalt.org/test/sic_5.htm. Accessed: 28/07/2022.

Siler, S. A., & Klahr, D. (2012). Detecting, classifying and remediating: Children's explicit and implicit misconceptions about experimental design. In R. W. Proctor & E. J. Capaldi (Eds.), *Psychology of science: Implicit and explicit processes* (pp. 137–180). Oxford University Press. https://doi.org/10.1093/acprof:oso/9780199753628.003.0007

Tschirgi, J. E. (1980). Sensible reasoning: A hypothesis about hypotheses. *Child Development, 51*(11), 1–10. https://doi.org/10.2307/1129583

Vincent-Ruz, P., & Schunn, C. D. (2021). Identity complexes and science identity in early secondary: Mono-topical or in combination with other topical identities. *Research in Science Education, 51*(S1), 369–390. https://doi.org/10.1007/s11165-019-09882-0

Vosniadou, S. (2019). The development of students' understanding of science. *Frontiers in Education, 4*, Article 32, 2368. https://doi.org/10.3389/feduc.2019.00032

Yen, W. M., & Fitzpatrick, A. R. (2006). Item response theory. In R. L. Brennan (Ed.), *Educational measurement* (4th ed., pp. 111–153). Praeger.

Zarkadis, N., Stamovlasis, D., & Papageorgiou, G. (2021). Effects of three neo piagetian constructs on students' portrayed representations of the atomic structure: A latent class analysis. *European Journal of Physics Education, 12*, 1–14. https://doi.org/10.20308/ejpe.v12i1.306

Martina Brandenburger works at the Department of Physics, University of Education Freiburg. She owns a PhD in Physics education from the University of Education Freiburg. In her research, she focuses on problem solving in Physics and utilize different latent variable methods like Rasch analysis, LCA or structural equation modelling.

Martin Schwichow currently works at the Department of Physics, University of Education Freiburg. He owns a PhD in Physics education from the University of Kiel. His research interests are teaching and measuring students' inquiry skills and particularly students' skills in designing controlled experiments. Thereby, he uses different methods like Rasch analysis, LCA or structural equation modelling.

Chapter 7
A Scientific Approach to Assessment: Rasch Measurement and the Four Building Blocks

Haider Ali Bhatti, Smriti Mehta, Rebecca McNeil, Shih-Ying Yao, and Mark Wilson

Abstract Rasch measurement, despite its many advantages, has not been widely utilized in the social sciences, likely due to diffuse adoption across disciplines and lack of an integrated, central framework. To address the challenge of applying Rasch measurement to science education research, this chapter describes the BEAR Assessment System (BAS; Wilson M, Sloane K, Appl Meas Educ 13(2): 181–208, 2000), a measurement framework built on four *building blocks* (Wilson M, Constructing measures: An item response modeling approach. Routledge, 2005, 2023). These building blocks—Construct Map, Items Design, Outcome Space, and Calibration Model—integrate Rasch measurement into a cohesive framework, forming a cyclical and iterative sequence that emulates the scientific process, making it an especially appropriate tool for assessment in science education. After describing this integrative approach to measurement, the chapter delineates an application of this framework by researchers at the UC Berkeley Evaluation and Assessment Research Center (BEAR) and Stanford University for assessing middle school students' proficiency with *Arguing from Scientific Evidence* (ARG).

7.1 Integrating the Four Building Blocks and the Scientific Process

The goal of science is to build more accurate explanations of the natural world and it is achieved through using empirical evidence to build scientific arguments. In proposing hypotheses, designing studies, collecting data, and interpreting the results,

H. A. Bhatti (✉) · S. Mehta · R. McNeil · M. Wilson
University of California, Berkeley, CA, USA
e-mail: haideralibhatti@berkeley.edu; smriti@berkeley.edu; rebeccamcneil@berkeley.edu; markw@berkeley.edu

S.-Y. Yao
Singapore University of Social Sciences, Singapore, Singapore
e-mail: syyao@suss.edu.sg

X. Liu, W. J. Boone (eds.), *Advances in Applications of Rasch Measurement in Science Education*, Contemporary Trends and Issues in Science Education 57,
https://doi.org/10.1007/978-3-031-28776-3_7

scientists are essentially piecing together arguments in favor of a theory. Scientific argumentation is thus at the core of the scientific enterprise. Scientific arguments involve three components: an idea (a hypothesis or theory), the expectations generated by that idea (frequently called predictions), and the actual observations relevant to those expectations (the evidence; Understanding Science, 2022). These components are always related in the same logical way:

1. If the hypothesis was true, what would be an observable consequence (i.e., what is our expected observation)?
2. What is the empirical observation?
3. Does the observation match the expectation?

Using the same logic, the BEAR Assessment System (BAS; Wilson & Sloane, 2000) offers a scientific approach to measurement that is dynamic and iterative. Just as in science, the emphasis in this approach is on theory development and the use of empirical data to assess whether one's hypothesis is supported. An early, informal example is the measurement of intelligence, which, in conjunction with statistical techniques like factor analysis, allowed researchers to develop, test, and refine theories about the nature of intelligence (e.g., one-factor vs. multiple-factor theories; Mackintosh, 2011).

The primary hypothesis in psychometrics is the validity of the instrument for measuring the construct in question, and it is tested through collecting data and interpreting the findings. Once this empirical analysis has been conducted, the results are then fed into the refinement of both the instrument and the scientific theory underlying the construct, creating a bidirectional relationship between theory and empirical investigation. The integration of Rasch measurement within the four building blocks results in a *scientific approach to assessment* that embodies the scientific process. In this approach, it is critical to differentiate two ideas about how science works. The *traditional scientific method* (TSM) and the *scientific process* are two contrasting, often conflated, ideas. In the context of science education, the scientific process has unfortunately long been associated with textbook representations of a linear, one-size-fits-all scientific method, or TSM. The TSM usually includes simplified, chronological steps:

1. Ask a question
2. Formulate hypothesis
3. Perform experiment
4. Collect data
5. Draw conclusions

Students copy these steps verbatim as they complete "experiments." Moreover, this erroneous association between the scientific process and the TSM is recapitulated throughout the typical structure of peer-reviewed scientific articles in which researchers state the problem, describe methods and materials, lay out the experiment, list results, and discuss conclusions (Padian, 2018). Thus, based on typical representations of science and the structure of scientific reports, students end up viewing the TSM as the only way and the "right" way to do science. This misconceived view of how science works has led to much critique of the TSM throughout science education research (Bauer, 1992; Hodson, 1996; Rudolph, 2005; Windschitl et al., 2008).

Though the TSM offers a standardized and seemingly objective structure, a better portrayal of science is one that represents it as a *process* that is nonlinear, dynamic, and iterative (Understanding Science, 2022). This representation of how science works is explained in *Understanding Science,* a resource developed by the University of California Museum of Paleontology, (https://undsci.berkeley.edu/). We reference this framework because of its widespread applicability to all areas and ages of science education (K-16), particularly when it comes to understanding the nature of science. *Understanding Science* also includes resources for teachers and educators to help elucidate the scientific process in ways that go beyond the TSM. Additionally, *Understanding Science* has been used in numerous contexts—by science education researchers to illustrate and bring students' science stories to life (Arlidge et al., 2017), by teachers as a module for teaching how science works (Price & Perez, 2018) and building the foundations for understanding the nature and process of science (Thanukos et al., 2010), and by scientists for improving public understanding of science (Scotchmoor et al., 2009).

In this chapter, we aim to connect the four building blocks measurement framework (Wilson, 2005, 2023) with portions of the "How Science Works" framework (Fig. 7.1) from *Understanding Science* to emphasize the scientific approach to assessment. The framework represents a nonlinear, dynamic, iterative portrayal of how science works and aligns with our approach to assessment through the four building blocks. Specifically, this chapter will explore the connections between two

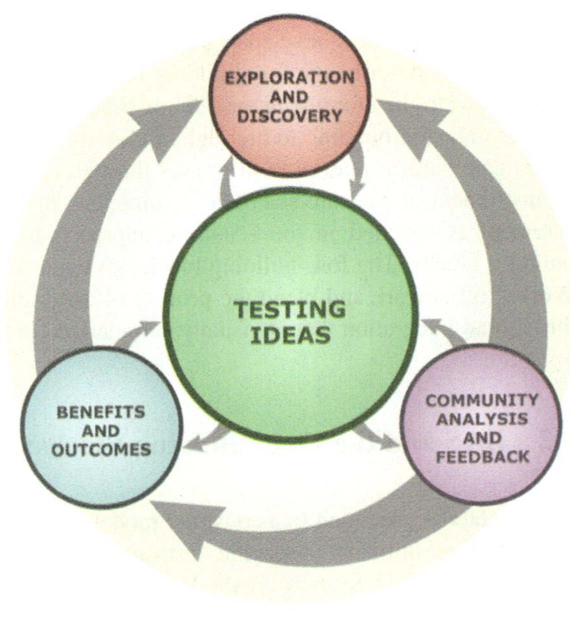

Fig. 7.1 The how science works framework. (Retrieved from https://undsci.berkeley.edu/)

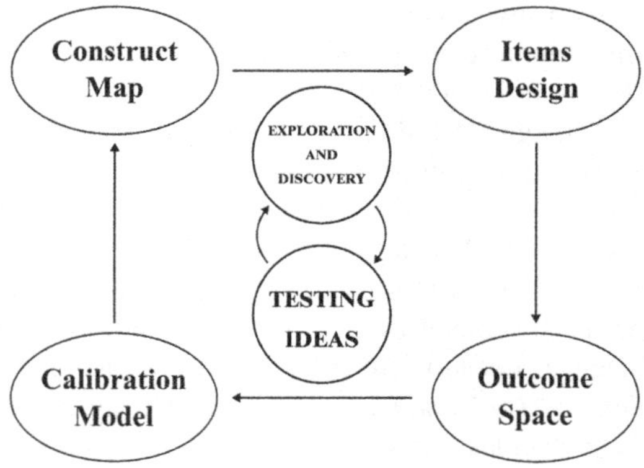

Fig. 7.2 The four building blocks in the BEAR assessment system (BAS) integrated with portions of the scientific process. (Adapted from Wilson, 2005)

portions of the framework—*Exploration and Discovery* and *Testing Ideas*—with the four building blocks.

Though we do not explore the connections between the four building blocks and other portions of the framework (*Benefits and Outcomes* and *Community Analysis and Feedback*), we recognize that these portions also influence how science works. However, for the purposes of this chapter, our scientific approach to assessment will be explained through a series of connections that integrate *Exploration and Discovery* and *Testing Ideas* with the four building blocks. Through these connections, we present a robustly *scientific* view of assessment, with epistemic features that match scientific knowledge such as being testable, revisable, explanatory, conjectural, and generative (Windschitl et al., 2008). Each of these epistemic features is integrated within the four building blocks approach to assessment. Additionally, these epistemic features are in stark contrast to the static, step-by-step, "cookbook" features of the TSM. The process of assessment in science education should match, epistemically and practically, the process of science, and the BAS framework helps achieve that goal.

Below, we describe the four building blocks of assessment within BAS and how each relates to the aforementioned "How Science Works" framework. We then use this theoretical grounding to highlight a case study that uses Rasch measurement to investigate student reasoning processes through the development and validation of an instrument to measure scientific argumentation. This application of Rasch measurement is couched in the scientific approach to assessment through the four building blocks. The four building blocks go hand-in-hand with the "How Science Works" framework and blend the process of assessment with the scientific process through an integration that is visually summarized in Fig. 7.2.

7.2 BAS and the Four Building Blocks: An Overview

This chapter is grounded in a construct modeling approach to measuring psychosocial variables utilizing a specific instrument development framework called the BEAR Assessment System (BAS; Wilson & Sloane 2000; Wilson, 2005, 2023).

The BAS uses four "building blocks" to address the challenges associated with measuring psychosocial variables. The four building blocks are: (a) Construct Map, (b) Items Design, (c) Outcome Space, and (d) Calibration Model. These building blocks are best represented as a cyclical (and often iterative) sequence during assessment development, as shown in Fig. 7.2. Importantly, Fig. 7.2 also centers the *Exploration and Discovery* and *Testing Ideas* portion of "How Science Works" framework within the four building blocks, an integration that will be further explained throughout the chapter.

7.3 ARG Case Study: Using Rasch Measurement to Investigate Student Reasoning Processes

Before discussing each of the four building blocks, it is important to establish the educational context in which this scientific approach to assessment takes place. The publication of the Framework for K-12 Science Education (National Research Council, 2012) and the Next Generation Science Standards (NGSS; NGSS Lead States, 2013) served as a call for reconceptualizing science education. These consensus reports frame science education with respect to three key areas—disciplinary core ideas (DCIs), scientific and engineering practices (SEPs), and cross-cutting concepts across science and engineering. In the following case study, we focus on the work of researchers at the UC Berkeley Evaluation and Assessment Research Center (BEAR) and Stanford University who used Rasch measurement, applied through the four building blocks approach, within an area of middle school science called *Arguing from Scientific Evidence*. As explained above, this scientific approach to assessment represents a robust application of Rasch measurement in the assessment of middle school SEPs, specifically the "Engaging in Argument from Evidence" SEP.

Arguing from scientific evidence requires *student reasoning*, which is a skill used in the construction of arguments and thus, represents an important component of the competency to argue (Osborne et al., 2012; Walton, 1990). Focusing on this component of argumentation competency, Osborne and colleagues adopted the Toulmin model of argument (Toulmin, 2003) as the basis for formulating their hypothesis of the development of reasoning ability. Under this model, there are six key elements for analyzing the pattern of an argument: claim, data, warrant, qualifier, rebuttal, and backing. As a skill, student reasoning can be developed and measured in the context of middle school science, specifically through an intersection of our scientific approach to assessment via the four building blocks and Rasch measurement practices. The research presented here stems from the *Learning Progressions in Middle School Science Instruction and Assessment* (LPS) project centered on the investigation of student reasoning processes in a middle school (grades 6–8) science domain called *Structure of Matter* (SOM). This work was conducted with middle school students from a large school district in the San Francisco Bay Area. More

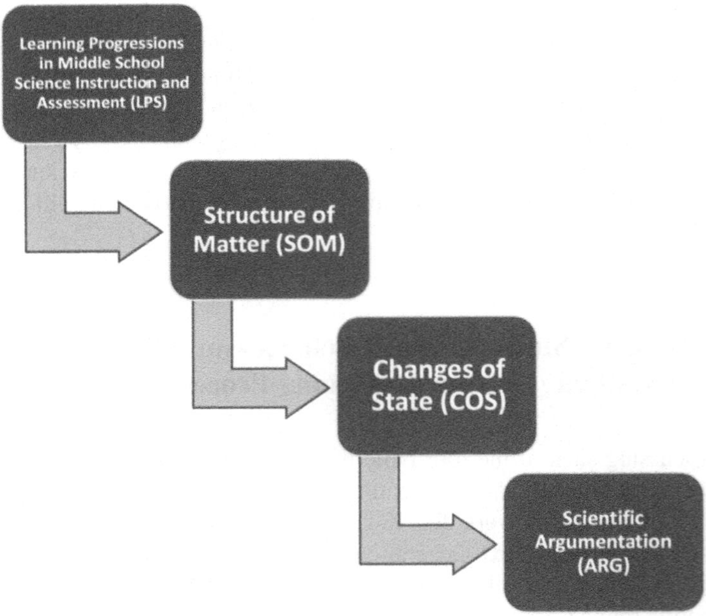

Fig. 7.3 Overview of ARG Case Study Context

specifically, in this case study, researchers focused on a particular scientific argumentation assessment contextualized in the *Changes of State* (COS) subtopic within the larger SOM domain. In other words, the assessment targeted middle school science students' understanding of the physical states of matter and phase changes between those states. Furthermore, within this COS context, researchers measured the *Argumentation* (ARG) construct rooted in students' applying reasoning processes as they engaged in the COS assessment tasks. The hierarchical organization of the ARG case study is summarized in Fig. 7.3. Overall, research from the LPS project utilized each of the four building blocks as means to exemplify the scientific approach to assessment coming to life in the context of an NGSS-aligned middle school science curriculum.

7.4 Building Block 1: Construct Map (*Exploration and Discovery*)

7.4.1 Defining the Construct

The first building block, the **Construct Map**, begins as an idea, a "latent rather than manifest conception" (p. 6, Wilson, 2005), which aligns with the *Exploration and Discovery* portion of the "How Science Works" framework (Understanding Science,

2022). When engaging in this first building block of assessment, the development of a construct map must be based on a clearly defined construct. To define a construct, researchers must make initial observations, ask questions, and explore the literature to develop the first inklings of a construct or attribute to be measured. These initial steps are similar to the early investigative stages of the scientific process. During the exploration and discovery phases of the process, it is crucial to recognize what is already known and what needs to be further explored. As described in *Understanding Science*, "An understanding of what other scientists have already figured out about a particular topic is critical to the process. This background knowledge allows scientists to recognize revealing observations for what they are, to make connections between ideas and observations, and to figure out which questions can be fruitfully tackled with available tools." Similarly, the construct map requires "variable clarification" steps in which the construct to be measured is distinguished from other, closely related, constructs (Wilson, 2005, 2023). In both the first building block and the *Exploration and Discovery* phase, such due diligence validates the novelty of both the assessment and the science to be done in later steps of the respective processes.

In the ARG case study, the first building block is based on the *Argumentation* (ARG) construct, a latent but measurable attribute in students. A review of the literature (i.e., engaging in an *Exploration and Discovery* step of the scientific process) provides a critical point of variable clarification that differentiates "argument" from "explanation." When science students participate in *explanation*, the goal is understanding, specifically by identifying dependent relationships and answering the causal question of why a particular phenomenon occurs (Osborne & Patterson, 2011). In contrast, when science students participate in *argumentation*, the goal is persuasion. In line with this goal, argumentation is defined in the literature as "a complex form of reasoning requiring domain-specific knowledge to construct and critique claims and their relation to any supporting evidence to persuade other members of the community of their validity" (p. 823, Osborne et al., 2016). Overall, student reasoning processes are situated within the larger construct of scientific argumentation and represent an ability for constructing arguments.

7.4.2 Building the Map

After conducting background research and defining the construct or attribute to be measured, the first building block also involves developing the construct into a "map" that represents an underlying continuum of the attribute of interest. This continuum contains the extremes of the attribute, from high to low, with qualitatively distinct "waypoints" between the extremes. For example, the continuum may map a particular skill that ranges from novice to expert, or a particular attitude that ranges from loathes to loves. Later, the quantitative underpinnings of this continuum are established through the other building blocks, but it is initially described qualitatively using waypoints. Thus, the development of this construct map represents an initial hypothesis with a theorized range and waypoints of a construct that can be measured. This process of construct map development is similar to the hypothesis generation that occurs within the scientific process. In the scientific process, a hypothesis is a proposed explanation for a narrow set of phenomena. New hypotheses are formulated through prior experience, background knowledge, preliminary observation, and logic (Understanding Science, 2022). When defining a specific construct of interest and developing a map that displays that construct along a continuum with theorized waypoints, similar procedures to hypothesis generation are undertaken. Overall, the construct of interest represents a theoretical ideal situated within a hypothetical map that is subject to change based on further experimentation. Thus, in a construct map, the "map" is a hypothesis-driven developmental perspective related to the construct.

In the ARG case study, the developmental perspective depicts students' progress from lesser to greater expertise when engaging in scientific argumentation. The construct map for argumentation (Table 7.1) originates from Osborne et al. (2012) and was further elaborated by Yao et al. (2015). It consists of three primary waypoints that map students' ability to construct and critique increasingly complex arguments. The lowest waypoint of the construct map, Waypoint 0 (zero degrees of coordination), maps students who can "identify individual core elements of arguments that are either being constructed or critiqued, but not necessarily *coordinate* such claims, warrants, and evidence together to form complete arguments" (p. 11, Yao et al., 2015). The coordination skill arises in Waypoint 1 (one degree of coordination), which includes the requirement to "coordinate a relationship between a claim and supporting evidence through explication of a warrant" (p. 11, Yao et al., 2015). Lastly, Waypoint 2 (two or more degrees of coordination) "requires the coordination of claim and evidence with explication of no less than two warrants" (p. 11, Yao et al., 2015). Readers can refer to Yao et al. (2015) for a complete description of construct waypoints, including the various sublevels (e.g., 0a, 1b, 2c).

Table 7.1 Hypothetical construct map for argumentation

Waypoint	Constructing arguments	Critiquing arguments
2c	Constructing a counter [argument]	
2b		Identifying relative significance of several pieces of evidence
2a	Making a comparative argument	
1b	Constructing a rebuttal	Identifying weak arguments
1a	Constructing an argument with data or reasons	
0c		Identifying data or a reason supporting a claim
0b		Identifying a claim
0a	Making a claim	

Adapted from Yao et al. (2015)

7.4.3 *Testing the Hypothesis*

It is important to reiterate that this construct map represents a *hypothesis* of proposed waypoints as part of an underlying continuum for the construct. Much like the scientific process, this hypothesis is a proposed explanation for a narrow set of phenomena (i.e., a proposed "map" for a well-defined construct ranging from high to low). As with all scientific hypotheses, the construct map hypothesis must be rigorously tested to determine if students do indeed fall somewhere on the proposed continuum based on their item responses and if so, where. In other words, the construct map must be "tested," much like scientific ideas are tested.

The testable nature of both scientific hypotheses and construct maps is an epistemic hallmark of scientific knowledge generation. In this sense, when something is testable, it is "advanced by proposing new hypotheses that express possible relationships between events, processes, or properties within these models or theories, and by using various domain-specific methods for gathering data aimed at evaluating these hypotheses" (p. 944, Windschitl et al., 2008). The four building blocks approach to assessment requires the construct map to be "tested" in some way, using various domain-specific methods (e.g., construct modeling, Rasch measurement), while the scientific process requires a more general approach to hypothesis testing. In both cases, there is a need for scientific testing and thus, we shift to the next building block, Items Design, and its relationship to the *Testing Ideas* portion of the framework.

7.5 Building Block 2: Items Design (*Testing Ideas*)

Testing Ideas is a critical component of the scientific process. It involves gathering data through experimentation and interpreting how the data link back to the hypothesis. This interpretation of how the data link back to the hypothesis requires an a priori understanding of what one would *expect* to observe if an idea were correct and comparing that expectation to what one *actually* observes (Understanding Science, 2022). In other words, testing ideas requires the application of reasoning about what one would logically expect to observe if a given hypothesis or theory was true by assessing whether the expectation is supported by empirical evidence. Observations that match the expectations generated by an idea are generally interpreted as supporting evidence and mismatches are considered contradictory evidence (Understanding Science, 2022).

These components of testing ideas in the scientific process connect with the second building block, the **Items Design**. This building block allows for the specification of a causal link between the construct underlying the construct map and the responses to items within an instrument. Items provide a way to reveal a particular construct within a respondent. The two-way flow of causality from the construct to the item responses and of inference from the items responses back to

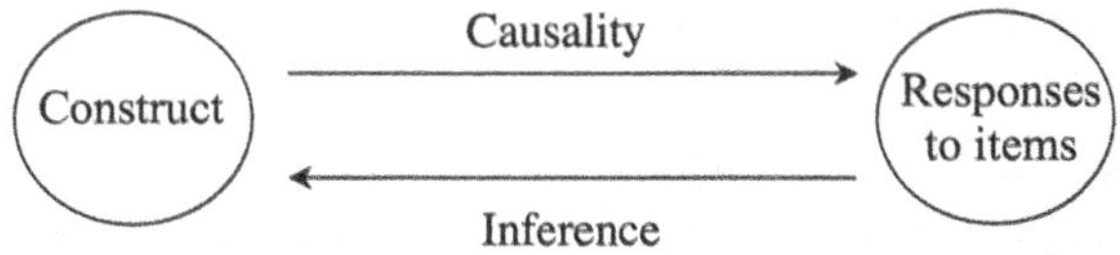

Fig. 7.4 The two-way flow of causality and inference. (Retrieved from Wilson, 2005)

construct is summarized in Fig. 7.4. As stated by Wilson (2005), "the measurer needs to build a structure that links the construct closely to the items—one that brings the inferences as close as possible to the observations" (p. 8). Establishing this linking structure embodies the "design" feature of this building block, which includes designing (1) a procedure that allows observations to be made under a set of standard conditions that span the intended range of the item contexts, and (2) a procedure for classifying those observations into a set of standard categories (Wilson, 2005, 2023).

In this case study, these design procedures result in the development of items that elicit student responses as they engage in scientific argumentation. Based on their responses, students' locations along the construct map for argumentation can be revealed once empirical data have been collected. Comparing response patterns to items with predictions from the hypothesized model (i.e., comparing expected versus actual results in the scientific process) is useful because it helps reveal discrepancies in the "theory," or alternatively, in the item design and implementation (Osborne et al., 2016). Ultimately, the purpose of this building block is to develop a causal link between the construct underlying the construct map and the responses to items within an instrument. This causal link requires the development of items (with a conscious and purposeful design) because items represent a way to reveal a particular construct within a respondent. Establishing such causality between the construct map and the items design emulates the causality that is expected in the scientific process when a hypothesis is tested through experimentation, yielding both expected and actual results.

The causality between the items design and the construct also aligns with the "explanatory" epistemic feature of scientific knowledge. Scientific knowledge generation is explanatory, meaning that it "provides causal accounts of events and processes, as opposed to accumulating descriptive detail about phenomena or merely seeking patterns" (p. 944, Windschitl et al., 2008). Causality in the scientific process comes from testing ideas, while casualty in the four building blocks approach to assessment comes (at least initially) from the link between the construct map and the items design. Additionally, scientific knowledge also has the epistemic feature of being conjectural, meaning that "causal accounts often involve theoretical or unobservable processes that can only be inferred from empirical observation [data] and that scientific argument aims to persuade others that explanations based on these inferences account most adequately for the observations" (p. 944, Windschitl et al., 2008). Previously, we discussed how a construct is a "latent rather than manifest conception" and a theoretical ideal (p. 6, Wilson, 2005). The inference that occurs from the theoretical to the actual (what is observed in respondents) is based on the application of the items design building block.

To summarize, the causal link between the construct map and items represents the manifestation of the once unobservable construct into an observable form, specifically through responses to items designed to measure the latent construct. Based on the responses to items, the measurer can infer the underlying latent construct. With these procedures in mind, in the case study, Yao et al. (2015) designed argumentation items that measured students' reasoning ability as they engaged in the Change of State (COS) topic within the SOM science domain. They developed an assessment that included 20 argumentation items (19 have been retained in the current analysis) bundled into four different COS tasks. All items were open-ended, with the exception of four items that required students to underline different components of arguments (e.g., a claim, data that supports an argument, etc.) directly from passages given in the prompts. More specifically, the four different tasks consisted of grouped items in which hypothetical evidence was presented and accompanied by one or more fictitious arguments purporting to address an overarching question (Yao et al., 2015; see Fig. 7.11 for task 1, *Desalinating Water*). Overall, these items tasked students to critique fictitious arguments and/or construct their own arguments that they believed to be superior within the COS context. These tasks are directly mapped onto waypoints within the previously described argumentation construct map (See Fig. 7.7 for waypoint mapping).

Additionally, as part of the iterative four building blocks cycle of assessment, the argumentation items went through rounds of think-aloud interviews and revisions. Based on this iterative procedure, Yao et al. (2015) developed a scoring guide for each argumentation item. Scoring guides were drafted and trialed with samples of empirical student responses and then revised based on the data. Readers can refer to Yao et al. (2015) for a complete description of the items' design and Fig. 7.11 for the complete *Desalinating Water* item bundle. Figure 7.7 contains the scoring guide for item A1-3e and item level analysis can be found in Appendix B. Overall, the argumentation construct manifests through an instrument that contains argumentation items and the format in which the items are presented to the respondent is the items design. This leads us into the third building block, the Outcome Space.

7.6 Building Block 3: Outcome Space (*Testing Ideas*)

As mentioned in the previous section, the *Testing Ideas* phase of the scientific process requires building a structure that links the construct to the item responses. The procedure that classifies the observed responses into standard categories is called the **Outcome Space**. This procedure involves decisions about what is included in the domain of possible responses and how these responses (known as local values or "sum scores"; Mari et al., 2021) will be scored. As the first step in the inference process, the Outcome Space ties together the hypothesis (the construct map), observation (the responses), and the measurement (the scores or ordered categories). This is part of the *Interpreting Data* stage in the scientific process—once the empirical data has been gathered, it needs to be interpreted to ascertain whether the original hypothesis (i.e., the Construct Map) is supported or refuted. Though data may seem complex and be represented by detailed graphs or complex statistical analyses, fundamentally, they are simply observations. And ideas are supported when observations (i.e., results) match expected observations and are contradicted when they do not.

The Outcome Space establishes the framework that allows the researcher to determine whether the data refutes or supports the idea by establishing categories into which any possible response can be categorized and then setting a criterion for scoring those categories (i.e., relating each category to a specific waypoint on the construct). An Outcome Space results in qualitatively distinct categories that are (a) ordered, (b) well-defined, (c) finite and exhaustive, (d) research-based, and (e) context-specific. For constructs assessed in an achievement context, a research-based model of cognition and learning should be the foundation for the definition of the construct, and hence also for the design of the outcome space and the development of items. For open-ended items, the outcome space is based on a qualitative assessment and categorization of student responses. The outcome spaces for fixed-response items (like Likert-type items) are the responses that are fixed a priori. Ideally, however, the fixed responses would be based on qualitative study of open-ended items first to create the fixed-response options (Wilson, 2005, 2023).

To be considered *well-defined*, the outcome space must come with clear descriptions of the construct and the ordered categories, example items and responses, and training materials (including scoring guides) for free-response items. The training materials are essential for establishing consensus among different judges on the same items so that categories are assigned consistently. The training materials should also include the relevant literature that has led to the development of the construct (*research-based*). The categories must be *ordered* to reflect proficiency on the construct, from least proficient to most proficient. In order to be useful, the categories must also be *specific to each context* to which they are applied, and finally, be limited in number so that any particular item response could be categorized into a specific category (*finite and exhaustive*).

7.6.1 Scoring Guide

Categorization and scoring are linked—once categorizations have been made, quantitative values can be assigned to the different waypoints according to the item's scoring guide. The scoring guide is a specific manifestation of the generalized Outcome Space that establishes clear and explicit performance criteria for a particular assessment. The scoring guides should be supplemented with annotated exemplar responses at each waypoint of every item. In measurement, as in science, the process of creating an outcome space and scoring guides is cyclical and initial iterations of this process will ideally lead to adaptation of the ordered categories and scoring systems.

There are multiple approaches to constructing an Outcome Space, and any appropriate approach is highly context-dependent. Nevertheless, owing to its flexibility and the possibility of applying it to many different contexts, an approach that is apt for science education research is the SOLO (Structure of the Observed Learning Outcome) taxonomy (Biggs & Collis, 1982). This framework allows for classification of learning outcomes in terms of complexity, thus allowing an assessment of students' work in terms of quality (the complexity of the reasoning process) and not quantity (the amount they got right). Additionally, it has already been applied to many science education contexts and the multistructural response (Figs. 7.5 and 7.6), due to multiple ways it can manifest, has been extended within the BAS framework (Wilson, 2005, 2023).

In summary, the Outcome Space is the first step in assessing the overlap between expectations and observations. A match between expectations and observations lends support to the underlying theory, whereas a mismatch potentially refutes

An *extended abstract* response is one that not only includes all relevant pieces of information, but extends the response to integrate relevant pieces of information not in the stimulus.

A *relational* response integrates all relevant pieces of information from the stimulus.

A *multistructural* response is one that responds to several relevant pieces of information from the stimulus.

A *unistructural* response is one that responds to only one relevant piece of information from the stimulus.

A *pre-structural* response is one that consists only of irrelevant information.

Fig. 7.5 The SOLO taxonomy. (Adapted from Wilson, 2005)

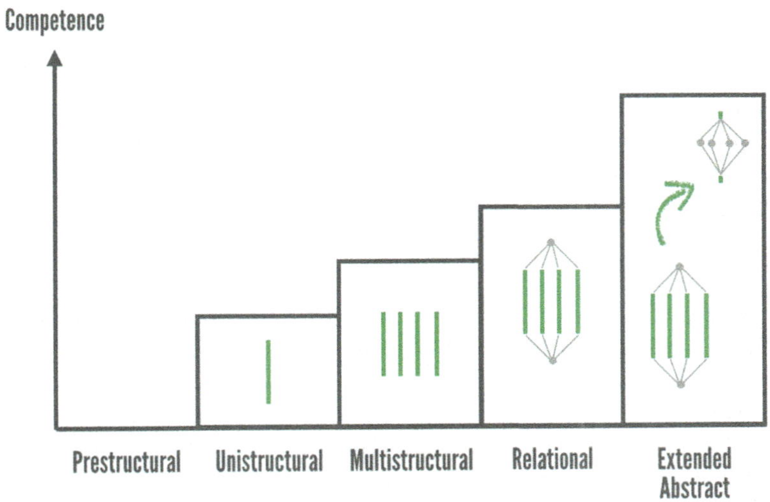

Fig. 7.6 Pictorial depiction of the SOLO taxonomy. (Diagram by Doug Belshaw, CC0)

it. The Outcome Space is a generalized version of the *scoring guide*, which is specific to individual items and is created by identifying the exhaustive categories for each item and aligning them to a waypoint on the construct map.

7.6.2 The ARG Outcome Space

The scientific enterprise aims to provide more accurate explanations of natural phenomena through the use of empirical evidence in service of building scientific arguments. Scientific argumentation begins with an idea or a hypothesis from which expected observations are derived. These expectations are then weighed against empirical observations made under specific conditions. If the idea/hypothesis generates expectations that hold true (are observed), then the idea is more likely to be accurate. If the idea generates expectations that don't hold true (are not observed), then we are less likely to accept the idea. Though the structure of this argument is consistent (hypothesis, then expectation, then actual observation), its pieces may be assembled in different orders. That is, one may come up with an idea (hypothesis) about what is true in the natural world *after* some observation (evidence).

The Scientific Argumentation (ARG) learning progression consists of three waypoints that are distinguished by an increasing degree of sophistication of student understanding, with each waypoint requiring greater coordination between claims and pieces of evidence (Osborne et al., 2016). Items intended to assess the various waypoints of each learning progression have been developed, tested, and validated using the BAS approach, which includes the application of the partial credit model (Masters, 1982), the focus in the current analysis, the Multidimensional Random Coefficients Multinomial Logit Model (MRCMLM; Adams et al., 1997), and Differential Item Functioning (DIF) analysis. The ARG progression was first validated in the context of physical science tasks (Osborne et al., 2016).

Scoring Guide To differentiate students' relative proficiency with scientific argumentation, an iterative procedure was employed to develop a scoring guide for each item. Initial scoring guides were drafted to describe and characterize expected student responses with respect to proficiency levels they demonstrated. These scoring guides were then used to score a sample of empirical student responses from pilot trials for each item. Scoring guides that were unclear or difficult to use were revised accordingly.

As an example, Fig. 7.7 presents the scoring guide for item A1-3e in the *Desalinating Water* item bundle (A1). This item asks students to select the best method for desalinating water and provide a justification for their choice. Each argumentation item is designed to assess one of the three main waypoints, i.e., 0, 1, or 2, in the hypothesized scientific argumentation construct. Item A1-3e was developed to assess waypoint 2 ability in the hypothesized argumentation construct: to identify evidence that contradicts a claim and build an alternative argument about why the claim may be flawed. Partial credit was given to responses where reasoning

Score	Descriptions	Empirical Examples	Waypoint
2	Claim + reason provided in the question prompt. *A good answer would state which they have chosen and provide a relevant reason – that is provide claim plus relevant reason*	41004: "I think the best method is to heat seawater so that it evaporates leaving the unwanted salt behind, because even if this method seems to use more energy, it doesn't matter, because just because method 2 uses less energy, you have to do more work to clean and take away." 41001: "I think that the second method would be the best way, it takes less energy but it has its cons too." 41003: "I think the membrane is the best method because it uses half the energy and you can probably turn the sludge into some kind of alternative fuel." 'I think the membrane method is best because although to have to manually dispose of sludge and quite often, you use less amounts of energy to desalinate water.	1a: Constructing an argument with data or reasons
1	Claim + reason not provided in the question prompt *Here they provide claim but the reason is not in the information provided in the question prompt*	40412: "Using membranes is more efficient than distilling water" (The kind of efficiency is not specified...energy efficiency? Time efficiency?) 41009: "I think the first method is better when you can obtain energy from solar panels."	0b: Identifying data or a reason supporting a claim
		41016: "I think using the one that takes energy is better because you can have artificial energy or sun energy." 40403: "Evaporation, I don't like to clean things."	
0	Claim only. *The student makes a claim without any justification or very irrelevant justification*	41005: "The best method." 41007: "To leave it there." 41006: "Both."	0a: Making a claim

Fig. 7.7 Scoring guide of item A1-3e in the desalinating water item bundle. (Adapted from Yao et al. 2015)

may not have been entirely correct. Responses with full credit, i.e., a score of 2 in the current example, are regarded as fully demonstrating the proficiency level an item is designed to assess, i.e., Waypoint 2 in this example. Alternatively, responses with partial credit, i.e., a score of 1 in the current example, are considered as demonstrating partial proficiency. The same logic applies to the scoring of all items.

7.7 Building Block 4: Calibration Model (*Testing Ideas*)

The next step involves relating the scored responses to the construct map by translating the scores into "locations" on the construct's continuum (Wilson, 2005, 2023). This part of the inferential process, the last building block, is the **Calibration Model**, and the Rasch model is a special case of the calibration model. The Rasch model is evaluated at both the item and the instrument level and models the *probability* of the observed responses. In this approach, individual response data is used to calibrate items, evaluate the empirical characteristics of the instrument, and assess how well the data fit the conceptual model. The central steps in the calibration model include item-level analysis, checking fit, and assessing the validity and reliability of the instrument.

In the Rasch model, the "amount" of construct that the items and students have work with each other such that the responses of the persons to the items depend on the *difference* between a student's ability (the amount of construct the student has) and the difficulty of the item (the amount of construct the item has). When the item and student have an equal amount of the construct, the probability of a correct response on that item is 0.5. When the student's ability is higher than the item difficulty, the probability of a correct response is higher than 0.5, and conversely, if the item has a higher amount of the construct (is more difficult) than the student, the probability of a correct response is lower than 0.5.

The simple Rasch model can thus be written as an equation for an individual's response, X_i (1 = correct, 0 = incorrect), on item i of a k-item test where the probability of a correct response ($X_i = 1$) is:

$$\text{Probability } (X_i = 1|\theta\ \delta_i) = \frac{\exp(\theta - \delta_i)}{1 + \exp(\theta - \delta_i)}$$

where θ is the student's ability level and δ_i is the difficulty of item i. The relationship between the probability of getting an item correct ($X_i = 1$) and the location on the construct map can be depicted graphically (Fig. 7.8).

For the model estimates to be interpretable, there needs to be evidence that the data approximates what is required by the model. These requirements include an assumption of unidimensionality (a single continuous latent construct), local independence, person-invariant item estimates, and item-invariant person estimates. The assumption of local or conditional independence is an important one as it implies nonnegative covariances between all monotone increasing functions of a set of item responses given any function of the remaining item responses (Bell et al., 1988; Rosenbaum, 1984).

$$P(X_1 = x_1, X_2 = x_2, \ldots, X_k = x_k|\theta) = \prod_i P(X_i = x_i|\theta)$$

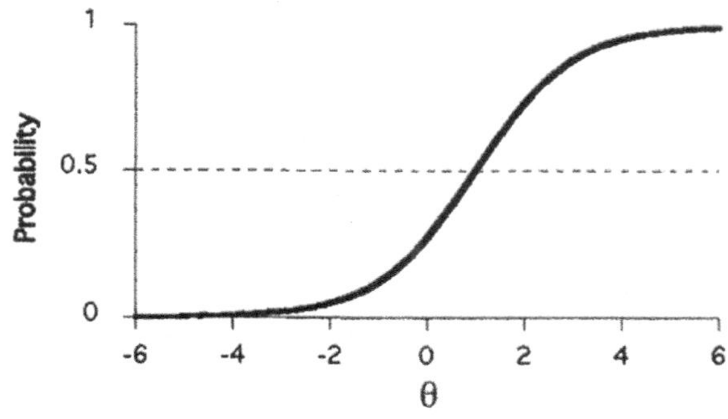

Fig. 7.8 Relationship between student ability, θ, and probability of a score of 1 for an item with difficulty, δ_i, 1. (Retrieved from Wilson, 2005)

One of the primary benefits of the Calibration Model is that it yields—provided the model fits the data well—a "scale," akin to a ruler, that allows researchers to compare items and persons on the same metric. This approach also makes it possible to graphically represent the "locations" of the items and respondents on either side of a single continuum according to their *estimated* location on the construct, thus placing both the items and respondents and their relative positions in a single visual display. This person-item map is also known as the Wright Map (named after a major proponent of Rasch measurement, Ben Wright; see below for an example in Fig. 7.9) and represents an empirical manifestation of the Construct Map.

The Wright Map displays Rasch-Thurstone thresholds (Linacre, 2010; Wu et al., 1998). Results are plotted by item and provide a visual representation of the relative difficulties of items aligned with the various waypoints of the learning progressions, thus allowing investigation of the hypothesized waypoints as "bands" in the Wright map. These bands allow educators to see which items are nearest to students' current ability levels. This, in turn, enables educators to assess their students' zones of proximal development (Vygotsky, 1978), which are the "sweet spots" between students' ability when working independently versus with the guidance of more capable peers or adults. This pedagogical tactic afforded by the BAS approach allows educators to provide important scaffolding by selecting tasks that may be challenging, but not too difficult, for students to complete.

7.7.1 *The ARG Calibration Model*

Model Formulation Since the data are obtained from a mix of dichotomous and polytomous items, the Rasch Partial Credit Model (Masters, 1982) was used to evaluate the measurement properties of the instrument and of each individual item. Using the Rasch Partial Credit Model, the probability of student p scoring x on item i can be formulated as follows:

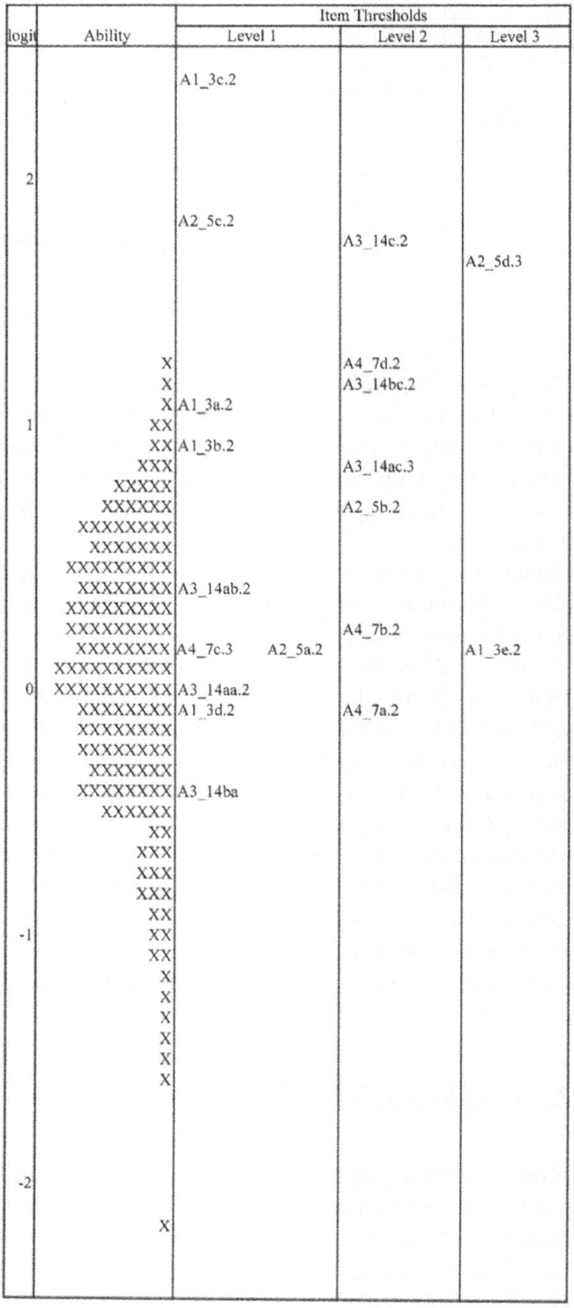

Fig. 7.9 Wright map with item thresholds of the full-credit response categories

$$P_{pix}(\theta) = \frac{exp\left[\sum_{j=0}^{x}\left(\theta_p - \delta_i - \tau_{ij}\right)\right]}{\sum_{r=0}^{m_i}\left[exp\sum_{j=0}^{r}\left(\theta_p - \delta_i - \tau_{ij}\right)\right]}, x = 0, \ldots m_i,$$

$$\text{where} \sum_{j=0}^{0}\left(\theta_p - \delta_i - \tau_{ij}\right) = 0$$

where θ stands for the latent ability of student p, δ represents the overall difficulty of item i, and τ refers to the additional step parameter associated with category j of item i.

The partial credit model was estimated using the software ConQuest (Adams et al., 2020) using the marginal maximum likelihood estimator (Gauss-Hermite Quadrature with 15 nodes), where ability parameter θ_p is assumed to be normally distributed with $N\left(\mu_\theta, \sigma_\theta^2\right)$. Item parameters, δ_i and τ_{ij}, are assumed to be fixed parameters. For model identification, the mean of the ability distribution is fixed as 0 and $\sum_j \tau_{ij} = 0$ for each item i. For ConQuest command codes, see Appendix C.

7.8 ARG Data

7.8.1 Instrument

The data for this case study comes from 19 argumentation items embedded in four tasks related to the Changes of State topic in grade eight physical science. Table 7.2 provides a summary of the information about the argumentation items. Figure 7.11 presents the A1 (*Desalinating Water*) argumentation item bundle used in the current study. Appendix B contains the individual item analysis output for item A1-3e from *Desalinating Water* Item Bundle. The 19 argumentation items were contextualized in four tasks (passages), each of which was followed by a set of items; all were open-ended except for four items that asked students to underline different components of arguments in the passages. Partial credit was given for most of the argumentation items, and thus the original data are a mix of dichotomous and polytomous items. As part of the iterative cycle for improving the validity of the instruments, some items were dropped from the currently presented analysis after review of empirical data due to misfit (Yao, 2013; Yao et al., 2015).

7.8.2 Sample

The data were collected from 347 grade eight students (54.97% female) from a local school district. Non-responses were coded as inaccurate responses. According to the results of the 2011 California Standards Test in Science (California Department of

Table 7.2 Information about argumentation items and corresponding waypoints

Item #	Bundle	Item name	Waypoint
1	A1.	A1-3a	1
2	*Desalinating water*	A1-3b	1
3		A1-3c	1
4		A1-3d	1
5		**A1-3e**	**3**
6	A2.	A2-5a*	1
7	*Sugar in the water*	A2-5b*	2
8		A2-5c*	1
9		A2-5d	3
10	A3.	A3-14aa	1
11	*Ice to water vapor*	A3-14ab	1
12		A3-14ac	2
13		A3-14ba	1
14		A3-14bc	2
15		A3-14c	2
16	A4.	A4-7a	2
17	*Something in the air*	A4-7b	2
18		A4-7c*	1
19		A4-7d	2

Note. Items with * are those with two parts: the first part asks students to select one of the choices given in the item prompt and the second part asks students to provide an explanation

Education, 2013), 195 of these students (57.02%) were classified into the "Advanced" category, 61 students (17.84%) were in the "Proficient" category, 57 students (16.67%) were in the "Basic" category, 20 students (5.85%) were in the "Below Basic" category, and nine students (2.63%) were in the "Far Below Basic" category.

7.8.3 *Validity and Reliability*

Among other psychometric properties, evidence on the validity of the assessment of scientific argumentation is of special interest. Validity refers to the degree to which the instrument measures the construct it was designed to measure, and thus leads to valid scientific inferences (AERA, APA, & NCME, 2014). In other words, validating an instrument involves gathering evidence that provides a scientific basis for interpreting the scores in the proposed way. It is also important to note that the concept of validity is embedded within a context, which in the current case is middle school science education and the purpose of the items to assess the level of sophistication students have reached in terms of their argumentation ability. The items were designed to measure a unidimensional construct, which has been supported by earlier empirical investigations (Yao, 2013; Yao et al., 2015).

However, the empirical pattern of item thresholds may not match completely with the hypothesized waypoints (see Wright Map section below).

Psychometric properties, such as test reliability and item discrimination, were also examined to evaluate the function of the items. Cronbach's Alpha was estimated as 0.73. The weighted likelihood estimate (WLE) person separation reliability was estimated as 0.77, and the Expected A Posteriori (EAP) reliability was estimated as 0.75. This suggests reasonable internal consistency among the items.

Item Discrimination The weighted mean square (MNSQ) statistic was used to evaluate the fit between the partial credit model and data on each item, which measures the degree of discrepancy between the expected and observed item scores. The weighted mean square statistic has an expected value of one. Items with values of the weighted mean square statistic larger than one are those with observed variance greater than the expected and contribute less toward the overall estimation of the latent variables. For the MNSQ fit statistics we provide a 95% confidence interval for the expected value of the MNSQ (value of 1 under the null hypothesis). If the MNSQ fit statistic lies outside that interval, we reject the null that the data conforms to the model. Conventionally, the weighted t statistic is used to test the statistical significance of the mean square (Wright & Masters, 1982). Items that have a weighted mean square statistic outside the range of 0.75 and 1.33 and have the absolute value of the weighted t statistic larger than 2 should be considered problematic (Wilson, 2005, p. 129). All the investigated argumentation items have weighted mean square statistics within the range of 0.75 and 1.33, which provides support for the use of the partial credit model. Table 7.3 below presents the item difficulties and mean square estimates for the argumentation items. Only one item, A2-5d, has a weighted t-statistic greater than 2 the weighted mean square statistic of this item is 1.19, which is close to the upper boundary of the acceptable range (i.e., 1.33). When the value of the weighted mean square statistic is larger than one, it suggests that the observed residual variance is greater than the expected. This is the only argumentation item that assessed students' ability to recognize uncertainty in the evidence. More items of this kind are needed in the future to investigate whether recognizing uncertainty is indeed more sophisticated than providing reasoning to back a claim as hypothesized.

7.8.4 *Wright Map*

As previously noted, an advantage of the BAS approach to studying learning progressions is the affordance of communicating findings using Wright maps (Wilson, 2005, 2023). Figure 7.9 presents a Wright map, which displays the student ability distribution and item thresholds on the same logit scale. The second column in the figure represents the distribution of the estimated student ability from most able at the top to least able at the bottom of the map, with mean equal to 0 and variance equal to 0.37. Each "x" represents approximately 0.6 students and the location of "x"

Table 7.3 Delta (difficulty) estimates and weighted MNSQ statistics

Item	Item name	□	MNSQ	CI	*t*
1	A1-3a	−0.048	1.05	(0.87, 1.13)	0.8
2	A1-3b	0.747	1.07	(0.87, 1.13)	1.0
3	A1-3c	1.308	0.94	(0.85, 1.15)	−0.7
4	A1-3d	−0.726	0.89	(0.87, 1.13)	−1.7
5	**A1-3e**	**−0.551**	**0.99**	**(0.88, 1.12)**	**−0.2**
6	A2-5a*	−0.653	0.91	(0.88, 1.12)	−1.5
7	A2-5b*	−0.432	0.91	(0.88, 1.12)	−1.5
8	A2-5c	0.350	0.97	(0.89, 1.11)	−0.5
9	A2-5d	0.695	1.19	(0.87, 1.13)	2.9
10	A3-14aa	−0.348	0.97	(0.88, 1.12)	−0.4
11	A3-14ab	−0.298	1.08	(0.88, 1.12)	1.3
12	A3-14 ac	0.315	0.98	(0.88, 1.12)	−0.4
13	A3-14ba	−0.938	1.04	(0.88, 1.12)	0.6
14	A3-14bc	0.434	0.91	(0.88, 1.12)	−1.5
15	A3-14c	1.509	1.00	(0.70, 1.30)	0.0
16	A4-7a	−0.649	1.08	(0.87, 1.13)	1.2
17	A4-7b	−0.654	1.07	(0.88, 1.12)	1.1
18	A4-7c	−1.654	0.90	(0.81, 1.19)	−1.0
19	A4-7d	−0.087	0.99	(0.87, 1.13)	−0.2

Note. The individual item analysis output for Item A1-3e (Desalinating Water), bolded above, is available in Appendix B

indicates the level of ability estimate. The third to the fifth columns present the Thurstone thresholds of the full-credit response category for each item (Wilson, 2005, 2023). The j_{th} Thurstonian threshold represents the point where the probability of scoring below j is equal to the probability of scoring j and above. In the current case, all item thresholds shown on the Wright map are the full-credit Thurstone thresholds. If the ability estimate of a student and the full-credit Thurstone threshold of an item are located at the same position on the map, the probability of this student obtaining full credit on the item is 0.5. More information about a student's proficiency can be obtained if locations of student ability and item thresholds are close. If items are too easy or too difficult for the student, very little information is gained. As a result, an ideal measurement situation is that the item threshold distribution covers the span of the student ability distribution so that appropriate ability estimates for students of all performance levels can be obtained. The current map suggests that there is a lack of items at the lower end and easier items are needed to avoid the issue of a floor effect.

Each item is hypothesized to measure one specific learning progression level. Items measuring a lower progression level are expected to have lower difficulty levels, compared to items measuring a higher progression level. Ideally, item thresholds tapping the same progression level will locate at similar positions on the Wright map and form a band, which represents the difficulty level of the progression level. The empirical ordering of the progression waypoints can then be compared to the theoretical ordering of the hypothesized learning progression. Figure 7.9 indicates that the expected structure is not present.

Figure 7.10 presents the banding of the Wright map based on the smaller set of items, where the four items with small factor loadings are excluded. Although not perfect, an ordinal structure among locations of item thresholds associated with

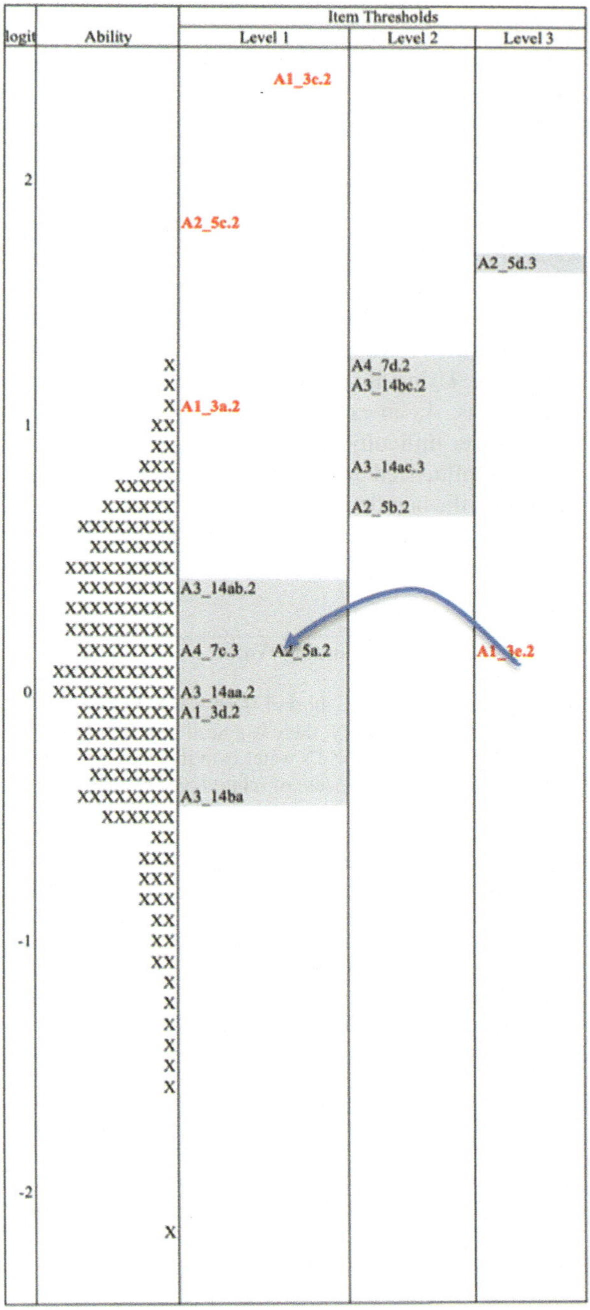

Fig. 7.10 Wright map with item thresholds of the full-credit response categories (4 items removed)

different progression levels emerges in Fig. 7.10. This structure could be made clearer by reclassifying item A1-3e from Waypoint 3 to Waypoint 1. For Waypoint 3, item A1-3e is surprisingly easy. Item A1-3e asks students to decide "*using only the information given above, which method do you think is the best one? Explain why you think this is the best method.*" It was hypothesized that this item assessed a student's ability to evaluate and compare competing arguments. However, according to the scoring guide, students get full credit if they successfully give a claim, i.e., choose one of the two desalination methods, and a reason provided in the prompt. Cognitively, succeeding on this item thus only requires the ability to identify a claim and a relevant piece of supporting evidence. We would therefore re-classify item A1-3e as Waypoint 1. This item is a good example of how the way an item is written may affect the cognitive demand it requires. Also, more Waypoint-3 items will be needed for a further examination of the "banding" and average difficulty level of the Waypoint-3 items.

This (lack of) evidence for the hypothesized structure denotes that some items did not perform the way they were intended to measure the latent argumentation proficiency. Using the empirical data, we can speculate about the items that violated expectations. As an example, item A1-3c was a Waypoint 1 item, and thus should have a lower difficulty level. However, a closer examination reveals that language may have influenced the difficulty level of the item. To illustrate, Fig. 7.11 shows the scenario of the bundle A1, along with all the items. Bundle A1 establishes a scenario with three paragraphs: the first paragraph presents the issue of California being short of fresh water, and each of the remaining two paragraphs presents a method of

Item Bundle A1. Desalinating Water

California is increasingly short of fresh water. Lack of freshwater limits the amount of food that can be grown. Clearly, there is a need to better manage this increasingly valuable resource. 97% of the world's water is in the oceans. The trouble is that it is salty and cannot be drunk. The process of removing salt from water is called *desalination*.

One method of desalination is to use energy to heat seawater so that it evaporates leaving the unwanted salt behind. The water vapor is then passed over a cold object where it turns back to pure water. A great deal of energy is lost during this process.

Another method involves pushing the seawater through a sheet with tiny holes in it called a membrane. Only the water particles pass through leaving all the salt on the other side. This uses half the energy but the membranes get clogged and have to be cleaned regularly. All the sludge produced then has to be taken away.

3a. What is the main claim that is being made in this piece?
3b. What are two pieces of data that are used to support the argument for desalination?
3c. What is the main argument against the standard method of desalinating water?
3d. What is the main argument against using membranes?
3e. Using only the information given above, which method do you think is the best one? Explain why you think this is the best method.

Fig. 7.11 Desalinating water item bundle A1

desalination to obtain fresh water from the oceans. Hypothetically, items A1-3c and A1-3d are expected to have similar difficulty levels, since they both ask students to identify an argument against a particular method of desalination. However, item A1-3c appears to be much more difficult than item A1-3d in the empirical data. One possible reason is the ambiguity of the language. Item A1-3d indicates clearly that the method students should identify an argument against is the one "using membranes," which helps students to locate the critical paragraph, i.e., the third paragraph, correctly. In contrast, item A1-3c refers to the method students should identify an argument against "the standard method of desalination," the corresponding paragraph of which is the second paragraph. Throughout the text, however, "the standard method of desalination," is not specified explicitly.

Evaluation of Item A1-3a (Fig. 7.11) provides another implication on the assessment of scientific reasoning. According to the scoring guide, students got full credit only if their responses referred to both (a) water shortage and (b) how the water shortage is to be addressed. 55% of the students mentioned only one of the two pieces in their responses and were awarded partial credit. These pieces of evidence suggest that identifying parts of a claim may be easy as hypothesized yet it is difficult to identify all pieces in a given claim. This finding calls for reflection on two general assessment issues: (a) what level of sophistication of student responses should be sufficient to be considered as demonstrating performance at a particular learning progression level, and (b) how item design can be improved to prompt students to demonstrate their understanding and proficiency level to the fullest.

7.9 Discussion

In this chapter, we presented the development and validation of a measurement instrument for assessing scientific argumentation using the four building blocks within the BAS framework. This development and validation process aligns with the "How Science Works" framework, namely *Exploration and Discovery* and *Testing Ideas*. Additionally, the epistemic features of scientific knowledge generation—testable, explanatory, conjectural, revisable, and generative (Windschitl et al., 2008)—are similarly embodied within this scientific approach to assessment. Importantly, we wish to emphasize that two epistemic features—revisable and generative—are especially relevant when considering the cyclical and iterative nature of the four building blocks and the "How Science Works" framework. As a process that involves revision, assessment methods (e.g., Items Design) can and should change in response to new evidence. And as a process that is generative, the four building blocks approach generates new ideas, hypotheses, and predictions for testing, rather than static "end-products" of inquiry (Windschitl et al., 2008).

The ARG case study in this chapter was based on research aimed at fulfilling the need for an explicit emphasis on scientific argumentation in science classrooms. Argumentation is an essential practice within science, but challenges about how to support and evaluate students' scientific argumentation proficiency in the classroom persist. Without valid argumentation assessments, science educators will not be able to readily diagnose their students' learning needs and provide instructional interventions. As a consequence, argumentation may not be adequately incorporated into science classrooms.

The motivation to develop an argumentation assessment also comes from the recognition that "Engaging in Argument from Evidence" is an explicit SEP within the NGSS. Just as scientific content knowledge in the form of disciplinary core ideas (DCIs) can be developed through validated learning progressions, SEPs can also develop through learning progressions built on valid assessments (Osborne et al., 2016). The ARG case study demonstrated the development and initial validation of a hypothesized construct map based on students' ability to engage in various argumentation tasks. Considering the revisable and generative nature of the assessment process, later work by Osborne et al. (2016) advanced the initial construct by Yao et al. (2015) into a more robust learning progression consisting of several construct maps linked together (see Table 1 in Osborne et al., 2016). This advancement demonstrates how a single construct map representing a unidimensional cognitive model that lays out students' proficiency levels on one single construct can develop into a more robust learning progression, or a multidimensional conceptual framework that depicts the relationships of two or more construct maps (Wilson, 2009; Yao et al., 2015).

7.10 Conclusion

Before scientists can observe patterns in nature, they need accurate instruments for observation. Valid and reliable measurement is thus the starting point of the scientific enterprise. And the starting point of measurement is an initial theory about the phenomenon being measured. Within the integrated Rasch-BAS framework, this theory is treated like a scientific hypothesis in its own right (The Construct Map), leading to the generation of an instrument (Items Design), a hypothesized structure (The Outcome Space), and a test of whether empirical data generated by the instrument supports that hypothesis (The Calibration Model). Just as in science, the empirical data sheds light on whether the empirical observations match what was hypothesized, and ideally results in updating the current theory and generating new hypotheses for testing. This integration of Rasch Measurement with the scientific BAS approach provides an especially suitable method for assessment in science education research.

Appendices

Appendix A: Argumentation Items

Changes of State Desalinating Water Item Bundle

Desalinating Water (A1)

California is increasingly short of fresh water. Lack of freshwater limits the amount of food that can be grown. Clearly, there is a need to better manage this increasingly valuable resource. 97% of the world's water is in the oceans. The trouble is that it is salty and cannot be drunk. The process of removing salt from water is called *desalination*.

One method of desalination is to use energy to heat seawater so that it evaporates leaving the unwanted salt behind. The water vapor is then passed over a cold object where it turns back to pure water. A great deal of energy is lost during this process.

Another method involves pushing the seawater through a sheet with tiny holes in it called a membrane. Only the water particles pass through leaving all the salt on the other side. This uses half the energy but the membranes get clogged and have to be cleaned regularly. All the sludge produced then has to be taken away.

Questions

3a. What is the main claim that is being made in this piece?

3b. What are two pieces of data that are used to support the argument for desalination?

3c. What is the main argument against the standard method of desalinating water?

3d. What is the main argument against using membranes?

3e. Using only the information given above, which method do you think is the best one?

Explain why you think this is the best method. **[See** Fig. 7.7 **for scoring guide]**

Appendix B: Item Analysis Output – Item A1-3e

Changes of State Desalinating Water Item Bundle

Item 5, (A1_3e)

Cases for this item: 335

Item-Rest Cor.: 0.35

Item-Total Cor.: 0.45

Item Threshold(s): −1.14, 0.03

Weighted MNSQ: 0.99

Item Delta(s): −0.77, −0.34

Label	Score	Count	% of tot	Pt Bis	t (p)	PV1Avg:1	PV1 SD:1
0	0.00	63	18.81	−0.43	−8.6(.000)	−0.508	0.574
1	1.00	108	32.24	0.14	2.6(.009)	0.122	0.544
2	2.00	164	48.96	0.20	3.7(.000)	0.161	0.555

Appendix C: ConQuest Codes

```
cos_arg_19_pcm_command.txt

Title Change of State - Argumentation 19 - Partial Credit Model;

datafile cos_arg_19_data.txt;

format id 1-10 responses 11-29;

labels << cos_arg_19_labels.txt;

codes 0,1,2,3;

set constraints = cases;

model item + item*step;

estimate! method = gauss, stderr=quick, nodes=15;

show !estimates=latent >> cos_arg_19_pcm_SHW.txt;
itanal >> cos_arg_19_pcm_ITN.txt;

show cases ! estimates=eap >> cos_arg_19_pcm_EAP.txt;
show cases ! estimates=wle >> cos_arg_19_pcm_WLE.txt;

export parameters >> cos_arg_19_pcm_PRM.txt;
show !tables=1:2:3:4:5:7,estimate=latent >> cos_arg_19_pcm_GIN.txt;

reset;
```

Data & output txt files available upon request.

References

Adams, R. J., Wilson, M., & Wang, W. C. (1997). The multidimensional random coefficients multinomial logit model. *Applied Psychological Measurement, 21*(1), 1–23.

Adams, R. J, Wu, M. L, Cloney, D., & Wilson, M. R. (2020). *ACER ConQuest: Generalised item response modeling software* [Computer software]. Version 5. Australian Council for Educational Research.

American Educational Research Association, American Psychological Association, & National Council on Measurement in Education (Eds.). (2014). *Standards for educational and psychological testing.*. American Educational Research Association.

Arlidge, S. M., Thanukos, A., & Bean, J. R. (2017). Using the understanding science flowchart to illustrate and bring students' science stories to life. *The Bulletin of the Ecological Society of America, 98*(3), 211–226. https://doi.org/10.1002/bes2.1330

Bauer, H. H. (1992). *Scientific literacy and the myth of the scientific method.* University of Illinois Press.

Bell, R. C., Pattison, P. E., & Withers, G. P. (1988). Conditional independence in a clustered item test. *Applied Psychological Measurement, 12*(1), 15–26.

Biggs, J. B., & Collis, K. F. (1982). *Evaluating the quality of learning: The solo taxonomy (Structure of the observed learning outcome).* Academic Press.

California Department of Education. (2013). *STAR CST blueprints.* Retrieved from https://www.cde.ca.gov/re/pr/star13rfl.asp

Hodson, D. (1996). Laboratory work as scientific method: Three decades of confusion and distortion. *Journal of Curriculum Studies, 28*(2), 115–135.

Lead States, N. G. S. S. (2013). *Next generation science standards: For states, by states.* The National Academies Press.

Linacre, J. M. (2010). Transitional categories and usefully disordered thresholds. *Online Educational Research Journal, 1*(3), 1–10.

Mackintosh, N. J. (2011). History of theories and measurement of intelligence. In *The Cambridge handbook of intelligence* (pp. 3–19).

Mari, L., Wilson, M., & Maul, A. (2021). *Measurement across the sciences.* Springer.

Masters, G. N. (1982). A Rasch model for partial credit scoring. *Psychometrika, 47*(2), 149–174.

National Research Council. (2012). *A framework for K-12 science education: Practices, crosscutting concepts, and core ideas.* National Academies Press.

Osborne, J. F., & Patterson, A. (2011). Scientific argument and explanation: A necessary distinction? *Science Education, 95*(4), 627–638.

Osborne, J. F., Henderson, B., MacPherson, A., & Szu, E. (2012, April). *Assessing scientific argumentation by middle school pupils and testing a learning progression for argumentation.* Paper presented at the annual meeting of the American Educational Research Association, Vancouver.

Osborne, J. F., Henderson, J. B., MacPherson, A., Szu, E., Wild, A., & Yao, S. Y. (2016). The development and validation of a learning progression for argumentation in science. *Journal of Research in Science Teaching, 53*(6), 821–846.

Padian, K. (2018). Narrative and "Anti-narrative" in science: How scientists tell stories, and don't. *Integrative and Comparative Biology.* https://doi.org/10.1093/icb/icy038

Price, R., & Perez, K. (2018). Research and teaching: Many paths toward discovery: A module for teaching how science works. *Journal of College Science Teaching, 047*(03).

Rosenbaum, P. R. (1984). Testing the conditional independence and monotonicity assumptions of item response theory. *Psychometrika, 49*(3), 425–435.

Rudolph, J. L. (2005). Epistemology for the masses: The origins of "the scientific method" in American schools. *History of Education Quarterly, 45*(3), 341–376.

Scotchmoor, J., Thanukos, A., & Potter, S. (2009). Improving the public understanding of science: New initiatives. *American Journal of Botany, 96*(10), 1760–1766. https://doi.org/10.3732/ajb.0900014

Thanukos, A., Scotchmoor, J. G., Caldwell, R., & Lindberg, D. R. (2010). Science 101: Building the foundations for real understanding. *Science, 330*(6012), 1764–1765. https://doi.org/10.1126/science.1186994

Toulmin, S. E. (2003). *The uses of argument*. Cambridge University Press.

Understanding Science. (2022). *University of California Museum of Paleontology*. 3 January 2022. http://www.understandingscience.org

Vygotsky, L. S. (1978). Zone of proximal development: A new approach. In *Mind in society: The development of higher psychological processes* (pp. 84–91).

Walton, D. N. (1990). What is reasoning? What is an argument? *The Journal of Philosophy, 87*(8), 399–419.

Wilson, M. (2005). *Constructing measures: An item response modeling approach*. Routledge.

Wilson, M. (2009). Measuring progressions: Assessment structures underlying a learning progression. *Journal of Research in Science Teaching, 46*, 716–730.

Wilson, M. (2023). *Constructing measures: An item response modeling approach*. Routledge.

Wilson, M., & Sloane, K. (2000). From principles to practice: An embedded assessment system. *Applied Measurement in Education, 13*(2), 181–208.

Windschitl, M., Thompson, J., & Braaten, M. (2008). Beyond the scientific method: Model-based inquiry as a new paradigm of preference for school science investigations. *Science Education, 92*(5), 941–967.

Wright, B. D., & Masters, G. N. (1982). *Rating scale analysis*. MESA Press.

Wu, M., Adams, R. J., & Wilson, M. (1998). *ACERConQuest* [computer program]. ACER.

Yao, S. Y. (2013). *Investigating the validity of a scientific argumentation assessment using psychometric methods*. (Doctoral dissertation, UC Berkeley).

Yao, S. Y., Wilson, M., Henderson, J. B., & Osborne, J. (2015). Investigating the function of content and argumentation items in a science test: A multidimensional approach. *Journal of Applied Measurement, 16*(2), 171–192.

Haider Ali Bhatti is a PhD candidate in the Graduate Group in Science and Mathematics Education (SESAME) program at the University of California, Berkeley. His research focuses on making undergraduate STEM education more inclusive, interpersonal, and interdisciplinary through the context of bioinspired design.

Smriti Mehta is a PhD candidate in the Social-Personality Area at the Department of Psychology at the University of California, Berkeley. Her research interests include psychosocial factors affecting academic motivation and achievement, measurement and psychometrics, and statistical education.

Rebecca McNeil holds a master's degree from the University of California, Berkeley with an emphasis in Social Research Methods. Her research foci include STEM learning and assessment for teachers and students in K-12 and higher education, culturally responsive assessment development, and program evaluation.

Shih-Ying Yao currently works as a Lecturer at the Teaching & Learning Centre, Singapore University of Social Sciences (SUSS). Before joining SUSS, Dr. Yao has conducted educational research at the University of California, Berkeley and the National Institute of Education, Singapore for years. Her research focuses on the application of psychometric methods to the development and validation of assessments and other research instruments. In addition to educational measurement, she is also interested in the application of innovative educational technologies in higher education.

Mark Wilson is a Distinguished Professor of Education at the University of California, Berkeley, and a Professor at the University of Melbourne. He teaches courses on measurement in the social sciences, multidimensional measurement, and applied statistics. His research focuses on the establishment of a framework for measurement practice informed by the philosophy of measurement, on statistical models that are aligned with scientific models of the construct, and on instruments to measure new constructs.

Chapter 8
Applying Rasch Modeling to a Global Climate Change Concept Knowledge Assessment for Secondary Students

Amanda A. Olsen, Silvia-Jessica Mostacedo-Marasovic, and Cory T. Forbes

Abstract This current research uses Rasch modeling to compare the person- and item-fit statistics from two commonly used estimators, the joint maximum likelihood (JML) estimator and the marginal maximum likelihood (MML) estimator using an assessment used to measure secondary students' climate science and global climate change epistemological knowledge. Data was provided by a module involving EzGCM (Easy Global Climate Modeling), a web-based climate modeling tool. Overall, there was not a statistically significant difference between the JML and MML estimators.

This work was supported by the National Science Foundation (NSF-DRK12 #1720838). Any opinions, findings, and conclusions or recommendations expressed in this material are those of the author(s) and do not necessarily reflect the views of the National Science Foundation. We especially thank the teachers and students who made this research possible.

A. A. Olsen (✉)
Department of Educational, School & Counseling Psychology, University of Missouri, Columbia, USA
e-mail: aolsen@missouri.edu

S.-J. Mostacedo-Marasovic
Department of Earth and Environmental Sciences, College of Science, University of Texas at Arlington, Arlington, TX, USA
e-mail: sjm3161@mavs.uta.edu

C. T. Forbes
Department of Curriculum and Instruction, College of Education, University of Texas at Arlington, Arlington, TX, USA

Department of Earth and Environmental Sciences, College of Science, University of Texas at Arlington, Arlington, TX, USA
e-mail: cory.forbes@uta.edu

X. Liu, W. J. Boone (eds.), *Advances in Applications of Rasch Measurement in Science Education*, Contemporary Trends and Issues in Science Education 57,
https://doi.org/10.1007/978-3-031-28776-3_8

8.1 Introduction

To study Earth's climate and make predictions about climatic patterns, scientists rely on robust, data-intensive, computer-based models. Modeling is a core scientific practice emphasized in the Next Generation Science Standards (NGSS) in which models, or working representations of natural systems, are used in collaborative sense-making about system-specific phenomena (National Research Council, 2013). To develop climate literacy, or the ability to engage with social, economic, and cultural issues around climate issues, students must develop knowledge of both the conceptual and epistemic underpinnings of climate science, the latter of which foregrounds understanding of how models are used to study the Earth's climate and global climate change (GCC). Growing climate education research shows that there are limited opportunities for students to learn about climate science and global climate models. Because scientific modeling remains underemphasized in K-12 science classrooms, there is a need to learn more about the design of science learning environments that promote students' model-based reasoning about climate.

8.2 Literature Review

Of the limited work completed on climate science and global climate change epistemological understanding, studies have used mixed methods analyses (Tasquier et al., 2016), qualitative methods (Liu & Roehrig, 2017), and quantitative methods such as text analysis (Bråten & Strømsø, 2009), confirmatory factor analysis (Yli-Panula et al., 2014), and t-tests (Rebich & Gautier, 2005), among others, to measure change and gage climate science understanding. However, one methodology that is underutilized in measuring climate science epistemological understanding is Rasch modeling. Broadly, Rasch modeling is a type of analysis that can measure latent traits such as ability and can determine the probability that an individual will answer an item correctly on an assessment. For example, Bae et al. (2021) used Rasch modeling to investigate the engagement and value of dialogic interactions on epistemic climate knowledge. Their results suggested that both engagement and value can be used to measure epistemic climate knowledge. Similarly, Quarderer et al. (2021) used Rasch modeling to determine connections between climate literacy and epistemic cognition for eighth grade students. They found a positive relationship between climate literacy and epistemic cognition, specifically that participants often show quasi-reflective judgement when justifying their beliefs about climate change.

However, of the research using Rasch models in climate science and global climate change research, few are using Rasch models to guide the development of climate change assessments and tests as suggested by Boone (2017), as both Bae et al. (2021) and Quarderer et al. (2021) used Rasch modeling to measure student perceptions of climate change. Of the research investigating assessments and tests, few to none are using applied data, with no study to our knowledge using applied data to better understand Rasch estimation models for climate science and global climate change research.

Research using Rasch estimation models are typically simulation studies where different estimation models are simulated with different person and item-fit indices. Two commonly used and tested Rasch estimation models are the joint maximum likelihood (JML) estimator and the marginal maximum likelihood (MML) estimator. When comparing these two estimators, with the "MML approach the person parameters are treated as random effects and are integrated out from the likelihood function, while in the JML approach the person parameters are treated as fixed effect parameters and kept in the likelihood function," (Chen et al., 2019, p. 125). Therefore, the evaluation of the numerical integrals in the MML approach is replaced by maximizing, with consideration to the person parameters in the JML estimator. However, there is limited consensus on which estimator may be best. For example, the JML estimator has been cited to produce inconsistent item parameter results for infinite data (Chen et al., 2019), since the number of parameters increase with each added individual, although research by Wright (1988) suggests that this bias is correctable with Embretson and Reise (2000) suggesting that even if the JML estimator is biased, this estimator still gives reasonable point estimations. In contrast to the JML estimator, the MML estimator can produce parameter estimates for questionnaires with large sample sizes, although assertions must be made about the sample distribution, meaning person parameters need to be calculated after estimating the item difficulty parameters. Typically, the sample distribution is modeled to include individuals with extreme scores (i.e., individuals with zero and perfect scores) (Linacre, 2004).

In addition, each estimation model may produce different results for person- and item-fit statistics. The purpose of person-fit statistics is two-fold. The person-fit statistics must be able to identify patterns that are both believable and improbable, and second, person-fit statistics must be able to identify misleading measures (Meijer & Sitsma, 2001). For example, if an individual answered all the items designated as easy incorrectly and answered all items determined to be difficult correctly, their response pattern on this exam would be questionable. Therefore, when there is person misfit in a measurement model, error can be introduced into the test score.

The second aspect of Rasch models includes item-fit statistics. Item-fit statistics include the outfit (outlier-sensitive fit), and infit (inlier-sensitive or information-weighted fit), mean-squares (shows the amount of distortion in the measurement), and standardized fit statistics. Outfit statistics are sensitive to the pattern of responses to items targeted on the person, and vice versa, while infit statistics are more sensitive to responses to item with difficulty far from a person. Mean-square fit statistics represent the size of randomness and standardized fit statistics represent t-tests of the hypothesis, specifically measuring whether data fit the model (perfectly). Standardized fit statistics are typically reported as z-scores (Linacre, 2002).

For example, Le and Adams (2013) used Monte Carlo simulations to compare the different item parameters for JML and MML estimation methods. They found that the MML estimator was best when the assumption of the ability distribution was matched. Similarly, Kim (2001) evaluated the Markov chain monte carlo method for estimating item and person parameters using the CML, MML, and the JML estimators. They found that the item parameter estimates for the different methods produced almost identical results. Work by Robitzsch (2021) also compared three different estimation models (MML, JML, conditional maximum likelihood (CML), and limited information methods) on simulated nonnormality, number of items, sample size, and distribution of item difficulties. Their results suggest that JML and LIM estimation models may be most useful in applied research settings.

Understanding how Rasch estimation models work with applied, non-simulated data is important, especially for climate science and global climate change research, given that Rasch modeling can correct erroneous conclusions that result from typical classical test theory methods. Specifically, work using applied data to better understand Rasch estimations models would advance the application of Rasch measurement and contribute to the field by helping researchers and practitioners design and better understand practical climate science and global climate change assessments that can be applied and implemented by others in the field.

8.3 The Current Study

We are currently engaged in a multi-year, mixed-methods, design-based research project to investigate classroom use of EzGCM (Easy Global Climate Modeling), a web-based climate modeling suite designed to provide non-scientists experiences with climate modeling. We have developed, piloted, studied, and revised a 6-week climate science module for secondary science classrooms designed around EzGCM (Bhattacharya et al., 2020, 2021a, b; Carroll-Steward et al., 2021; Forbes et al., 2020). Overall, the module has supported students' understanding of climate models, including the data involved in them, and their temporal and spatial components, allowing students to develop a scientific understanding of climate (Battacharya et al., 2020). It has also provided instructors with a curriculum that can be adapted in response to students' educational needs and contextual factors (Carroll-Steward et al., 2021).

Despite documented project-specific successes in this domain, supporting students' model-based reasoning about natural phenomena, including climate change, remains a challenge. Adequately assessing the epistemic dimensions of their learning is even more challenging. Although a number research-based climate concept inventories have been developed and tested to provide measures of students' conceptual understanding (Aksit et al., 2018; Libarkin et al., 2018), no such measures currently exist that afford a measure of their epistemic understanding. As part of this project, including the broader project research, we have developed an assessment administered to students before and after the module. This assessment includes two subsets of items: one providing a measure of their conceptual understanding of Earth's climate and GCC and another providing a measure of their epistemic understanding. While the conceptual component of the assessment is based on existing, research-based assessment instruments in the field, the epistemic measure is a novel, project-developed instrument employed here for the first time. Therefore, the aim of the chapter is to perform a Rasch analysis to evaluate the level of difficulty of each of the assessment's items, as well as to compare the variations in students' understanding of the epistemic dimensions of Earth's climate and climate change. We also investigate two different estimation models, specifically the joint maximum likelihood (JML) estimator and the marginal maximum likelihood (MML) estimator, to see whether results varied on the epistemic dimensions of an Earth's climate and climate change assessment.

8.4 Method

8.4.1 Defining and Developing the Construct

To provide a measure of students' knowledge of climate-related concepts and define the construct, we used two research-based climate assessments (see Aksit et al., 2018; Breslyn et al., 2017), that were recently developed to measure understanding of climate concepts validated and shown to be reliable with secondary student populations. Note that Aksit et al. (2018), used work from Libarkin et al. (2015), McNeal, Libarkin, et al. (2014a), and McNeal, Spry et al. (2014b). To provide a measure of their knowledge of epistemic dimensions of scientific modeling, students completed an assessment designed by the project team, again using two research-based climate assessments as a guide.

8.4.2 Presenting the Item Construction Process

When developing the items used to measure the epistemic domain, we engaged in the process of evidence-centered design (ECD; Mislevy & Haertel, 2006), which involves the: (1) analysis of the content domain (domain analysis); (2) elaboration of the content (domain modeling), and; (3) articulation of a conceptual assessment framework. In the initial domain analysis stage, each epistemic dimension was unpacked to identify core elements and components. In our underlying theoretical framework, the epistemic dimension of students' understanding was defined by two primary subcategories: representational affordances and computational affordances. Representational affordances refer to the various ways in which EzGCM enables students to visualize fundamental climate data. Computational affordances refer to the various data transformations EzGCM affords within the interface. In the domain modeling stage, each epistemic dimension was then used as a primary claim about students' performance to articulate epistemic knowledge, types of work products that would elicit targeted performances, and features of these student items. This work led to the articulation of an assessment design space for each targeted outcome, meaning articulation of specific criteria for both representational and computational affordances, for each of which an assessment item was developed. In the third step, information from the previous two steps was used to articulate levels of student understanding, identify appropriate assessment items, and make decisions about how to evaluate evidence of students' thinking about each epistemic dimension. These stages resulted in the development of initial assessment items that are aligned with epistemic dimensions of model-based learning about Earth's climate and global climate change.

8.4.3 Presenting Sample Items

The 27 sample items that emerged from the evidence-centered design process to measure the epistemic domain are provided below. The questionnaire contained multiple choice items and "select all that apply" items that were dichotomously scored, where 1 meant the student answered the item correctly and zero meant the student answered the item score incorrectly. An individual who scored a perfect on the exam would have a total score of 27 points, whereas an individual who answered all items incorrectly would have scored zero points.

8.4.4 Sampling and Data Collection/Pilot Testing

During the pilot testing, 211 ninth grade high school students completed a 27-item questionnaire that measured student content knowledge (the epistemic dimension) regarding global climate change and environmental science. This assessment was completed by students who were enrolled in two secondary science teachers' geoscience classes in Nebraska who were participating in other components of the larger study. The geoscience course was compulsory for the district, meaning no self-selection into the course occurred. This research was approved by the University of Nebraska's institutional review board and participating students and teachers were free to withdraw at any time.

8.4.5 Rasch Analysis

Estimation Models

Since items were scored as either correct or incorrect, a dichotomous Rasch model was conducted using the Test Analysis Modules (TAM) package (Robitzsch et al., 2022) in R (R Core Team, 2022). In the TAM package, the MML estimator was computed using the tam.mml function which was based off work by Adams et al.

(1997) and the JML estimation was computed using the tam.jml function which was based off work by Linacre (1994). We recognize that there are many options to run Rasch modeling. Specifically, the JML estimator is commonly used and is available on WINSTEPS (Linacre, 2022; see https://www.winsteps.com/), and the MML estimator can be found in software such as BILOG-MG (Zimowski et al., 1996; see https://ssicentral.com/index.php/products/bilogmg-gen/) and ConQuest (Wu et al., 2007; see https://www.acer.org/au/conquest).

Below we provide the basic equation for a Rasch model with dichotomous data

$$P(X=1|B_n,D_i)=\frac{e^{(B_n-D_i)}}{1+e^{(B_n-D_i)}}$$

where B_n is the ability of person n and D_i is the difficulty of item i. Since this is for dichotomous data, $X=1$, would mean the item was scored as correct whereas $X=0$ would mean the answer was scored as incorrect. P is the probability that person n with ability B_n would answer the item correctly.

For each estimator, items were analyzed using item difficulty in logits and the standard error for each item. Standard errors were used to determine whether the item difficulties may overlap. Note, the mean of the item difficulties was constrained to zero.

Person- and Item-Fit Statistics

Person- and item-fit statistics were also calculated using both estimation methods in the TAM package (Robitzsch et al., 2022). For this analysis the item- and person-fit included the mean square error and the standardized (t) versions of the outfit and infit statistics. Mean-squares near 1 means there is little distortion, whereas mean-squares greater than one or below one suggests distortion. Standardized values above three meant these data were unexpected, and standardized values below negative two meant these data were too predictable (Linacre, 2002). For more information about person- and item-fit statistics we recommend visiting https://www.rasch.org/rmt/rmt162f.htm. In addition, we used Warm's likelihood estimation (WLE) for the person ability statistics (Warm, 1989), which was selected as WLE can provide more central estimates compared to other person ability statistics such as maximum likelihood estimates (Linacre, 2009).

Since item- and person-t statistics are measures on the same logit or log-odds unit scale (see https://www.rasch.org/rmt/rmt72e.htm), item parameter and person parameter results from the JML estimator and the MML estimator can be correlated using Pearson's correlation. Correlations were interpreted from zero to plus/minus one, where one equaled a perfect relationship between the estimators and zero represented no relationship between the estimators. The plus or minus represents the direction of the relationship.

8.5 Results

Using a dichotomous Rasch model, the calibration of secondary students ($N = 211$) and items ($N = 27$) were summarized using logit-scale calibrations, standard errors, and model-fit statistics. Results from the MML estimator and JML estimator item difficulty and model summary are provided in Tables 8.1 and 8.2, respectively. Please note that the mean item difficulty was restrained to zero, where an item difficulty below zero stands for an easier than the mean item difficulty and an item difficulty above zero stands for a more difficult than the mean item difficulty.

As seen in Table 8.2, item difficulty and standard errors between the two estimation methods were similar and negligible as seen by the mean and standard deviation statistics. Results from two t-tests (assuming normality), one analyzing the differences in means ($t(52) = .005$, $p = .996$) and one analyzing differences in standard errors ($t(52) = = 1.102$, $p = .275$), found no statistically significant differences between the two groups. On average, descriptively, items estimated using the JML estimator had a slightly higher (one, one hundredth) average and standard error (three one hundredths) compared to the MML estimator. The hardest items were item one "Which of the following is the best definition of a greenhouse gas?" and item 27 "Not every action taken by humans contributes to climate change. Which of the following human activities does NOT contribute to climate change?" The easiest item was item nine, "Which is the best description of the differences between climate and weather?" Of the 27 items, 10 items had difficulties below zero suggesting they were easier than the ability of an average person in the dataset, and 17 items had difficulties above zero suggesting they were more difficult than the ability of an average person in the dataset. This is visually shown in Fig. 8.1, which is a Wright Map of the climate assessment. Since results from the MML and JML estimator were virtually the same, we decided to only provide the Wright Map for the MML estimator.

In addition to comparing item difficulty and standard errors by estimator, we have also provided a model summary by estimation method for both item and person fit parameters. When comparing the two estimators, the logit scale location, standard error, outfit MSE and infit MSE were for the most part, very similar. For both estimators, on average students were located lower on the logit scale (JML; $M = 0.00$, $SD = 0.84$ and MML; $M = 0.00$, $SD = 0.80$) compared to the items (JML; $M = 0.28$, $SD = 0.69$ and MML; $M = 0.28$, $SD = 0.68$). This suggests that some of the items may have been slightly difficult for the sample of students who participated on Earth climate and GCC assessment. Further, the mean standard error was larger for the persons compared to the items for both the JML and MML estimator, respectively (JML; $M = 0.43$, $SD = 0.04$ and MML; $M = 0.43$, $SD = 0.04$). We note that the assessment was validated with high school and college students, and our sample included ninth grade students, perhaps suggesting the assessment may have been too difficult. However, since the MSE were near 0 that suggests that there was little distortion with these data (Linacre, 2002). The largest differences were seen between the standardized outfit mean and standard deviation,

Table 8.1 Sample questionnaire items

1	Which of following is the best definition of a greenhouse gas?
2	How has the amount of carbon dioxide in the atmosphere changed since the start of the industrial revolution 150 years ago?
3	Which of the following contributes to the transfer of thermal energy from place to place around the earth?
4	How does sunlight affect temperature on earth?
5	Which is the best definition of a positive feedback loop in the climate system?
6	Which of the following will occur if the amount of ice floating in the ocean decreases?
7	Which of the following best describes how plants take in carbon dioxide?
8	Which of the following would most likely occur if the oceans stopped absorbing carbon dioxide?
9	Which is the best description of the differences between climate and weather?
10	Which of the following statements about global warming over the past 50 years is most accurate?
11	Which of the following statements about air temperature change over the past millions of years is most accurate?
12	How much incoming sunlight do greenhouse gasses absorb?
13	What do greenhouse gases do?
14	Which of the following could cause the Earth's surface temperature to change?
15	What information do ice cores from glaciers contain about earth?
16	Which statements about non-greenhouse gasses are accurate?
17	Which of the following would cause Earth's average global temperature to rise?
18	Over the past several decades, the earth has warmed faster than any other time period. What best explains this increase?
19	There is strong evidence that there is more carbon dioxide in the atmosphere now than in the past several hundred years. What is the most likely cause of the current increase in carbon dioxide?
20	Scientists believe that global temperatures are rising primarily because of...
21	Climate change projections for the future are...
22	How does the rate that humans produce greenhouse gases relate to how quickly they are being removed by plants?
23	What is the relationship between Earth's temperature and the Earth's atmosphere?
24	Data collected by scientists indicate that the average global temperature is rising and will continue to rise in the foreseeable future. What actions could people in your community take to reduce the negative impacts of climate change?
25	How is carbon dioxide removed from the atmosphere?
26	Where can scientists see evidence of climate change?
27	Not every action taken by humans contributes to climate change. Which of the following does NOT contribute to climate change?

Note: Items were adapted from work by Aksit et al. (2018), Breslyn et al. (2017)

in addition to the standardized infit mean and standard deviation. However, since these standardized data were between −2 and 3, that suggests that these data were not too unexpected or predictable. This suggests that there was overall adequate model fit for the item and person calibrations on the logit scale. In addition, results

Table 8.2 Item difficulty by estimation method

	JLM estimator		MML estimator	
Item	Item difficulty	SE	Item difficulty	SE
1	1.562	0.183	1.561	0.181
2	−0.800	0.157	−0.801	0.154
3	−0.616	0.153	−0.617	0.150
4	0.682	0.155	0.681	0.152
5	1.317	0.172	1.316	0.170
6	0.479	0.151	0.478	0.148
7	0.178	0.149	0.177	0.146
8	−0.090	0.149	−0.091	0.146
9	−0.848	0.158	−0.849	0.155
10	−0.224	0.149	−0.225	0.146
11	0.964	0.162	0.963	0.158
12	1.123	0.166	1.122	0.163
13	0.944	0.161	0.943	0.158
14	−0.661	0.153	−0.662	0.151
15	0.030	0.148	0.029	0.145
16	0.919	0.160	0.918	0.157
17	0.220	0.149	0.219	0.146
18	−0.245	0.149	−0.246	0.146
19	0.306	0.150	0.305	0.146
20	0.030	0.148	0.029	0.145
21	0.199	0.149	0.198	0.146
22	−0.033	0.148	−0.034	0.145
23	−0.033	0.148	−0.034	0.145
24	0.705	0.155	0.704	0.152
25	0.135	0.149	0.134	0.145
26	−0.224	0.149	−0.225	0.146
27	1.562	0.183	1.561	0.181
Mean(SD)	0.281(0.684)	0.156(0.010)	0.280(0.684)	0.153(0.010)

Note: JML means joint maximum likelihood estimation. MML means marginal maximum likelihood estimation. SE means standard error and SD means standard deviation

from the WLE reliability indicate acceptable person separation reliability. Although this statistic is not equivalent to Cronbach's (1951) alpha, it is often interpreted using Cronbach's alpha benchmarks ($\propto \geq 0.9$ = excellent, $0.9 > \propto \geq 0.8$ = good, $0.8 > \propto \geq 0.7$ = acceptable, $0.7 > \propto \geq 0.6$ = questionable, $0.6 > \propto \geq 0.5$ poor, and $0.50 > \propto$ = unacceptable), therefore, our assessment had "acceptable" person separation reliability.

To compare the differences between the item parameter and person parameter results from the JML estimator and MML estimator, a Pearson's correlation was completed. Results from the correlations between the JML estimator and MML estimator were strong and positive for both the item-fit parameters ($r = 0.99$) and the person-fit parameters ($r = 0.99$). Since 1 is representative of a perfect, positive

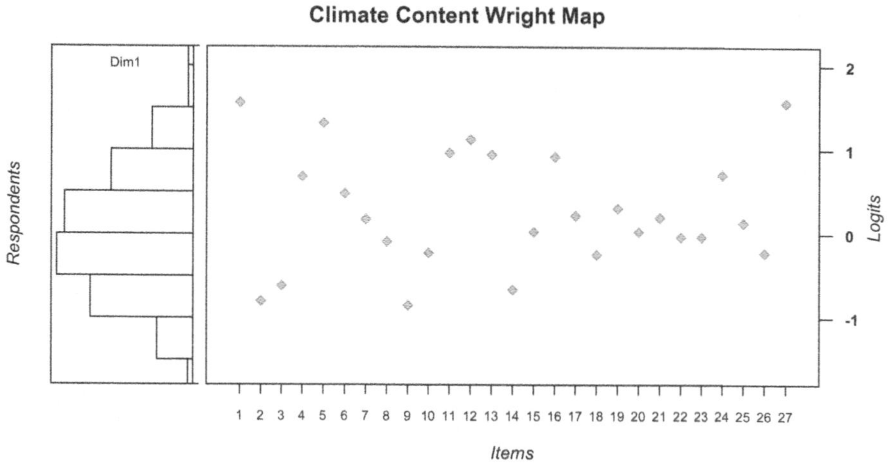

Fig. 8.1 Climate assessment wright map for the MML estimator

relationship, arguably we can state that the JML estimator and MML estimator models produced almost identical item-fit and person-fit statistics.

8.6 Discussion

The purpose of the current study was to determine whether there were differences between two commonly used estimators, the MML estimator and the JML estimator on item difficulty, and person- and item-fit statistics with the goal of better understanding the complexities of a research-based climate assessment that measured different epistemic dimensions of model-based learning about Earth's climate the GCC. Overall, as stated in the results section, the JML estimator and the MML estimator produced almost identical item-fit and person-fit statistics. This does not align with previous research suggesting that the JML estimator is biased due to our non-significant results when comparing results from the two estimators (Embretson & Reise, 2000; Wright, 1988). However, we note that these results are only valid for the current dataset.

8.6.1 Comparison Between Item Difficulty

Overall, there were minimal differences between the JML and the MML estimators on item difficulty, with the JML estimator having a slightly higher mean average and standard error than the MML estimator. Items greater than 0 were considered to be difficult, and items lower than 0 were considered to be easy. When looking at the different items that were considered difficult (items 1, 4, 5, 6, 7, 11, 12, 13, 15, 16, 17, 19, 20, 21, 24, 25, 27) and easy (items 2, 3, 8, 9, 10, 14, 18, 22, 23, 26) there did not appear to be a pattern between the item content or item style (i.e., select all of the above, which of the following, etc.). However, we did notice that some concepts may have been more difficult or easy for some students.

When considering the items with the highest logit digits (1 to 2) the items in which students had most difficulties were at indicating the role of greenhouse gases (GHG) in radiation absorption, what a positive feedback loop does, the amount of solar radiation absorbed by GHG, and at separating climate change from ozone depletion processes (Items 1, 27, 5, and 12). Next, the items with logit digits from 0 to 1 with which students had difficulties to respond were related to identifying the roles of natural processes and human activities on climate change, identifying the role of GHG and non-GHG in radiation processes, identifying actions people can adopt to reduce climate change, indicating the role of solar radiation in land and air surface temperatures, identifying impacts that ice reduction would have on solar radiation and air temperature, the impact of the increase of use of fossil fuels on CO^2, how climate change projections are based on observations and may differ from what was estimated, and the role of plants to remove CO_2. Overall, these items denote interactions between processes and correlations, where it seems that those related to radiation processes and GHG may have been most difficult. The items with logit numbers between zero to negative one with which students had least difficulties were related to indicating the role of the atmosphere to maintain energy and warm the Earth, the role of oceans as CO_2 storage, identification of evidence of climate change in various locations, the role of people in global warming when compared to other natural processes, the importance of ocean water to move energy, indicating that CO_2 has increased since the Industrial Revolution, and describing that climate changes over long periods of time whereas weather changes over short periods of time. These items focus more on factual evidence. Overall, we recommend that either the JML or the MML estimator could be used to estimate item difficulty, due to the similarities between the coefficients of these two models. However, we recommend that this is explored more thoroughly using simulation data.

8.6.2 Comparison Between Item- and Person-Fit Statistics

As seen in Table 8.3, for both estimators, on average students were located lower on the logit scale compared with the items. This suggests that on this particular assessment, items may have been slightly too difficult for the sample of students that participated. This is possible as the assessment was taken by ninth grades and the assessment was validated for high school and college students. We would recommend potentially eliminating items that have high item difficulty or rewriting

Table 8.3 Model summary by estimation method

	JML estimator		MML estimator	
Statistic	Items	Persons	Items	Persons
Logit Scale Location Mean	0.28	0.00	0.28	0.00
Logit Scale Location SD	0.69	0.84	0.68	0.80
Standard Error Mean	0.16	0.43	0.15	0.43
Standard Error SD	0.01	0.04	0.01	0.04
Outfit MSE Mean	1.01	1.01	1.02	1.00
Outfit MSE SD	0.20	0.24	0.15	0.24
Infit MSE Mean	1.00	0.99	1.00	0.98
Infit MSE SD	0.14	0.16	0.10	0.16
Std. Outfit Mean	−0.35	0.00	−0.18	−0.05
Std. Outfit SD	2.21	1.13	2.35	1.14
Std. Infit Mean	−0.41	−0.07	−0.30	−0.13
Std. Infit SD	2.22	1.05	1.75	1.07
WLE reliability	0.71	0.71	0.73	0.71

Note: JML means joint maximum likelihood estimation. MML means marginal maximum likelihood estimation. SE means standard error. SD means standard deviation. Std. means standardized and MSE means mean square error

these items for clarity. There were also large mean standard errors for persons compared to items for both estimators. We recommend potentially seeing whether items can be adjusted to better suit the students, or determining whether recruiting different students may fit the assessment better.

However, based on MSE and the standardized infit and outfit statistics, results show that either the MML estimator or the JML estimator are suitable to analyze either person- or item-fit statistics to better understand complex science learning outcomes on a global climate change and environmental science questionnaire with a sample size around 200 individuals. This adds to the literature as there is unclear consensus on which person- and item-fit statistics may be best suited to understand these types of science learning outcomes. However, we note that this should more thoroughly be explored using simulation data, as it is possible that our data happened to be well suited for Rasch modeling.

In addition, our results showed that there was "acceptable" person separation reliability using Cronbach's alpha as a benchmark. However, the assessment could possibly be revised to increase reliability for both the JML and MML estimator. As discussed earlier, it is possible that the assessment was too difficult for the students completing it, therefore, by potentially creating an assessment more geared towards ninth and tenth grade high school students compared to high school and college students, reliability may increase.

8.6.3 *Implications/Significance for Global Climate Change and GCC Inventories/Assessments*

There are multiple implications the use of Rasch modeling has for science assessments. First, Rasch modeling can help structure assessments with items of varying levels of difficulty that can help discriminate different levels of understanding of science contents, in this case epistemic aspects of GCC. Second, Rasch modeling can also help develop assessments when there aren't instruments available yet, or these are at an early stage of development, like the assessment used in this study. This is important as current environmental issues continue to be addressed in school but are relatively new when compared to other fields of science. Third, when assessments are available, this methodology can also help identify items that may need to be re-defined to identify if the items are helping evaluate the learning objectives of interest.

Future research can continue comparing and contrasting different estimation models to help the scientific community best understand which estimation models should be used. Perhaps different types of items or items with different types of science content areas may fit better with a specific type of estimator. There are also different estimators that could be explored in future studies such as the conditional maximum likelihood estimator and the limited information methods (Robitzsch, 2021).

Appendix

Annotated R Code

```
dat_0 <- data[1:211, 1:29]
dat_1 <- data[1:211, 3:29]


View(dat_1)


library(TAM)
library(WrightMap)
summary(dat_1)


rasch=tam(dat_1)
summary(rasch)


IRT.WrightMap(rasch, show.thr.lab=FALSE, main.title = "Climate
Assessment Wright Map")
```

```
difficulty <- rasch$xsi
difficulty


summary(difficulty)
sd(difficulty$xsi)
mean(difficulty$xsi)
sd(difficulty$se.xsi)
mean(difficulty$se.xsi)


item.fit <- tam.fit(rasch)
item.fit=as.data.frame(item.fit$itemfit)
summary(item.fit)


content=as.data.frame(tam.wle(rasch))
summary(content)


person.fit=tam.personfit(rasch)
summary(person.fit)


summary.table.statistics <- c("Logit Scale Location Mean",
                  "Logit Scale Location SD",
                  "Standard Error Mean",
                  "Standard Error SD",
                  "Outfit MSE Mean",
                  "Outfit MSE SD",
                  "Infit MSE Mean",
                  "Infit MSE SD",
                  "Std. Outfit Mean",
                  "Std. Outfit SD",
                  "Std. Infit Mean",
                  "Std. Infit SD",
                  "Reliability of Separation")


item.summary.results <- rbind(mean(difficulty$xsi),
                  sd(difficulty$xsi),
                  mean(difficulty$se.xsi),
                  sd(difficulty$se.xsi),
                  mean(item.fit$Outfit),
                  sd(item.fit$Outfit),
                  mean(item.fit$Infit),
                  sd(item.fit$Infit),
                  mean(item.fit$Outfit_t),
                  sd(item.fit$Outfit_t),
                  mean(item.fit$Infit_t),
                  sd(item.fit$Infit_t),
                  rasch$EAP.rel)
```

```
person.summary.results <- rbind(mean(content$theta),
                  sd(content$theta),
                  mean(content$error),
                  sd(content$error),
                  mean(person.fit$outfitPerson),
                  sd(person.fit$outfitPerson),
                  mean(person.fit$infitPerson),
                  sd(person.fit$infitPerson),
                  mean(person.fit$outfitPerson_t),
                  sd(person.fit$outfitPerson_t),
                  mean(person.fit$infitPerson_t),
                  sd(person.fit$infitPerson_t),
                  mean(content$WLE.rel))


item.summary.results_rounded <- round(item.summary.results, digits
= 2)


person.summary.results_rounded <- round(person.summary.results,
digits = 2)


Table1 <- cbind.data.frame(summary.table.statistics,
                  item.summary.results_rounded,
                  person.summary.results_rounded)


# add descriptive column labels:
names(Table1) <- c("Statistic", "Items", "Persons")


####


# Calculate the proportion correct for each task:
TaskCorrect <- apply(dat_1, 2, sum)
PropCorrect <- (TaskCorrect/nrow(dat_1))


# Combine item calibration results in a table:
Table2 <- cbind.data.frame(item.fit$parameter,
                  PropCorrect,
                  difficulty$xsi,
                  difficulty$se.xsi,
                  item.fit$Outfit,
                  item.fit$Outfit_t,
                  item.fit$Infit,
                  item.fit$Infit_t)
names(Table2) <- c( "Task ID", "Proportion Correct", "Item
Location","Item SE","Outfit MSE","Std. Outfit", "Infit MSE","Std. Infit")
```

```
# Sort Table 2 by Item difficulty:
Table2 <- Table2[order(-Table2$`Item Location`),]


# Round the numeric values (all columns except the first one) to 2 digits:
Table2[, -1] <- round(Table2[,-1], digits = 2)
Table2


#########


# Calculate proportion correct for persons:
PersonPropCorrect <- content$PersonScores / content$PersonMax


# Combine person calibration results in a table:
Table3 <- cbind.data.frame(content$pid,
                 PersonPropCorrect,
                 content$theta,
                 content$error,
                 person.fit$outfitPerson,
                 person.fit$outfitPerson_t,
                 person.fit$infitPerson,
                 person.fit$infitPerson_t)


names(Table3) <- c("Person ID", "Proportion Correct", "Person
Location","Person SE","Outfit MSE","Std. Outfit", "Infit MSE","Std.
Infit")


# Round the numeric values (all columns except the first one) to 2 digits:
Table3[, -1] <- round(Table3[,-1], digits = 2)
Table3


######


#JMLE Analysis


jmle.dichot.transreas <- tam.jml(dat_1)
summary(jmle.dichot.transreas)


jmle.difficulty <- jmle.dichot.transreas$xsi
theta <- jmle.dichot.transreas$theta
```

```
wrightMap(thetas = theta, thresholds = jmle.difficulty,
      show.thr.lab=FALSE, main.title = "Content Wright Map: JMLE")


jmle.item.se <- jmle.dichot.transreas$errorP
jmle.item.estimates <- cbind.data.frame(c(1:27),
                 jmle.difficulty,
                 jmle.item.se)
names(jmle.item.estimates) <- c("Item ID", "Item Location", "Item
Location SE")
summary(jmle.item.estimates)


sd(jmle.item.estimates$`Item Location`)
hist(jmle.item.estimates$`Item Location`, main = "Histogram of Item
Location Estimates for Science Content Data: JMLE", xlab = "Item
Location Estimates in Logits")


jmle.person.se <- jmle.dichot.transreas$errorWLE


jmle.person.estimates <- cbind.data.frame(dat_0$`Student ID`, theta,
jmle.person.se)
names(jmle.person.estimates) <- c("Student ID", "Theta", "Theta SE")


summary(jmle.person.estimates)
sd(jmle.person.estimates$Theta)


jmle.fit <- tam.jml.fit(jmle.dichot.transreas)


jmle.item.fit <- as.data.frame(jmle.fit$fit.item)
jmle.item.fit
summary(jmle.item.fit)


jmle.person.fit <- as.data.frame(jmle.fit$fit.person)
# View(jmle.person.fit)
summary(jmle.person.fit)


jmle.summary.table.statistics <- c("Logit Scale Location Mean",
                 "Logit Scale Location SD",
                 "Standard Error Mean",
                 "Standard Error SD",
                 "Outfit MSE Mean",
                 "Outfit MSE SD",
```

```
                "Infit MSE Mean",
                "Infit MSE SD",
                "Std. Outfit Mean",
                "Std. Outfit SD",
                "Std. Infit Mean",
                "Std. Infit SD",
                "Reliability of Separation")


jmle.item.summary.results <- rbind(mean(jmle.item.estimates$`Item
Location`),
                sd(jmle.item.estimates$`Item Location`),
                mean(jmle.item.estimates$`Item Location SE`),
                sd(jmle.item.estimates$`Item Location SE`),
                mean(jmle.item.fit$outfitItem),
                sd(jmle.item.fit$outfitItem),
                mean(jmle.item.fit$infitItem),
                sd(jmle.item.fit$infitItem),
                mean(jmle.item.fit$outfitItem_t),
                sd(jmle.item.fit$outfitItem_t),
                mean(jmle.item.fit$infitItem_t),
                sd(jmle.item.fit$infitItem_t),
                jmle.dichot.transreas$WLEreliability)


jmle.person.summary.results <- rbind(mean(jmle.person.estimates
$Theta),
                sd(jmle.person.estimates$Theta),
                mean(jmle.person.estimates$`Theta SE`),
                sd(jmle.person.estimates$`Theta SE`),
                mean(jmle.person.fit$outfitPerson),
                sd(jmle.person.fit$outfitPerson),
                mean(jmle.person.fit$infitPerson),
                sd(jmle.person.fit$infitPerson),
                mean(jmle.person.fit$outfitPerson_t),
                sd(jmle.person.fit$outfitPerson_t),
                mean(jmle.person.fit$infitPerson_t),
                sd(jmle.person.fit$infitPerson_t),
                mean(content$WLE.rel))


jmle.item.summary.results_rounded <- round(jmle.item.summary.
results, digits = 2)


jmle.person.summary.results_rounded <- round(jmle.person.summary.
results, digits = 2)
```

```
jmle.Table1 <- cbind.data.frame(jmle.summary.table.statistics,
                    jmle.item.summary.results_rounded,
                    jmle.person.summary.results_rounded)


# Add descriptive column labels:
names(jmle.Table1) <- c("Statistic", "Items", "Persons")


# Calculate the proportion correct for each task:


TaskCorrect <- apply(dat_1, 2, sum)
PropCorrect <- (TaskCorrect/nrow(dat_1))


jmle.Table2 <- cbind.data.frame(jmle.item.fit$item,
                    PropCorrect,
                    jmle.difficulty,
                    jmle.item.se,
                    item.fit$Outfit,
                    item.fit$Outfit_t,
                    item.fit$Infit,
                    item.fit$Infit_t)
names(jmle.Table2) <- c("Task ID", "Proportion Correct", "Item
Location","Item SE","Outfit MSE","Std. Outfit", "Infit MSE","Std. Infit")


# Sort the table by Item difficulty:
jmle.Table2 <- jmle.Table2[order(-jmle.Table2$`Item Location`),]


# Round the numeric values (all columns except the first one) for
presentation in a table:
jmle.Table2[, -1] <- round(jmle.Table2[, -1], digits = 2)


# calculate proportion correct for persons:
PersonPropCorrect <- content$PersonScores / content$PersonMax


jmle.Table3 <- cbind.data.frame(jmle.person.estimates$`Student ID`,
                    PersonPropCorrect,
                    jmle.person.estimates$Theta,
                    jmle.person.estimates$`Theta SE`,
                    jmle.person.fit$outfitPerson,
                    jmle.person.fit$outfitPerson_t,
                    jmle.person.fit$infitPerson,
                    jmle.person.fit$infitPerson_t)
```

```
names(jmle.Table3) <- c("Person ID", "Proportion Correct", "Person
Location","Person SE","Outfit MSE","Std. Outfit", "Infit MSE","Std.
Infit")


# Round the numeric values (all columns except the first one) for
presentation in a table:
jmle.Table3[, -1] <- round(jmle.Table3[, -1], digits = 2)


plot(difficulty$xsi, jmle.difficulty,
  pch=16,
  xlab= "MMLE Estimate",
  ylab= "JMLE Estimate",
  main="Item Parameter Estimate Comparison")


cor(difficulty$xsi, jmle.difficulty)


plot(content$theta, jmle.person.estimates$Theta,
  pch=16,
  xlab= "MMLE Estimate",
  ylab= "JMLE Estimate",
  main="Person Parameter Estimate Comparison")


cor(content$theta, jmle.person.estimates$Theta)
```

References

Adams, R. J., Wilson, M., & Wang, W.-C. (1997). The multidimensional random coefficients multinomial logit model. *Applied Psychological Measurement, 21*(1), 1–23. https://doi.org/10.1177/0146621697211001

Aksit, O., McNeal, K. S., Gold, A. U., Libarkin, J. C., & Harris, S. (2018). The influence of instruction, prior knowledge, and values on climate change risk perception among undergraduates. *Journal of Research in Science Teaching, 55*(4), 550–572.

Bae, Y., Fulmer, G. W., & Hand, B. M. (2021). Developing latent constructs of dialogic interaction to examine the epistemic climate: Rasch modeling. *School Science and Mathematics, 121*(3), 164–174. https://doi.org/10.1111/ssm.12460

Bhattacharya, D., Chandler, M., Carroll-Steward, K., & Forbes, C. T. (2020). Investigating the phenomenon of increasing surface air temperatures using a global climate modeling approach. *The Science Teacher, 88*(1), 58–66. https://www.jstor.org/stable/27048205

Bhattacharya, D., Carroll-Steward, K., & Forbes, C. T. (2021a). Climate education in secondary science: Comparison of model-based and non-model-based investigations of Earth's climate. *International Journal of Science Education, 43*(13), 2226–2249. https://doi.org/10.1080/09500693.2021.1958022

Bhattacharya, D., Carroll-Steward, K., & Forbes, C. T. (2021b). Empirical research on K-12 climate education: A systematic review of the literature. *Journal of Geoscience Education, 69*(3), 223–247. https://doi.org/10.1080/10899995.2020.1838848

Boone, W. J. (2017). Rasch analysis for instrument development: Why, when, and how? *CBE-Life Sciences Education, 15*(4), 1–7. https://doi.org/10.1187/cbe.16-04-0148

Bråten, I., & Strømsø, H. I. (2009). Effects of task instruction and personal epistemology on the understanding of multiple texts about climate change. *Discourse Processes, 47*(1), 1–31. https://doi.org/10.1080/01638530902959646

Breslyn, W., Drewes, A., McGinnis, J. R., Hestness, E., & Mouza, C. (2017). Development of an empirically-based conditional learning progression for climate change. *Science Education International, 28*(3), 214–223. https://doi.org/10.33828/sei.v28.i3.5

Carroll-Steward, K., Bhattacharya, D., Chandler, M. C., & Forbes, C. T. (2021). Secondary science teachers' implementation of a curricular intervention when teaching with global climate models. *Journal of Geoscience Education, 70*, 474–489. https://doi.org/10.1080/10899995.2021.1980706

Chen, Y., Li, X., & Zhang, S. (2019). Joint maximum likelihood estimation for high-dimensional exploratory item factor analysis. *Psychometrika, 84*(1), 124–146. https://doi.org/10.1007/s11336-018-9646-5

Cronbach, L. J. (1951). Coefficient alpha and the internal structure of tests. *Psychometrika, 16*, 297–334. https://doi.org/10.1007/BF02310555

Embretson, S. E., & Reise, S. P. (2000). *Item response theory for psychologists*. Lawrence Erlbaum Associates Publishers.

Forbes, C. T., Chandler, M., Blake, J., Bhattacharya, D., Carroll-Steward, K., Johnson, V., DeGrand, T., Mason, W., & Murrow, B. (2020). Fostering climate literacy with global climate models in secondary science classrooms: Insights from a collaborative partnership. In J. Henderson & A. Drewes (Eds.), *Teaching climate change in the United States* (pp. 29–43). Routledge.

Kim, S.-H. (2001). An evaluation of a Markov chain Monte Carlo method for the Rasch model. *Applied Psychological Measurement, 25*(2), 163–176. https://doi.org/10.1177/01466210122031984

Le, L. T., & Adams, R. J. (2013). *Accuracy of Rasch model item parameter estimation*. https://research.acer.edu.au/ar_misc/13.

Libarkin, J. C., Thomas, S. R., & Ording, G. (2015). Factor analysis of drawings: Application to college student models of the greenhouse effect. *International Journal of Science Education, 37*(13), 2214–2236. https://doi.org/10.1080/09500693.2015.1074757

Libarkin, J. C., Gold, A. U., Harris, S. E., McNeal, K. S., & Bowles, R. P. (2018). A new, valid measure of climate change understanding: Associations with risk perception. *Climatic Change, 150*(3), 403–416.

Linacre, J. M. (1994). *Many-facet Rasch measurement*. MESA Press.

Linacre, J. M. (2002). What do infit and outfit, mean-square and standardized mean? *Rasch Measurement Transactions, 16*, 878.

Linacre, J. M. (2004). Rasch model estimation: Further topics. *Journal of Applied Measurement, 5*(1), 95–110.

Linacre, J. M. (2009). The efficacy of Warm's weighted mean likelihood estimate (WLE) correction to maximum likelihood estimate (MLE) bias. *Rasch Measurement Transactions, 23*(1), 1188–1189.

Linacre, J. M. (2022). Winsteps® (version 5.2.3) [computer software]. Winsteps.com. Available from https://www.winsteps.com/

Liu, S., & Roehrig, G. (2017). Exploring science teachers' argumentation and personal epistemology about global climate change. *Research in Science Education, 49*, 173–189. https://doi.org/10.1007/s11165-017-9617-3

McNeal, K. S., Libarkin, J. C., Ledley, T. S., Bardar, E., Haddad, N., Ellins, K., & Dutta, S. (2014a). The role of research in online curriculum development: The case of earthlabs climate change and earth system modules. *Journal of Geoscience Education, 62*(4), 560–577. https://doi.org/10.5408/13-060.1

McNeal, K. S., Spry, J. M., Mitra, R., & Tipton, J. L. (2014b). Measuring student engagement, knowledge, and perceptions of climate change in an introductory environmental geology course. *Journal of Geoscience Education, 62*(4), 655–667. https://doi.org/10.5408/13-111.1

Meijer, R. R., & Sijtsma, K. (2001). Methodology review: Evaluating person fit. *Applied Psychological Measurement, 25*(2), 107–135. https://doi.org/10.1177/01466210122031957

Mislevy, R. J., & Haertel, G. D. (2006). Implications of evidence-centered design for educational testing. *Educational Measurement, 25*(4), 6–20. https://doi.org/10.1111/j.1745-3992.2006.00075.x

National Research Council. (2013). *Next generation science standards: For states, by states*. The National Academies Press. https://doi.org/10.17226/18290

Quarderer, N. A., Fulmer, G. W., Hand, B., & Neal, T. A. (2021). Unpacking the connections between 8th graders' climate literacy and epistemic cognition. *Journal of Research in Science Teaching, 58*(10), 1527–1556. https://doi.org/10.1002/tea.21717

R Core Team. (2022). *R: A language and environment for statistical computing*. R Foundation for Statistical Computing. https://R-project.org

Rebich, S., & Gautier, C. (2005). Concept mapping to reveal prior knowledge and conceptual change in a mock summit course on global climate change. *Journal of Geoscience Education, 53*(4), 355–365. https://doi.org/10.5408/1089-9995-53.4.355

Robitzsch, A. (2021). A comprehensive simulation study of estimation methods for the Rasch model. *Stat, 4*, 814–836. https://doi.org/10.3390/stats4040048

Robitzsch, A., Kiefer, T., & Wu, M. (2022). *TAM: Test analysis modules*. R package version 4.0-16. https://CRAN.R-project.org/package=TAM

Tasquier, G., Levrini, O., & Dillion, J. (2016). Exploring students' epistemological knowledge of models and modeling in science: Results from a teaching/learning experience on climate change. *International Journal of Science Education, 38*(4), 539–563.

Warm, T. A. (1989). Weighted likelihood estimation of ability in item response theory. *Psychometrika, 54*, 427–450. https://doi.org/10.1007/BF02294627

Wright, B. (1988). The efficacy of unconditional maximum likelihood bias correction: Comment on Jansen, van den Wollenberg, and Wierda. *Applied Psychological Measurement, 12*(3), 315–318.

Wu, M. L., Adams, R. J., Wilson, M. R., & Haldane, S. (2007). *ACER ConQuest 2.0: General item response modeling software [computer software]*. Australian Council for Educational Research.

Yli-Panula, E., Laakkonen, E., & Vauras, M. (2014). High school students' topic-specific epistemic beliefs about climate change: An assessment related study. *Education Sciences, 11*(8), 440. https://doi.org/10.3390/educsci11080440

Zimowski, M. F., Muraki, E., Mislevy, R. J., & Brock, R. D. (1996). *BILOG-MG: Multiple group IRT analysis and test maintenance for binary items [computer software]*. Scientific Software International.

Amanda A. Olsen, Ph.D., is an Associate Professor of Statistics, Measurement, and Evaluation in Education at the University of Missouri. She is an applied, quantitative methodologist with substantive interests in school climate, equity, and interdisciplinary collaborations.

Silvia-Jessica Mostacedo-Marasovic is a PhD candidate from the Department of Earth and Environmental Sciences at the University of Texas at Arlington. Master of Sustainable Development Practice with specializations on water resources, climate sciences, and higher education.

Cory T. Forbes, Ph.D., is Fenton Wayne Robnett Professor of Science Education and Chair of the Department of Curriculum and Instruction in the College of Education at the University of Texas at Arlington with a secondary appointment at UTA's Department of Earth and Environmental Sciences. Director of the STEM Education Research Collaboratorium and Resource Center.

Chapter 9
Validating the Progression of Chemistry-Based Health Literacy: An Application of Rasch Analysis

Jonathan M. Barcelo and Nona Marlene B. Ferido

Abstract Current reforms in teaching science to allied health science students emphasize the evidence-based approach to decision-making. To adopt this approach in chemistry subjects, the construct *chemistry-based health literacy* was developed, and four levels were proposed to describe an increasing ability to support health decisions with chemistry concepts. Rasch analysis provided validity evidence of the chemistry-based health literacy test and the rating scale functioning that corresponds to the four levels of the construct. The results can guide science educators in designing learning activities and teaching interventions that promote the progression of chemistry-based health literacy in health science students.

9.1 Introduction

An allied health science student is expected to apply science concepts to health issues, whether the context is related to making health decisions, or understanding the scientific rationale of clinical decisions and interventions. However, applying scientific knowledge to health contexts is difficult and requires more than simple recall of scientific information and algorithmic reasoning, since the decision-making process is essentially context-bound (Thomas & Ellaway, 2021). Making clinical decisions also requires an evidence-based approach (Lehane et al., 2018) and suggests a propensity toward using an evidence-based curriculum in health program education (Asokan, 2012; Patelarou et al., 2017). With this reform, teaching science

J. M. Barcelo (✉)
School of Nursing, Allied Health and Biological Sciences, Saint Louis University, Baguio City, Philippines

University of the Philippines Open University, Los Baños, Laguna, Philippines
e-mail: jmbarcelo@slu.edu.ph

N. M. B. Ferido
Assessment, Curriculum, & Technology Research Centre (ACTRC), Quezon City, Philippines

X. Liu, W. J. Boone (eds.), *Advances in Applications of Rasch Measurement in Science Education*, Contemporary Trends and Issues in Science Education 57,
https://doi.org/10.1007/978-3-031-28776-3_9

to health science students is expected to promote the meaningful application of scientific evidence to various health contexts.

The current situation in chemistry education in health science programs calls for reforms in increasing the context and identifying the topics that best address the learning outcomes for future healthcare professionals. Chemical knowledge application should have a specific context (Cooper & Stowe, 2018; King, 2012), and scientific knowledge should guide individuals when making decisions (von Winterfeldt, 2013). Hence, health contexts related to health promotion, disease prevention, and curative interventions were selected. These topics include nutrition, diagnostics, and pharmacology, since chemistry concepts are ubiquitous, and chemistry subjects are common prerequisites.

Developing a construct to measure when teaching science to allied health science students is critical because health contexts require a critical understanding of several topics. To address the evidence-based framework desired in allied health science programs, we contended that the construct should use science concepts related to health issues. Hence, we developed a construct that links chemistry concepts with health decisions – *chemistry-based health literacy*. Using raw scores to detect the progression of this construct does not ensure measurement reliability, thereby necessitating the application of Rasch analysis. Aside from validating the research instrument or tool that measures the construct, Rasch analysis also provides validity evidence for a hypothesized progression, allowing a more robust quantitative interpretation of results that may inform the reframing of science education in health science programs.

This study demonstrates how Rasch analysis provides evidence of the validity of the *Chemistry-based Health Literacy Test*, a tool that measures chemistry-based health literacy of health science students in nutrition, diagnostics, and pharmacology. By adopting a Rasch analysis-based approach, it is possible to develop an assessment and evaluation tool that measures the application of chemistry to health contexts, which is the main thrust in science education for allied health science programs.

9.2 Defining Chemistry-Based Health Literacy

Chemistry-based health literacy is defined in this study as the *ability to choose a correct health decision by applying relevant and contextualized chemical knowledge and avoiding heuristic-based judgment*. Based on this definition, it is hypothesized that the selection of a correct health decision should be supported by chemistry-related reasoning that is accurate and contextualized to a health context. Scientific knowledge influences beliefs and values which guide decision-making processes (von Winterfeldt, 2013). While the decision-making process in socio-scientific issues is not entirely dependent on scientific knowledge (Bell & Lederman, 2003), using scientific evidence in supporting decisions related to health enables students to distinguish between scientific and pseudoscientific information and allows a re-evaluation of whether heuristics are applicable in a health-related decision or not. Since chemistry subjects are prerequisites to learning professional subjects such as nutrition and diet therapy, clinical chemistry, and pharmacology, promoting the application of chemistry concepts to health contexts can make chemistry concepts more relevant in allied health science professions.

This study posited that chemistry-based health literacy is not entirely similar to health literacy or chemical literacy, as the construct involves the application of scientific knowledge to health decisions. Health literacy is based on the assessment of an individual's self-belief in their ability to access adequate and correct health information, understand information, or navigate through the healthcare system (Abel et al., 2014; Hawkins et al., 2017; Mullan et al., 2017), suggesting that health literacy requires a multidisciplinary approach. Chemical literacy, on the other hand, is defined as the ability to comprehend and apply the knowledge of chemistry in everyday life in terms of understanding three aspects – knowledge, awareness, and the application of chemistry in daily life appropriately and effectively (Thummathong & Thathong, 2018). We acknowledge that chemistry-based health literacy is a construct suited for classroom assessment and evaluation for allied health science students.

Chemistry-based health literacy is based on the decision-making framework described by von Winterfeldt (2013). According to the framework, scientific knowledge influences values and beliefs, which, in turn, guides decisions. This study supports the claim that science has a tremendous effect on the decision-making process (Lee, 2012), both on a personal and societal level. In health science professions, decision-making is sometimes based on heuristic reasoning (Blumenthal-Barby & Krieger, 2014; Featherston et al., 2020), although such decisions may not always be rational. Similarly, heuristic reasoning was also observed among students when learning chemistry concepts and based on the familiarity of the concept, perception of similarity of a new concept to a known concept, or using a single differentiating characteristic as a basis for making inferences (Maeyer & Talanquer, 2010; Talanquer, 2014). Furthermore, pseudoscientific information has negatively impacted decision-making related to health, requiring the need for scientifically validated claims remains to be a priority when teaching students in allied health science programs.

We evaluated the explanations that support a correct health decision using four competencies: (a) the *ability to link chemistry concepts and health decisions (or claims)*; (b) *the ability to appraise chemical information*; (c) *the ability to generate inference from chemical data*; and, (d) *the ability to apply latent and explicit chemistry concepts in each situation*. Students with high chemistry-based health literacy should select correct health decisions and support the decision with accurate, contextualized, and relevant chemical information. In addition, a student with a high level of chemistry-based health literacy should also avoid using simplistic or heuristic-based reasoning when choosing health decisions. On the other hand, students with low chemistry-based health literacy are unable to select appropriate health decisions, exhibit no knowledge or misconceptions about chemical concepts, and are unable to apply chemical knowledge.

Using chemistry-based health literacy as a construct to measure the application of chemistry concepts in health decisions is suited to the learning needs of health science students in the Philippines. Aside from empowering students to increase the health literacy of Filipinos (Agosto et al., 2018; Camiling, 2019), chemistry educators can design teaching interventions that promote the progression of chemistry-based health literacy to allied health science students, making the curriculum more cohesive and relevant to allied health science programs.

9.3 Factors Affecting Chemistry-Based Health Literacy

When applying chemistry concepts to health contexts, educators consider the student's prior knowledge about a chemistry concept, and ability to relate information to a different situation. Students should demonstrate the ability to apply their prior knowledge in new situations to strengthen conceptual connections. Compared to novice learners, expert learners use the scientific method and a logical approach when solving problems related to open-ended questions in chemistry (Randles & Overton, 2015). An expert chemistry learner navigates the relationship between the macroscopic, sub-microscopic, and symbolic domains of chemical representations; a novice learner focuses on the surface features of these representations (Kozma, 2003). Given these characteristics, students with adequate prior knowledge of chemistry concepts can interpret chemical information, make inferences, and apply chemical knowledge in supporting health decisions.

The ability to generate explanations that support a health decision requires skillful recognition of the type of embedded chemical information in each situation. Chemical information is derived from *in vitro*, *in vivo*, or *in silico* methods (Benfenati et al., 2010). Data from *in vivo* experiments may involve complex metabolic pathways (Pang et al., 2014), requiring a careful interpretation of the results. In most cases, the information from chemical experiments is extrapolated (Rehberger et al., 2018) to determine the effects of chemical substances on humans. Doing this, however, requires a skillful and critical interpretation of data. Supporting health decisions with any chemical information requires a critical evaluation of the applicability of the information to health contexts, and may require a high cognitive load, especially if the context involves *in vivo* chemical systems.

9.4 Method

9.4.1 Development of the Chemistry-Based Health Literacy Test (CbHLT)

The chemistry-based health literacy test is a 28-item two-tiered test using various contexts in nutrition (food choice, food preparation), pharmacology (food-drug interaction, drug action, natural products), and diagnostics. The summary of items is shown in Table 9.1 The topics selected are those that rely on prior knowledge of

Table 9.1 Summary of assessment items in the Chemistry-based Health Literacy Test

Item	Code	Context	Embedded chemistry concept	Hypothesized difficulty
1	N1	Reusing frying oil	Oxidation of fatty acids (in vitro)	Easy
2	P1	Mixing medicine with food	Metal complexation (in vitro)	Easy
3	N2	Formula milk preparation	Protein denaturation (in vitro)	Easy
4	NP1	Food as replacement of medicine	Chemical property (in vitro)	Easy
5	N3	Noodle product	Chemical property (in vitro)	Easy
6	P2	Anti-asthma medication	pH and buffers (in vitro)	Moderately difficult
7	D1	Urinalysis	Chemical reactions (in vitro)	Easy
8	N4	Dietary antioxidants	Extraction of substance (in vitro)	Easy
9	N5	Cholesterol-rich diet	Structure-property relationship (in vivo)	Difficult
10	P3	Vasodilatory medication	Chemical property (in vitro)	Moderately difficult
11	P4	Fluid and electrolyte imbalance	Chemical property (in vitro)	Moderately difficult
12	NP2	Antifungal medication	Chemical property (in vitro)	Easy
13	N6	Frying of potato	Chemical reactions (in vitro)	Moderately difficult
14	D2	Use of alternative diag-nostic kits	Chemical reactions (in vitro)	Moderately difficult
15	P5	Herbal antioxidant	Air oxidation (in vitro)	Moderately difficult
16	D3	Enzyme-based diagnostic kit	Chemical reactions (in vitro)	Moderately difficult
17	P6	Alternative antibacterial drug	Structure-property relationship (in vitro)	Moderately difficult
18	N7	Health benefits of fruit extracts	Chemical reactions (in vitro)	Moderately difficult
19	N8	Food as nutraceutical products	Chemical reactions (in vivo)	Difficult
20	P7	Side effects of medications	Chemical reactions (in vivo)	Difficult
21	N9	Food supplement	Structure-property relationship (in vivo)	Difficult
22	N10	Food preparation	Chemical reactions (in vitro)	Easy
23	N11	Food preservatives	Chemical reactions (in vivo)	Difficult
24	N12	Alkaline water	pH and buffers (in vivo)	Difficult
25	N13	Fruit extracts	Neutralization (in vivo)	Difficult
26	D4	Blood glucose measurement	Chemical reactions (in vitro)	Difficult

(continued)

Table 9.1 (continued)

Item	Code	Context	Embedded chemistry concept	Hypothesized difficulty
27	D5	Detection of analyte	Chemical reactions (in vitro)	Moderately difficult
28	D6	Stool analysis	Chemical reactions (in vitro)	Moderately difficult

Note: *N* nutrition, *D* diagnostics, *P* pharmacology, *NP* natural products

chemistry concepts. Next, we screened the topics highlighting health promotion, disease prevention, and curative interventions, following the thrust of the Department of Health. Eight chemistry teachers with more than five years of experience teaching chemistry subjects to students in allied health science programs deliberated the final topics in the test. The difficulty of items was hypothesized based on the embedded chemistry concept and the type of chemical environment involved in the situation (*in vitro* versus *in vivo* chemical systems). We hypothesized that items with embedded concepts in general inorganic chemistry and organic chemistry concepts were less difficult compared to items in analytical chemistry and biochemistry. We also postulated that achieving high chemistry-based health literacy is more difficult when the context involves *in vivo* chemical systems.

9.4.2 Validation of the Chemistry-Based Health Literacy Test

Four chemistry professors and one chemistry education professor evaluated the first draft of the chemistry-based health literacy test. We determined the content validity index and kappa statistics and revised the test based on the recommendations. Table 9.2 shows that the item content validity indices ranged from 0.42 to 1.00 and are interpreted as fair to excellent (Polit & Beck, 2006). However, the I-CVI results implied that some items need to be revised.

Students may select health decisions based on simple chemistry rules or heuristics, so the situations were designed to have an embedded chemistry topic that students can use when providing chemistry explanations. To link the test to the progression of chemistry-based health literacy, we also hypothesized that each level of chemistry-based health literacy should reflect an increasing ability to choose a correct claim or health decision and display a progressive aptitude to support the decision with an accurate, relevant, and contextualized explanation that is related to the embedded chemistry concept. We presumed that the progression is similar in items related to nutrition, diagnostics, and pharmacology topics.

To determine the hypothesized difficulty of the items, we considered the prior knowledge of students and the context of the situation related to nutrition, diagnostics, and pharmacology. The topics that involved the simple recall of general inorganic chemistry, organic chemistry, and analytical chemistry were presumed to be easier for students due to familiarity. These topics included chemical properties,

Table 9.2 Item content validity of Chemistry-based Health Literacy Test

Item	Code	I-CVI	I-CVI interpretation	pc	k	Interpretation of k
1	P1	1.00	Appropriate	0.03	1.00	Excellent
2	N2	0.80	Revise	0.16	0.76	Excellent
3	NP1	1.00	Appropriate	0.03	1.00	Excellent
4	N3	0.80	Revise	0.16	0.76	Excellent
5	P2	0.80	Revise	0.16	0.76	Excellent
6	D1	0.80	Revise	0.16	0.76	Excellent
7	N4	0.80	Revise	0.16	0.76	Excellent
8	N5	0.80	Revise	0.16	0.76	Excellent
9	P3	1.00	Revise	0.03	1.00	Excellent
10	P4	0.80	Revise	0.16	0.76	Excellent
11	NP2	1.00	Appropriate	0.03	1.00	Excellent
12	N6	0.80	Appropriate	0.16	0.76	Excellent
13	D2	1.00	Appropriate	0.03	1.00	Excellent
14	P5	1.00	Appropriate	0.03	1.00	Excellent
15	D3	0.80	Revise	0.16	0.76	Excellent
16	P6	0.80	Revise	0.16	0.76	Excellent
17	N7	0.80	Revise	0.16	0.76	Excellent
18	N8	0.80	Revise	0.16	0.76	Excellent
19	P7	0.60	Revise	0.31	0.42	Fair
20	N9	0.80	Revise	0.16	0.76	Excellent
21	N10	0.80	Revise	0.16	0.76	Excellent
22	N11	0.80	Revise	0.16	0.76	Excellent
23	N12	1.00	Appropriate	0.03	1.00	Excellent
24	N13	0.80	Revise	0.16	0.76	Excellent
25	D4	0.80	Revise	0.16	0.76	Excellent
26	D5	1.00	Revise	0.03	1.00	Excellent
27	D6	0.80	Revise	0.16	0.76	Excellent
28	N1	0.60	Revise	0.31	0.42	Fair

Note: *N* nutrition, *D* diagnostics, *P* pharmacology, *NP* natural products

metal complexation, and basic laboratory procedures (extraction). Situations that highlight chemical reactions and structure-property relationships and involve *in vitro* chemical systems are classified as moderately difficult items. Topics that require analysis of *in vivo* chemical systems (cellular or physiologic environment) are ranked difficult. *In vivo* chemical systems are more complex due to the involvement of metabolic pathways. Using *in vivo* evidence may also involve a higher cognitive load since most metabolic pathways are still incompletely understood (Pang et al., 2014).

Each item has three parts – a situation, four claims or decisions, and a blank box where students will write their explanations. Furthermore, the embedded chemistry concepts are not explicitly described to test whether students recognize the embedded chemistry concept in the situation. Embedded chemistry concepts are not presented directly to avoid "teaching the correct explanation" and allow the students to figure out the chemistry concepts that are applied to the situation independently.

An example of a nutrition-related item in CbHLT is shown below.

Situation

A citrus-flavored noodle product was advertised that it has been enriched with a "concentrated citrus flavor." You were asked by *Onat*, a 30-year-old father if it is alright to feed this product to his four-year-old son. One of his neighbors, Nitoy, stated that the noodle is rich in fruit vitamins because it has a citrus flavor. What is an APPROPRIATE response?

Claim or Decision

A. "Since the noodle product has a citrus flavor, the product is rich in vitamins."
B. "The noodle is not a source of vitamins because it contains ascorbic acid only."
C. "The noodle is a good nutrient source because it contains concentrated citrus flavor."
D. "A product which has a citrus flavor does not necessarily mean that it has vitamins." ➔ **correct answer**

Explanation

9.4.3 Criteria for Classifying Types of Explanations

An initial survey was conducted in 2019 to a total of 100 students (45% first-year students, 55% second-year students) in a tertiary institution in Baguio City, Philippines. Students were asked to explain their answers to questions related to nutrition (*food choice*, *food preparation*), diagnostics (*principles of diagnostic tests*), and pharmacology (*drug interaction*, *evaluation of natural products*, and *food and drug interaction*). The explanations were analyzed by eight science teachers with more than five years of experience in teaching general inorganic chemistry, organic chemistry, analytical chemistry, and biochemistry to health science students. The general types of answers were:

- Correct health decisions supported by complete chemical information or evidence. A correct generalization is given. The generalization is supported by sufficient, correct, and related evidence. Furthermore, there is no over-generalization or simplification of chemical information. The generalization has a similar context to the pieces of evidence.

- Correct health decision supported by correct but incomplete chemical information or evidence. A correct claim or decision is given and the evidence is accurate but incomplete. Some pieces of evidence are assumed to be true to the context.
- Correct health decision that is unsupported by chemical information or evidence. A correct generalization but the information or evidence to support the claim or decision is not relevant, inaccurate, or unrelated.
- Incorrect health decisions based on correct and complete chemical information. An incorrect generalization is given from sufficient, correct, and related evidence.
- Incorrect health decisions based on correct but incomplete chemical information. An incorrect claim or decision is given. The pieces of evidence are accurate but incomplete. Some information used to justify the decision is misinterpreted. The context of the evidence or information differs from the context of the claim or decision.

    This type of answer included the following subtypes:

    (a) Hasty Generalization. An incorrect claim or decision is used as a general rule or generic conclusion from insufficient but correct evidence derived from a specific or contextualized situation.
    (b) Sweeping Generalization – An incorrect generalization of a specific or contextualized situation based on insufficient but correct proof adapted from a general or generic situation.

- Incorrect health decisions based on incorrect or absent chemical information or pieces of evidence. An incorrect claim or decision is given. The pieces of evidence are incorrect and unrelated.

Gotwals and Songer (2013) evaluated argument-based explanations in ecology based on the ability to articulate the claim, the use of appropriate and sufficient evidence to support these claims, the and use of reasoning that draws on scientific principles to explicitly link the evidence to the claim or decision. The explanations of students varied greatly, particularly on the selection of evidence. Students sometimes used simplified rules (e.g., *water is a universal solvent*), non-chemistry explanations (e.g., *it is the role of the doctor to inform patients on what medicine to take*), or heuristics (e.g., *the drug is safe because it comes from plants, and plants are safe to eat*).

Some arguments involved preconceived notions or unrelated claims. We concurred that students with adequate *chemistry-based health literacy* should select appropriate chemistry-related cues from the situation, develop correct inferences based on the provided evidence(s) and evaluate the credibility of the chemical information. With these considerations, we selected four criteria for classifying the type of explanation in the chemistry-based health literacy test: a) the ability to link chemistry concepts and health decisions (or claims); b) the ability to appraise chemical information; c) the ability to generate inference from chemical data; and, d) the ability to apply latent and explicit chemistry concepts in a given situation.

9.4.4 Rubric for the Evaluation of Explanations

The scoring of the answers involved the evaluation of the student's claim or decision and explanation, based on the learning progression-based assessment described by Gotwals and Songer (2013). However, a difference in the scoring of answers in the chemistry-based health literacy test is the determination of the type of explanation based on a scoring rubric. The students' claim or decision was evaluated based on accuracy, and is coded as "0″ if it is incorrect, and "1″ if it is correct.

The explanations, on the other hand, were evaluated based on accuracy and type. The accuracy of the explanation was coded as "0" if it is incorrect, and "1" if it is correct. Then, we determined the type of explanation by obtaining the average score using the rubric (Table 9.3). Explanations were classified as Type 1, Type 2, and Type 3 based on a range of scores (Type 1 = average score between 0.00 and 0.75; Type 2 = average score between 1.51 and 2.25; Type 3 = average score between 2.26 and 3.00).

9.4.5 Scoring System and Hypothesized Progression of Chemistry-Based Health Literacy

The scoring system shown in Table 9.4 shows that the basis for the levels of chemistry-based health literacy is the accuracy of a claim (or decision), accuracy of explanation, and type of explanation. The identified levels of chemistry-based health literacy were the basis for the proposed learning progression of allied health science students. We argue that Level 0 characterizes a student who cannot choose correct claims or decisions. Even if the explanation is correct, the student does not exhibit chemistry-based health literacy if the claim is incorrect.

Answers involving incorrect claims and a correct or incorrect explanation are classified as Level 1. In previous sections, we discussed that health decisions are not entirely dependent on chemistry knowledge, and non-chemistry explanations, heuristics, or simplified rules may still guide students in selecting correct claims. Hence, we considered Level 1 chemistry-based health literacy as an ability to choose correct claims or decisions and an attempt to link explanations, albeit incomplete, non-chemistry-related or simplistic. Assignment of levels is limited by these basic premises, but we promote the idea that the "middle" levels of chemistry-based health literacy are areas that may require future investigations. For now, the middle levels proposed in the study help describe the progression of chemistry-based health literacy in allied health science students.

Using the scoring system, four levels of chemistry-based health literacy are proposed (Fig. 9.1). The first level (*insufficient* chemistry-based health literacy) corresponds to the inability to select the correct claim or health decision. The second level (*basic* chemistry-based health literacy) reflects an ability to choose a correct claim or decision and an attempt to support the claim with an explanation, even if

Table 9.3 Rubric for the evaluation of explanations in chemistry-based health literacy test

Criterion	Score	Description
Linking of chemistry concepts and health decision	**3**	Linking of chemical concepts as a basis to the health decision is based on multiple contexts on how chemistry concepts are linked to health concepts in the situation.
	2	Linking of chemical concepts as a basis for health decisions is based on a linear, one-dimensional context on how chemistry concepts are linked to health concepts in the situation.
	1	Linking of chemical concepts as a basis for health decisions is based on the similarity of concepts in chemistry and the chemistry-based concepts in the situation.
	0	Linking of chemical concepts to health decisions is based on the definition of chemical terms or description of chemistry concepts; no attempt to link chemistry concepts to health decisions
Appraisal of chemical information	**3**	Appraisal of chemistry information involves the recognition of limitations and strengths of chemical data which are relevant to the context of the health situation.
	2	Appraisal of chemistry information involves recognition of the limitations and strengths of chemical data without involving the context of the situation.
	1	Appraisal of chemistry information is based on the similarity of the chemical concept which was previously learned to the health situation but without involving the context of the situation.
	0	Appraisal of chemistry information is based on familiarity with definitions of chemical terms or concepts only; use of heuristic-based judgment; no attempt to appraise chemical information
Inference from chemical data	**3**	Inferences from chemical data are based on the nature of the chemical information and its suitability to be applied in the situation.
	2	Inferences from chemical data are based on the nature of chemical information and its similarity to the chemical concepts related to the situation.
	1	Inferences are based on simplified rules in chemistry.
	0	Inferences are limited and are based on the definition of chemical terms and concepts only; inferences are based on non-chemistry concepts or heuristics
Application of latent and explicit chemistry concept in the situation	**3**	Application of related macroscopic and abstract chemistry concepts (sub-microscopic or symbolic) which are not stated in the situation in addition to explicitly stated chemistry concepts.

(continued)

Table 9.3 (continued)

Criterion	Score	Description
	2	Application of macroscopic level or chemistry rules which are stated in the situation.
	1	Application of simplified chemistry rules or concepts that are not stated in the situation.
	0	Utilization of chemistry concepts and definitions without applying them to the health situation; use of non-chemistry-related explanations; no explanation is provided

Table 9.4 Scoring system and level of chemistry reasoning in CbHLT

Accuracy of Claim or Decision	Accuracy of Explanation	Type of Explanation	Final Score	Level of chemistry-based health literacy
Correct	Correct	Type 3	3	**Level 3**
Correct	Correct	Type 2	2	**Level 2**
Correct	Correct or incorrect	Type 1	1	**Level 1**
Incorrect	Correct or incorrect	Type 1	0	**Level 0**

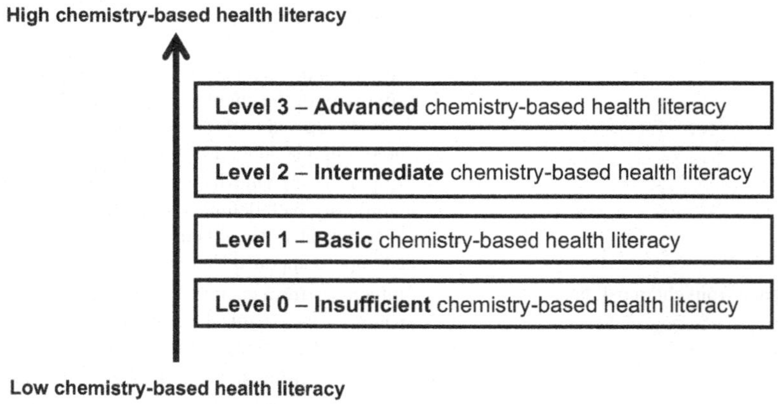

Fig. 9.1 Hypothesized progression of chemistry-based health literacy

there is a lack of contextualization, a lack of inference of chemical information, or utilization of heuristic reasoning.

The third level (*intermediate* chemistry-based health literacy) describes an ability to support correct claims or decisions with accurate explanations but exhibits deficiencies in some criteria (incomplete appraisal of information, incomplete inference of chemical data, or incomplete contextualization). The last level (*advanced* chemistry-based health literacy) is congruous to an ability to choose a correct claim or decision and support it with correct and contextualized chemistry explanations, exhibit adequate appraisal of chemical information, and display inference skills based on the given chemical data.

9.4.6 Ethics Approval

The research protocol for pilot testing of the chemistry-based health literacy test was reviewed last February 18, 2019, approved last March 21, 2019, and was assigned with Protocol Number SLU-REC 2019-023. After ethics approval, letters of communication were sent to various tertiary institutions in the Philippines offering the BS Medical Laboratory Science (Medical Technology) program to coordinate the schedule of data gathering.

9.4.7 Data Collection

The test was administered to 517 students from six tertiary education institutions in the Philippines from April 2019 to January 2020. This population consists of 97 second-year students from a tertiary institution in Baguio City), 52 incoming second-year students from a university in Manila (completed general inorganic and organic chemistry and analytical chemistry), 92 first-year students from another institution in Metro Manila (currently enrolled in general inorganic and organic chemistry), 107 second-year students from another university in Metro Manila (completed general inorganic chemistry, organic chemistry, analytical chemistry, and biochemistry), 65 first-year medical technology students from a university in Pampanga (currently enrolled in general inorganic and organic chemistry), and 104 incoming third-year students in Davao City (completed general inorganic chemistry, organic chemistry, analytical chemistry, and biochemistry).

9.4.8 Interrater Reliability

The encoded explanations were sent to two raters who had sufficient knowledge in chemistry, nutrition, diagnostics, and pharmacology (one sophomore medical doctor and an MS Science graduate who finished a health science undergraduate program). The two raters underwent four hours of training in using the rubric (Table 9.2) to classify the types of explanations. The two raters evaluated a total of 168 explanations (6 explanations per item) independently for three weeks. The interrater reliability was $Kappa = 0.82$ ($\rho < 0.001$), which implies a good agreement (Landis & Koch, 1977; McHugh, 2012). For items that received different scores from the raters, we conducted a separate meeting until a consensus was reached.

9.4.9 Data Analysis

After data entry, we applied Rasch analysis to determine the validity of the hypothesized progression of chemistry-based health literacy and to evaluate the item functioning of the chemistry-based health literacy test. All answers were rated using the same rubric and scoring system since the progression of chemistry-based health literacy was assumed to be similar in items related to nutrition, diagnostics, and pharmacology. Hence, the Rasch rating scale model is considered appropriate. In a rating scale model, each item has the same threshold for the rating scale categories (Planinic et al., 2019). The Rasch rating scale model is more appropriate in this study since all items were assumed to share the same rating scale structure. The rating categories used in this study corresponded to the levels of chemistry-based health literacy and reflected a hypothesized progression of the construct.

Four general Rasch indices – person and item reliability, and person and item separation – were evaluated to determine the instrument quality and item hierarchy. Next, the items' misfits through the mean square (MNSQ) and standardized values (ZTSD) outfit and infit summary statistics were explored to investigate instrument functioning. The MNSQ outfit was regarded as sensitive to outliers (Neumann et al., 2013), so the MNSQ infit was given priority. Since the test is designed for the assessment and evaluation of allied health science students enrolled in chemistry subjects, we selected MNSQ values between 0.70 and 1.30 and ZTSD values below between −2.0 and +2.0 to be acceptable. The point-measure correlation was also investigated to determine whether an item contributed to the construct measured by the instrument (Arnold et al., 2018). Unidimensionality was evaluated by using principal component analysis (PCA) on standardized residuals. Rasch analysis was performed in Winsteps 4.4.5.

The Rasch rating scale model allows the evaluation of the validity of the progression of chemistry-based health literacy, based on the assumption that the level should increase with an increasing ability to apply contextualized chemistry explanations to correct claims or health decisions. The average measure and Andrich thresholds were investigated to evaluate this in the rating scale model. In a similar manner, Todd and Romine (2016) have applied the Rasch rating scale model to evaluate the validity with respect to the progression of the construct "genes code for proteins" in the assessment of modern genetics among college students.

9.4.10 Unidimensionality

According to Linacre (2012), the unexplained variance by the first contrast should be below 5% (less than 2 eigenvalues) if the items are exhibiting unidimensionality. The results of the analysis reported an eigenvalue = 1.91 in the first contrast, which satisfied the condition for unidimensionality. The result indicates that the items forming the chemistry-based health literacy test are measuring the same construct, and the response given to one item is independent of the response to other items in the instrument.

9.4.11 Item Fit Statistics

Table 9.5 shows the item difficulty estimates and fit statistics of the items in the chemistry-based health literacy test. Three items (item N1, item D1, and item N6) had MNSQ infit values which are not within 0.70 to 1.30. Item N12, on the other hand, had an MNSQ outfit >1.30. The ZTSD values were least prioritized in the

Table 9.5 Item fit statistics for CbHLT items

				Infit		Outfit		
Item Number	Label	Measure	SE	MNSQ	ZSTD	MNSQ	ZSTD	Pt. Meas. Correl.
1	N1	0.12	0.05	1.50	8.45	1.49	8.02	0.48
2	P1	0.04	0.05	1.14	2.53	1.13	2.40	0.44
3	N2	−0.09	0.05	1.03	0.55	1.04	0.75	0.46
4	NP1	0.12	0.05	0.95	−0.96	0.93	−1.35	0.51
5	N3	−0.49	0.05	1.09	1.61	1.06	1.11	0.54
6	P2	1.13	0.06	0.99	−0.12	1.02	0.36	0.33
7	D1	−0.34	0.05	0.60	−8.87	0.61	−8.40	0.52
8	N4	0.77	0.05	0.89	−2.06	0.95	−0.74	0.39
9	N5	1.60	0.07	1.15	1.96	1.27	3.02	0.25
10	P3	0.40	0.05	0.95	−0.90	0.94	−1.02	0.49
11	P4	0.94	0.06	0.77	−4.23	0.81	−3.22	0.37
12	NP2	0.82	0.05	1.14	2.39	1.09	1.37	0.49
13	N6	1.13	0.06	1.32	4.67	1.25	3.36	0.39
14	D2	0.40	0.05	0.80	−4.17	0.78	−4.19	0.54
15	P5	1.04	0.06	1.02	0.41	0.98	−0.25	0.44
16	D3	1.01	0.06	0.80	−3.65	0.82	−2.81	0.37
17	P6	0.80	0.05	0.97	−0.58	0.97	−0.44	0.44
18	N7	1.18	0.06	0.95	−0.73	0.94	−0.91	0.33
19	N8	1.31	0.06	1.19	2.78	1.17	2.15	0.36
20	P7	1.00	0.06	0.98	−0.38	0.95	−0.72	0.50
21	N9	1.10	0.06	0.83	−2.92	0.81	−2.93	0.39
22	N10	0.71	0.05	0.82	−3.44	0.80	−3.63	0.49
23	N11	1.65	0.07	1.00	0.03	0.97	−0.28	0.36
24	N12	1.96	0.08	1.22	2.49	1.32	3.08	0.19
25	N13	1.81	0.07	1.26	3.12	1.25	2.61	0.21
26	D4	1.63	0.07	0.87	−1.80	0.91	−1.09	0.32
27	D5	1.80	0.07	0.99	−0.08	1.08	0.93	0.21
28	D6	0.98	0.06	0.88	−2.04	0.89	−1.72	0.43

Note: Type of topics – *N* nutrition, *D* diagnostics, *NP* natural products, *P* pharmacology

analysis since ZTSD measures can be ignored if MNSQ values are within the acceptable range (Boone et al., 2015). The point measure correlation ranged from 0.19 to 0.54, indicating that all items are contributing to the construct that is measured. When the items fit the Rasch model, the items are assumed to measure the underlying construct (Oon et al., 2016).

Based on the results, there were four misfitting items. A Rasch analysis involving the remaining 24 items showed adequate reliability (person reliability = 0.99, item reliability = 0.82), and separation (person separation 2.17, item separation = 10.53). The instrument also exhibited unidimensionality (first eigenvalue = 1.42). All MNSQ infit and outfit values were within 0.70 to 1.30, and the ZTSD values were within −2.0 to +2.0. The point measure correlations ranged from 0.34 to 0.53, indicating that all items contribute to the measurement of the construct. These results suggest that the revised chemistry-based health literacy test fits the Rasch model.

9.4.12 Validity of Progression of Chemistry-Based Health Literacy

The four levels of the chemistry reasoning progression were used in all items in nutrition, diagnostics, and pharmacology in the chemistry-based health literacy test. Table 9.6 shows that the four categories, corresponding to the four levels of chemistry-based health literacy, have acceptable infit and outfit MNSQ. With an increase in the category score (0–1 to 2–3), the category measures also increased from −2.27 to 2.25, indicating that the rating scale is functioning as intended.

9.4.13 Item Category Function

Figure 9.2 shows the category probability of the four levels of chemistry-based health literacy using the Rasch rating scale model. Based on the figure, in each category, the Andrich thresholds are ordered in correspondence with the rating category, which corresponds to the four levels of chemistry-based health literacy. Since the thresholds increase monotonically for adjacent categories, it means that the

Table 9.6 Summary of the rating scale category

Category	Observed Count	Observed %	Infit MNSQ	Outfit MNSQ	Category Measure
0	5202	42	1.01	1.01	−2.27
1	4108	33	0.99	0.99	−0.63
2	2229	18	1.00	0.99	0.64
3	869	7	1.00	0.99	2.25

Note: Category 0 stands for “Level 1”; Category 1 stands for “Level 2”; Category 2 stands for “Level 3”; Category 3 stands for “Level 4”

```
P      -+---------+---------+---------+---------+---------+---------+-
R  1.0 +                                                             +
O      |                                                             |
B      |0                                                           3|
A      | 0000                                                  3333  |
B   .8 +     000                                            333      +
I      |        00                                        33         |
L      |          00                                    33           |
I      |            00                                33             |
T   .6 +              00                            33               +
Y      |                0                         33                 |
    .5 +                 00                      3                   +
O      |                   00                  33                    |
F   .4 +                  111*11111111  2222222*22                   +
       |               111    00    2**1     33  2222                |
R      |            111        0  22    11 3         222             |
E      |         111           2**       3*1            222          |
S   .2 +      111             22   00   33   11             222      +
P      |  1111            222       0*3       111           2222   |
O      |11              222        333 000       1111              22|
N      |         2222222    333333       000000      111111          |
S   .0 +*********333333333333                   000000000000*********+
E      -+---------+---------+---------+---------+---------+---------+-
       -3        -2        -1          0         1         2         3
        Person [MINUS] Item MEASURE
```

Fig. 9.2 Category probability curve of the four-level rating scale of chemistry-based health literacy test

levels of chemistry-based health literacy are functioning as intended in all items in nutrition, diagnostics, and pharmacology.

9.4.14 Using CbHLT for Dialogistic Assessment and Teaching Evaluation

Table 9.7 shows the conversion table of raw scores to Rasch scale scores. The scores ranged from 1 to 72 based on the scoring system that measures chemistry-based health literacy. By using the table as a reference, teachers can estimate the Rasch scale score from the raw score without performing a Rasch analysis. For example, if a student's raw score in CbHLT is 40, the score of the student on the Rasch scale is 0.24. Using the conversion table aids in the estimation of the student's ability on a Rasch scale without performing Rasch analysis.

The table (Table 9.7) also allows the estimation of the student's ability across all items of the chemistry-based health literacy test. Figure 9.3 shows the probable progression of levels in each item for a student's logit measure. The items on the right show the arrangement of items based on difficulty. The item at the bottom (item N3) is the easiest, while the item at the top (item N13) is the most difficult. The numbers in each item correspond to the predicted level of progression of chemistry-based health literacy (level 0, level 1, level 2, and level 3). We also

Table 9.7 Conversion table of raw scores to Rasch scale scores

Score	Ability estimate	SE	Score	Ability estimate	SE	Score	Ability estimate	SE
1	−4.33	1.01	25	−0.64	0.25	49	0.79	0.25
2	−3.62	0.72	26	−0.58	0.25	50	0.85	0.26
3	−3.2	0.59	27	−0.51	0.25	51	0.92	0.26
4	−2.9	0.52	28	−0.45	0.25	52	0.99	0.26
5	−2.65	0.47	29	−0.39	0.24	53	1.06	0.27
6	−2.45	0.43	30	−0.33	0.24	54	1.13	0.27
7	−2.28	0.4	31	−0.27	0.24	55	1.21	0.28
8	−2.13	0.38	32	−0.22	0.24	56	1.29	0.28
9	−1.99	0.36	33	−0.16	0.24	57	1.37	0.29
10	−1.87	0.34	34	−0.10	0.24	58	1.45	0.30
11	−1.76	0.33	35	−0.04	0.24	59	1.55	0.31
12	−1.65	0.32	36	0.01	0.24	60	1.64	0.32
13	−1.55	0.31	37	0.07	0.24	61	1.74	0.33
14	−1.46	0.3	38	0.13	0.24	62	1.85	0.34
15	−1.37	0.29	39	0.19	0.24	63	1.97	0.35
16	−1.28	0.29	40	0.24	0.24	64	2.11	0.37
17	−1.20	0.28	41	0.30	0.24	65	2.25	0.40
18	−1.12	0.28	42	0.36	0.24	66	2.42	0.42
19	−1.05	0.27	43	0.42	0.24	67	2.62	0.46
20	−0.97	0.27	44	0.48	0.24	68	2.85	0.51
21	−0.90	0.26	45	0.54	0.25	69	3.15	0.59
22	−0.84	0.26	46	0.60	0.25	70	3.57	0.71
23	−0.77	0.26	47	0.66	0.25	71	4.27	1.01
24	−0.70	0.25	48	0.72	0.25	72	5.48E	1.83

observed that our hypothesized item ordering based on item difficulty was valid. Based on the item difficulty, students had more difficulty in items that involve *in vivo* chemical systems or complex biochemistry topics. The distribution of items also suggests that chemistry concepts are applied differently in items related to nutrition, diagnostics, and pharmacology.

For example, if a student obtained a pretest score of 0.01 logit (raw score = 36), the student is more likely to exhibit level 1 (basic) chemistry-based health literacy in items N3, N2, NP1, P1, D2, and D3. Additionally, the same student is predicted to display level 0 (insufficient) chemistry-based health literacy in the remaining items. Based on the scoring rubric, a Level 1 chemistry-based health literacy indicates that the student can identify a correct health decision, but the explanation may be insufficient. Hence, the science teacher can be guided to improve the skills of students along the construct of *chemistry-based health literacy*. The teacher can focus on improving the accuracy of the student's explanation, ability to link chemistry concepts and health decisions (or claims), ability to appraise chemical

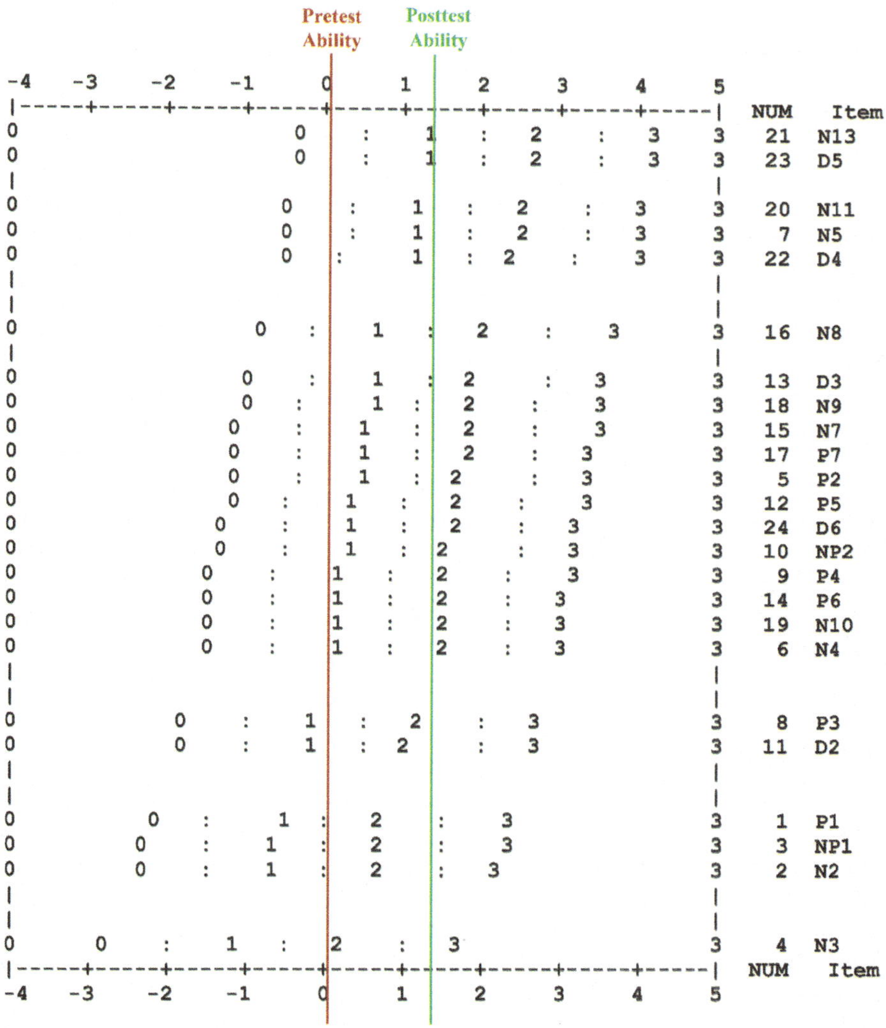

Fig. 9.3 Chemistry-based health literacy test variable map. The red and green lines indicate the student's ability and the possible level of chemistry-based health literacy in each item during the pretest and posttest, respectively. Items are arranged based on increasing difficulty from bottom to top

information, ability to generate inference from chemical data, and ability to apply latent and explicit chemistry concepts.

On the other hand, obtaining Level 0 in the items indicates that the student is unable to choose the correct claim or health decision. Hence, the teacher can focus on supporting the student's ability to select the most plausible claim or health decision, based on the context of the given situation. Since each topic has a specific embedded chemistry concept (Table 9.1), the science teacher can also focus on enhancing the knowledge of students on this chemistry topic to improve the ability of students to utilize chemical knowledge in explaining claims or health decisions.

The effectiveness of a teaching intervention in the chemistry subjects offered to health science students can also be evaluated using the students' ability to measure and the category probabilities per item. For instance, a science teacher introduced a new teaching approach in chemistry. After a teaching intervention, this student obtained a posttest score of 1.37 (raw score = 57). If the pretest score was initially 0.01, the science teacher can infer that the student's level of chemistry-based health literacy in items N3, N2, NP1, P1, D2, and D3 has progressed to Level 2 while the chemistry-based health literacy of the student in the remaining items increased from Level 0 to Level 1.

As described earlier, level 2 (intermediate chemistry-based health literacy) is characterized by the ability to choose a correct claim or health decision and utilize a correct chemistry-based explanation, but the explanations may lack contextualization, lack of inference from chemical information, or use of incorrect heuristic reasoning. The changes in the scores during the pretest and the posttest can be used to determine whether the teaching intervention was effective or not and whether the teaching intervention influenced the ability of students to select the correct claim or health decision or improve the student's ability to link correct and contextualized chemistry explanations to the claim or health decision.

9.5 Discussion

Allied health science programs promote an evidence-based approach in their curriculum (Asokan, 2012; Patelarou et al., 2017). With this thrust, science education in allied health science programs should also focus on teaching students to utilize scientific information as pieces of evidence. Previous studies that have integrated science concepts into relevant real-life issues such as climate change (Eggert et al., 2016; Flener-Lovitt, 2014) and biofuels (Dishadewi et al., 2020) have also highlighted the role of chemistry concepts in socio-scientific discourses.

Similarly, chemistry concepts serve as a cognitive lens that guides student reasoning when choosing decisions or claims. There are several socio-scientific issues related to health (Lee, 2012), and a holistic approach requires scientific evidence in decision-making. By applying Rasch analysis, we demonstrate that it is possible to establish the validity of the chemistry-based health literacy test and use the quantitative measurement of student ability. Furthermore, Rasch analysis supported the proposed item order, thereby providing a basis for the design of appropriate learning activities that can enhance the ability of students to select a health decision and use a correct and contextualized chemistry explanation as evidence to support the decision.

In this study, Rasch analysis has been an indispensable tool in validating the progression of chemistry-based health literacy and determining the validity of the chemistry-based health literacy test that measures the construct. In this study, the Rasch rating scale model was used to validate the raring scale corresponding to the levels of chemistry-based health literacy. As defined, chemistry-based health literacy describes the ability to support correct health decisions with accurate and contextualized explanations based on chemistry concepts and avoid heuristic-based thinking. As long as the hypothesized construct is sufficiently conceptualized along with theoretical considerations, it is possible to revise or refine the instrument and change the rating scale (Boone, 2016).

The validity of the rating scale corresponding to the four levels of chemistry-based health literacy has supported our hypothesis that an increasing level reflects the ability of students to choose a health decision and then rationalize the decision with correct and contextualized explanations. Our postulate that a similar pattern of reasoning in items related to nutrition, diagnostics, and pharmacology is also validated. With a rating scale model, the sufficiency of the rating scale options corresponding to the levels of chemistry-based health literacy is validated, allowing a quantitative analysis of results. However, the proposed levels may apply to students in the allied health science programs, particularly first-year and second-year medical laboratory science students. At this point, we divulge that the purpose of the chemistry-based health literacy test was to evaluate the effectiveness of a proposed teaching intervention in biochemistry for allied health science programs.

Aside from testing the validity of the levels of chemistry-based health literacy, Rasch analysis has also provided validity evidence of the hypothesized difficulty of items and helps in designing appropriate teaching interventions to improve the expected competency of the students. Indeed, it was more difficult for students to choose a health decision and apply correct and contextualized chemistry explanations in items that involved *in vivo* chemical systems, especially in nutrition-related topics (Fig. 9.3). These items include item N13 (fruit extracts), item N11 (food preservatives), item N5 (cholesterol-rich diet), and item N8 (food as nutraceutical products). Choosing correct health decisions or claims related to nutrition items is difficult because there are several considerations to include. In analyzing nutrition contexts, we consider the type of nutrient or contaminant in food, food preparation or processing, the number of nutrients required to expect health benefits, and the metabolism of the nutrients or contaminants in the human body. Hence, students were expected to have difficulties extrapolating the health effects based on available chemical information. In addition, decision-making related to the health valuation of food is challenged by the proliferation of pseudoscience and misinformation (Neale & Tapsell, 2019). In line with measuring the chemistry-based health literacy of students, the analysis of answers to items related to nutrition requires caution, and effort to disprove inaccurate claims and pseudoscientific information should utilize scientific evidence.

On the other hand, items with the lower difficulty involved *in vitro* systems and required recall of chemistry concepts from general inorganic chemistry, organic chemistry, and analytical chemistry. The results highlight the role of prior knowledge of chemistry concepts in supporting inferences and claims in health contexts and affirm previous reports that chemical sciences are instrumental in understanding complex health topics (Armstrong & Poë, 2020; Matlin et al., 2017). However, of equal importance is to highlight the limitations of inferences from chemical reactions observed in chemistry laboratories in academic settings because food, natural products, medicines, and biological samples are complex chemical matrices (Beutler, 2019; Peng et al., 2016).

The results of Rasch analysis can also inform the science teacher about the expected competencies of allied health science students when they complete their chemistry subjects. In the medical laboratory science program, general inorganic chemistry, organic chemistry, and analytical chemistry are offered during the first year while biochemistry is offered during the second year. Students in the first year are expected to finish all chemistry subjects before enrolling in professional subjects such as clinical chemistry. With the results, first-year medical laboratory science students are not expected to answer items that involve *in vivo* chemical systems but may show adequate competencies when answering items based on *in vitro* chemical systems. In contrast, students who have completed all chemistry prerequisites, including biochemistry, are expected to exhibit adequate competency to support health decisions with chemistry explanations involving *in vivo* chemical systems. By introducing previously learned concepts and highlighting the unique characteristics of an *in vivo* chemical system, students can learn the limitations of *in vitro* chemical information and may become more critical when making decisions.

The variable map allows a qualitative measure of the approximate level of chemistry-based health literacy in all items. The map provides an opportunity to directly evaluate the effectiveness of teaching interventions based on the changes in the student's ability. Along the hypothesized progression, students can be given a diagnosis. As discussed, the Rasch scale scores can be compared during the pretest and posttest without performing Rasch analysis by referring to the conversion table (Table 9.7). Changes in student ability can inform science teachers whether the teaching intervention promotes a higher level of chemistry-based health literacy or not. Since the levels of chemistry-based health literacy reflect an increase in the competencies presented in Table 9.3, chemistry teachers may be guided in selecting learning activities that focus on improving the ability of students to appraise chemical information, generate inferences from chemical data, and apply latent and explicit chemistry concepts. However, we caution teachers not to "teach correct health decisions" but focus on developing a thinking process that is required to guide the selection of a correct health decision.

Determining chemistry-based health literacy using the rubric for the evaluation of explanations may be time-consuming, and a reduction of items may be warranted for future studies. An ordered multiple-choice format may provide advantages in making the test easier for teachers to evaluate, as long as the options reflect the hypothesized levels of chemistry-based health literacy. When all items fit the Rasch model, the mean ability of students in items related to nutrition, diagnostics, and pharmacology can be determined and analyzed by simple parametric tests such as analysis of variance. Similarly, the mean ability of students in a class can be compared on topics that involve *in vivo* chemical systems, allowing chemistry educators to diagnose the difficulties of students based on the expected level of chemistry-based health literacy.

9.6 Conclusion

In this chapter, we described the application of Rasch modeling in validating the progression of chemistry-based health literacy, a construct that reflects the ability to support correct health decisions with correct and contextualized chemistry explanations in decisions involving nutrition, diagnostics, and pharmacology. Additionally, this study demonstrated that the Rasch rating scale model provides evidence of the validity of the rating scale corresponding to the increasing levels of chemistry-based health literacy. The application of Rasch analysis promotes a reliable measurement of student ability along the hypothesized progression of the construct and guides chemistry educators in designing appropriate learning activities and teaching interventions that promote chemistry concepts as meaningful and relevant sources of evidence in achieving the expected competencies of students in allied health science programs.

Acknowledgments We would like to acknowledge the Department of Science and Technology Science Education Institute (DOST-SEI) through the Capacity Building Program in Science and Mathematics Education (CBPSME) for providing support during the conduct of the study.

Appendix

Code

```
&INST
 TITLE = 'CbHLT.xlsx'
 PERSON = Person ; persons are ...
 ITEM = Item ; items are ...
 ITEM1 = 1 ; column of response to first item in data record
 NI = 24 ; number of items
 NAME1 = 26 ; column of first character of person identifying label
NAMELEN = 6 ; length of person label
 XWIDE = 1 ; number of columns per item response
 CODES = 0123 ; valid codes in data file
 UIMEAN = 0 ; item mean for local origin
 USCALE = 1 ; user scaling for logits
 UDECIM = 2 ; reported decimal places for user scaling
 ; Excel file created or last modified: 'date of analysis'
 ; Worksheet
 ; Excel Cases processed = 517
 ; Excel Variables processed = 26
 ; GROUPS = 0 Partial Credit model: in case items have different rating
```

```
scales
TOTALSCORE = Yes ; Include extreme responses in reported scores
; Person Label variables: columns in label: columns in line
@No. = 1E3 ; $C26W3
@Year-level = 5E5 ; $C30W1

&END
P1
N2
NP1
N3
P2
N4
N5
P3
P4
NP2
D2
P5
D3
P6
N7
N8
P7
N9
N10
N11
N13
D4
D5
D6
END LABELS
1213331312101300212011211 2
...
*WINSTEPS codes
```

References

Abel, T., Hofmann, K., Ackermann, S., Bucher, S., & Sakarya, S. (2014). Health literacy among young adults: A short survey tool for public health and health promotion research. *Health Promotion International, 30*(3), 725–735. https://doi.org/10.1093/heapro/dat096

Agosto, H. G. C., Briones, M. V. A., & Palatino, M. C. (2018). Correlates of health literacy among Filipinos aged 50–70 years old belonging to low-income families in a selected community. *Acta Medica Philippina, 52*(3), 239–244. https://doi.org/10.47895/amp.v52i3.397

Armstrong, D., & Poë, J. (2020). The science of human health—A context-based chemistry course for non-science majors incorporating systems thinking. *Journal of Chemical Education, 97*(11), 3957–3965. https://doi.org/10.1021/acs.jchemed.0c00887

Arnold, J., Boone, W., Kremer, K., & Mayer, J. (2018). Assessment of competencies in scientific inquiry through the application of Rasch measurement techniques. *Education Sciences, 8*(4), 184. https://doi.org/10.3390/educsci8040184

Asokan, G. (2012). Evidence-based practice curriculum in allied health professions for teaching-research-practice nexus. *Journal of Evidence-Based Medicine, 5*(4), 226–231. https://doi.org/10.1111/jebm.12000

Bell, R., & Lederman, N. (2003). Understandings of the nature of science and decision making on science and technology-based issues. *Science Education, 87*(3), 352–377. https://doi.org/10.1002/sce.10063

Benfenati, E., Gini, G., Hoffmann, S., & Luttik, R. (2010). Comparing in vivo, in vitro and in Silico methods and integrated strategies for chemical assessment: Problems and prospects. *Alternatives to Laboratory Animals, 38*(2), 153–166. https://doi.org/10.1177/026119291003800201

Beutler, J. (2019). Natural products as a Foundation for Drug Discovery. *Current Protocols in Pharmacology, 86*(1). https://doi.org/10.1002/cpph.67

Blumenthal-Barby, J., & Krieger, H. (2014). Cognitive biases and heuristics in medical decision making. *Medical Decision Making, 35*(4), 539–557. https://doi.org/10.1177/0272989x14547740

Boone, W. (2016). Rasch analysis for instrument development: Why, when, and how?. CBE-life sciences. *Education, 15*(4), rm4. https://doi.org/10.1187/cbe.16-04-0148

Boone, W., Townsend, J., & Staver, J. (2015). Utilizing multifaceted Rasch measurement through FACETS to evaluate science education data sets composed of judges, respondents, and rating scale items: An exemplar utilizing the elementary science teaching analysis matrix instrument. *Science Education, 100*(2), 221–238. https://doi.org/10.1002/sce.21210

Camiling, M. (2019). eHealth literacy of high school students in The Philippines. *IAFOR Journal of Education, 7*(2), 69–87. https://doi.org/10.22492/ije.7.2.04

Cooper, M., & Stowe, R. (2018). Chemistry education research—From personal empiricism to evidence, theory, and informed practice. *Chemical Reviews, 118*(12), 6053–6087. https://doi.org/10.1021/acs.chemrev.8b00020

Dishadewi, P., Wiyarsi, A., Prodjosantoso, A., & Nugraheni, A. (2020). Chemistry-based socio-scientific issues (SSis) as a learning context: An exploration study of biofuels. *Journal of Physics: Conference Series, 1440*(1), 012007. https://doi.org/10.1088/1742-6596/1440/1/012007

Eggert, S., Nitsch, A., Boone, W., Nückles, M., & Bögeholz, S. (2016). Supporting students' learning and Socioscientific reasoning about climate change—The effect of computer-based concept mapping scaffolds. *Research in Science Education, 47*(1), 137–159. https://doi.org/10.1007/s11165-015-9493-7

Featherston, R., Downie, L., Vogel, A., & Galvin, K. (2020). Decision making biases in the allied health professions: A systematic scoping review. *PLoS One, 15*(10), e0240716. https://doi.org/10.1371/journal.pone.0240716

Flener-Lovitt, C. (2014). Using the Socioscientific context of climate change to teach chemical content and the nature of science. *Journal of Chemical Education, 91*(10), 1587–1593. https://doi.org/10.1021/ed4006985

Gotwals, A., & Songer, N. (2013). Validity evidence for learning progression-based assessment items that fuse Core disciplinary ideas and science practices. *Journal of Research in Science Teaching, 50*(5), 597–626. https://doi.org/10.1002/tea.21083

Hawkins, M., Gill, S., Batterham, R., Elsworth, G., & Osborne, R. (2017). The health literacy questionnaire (HLQ) at the patient-clinician interface: A qualitative study of what patients and clinicians mean by their HLQ scores. *BMC Health Services Research, 17*(1). https://doi.org/10.1186/s12913-017-2254-8

King, D. (2012). New perspectives on context-based chemistry education: Using a dialectical sociocultural approach to view teaching and learning. *Studies in Science Education, 48*(1), 51–87. https://doi.org/10.1080/03057267.2012.655037

Kozma, R. (2003). The material features of multiple representations and their cognitive and social affordances for science understanding. *Learning and Instruction, 13*(2), 205–226. https://doi.org/10.1016/s0959-4752(02)00021-x

Landis, J., & Koch, G. (1977). The measurement of observer agreement for categorical data. *Biometrics, 33*(1), 159. https://doi.org/10.2307/2529310

Lee, Y. (2012). Socio-scientific issues in health contexts: Treading a rugged terrain. *International Journal of Science Education, 34*(3), 459–483. https://doi.org/10.1080/09500693.2011.613417

Lehane, E., Leahy-Warren, P., O'Riordan, C., Savage, E., Drennan, J., O'Tuathaigh, C., et al. (2018). Evidence-based practice education for healthcare professions: An expert view. *BMJ Evidence-Based Medicine, 24*(3), 103–108. https://doi.org/10.1136/bmjebm-2018-111019

Linacre, J. M. (2012). *Winsteps help for Rasch analysis*. Retrieved from Winsteps website: http://www.winsteps.com.

Maeyer, J., & Talanquer, V. (2010). The role of intuitive heuristics in students' thinking: Ranking chemical substances. *Science Education, 94*(6), 963–984. https://doi.org/10.1002/sce.20397

Matlin, S., Mehta, G., Krief, A., & Hopf, H. (2017). The chemical sciences and health: Strengthening synergies at a vital interface. *ACS Omega, 2*(10), 6819–6821. https://doi.org/10.1021/acsomega.7b01463

McHugh, M. (2012). Interrater reliability: The kappa statistic. *Biochemia Medica*, 276–282. https://doi.org/10.11613/bm.2012.031

Mullan, J., Burns, P., Weston, K., McLennan, P., Rich, W., Crowther, S., et al. (2017). Health literacy amongst health professional university students: A study using the health literacy questionnaire. *Education Sciences, 7*(2), 54. https://doi.org/10.3390/educsci7020054

Neale, E., & Tapsell, L. (2019). Perspective: The evidence-based framework in nutrition and dietetics: Implementation, challenges, and future directions. *Advances in Nutrition, 10*(1), 1–8. https://doi.org/10.1093/advances/nmy113

Neumann, K., Viering, T., Boone, W. J., & Fischer, H. E. (2013). Towards a learning progression of Energy. *Journal of Research in Science Teaching, 50*(2), 162–188. https://doi.org/10.1002/tea.21061

Oon, P., Spencer, B., & Kam, C. (2016). Psychometric quality of a student evaluation of teaching survey in higher education. *Assessment and Evaluation in Higher Education, 42*(5), 788–800. https://doi.org/10.1080/02602938.2016.1193119

Pang, G., Xie, J., Chen, Q., & Hu, Z. (2014). Energy intake, metabolic homeostasis, and human health. *Food Science and Human Wellness, 3*(3-4), 89–103. https://doi.org/10.1016/j.fshw.2015.01.001

Patelarou, A., Kyriakoulis, K., Stamou, A., Laliotis, A., Sifaki- Pistolla, D., Matalliotakis, M., et al. (2017). Approaches to teach evidence-based practice among health professionals: An overview of the existing evidence. *Advances in Medical Education and Practice, 8*, 455–464. https://doi.org/10.2147/amep.s134475

Peng, J., Tang, F., Zhou, R., Xie, X., Li, S., Xie, F., et al. (2016). New techniques of on-line biological sample processing and their application in the field of biopharmaceutical analysis. *Acta Pharmaceutica Sinica B, 6*(6), 540–551. https://doi.org/10.1016/j.apsb.2016.05.016

Planinic, M., Boone, W., Susac, A., & Ivanjek, L. (2019). Rasch analysis in physics education research: Why measurement matters. *Physical Review Physics Education Research, 15*(2). https://doi.org/10.1103/physrevphyseducres.15.020111

Polit, D., & Beck, C. (2006). The content validity index: Are you sure you know what's being reported? Critique and recommendations. *Research in Nursing and Health, 29*(5), 489–497. https://doi.org/10.1002/nur.20147

Randles, C., & Overton, T. (2015). Expert vs. novice: Approaches used by chemists when solving open-ended problems. *Chemistry Education Research and Practice, 16*(4), 811–823. https://doi.org/10.1039/c5rp00114e

Rehberger, K., Kropf, C., & Segner, H. (2018). In vitro or not in vitro: A short journey through a long history. *Environmental Sciences Europe, 30*(1). https://doi.org/10.1186/s12302-018-0151-3

Talanquer, V. (2014). Chemistry education: Ten heuristics to tame. *Journal of Chemical Education, 91*(8), 1091–1097. https://doi.org/10.1021/ed4008765

Thomas, A., & Ellaway, R. (2021). Rethinking implementation science for health professions education: A manifesto for change. *Perspectives on Medical Education, 10*(6), 362–368. https://doi.org/10.1007/s40037-021-00688-3

Thummathong, R., & Thathong, K. (2018). Chemical literacy levels of engineering students in Northeastern Thailand. *Kasetsart Journal of Social Sciences, 39*(3), 478–487. https://doi.org/10.1016/j.kjss.2018.06.009

Todd, A., & Romine, W. (2016). Validation of the learning progression-based assessment of modern genetics in a college context. *International Journal of Science Education, 38*(10), 1673–1698. https://doi.org/10.1080/09500693.2016.1212425

von Winterfeldt, D. (2013). Bridging the gap between science and decision making. *Proceedings of the National Academy of Sciences, 110*(supplement_3), 14055–14061. https://doi.org/10.1073/pnas.1213532110

Jonathan M. Barcelo, Ph.D., is a professor at Saint Louis University, Baguio City, Philippines, and a graduate of the University of the Philippines (UP) Open University, Los Baños, Laguna, Philippines. He teaches general inorganic chemistry, organic chemistry, and biochemistry subjects to allied health science students.

Nona Marlene B. Ferido, Ph.D., is the deputy director of the Assessment, Curriculum, & Technology Research Centre. Some of her research involvements in the Centre were: progress of students through the science curriculum, curriculum review of the Philippine K to 12 curriculum, feasibility study on micro-certification for the alternative learning system (ALS) in the Philippines, and the development and conduct of the professional development (PD) program for science, mathematics and reading teachers on assessment and emerging literacies. She was an Affiliate Associate Professor of the Faculty of Education of the University of the Philippines (UP) Open University, and served as a Science Education Specialist of the University of the Philippines National Institute for Science & Mathematics Education Development.

Chapter 10
Using Rasch Measurement to Develop Model-Based Reasoning Assessment Tasks

Cari F. Herrmann-Abell, Molly A. M. Stuhlsatz, Christopher D. Wilson, Jeffery Snowden, and Brian M. Donovan

Abstract This chapter details the role that Rasch measurement played in the development of an assessment instrument that can be used to measure the complex science learning described in the *Next Generation Science Standards* during a large-scale efficacy study. Four model-based reasoning (MBR) tasks were developed and tested along with content-focused (CF) items with high-school biology students, high-school biology teachers, and crowd-sourced adults. Rasch modeling was used to investigate the relative difficulties of the items within the tasks, explore the relationship between performance on the MBR tasks and the CF items, and compare the performance of students, teachers, and adults.

This chapter describes an assessment development effort that is part of a project to examine the efficacy of high school biology instructional materials. The Model-Based Educational Resources (MBER) materials were designed to provide students opportunities to engage in modeling to generate scientific understanding. MBER supports teachers' understanding and practice of model-based reasoning as an approach to support high school students in developing an integrated, multidimensional understanding of science. The MBER instructional materials are built around a clear storyline with explicit connections between biological ideas across the full year biology sequence and within each instructional unit. They provide teachers with pedagogical supports that outline how to engage students in the intellectual work of the classroom. This includes the use of an instructional framework that highlights the connections between the three main components of any modeling lesson: the phenomenon, the question about the phenomenon, and the model that explains the phenomenon. Finally, the curriculum gives teachers the freedom to modify the sequence to accommodate their needs. Having previously demonstrated significant promise of efficacy, the MBER materials are currently being studied in a large-scale quasi-experiment, which necessitated the development

C. F. Herrmann-Abell (✉) · M. A. M. Stuhlsatz · C. D. Wilson · J. Snowden · B. M. Donovan
BSCS Science Learning, Colorado Springs, CO, USA
e-mail: cabell@bscs.org

X. Liu, W. J. Boone (eds.), *Advances in Applications of Rasch Measurement in Science Education*, Contemporary Trends and Issues in Science Education 57,
https://doi.org/10.1007/978-3-031-28776-3_10

of teacher and student achievement measures. The goal of the assessment development effort is to create tasks that can be used to measure students' and teachers' ability to use modeling and model-based reasoning to make sense of biological phenomena. Rasch measurement was used to investigate the extent to which the items that make up the instrument form a unidimensional scale and the extent to which the scale formed by the instrument covers the full range of student and teacher performance.

10.1 Theoretical Framework

10.1.1 Multidimensional Science Learning

The *Next Generation Science Standards* (NGSS Lead States, 2013) organizes goals for student learning along three dimensions: (1) *scientific and engineering practices* includes asking questions and defining problems; developing and using models; planning and carrying out investigations; analyzing and interpreting data; using mathematics and computational thinking; constructing explanations and designing solutions; engaging in argument from evidence; and obtaining, evaluating, and communicating information; (2) *crosscutting concepts* includes patterns; cause and effect; scale, proportion, and quantity; systems and system models; energy and matter; structure and function; and stability and change; and (3) *disciplinary core ideas* includes foundational ideas from the life, physical, and earth sciences and engineering and technology. To further emphasize the vision of integrated learning and to guide instruction and assessment in science, the NGSS is organized around a set of *performance expectations* that combine what students should know and be able to do along the three dimensions.

10.1.2 Multidimensional Science Assessment

In *Developing Assessments for the Next Generation Science Standards* (NRC, 2014), the National Research Council has offered a set of recommendations calling for assessments that are "significantly different from those in current use" (p. 3). They state that assessing students' three-dimensional science understanding will require "assessment tasks that examine students' performance of scientific and engineering practices in the context of crosscutting concepts and disciplinary core ideas" (NRC, 2014, p. 44). Additionally, they advocate for the use of sets of interrelated items where the individual items may target discrete core ideas, practices, or crosscutting concepts but when taken as a whole, provide a complete picture of students' three-dimensional science understanding.

10.1.3 Assessment Development

We took a construct-centered approach to assessment development as recommended by the NRC (2014). This approach draws from frameworks such as Evidence Centered Design (Mislevy et al., 2003) and Construct Modeling (Wilson, 2005) as well as work done by AAAS Project 2061 (DeBoer, et al., 2008). The foundation of this approach is the development of a detailed description of the content and practices to be assessed. Developing clear descriptions of the content and practices brings the targeted construct to the forefront and helps to ensure that the resulting assessments actually assess the construct and provide the evidence needed to describe students' performance. The assessment development process consists of four steps. First, the construct is defined to make explicit what students are expected to know and do. Then, the construct definition is used to design items that require students to engage with data and information about real-world phenomena. Next, the outcome space is described, which includes developing rubrics and rules for scoring. Finally, the measurement model, which describes how results will be interpreted and used, is selected.

We also utilized the framework that resulted from the Task Annotation Project in Science (Achieve, 2019). In this project, Achieve collected tasks that were designed to be aligned to the NGSS and convened a panel of experts to annotate the tasks. After analyzing the annotations, the panel identified five characteristics that all three-dimensional science assessments should possess: (1) a focus on real-world phenomena, (2) requiring students to engage in sense making, (3) requiring students to use both disciplinary core ideas and science practices, (4) being comprehensible to students, and (5) supporting the intended purpose and use of the assessment.

10.1.4 Model-Based Reasoning

Within the NGSS, developing and using models is articulated as one of eight science and engineering practices through which students learn the disciplinary core ideas and crosscutting concepts of science. We define models as sets of ideas about the relationships between real and hypothetical objects and processes that can be used to make sense of natural phenomena (Gouvea & Passmore, 2017; Passmore et al., 2014). There are several aspects of model-based reasoning in science; Model formation, Model use, Model elaboration, Model articulation, Model evaluation, Model revision, and Model-based inquiry (Mislevy et al., 2017). Many of the science and engineering practices are organized around these aspects of modeling. For example, observation motivates the need to form a model or revise or elaborate an existing one; investigation is a way to evaluate model ideas; argumentation is a discursive strategy used to defend or critique a model; explanation is the use of a model to make sense of a particular data pattern or phenomenon. In our assessment development effort, we chose to focus on Model formation, use, evaluation, and revision because they were most closely aligned with the activities in which the students engage during instruction.

10.1.5 Model Ideas in Biology

The construct targeted by the instrument focuses on ideas in two areas of high school biology, natural selection and genetics. The following ideas are the basis of the contemporary scientific model of natural selection:

- Populations have the potential to grow exponentially, but they typically don't because of limited resources or environmental factors.
- Variation in traits naturally exists in populations.
- Depending on the environment, traits can be advantageous or disadvantageous.
- Individuals with more advantageous traits are more likely to survive and reproduce.
- Many traits are heritable, so offspring tend to resemble their parents.
- In each generation, the number born with the advantageous trait will increase and the number with the disadvantageous trait will decrease.
- Over time the distribution of traits in the population will change.

Regarding genetics, the instrument targets ideas about Mendelian genetics and molecular genetics. The ideas related to Mendelian genetics include

- Traits are characteristics of organisms. Traits can be physical or behavioral.
- Sexually reproducing organisms have two alleles that determine each trait, one from each parent. A parent passes only one of his/her two genes for a trait to each offspring. Random chance determines which of the two alleles is passed to each offspring.
- A gene for a trait can occur in different forms called alleles.
- When there are two variations for a trait (phenotypes) in a population, there are two alleles (1 and 2) and three possible allele combinations (genotypes): (1,1), (2,2) and (1,2).
- If (1,1) and (1,2) have one phenotype and (2,2) has the other, then 1 is the dominant allele. It always shows when present. 2 is the recessive allele and will only show if no dominant allele is present.

The ideas related to molecular genetics include

- DNA is the hereditary material and it codes for the protein.
- DNA is in a code, and it is read by the base pairing rules. This base pairing allows the cell to make an exact copy of DNA to pass along to daughter cells.
- Because DNA is in the cell's nucleus and ribosomes (where proteins are made) are found in the cytoplasm, a messenger RNA molecule transports the code from the nucleus to the ribosomes where translation of the code occurs.

- The messenger RNA is read in groups of 3 bases, that corresponds to an amino acid, the ribosome then strings the amino acids together to make a protein.
- Depending on the function of the protein, you will see different patterns of phenotypes in heterozygotes.
- Errors in the processing of DNA can lead to changes in the genetic material and are called mutations.
- Mutations are a source of genetic variation.

10.1.6 Task Structure

We developed a template for the structure of the tasks that uses the targeted aspects of model-based reasoning, as illustrated in Fig. 10.1. Each task begins with a description of a real-world phenomenon. After the phenomenon is described, a dataset is provided, and the respondent is asked to analyze those data and propose an initial model that could explain the phenomenon (model formation). Next, respondents are presented with an alternative model and additional data. They consider the proposed model and data and evaluate whether these new data support the model (model evaluation). Respondents are then asked to use their model to identify patterns or make predictions (model use). In some tasks, respondents are provided with data that contradict the proposed model and asked whether these data support the model (model evaluation). At the end of the task, respondents are prompted to write a revised model that fits all the data presented in the task (model revision).

10.1.7 Challenge of Instrument Development

Multidimensional assessments move away from the reliance on the efficiency and affordability of multiple-choice items used in traditional science assessments, and towards the use of more authentic tasks that tend to involve significant student

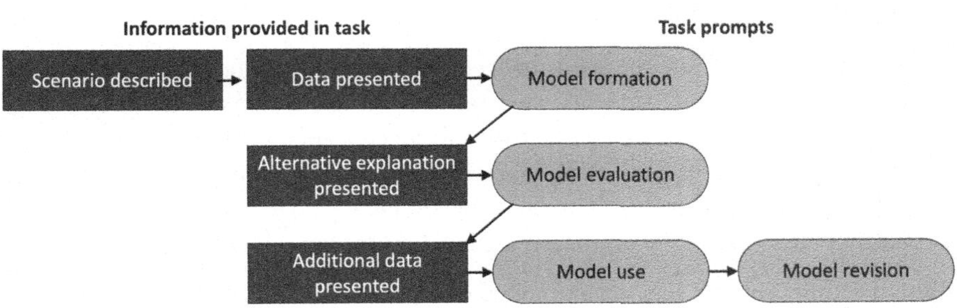

Fig. 10.1 Model-based reasoning task template

written work. Such assessments can be both time intensive for students and resource intensive to score. This presents a challenge for the development of a measurement instrument that will be used as a pre- and post-test in a large-scale efficacy study with thousands of students. We therefore needed to balance the use of different assessment formats to capture the richness of student thinking most effectively in a feasible and efficient manner.

The goal of the analyses described in this chapter was to design an instrument that (1) efficiently assesses high school students' and teachers' multidimensional understanding of natural selection and genetics, two essential content areas within the high school biology curriculum, with a focus on model-based reasoning and (2) can be used as an outcome measure in a large-scale efficacy trial. In order to reach these goals, the instrument must be made up of items that all measure the target construct and that span a range of difficulty appropriate for both students and teachers. Therefore, we sought to use Rasch measurement to answer the following research questions:

1. To what extent does an instrument made up of model-based reasoning tasks and content-focused items form a unidimensional scale?
2. To what extent does the scale formed by the instrument cover the full range of student and teacher performance?

10.2 Methods

10.2.1 Task Development

Using the construct-centered approach described above, we developed a set of four model-based reasoning tasks (MBR tasks). Each task follows the task structure template and begins with a biological phenomenon aligned with high school biology content, followed by a series of constructed-response and multiple-choice items that require students to use model-based reasoning. Two of the tasks were aligned to natural selection ideas using the contexts of rock pocket mice and lactose tolerance. The other two tasks assess students' understanding of genetics with one using Tay-Sachs disease as a context targeting both Mendelian and molecular genetics and the other using polydactyl cats as a context targeting only Mendelian genetics.

NGSS Alignment While the focus of the assessments was model-based reasoning using biology core ideas, students were required to use aspects of all three NGSS dimensions (disciplinary core ideas, cross-cutting concepts, science and engineering practices) when completing the tasks as listed in Fig. 10.2. For example, to complete the task, students must analyze graphical data to identify patterns that could be used as evidence to support a model-based explanation for how a trait is inherited.

Pilot Testing During the development of the tasks, the COVID-19 pandemic interrupted our plans to pilot test draft tasks with high school students. Because it

Disciplinary Core Ideas	• LS3.A Inheritance of Traits • LS4.B Natural Selection
Science and Engineering Practices	• Developing and Using Models • Analyzing and Interpreting Data • Constructing Explanations • Engaging in Argument from Evidence
Crosscutting Concepts	• Patterns • Cause and Effect

Fig. 10.2 NGSS dimensions utilized in the MBR assessment tasks

was necessary to collect pilot data to inform revisions to the assessment, we utilized CloudResearch.com (Litman et al., 2017) and Amazon's Mechanical Turk (MTurk) to collect responses to the assessment tasks from adults with at least a high school diploma from across the U.S. This method has been demonstrated to be a successful way to collect data to evaluate middle school science items (Sadler et al., 2016). MTurk is a crowdsourcing marketplace that connects businesses with individuals from around the country willing to do "human intelligence tasks" (HITs) for a fee.

A total of 234 MTurk respondents participated in the pilot test and each respondent was randomly assigned to two tasks. The participants ranged in age from 18 to 31 with an average age of 22. We used these responses to inform revisions to the tasks and evaluate scoring rubrics.

External Review Draft versions of all four tasks were reviewed by the project's external evaluator. One of the members of the evaluator's team is a content expert in biology, and they provided feedback on the scientific accuracy of the tasks. All the evaluator feedback was used to revise the tasks.

10.2.2 Instrument Development

We developed four versions of the instrument to reduce the testing burden on the respondents, yet still address a range of science ideas, practices, and contexts in the assessment. Each version included two MBR tasks and two sets of six content-focused (CF), multiple-choice items. The CF items were selected from the Conceptual Inventory of Natural Selection (CINS) (Anderson et al., 2002) and the Genetics Literacy Assessment Instrument (GLAI) (Bowling et al., 2008) and used as is. For the CINS item sets, we selected the lizard items and the finch items. The finch

	Version 1	Version 2	Version 3	Version 4
MBR Task 1	Rock Pocket Mice	Lactose Tolerance	Polydactyl Cats	Tay-Sachs Disease
CF Items	GLAI Item Set 1	GLAI Item Set 2	CINS Finch Items	CINS Lizard Items
MBR Task 2	Tay-Sachs Disease	Polydactyl Cats	Rock Pocket Mice	Lactose Tolerance
CF Items	CINS Lizard Items	CINS Finch Items	GLAI Item Set 1	GLAI Item Set 2

Fig. 10.3 Illustration of item distribution on the different versions of the instrument. Gray boxes represent natural selection items and black boxes represent genetics items

context is used within the intervention and serves as a proximal assessment. The lizard context is not used in the focal intervention and could be considered a more distal assessment. We selected 12 of the GLAI items based on the content alignment of the items and item difficulty data from a prior project that used the same items in an assessment of high school genetics knowledge.

Common Item Equating The test blueprint in Fig. 10.3 illustrates the distribution of MBR tasks and CF item sets across the four instrument versions and the order that they were administered (top to bottom). Each MBR task and set of CF items appeared on two of the versions to ensure linking across versions. For example, the Cats task appeared on versions 2 and 3, and the Mice task appeared on versions 1 and 3. Care was taken to ensure that all four versions were linked by common items and tasks so that the data from all versions could be combined during the Rasch analysis (Boone et al., 2014). The MBR tasks and CF item sets were also distributed so that each version covered both content areas.

10.2.3 Scoring

Each of the multiple-choice items were scored dichotomously. Respondents selecting the correct answer choice received one point, and respondents selecting any of the incorrect answer choices received no points. To define rubric criteria during the development of the item rubrics for constructed-response items, we started with ideal student responses. The ideal responses were broken down into statements that represented the components and interactions of the natural selection or genetics model being assessed by the items within the task. A sample rubric from an item in a natural selection task is shown in Fig. 10.4 and a sample rubric from an item in a genetics task is shown in Fig. 10.5.

The responses were rated on whether they included each statement in the rubric. The score for the item was based on the number of statements present in the response. For most items, the score is the sum of the statements present. For some items that required the use of evidence, the rubric included multiple evidence statements but only one piece of evidence was required to receive full credit on

Scientists think that there is an explanation for the patterns found at both Location A (light-colored rock) and Location B (dark-colored rock). Describe how scientists would explain the two patterns. *Ideal response:* *Natural selection would explain the patterns. In Location B, dark mice have an advantage over light colored mice that cause them to be more likely to survive and reproduce. Hair color is heritable so dark colored mice will have dark colored offspring. Over time, there will be more dark-colored mice and fewer light-colored mice at Location B.* *The proportion of dark-colored mice at Location A does not change because there was no change in the color of the land that gave the dark-colored mice an advantage.*	Elements of a correct response: • **Variation** = there are different colored mice in the population • **Predation** = the mice that have a similar color as their environment are the ones that will be less likely to be eaten by predators. • **Reproductive advantage** = dark-colored mice blend into their surroundings, making them more likely than light-colored mice to survive **and** reproduce in Location B. • **Heritable** = the fur color trait is heritable so dark colored mice will have dark colored offspring. • **Time (Location B)** = over time there will be more dark-colored mice and fewer light-colored mice at Location B • **Time (Location A)** = over time there will be no change in distribution because there is no environmental change that would give the dark-colored mice a reproductive advantage.

Fig. 10.4 Sample rubric from the rock pocket mice task. The prompt and ideal response are shown on the left and the rubric statements are listed on the right

the item (see Fig. 10.5, for example). As a result, the maximum scores for the constructed-response items ranged from two to six.

Interrater Reliability Two researchers initially scored 20 responses from each task. Interrater reliabilities were calculated. Rubric statements that had an interrater reliability of less than 0.75 (Koo & Li, 2016) were discussed to resolve disagreements. For elements that had an interrater reliability of 0.75 or higher, the scores from one researcher were used. After the discussion, the rubrics were revised, and one of the researchers scored the remaining responses for each task.

10.2.4 Rasch Analysis

To prepare for the Rasch analysis, data from the different test versions were combined into one file. We were able to combine the data because the common items and tasks that appeared on multiple versions provided sufficient linking across the four versions of the test. We used the software package WINSTEPS (Linacre, 2022) to estimate item and person measures, investigate the relative difficulties of the

How do you think the polydactyl trait is inherited? Be sure to cite evidence and use reasoning with ideas about genetics to describe how the evidence supports your claim.	Elements of a correct response:
Ideal response: *A kitten inherits one allele from each parent. The combination of alleles the kitten inherits will dictate whether or not the kitten is polydactyl. The evidence indicates that the polydactyl allele is dominant meaning that if the kitten inherits at least one polydactyl allele it will be polydactyl. Evidence 1 supports a dominant explanation because if one parent is polydactyl and polydactyl is dominant, you would predict that at least half of the kittens would be polydactyl regardless of whether the other parent is polydactyl. Evidence 2 supports this because it says that when both parents are polydactyl some are polydactyl, and some are not. If polydactyl was recessive, none of the cats would not be polydactyl in this case.*	• Offspring **inherit** one allele/gene from **each** parent. • The combination of alleles/genes dictate which trait the offspring has. • A gene has different alleles, in this case, two (dominant and recessive). • Offspring that inherit at least one dominant gene will have the dominant trait. • In this case, polydactyl is dominant. • **Evidence 1** supports the idea that polydactyl is dominant because if one parent is polydactyl and polydactyl is dominant you would predict that at least half of the kittens would be polydactyl regardless of whether the other parent is polydactyl. -OR- • **Evidence 2** supports the idea that polydactyl is dominant because it says that when both parents are polydactyl some are polydactyl, and some are not. If polydactyl was recessive, none of the cats would not be polydactyl in this case.

Fig. 10.5 Sample rubric from the Cats task. The prompt and ideal response are shown on the left and the rubric statements are listed on the right

items within the tasks, and explore the relationship between the performance on the MBR tasks and the CF items. Item and person measures are reported in logits, with zero representing the average item difficulty. Measures greater than zero represent more difficult items and higher performing respondents, and measures less than zero represent less difficult items and lower performing respondents.

Because the items within the MBR tasks are polytomously scored, the Rasch partial credit model was applied (Masters, 1982). The partial credit model is an extension of the dichotomous model and allows for a range of correctness from "not at all correct" to "completely correct." Additionally, different items are allowed to have different numbers of levels of correctness.

Data Fit Fit to the Rasch model was evaluated using fit statistics including person and item separation indices and reliabilities, infit and outfit mean-squares, and standard errors (Bond & Fox, 2007; Boone et al., 2014). Separation indices should be greater than 2.0. Ideally, infit and outfit mean-squares should be between 0.6 and 1.4. Standard errors should be close to zero.

Dimensionality A principal component analysis of the Rasch residuals was conducted to examine the dimensionality of the data and evaluate whether the CF items and MBR tasks were measuring the same overall construct. If the data was truly unidimensional, the first component of the correlation matrix of the residuals would be small. Simulation studies have shown that components less than 2 are at the random "noise" level (Wilson, 1994) with components less than 3 generally being considered small and indicating the test is largely unidimensional but measuring a "broad" dimension.

Differential Item Functioning Differential item functioning (DIF) was used to investigate whether the tasks functioned differently for respondents based on gender. In our analysis of gender DIF, an item would be considered to have DIF if the absolute value of the DIF contrast value was greater than or equal to 0.43 logits and the p value was less than .05 (Linacre, n.d.).

Wright Maps Wright maps plot the distribution of item difficulties and person measures on the same scale. The maps were used to compare the relative difficulties of items within the tasks, the relative difficulties of the MBR items and the CF items, and the relative performance levels of the various groups of respondents. The maps were also used to determine how well the set of items cover the range of difficulty (difficult to easy) and how well the distribution of item difficulties match the range of the person measures.

Cross-Plots We used cross-plots of the person measures to visually investigate correlation between performance on the MBR tasks and CF items. To construct the cross-plots, two person measures for each respondent were estimated: one using the responses from the MBR tasks and the other using the responses from the CF items. Pearson correlation coefficients and the statistical significance of the coefficients were also calculated.

10.3 Data Sources

This study includes data from three sources: high school students from two schools in the western U.S., high school biology teachers from the western U.S., and crowd-sourced adults from across the U.S. The high school students were pilot test participants in the efficacy study that prompted the development of this instrument. Students were randomly assigned a version of the instrument as a pre-test and a post-test. The teacher population includes treatment and comparison participants from the efficacy study. All the teachers responded to Version 4 of the instrument because we expected it to be the most challenging version. We predicted the Tay-Sachs task would be more difficult than the Cats task because of the inclusion of molecular genetics ideas. We predicted the Lactose task would be more difficult than the Mice task because it contained more explanation items.

Table 10.1 Distribution of respondents in the data set

	Percentage	Number
Pre-test high school students	31%	107
Post-test high school students	28%	97
High school biology teachers	19%	66
MTurk student equivalent	11%	40
MTurk teacher equivalent	11%	37

The crowd-sourced adult samples were obtained using a second round of HITs with Amazon's Mechanical Turk (MTurk) and CloudResearch.com (Litman et al., 2017). We obtained two samples, one made up of adults ages 18–20 with some high school experience (40) and one made up of adults ages 21–25 with at least a bachelor's degree (37). The younger group was selected to be roughly equivalent to the intended high school population and the older group was roughly equivalent to the teacher population and was likely to give us more sophisticated responses at the upper end of the distribution.

Because this study is focused more on the performance of the items than on the performance of the respondents, we cleaned the data set so that it included only responses from those who (1) answered at least one MBR task and one CF item set and (2) provide relevant responses. Nine respondents were removed because they did not respond to at least one MBR task and one set of CF items. Nineteen respondents who provided irrelevant responses (e.g., typing random strings of letters or writing "bio is cool") for all the items on the task were removed from the data set. Twenty respondents provided irrelevant responses to one MBR task and relevant responses for the other, perhaps because they were running out of time during the testing period. For these respondents, we replaced their irrelevant responses with missing data and kept their responses for the task to which they provided relevant responses. Table 10.1 displays the distribution of the 347 respondents in the final data set.

10.4 Results

10.4.1 Data Fit

Table 10.2 shows the initial fit statistics of the Rasch model to the data set.

Four items were found to have outfit mean-square values greater than 1.4, indicating unexpected responses to these items (Bond & Fox, 2007; Boone et al., 2014). This suggests that for these items some respondents were likely guessing and unexpectedly responded correctly. One way of addressing guessing is to assume guessing is a function of the difficulty of the item and the proficiency of the student. That is, if a low proficiency person gets a difficult item correct, it could be a sign that the person guessed correctly. Using the Rasch model, this information is captured in students' z-residuals for items, where a high z-residual indicates the person was

Table 10.2 Summary of initial fit statistics

	Item			Person		
	Min	Max	Median	Min	Max	Median
Standard error	0.07	0.25	0.16	0.22	1.08	0.34
Infit mean-square	0.74	1.65	1.00	0.30	2.74	0.92
Outfit mean-square	0.52	3.12	1.01	0.29	4.32	0.87
Point-measure correlation	0.26	0.71	0.51			
Separation index (reliability)	7.16 (0.98)			2.72 (0.88)		

Table 10.3 Summary of final fit statistics

	Item			Person		
	Min	Max	Median	Min	Max	Median
Standard error	0.08	0.27	0.16	0.25	1.86	0.36
Infit mean-square	0.77	1.40	0.98	0.35	2.65	0.94
Outfit mean-square	0.50	1.41	1.03	0.16	2.94	0.88
Point-measure correlation	0.27	0.72	0.49			
Separation index (reliability)	7.35 (0.98)			2.73 (0.88)		

unlikely to answer the item correctly, but they did. To decrease the influence of the unexpected responses on our item measures, we used an approach outlined by Andrich et al. (2012) in which a tailored data set is created by replacing student responses with large z-residual values with missing data. We replaced all student responses with z-residuals greater than or equal to 4 for the four misfitting items with missing data. This resulted in a total of 21 responses being replaced with missing data. Table 10.3 shows the fit statistics after these responses were replaced.

The tailored data set fit the partial credit Rasch model well. Table 10.4 presents the final individual item statistics. The person and item separation indices were both greater than two, and the corresponding reliabilities were high, 0.88 and 0.98, respectively. All items had Infit and Outfit statistics less than or equal to 1.4 and had positive point-measure correlations. The average person measure was close to zero (−0.04), indicating that the items were well matched to the participants.

10.4.2 Dimensionality

We conducted a principal component analysis of residuals to explore the dimensionality of the data set. Approximately 56% of the raw variance in the data was explained by the model, and the eigenvalue of the first contrast was slightly over two (2.11). This indicates that the items within the MBR tasks and CF items form a predominantly unidimensional scale, although the construct being measured may be broad.

Table 10.4 Rasch item statistics

Entry	Measure	Count	Score	Model S.E.	Infit MNSQ	Outfit MNSQ	PTMA	Item Name
1	−0.45	158	86	0.18	1.05	1.04	0.37	miceQ1
2	−2.44	158	137	0.25	0.92	0.72	0.38	miceQ2
3	−0.33	158	166	0.10	0.89	0.74	0.64	miceQ4Q5
4	−0.69	150	273	0.09	1.31	1.39	0.53	miceQ6Q7
5	0.33	158	177	0.10	0.87	0.85	0.69	miceQ8Q9
6	0.75	158	297	0.08	1.19	1.11	0.69	mice310
7	−2.15	158	132	0.23	0.86	0.63	0.46	miceQ11
8	1.08	211	312	0.09	1.40	1.41	0.57	TaySachsQ1
9	−2.77	211	195	0.27	0.87	0.51	0.38	TaySachsQ2
10	1.90	211	165	0.08	0.90	0.72	0.71	TaySachsQ3
11	−0.02	211	240	0.11	0.95	0.93	0.67	TaySachsQ4
12	0.32	211	213	0.11	1.04	1.014	0.65	TaySachsQ5
13	0.36	211	105	0.16	0.82	0.75	0.63	TaySachsQ6
14	2.20	211	232	0.08	1.23	1.09	0.70	TaySachsQ7
15	1.90	217	140	0.10	0.84	0.70	0.69	LactoseQ1
16	2.12	217	80	0.13	0.77	0.50	0.64	LactoseQ2
17	1.99	217	163	0.09	1.07	1.19	0.61	LactoseQ3
18	1.85	217	265	0.08	1.32	1.33	0.62	LactoseQ4
19	0.18	145	103	0.12	0.93	0.89	0.60	CatsQ2Q3
20	1.30	145	28	0.23	1.09	1.04	0.31	CatsQ4
21	−1.75	99	286	0.14	1.07	1.22	0.47	CatsQ5
22	1.02	145	146	0.09	0.79	0.71	0.72	Cats16
23	−1.59	155	115	0.20	0.85	1.09	0.49	GLAI30
24	−0.63	155	87	0.18	0.90	0.84	0.50	GLAI7
25	−0.38	155	79	0.18	1.06	1.05	0.35	GLAI16
26	0.41	225	107	0.16	1.02	1.10	0.48	GLAI8
27	0.38	225	108	0.16	0.86	0.80	0.60	GLAI4
28	0.38	225	108	0.16	0.77	0.74	0.65	GLAI22
29	−1.34	150	100	0.19	1.00	1.12	0.39	GLAI1
30	−0.24	220	127	0.16	1.08	1.13	0.48	GLAI20
31	0.72	220	90	0.16	0.87	0.85	0.61	GLAI25
32	0.45	150	48	0.20	0.88	0.82	0.53	GLAI15
33	0.15	220	112	0.16	0.83	0.78	0.64	GLAI6
34	−0.69	150	81	0.18	0.90	0.88	0.51	GLAI28
35	−2.20	215	189	0.23	0.93	0.89	0.38	Lizard1
36	0.79	221	94	0.16	1.02	1.30	0.48	Lizard2
37	−0.86	221	158	0.17	1.00	0.93	0.47	Lizard3
38	0.62	221	101	0.16	1.06	1.06	0.48	Lizard4
39	−0.34	221	139	0.16	1.03	1.11	0.47	Lizard5
40	0.22	221	117	0.16	1.05	1.06	0.48	Lizard6
41	−1.60	149	108	0.20	1.06	1.17	0.36	Finch1
42	−0.59	152	83	0.18	1.22	1.31	0.27	Finch2

(continued)

Table 10.4 (continued)

Entry	Measure	Count	Score	Model S.E.	Infit MNSQ	Outfit MNSQ	PTMA	Item Name
43	−0.76	152	88	0.18	1.08	1.07	0.39	Finch3
44	−0.56	152	82	0.18	0.96	0.93	0.48	Finch4
45	−0.56	151	82	0.18	1.14	1.24	0.32	Finch5
46	1.49	149	28	0.23	0.97	1.12	0.37	Finch6

Table 10.5 Summary of item difficulties by task or set

			# of items	Minimum difficulty	Maximum difficulty	Mean difficulty
Genetics	MBR tasks	Cats	4	−1.75	1.30	0.19
		Tay-Sachs	7	−2.77	2.20	0.44
	CF items	GLAI	12	−1.59	0.72	−0.20
Natural selection	MBR tasks	Lactose	4	1.85	2.12	1.97
		Mice	7	−2.44	0.75	−0.71
	CF items	Lizards	6	−2.20	0.79	−0.30
		Finches	6	−1.60	1.49	−0.43

10.4.3 *Item Difficulties*

Table 10.5 presents a summary of the item difficulties for the items within each MBR task and the sets of CF items from the GLAI and CINS. For the natural selection MBR tasks, the items that made up the Lactose tasks were, on average, more difficult than the items in the Mice task, 1.97 and − 0.71, respectively. For the genetics MBR tasks, the items that made up the Tay-Sachs task were, on average, more difficult than the items that made up the Cats task, 0.44 vs. 0.19, respectively.

Next, we examined the Wright map in Fig. 10.6 to investigate the relative differences in item difficulty. The map shows the person measures (on the left) and the item difficulties (on the right) on a single scale ranging from easier items and lower performing people at the bottom and harder items and higher performing people at the top. The easiest items across all the MBR tasks are items that ask respondents to use models to identify or predict patterns (shaded dark gray on the map). The items that required respondents to write model-based explanations (shaded black) were the most difficult items. Items that asked respondents to evaluate whether data support a model (shaded light gray) fall in between the explanation items and the pattern items.

These patterns in difficulty are not surprising. Most of the identify or predict patterns questions were multiple-choice items that require lower cognitive demand than constructed-response items. Multiple-choice items provide a set of answers from which respondents choose the correct one, while constructed-response items require the respondents to develop the answer on their own. However, one item on

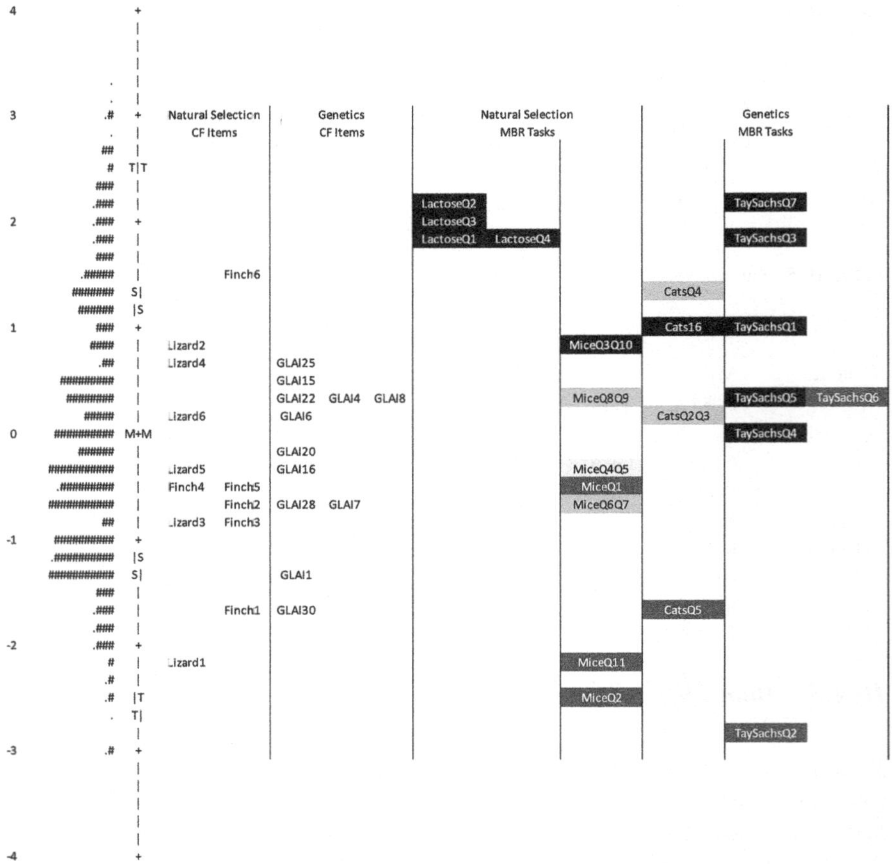

Fig. 10.6 Wright map showing the item difficulties (right side) and person measures (left side). Each # is 2 people and each "." is one person

the Cats task that asked respondents to evaluate whether data support a model was a multiple-choice item and is as difficult or more difficult than the constructed-response explanation items. For this item, the reasons presented in the answer choices for why the data support or refute the model are based on common alternative mental models about the inheritance of recessive and dominant traits. About 30% of the respondents selected the answer choice that indicated that offspring only inherit a recessive trait if both parents have the recessive trait. These respondents are not thinking about the possibility of two heterozygous parents producing a homozygous offspring with the recessive trait.

We were surprised about the difference in difficulty of the explanation items in the natural selection tasks. These items were designed to measure a similar concept, model-based reasoning with natural selection science ideas. Therefore, we expected to observe more similarity between the explanation items in both tasks. However, we found that the explanation items in the Lactose task are more difficult than the explanation item in the Mice task. It is possible that the contexts of the tasks may be playing a role in the differences in difficulty. Penuel et al. (2019) found similar patterns in which student performance depended significantly on the combination of tasks to which the students responded. When we looked more carefully at what students were writing for the different contexts, we observed that many students did not apply the natural selection model to the lactose context because they did not see drinking milk as essential for survival. One student wrote "This isn't a life or death trait in our world today so natural selection isn't a part of it." On the other hand, in the Mice task, being able to blend into the environment was seen as essential for survival.

10.4.4 Differential Item Functioning

DIF analysis was investigated for the gender groups where participants self-identified themselves as men or women. There were 137 participants who identified themselves as a boy or man and 143 participants who identified themselves as a girl or woman. We looked for DIF contrast values greater than or equal to 0.43 and found that 13 items met this criterion. Of the 13 items with large DIF contrast, only one had a p-value lower than 0.05, indicating statistically significant DIF. An analysis of the item did not surface any obvious explanation for why the item would behave differently for men and women. A closer look at the response patterns for this item revealed that it had the fewest responses compared to any other item. Thus, the DIF analysis only included 46 men and 33 women. It is possible that this sample is too small to conclusively judge DIF. We decided that we should retain the item and complete an additional DIF analysis when we have a larger set of responses.

10.4.5 Person Measures

A summary of the person measures by category is shown in Table 10.6. Figure 10.7 shows a Wright map which displays the item difficulties (Xs on the left) and person measures (numbers on right). The horizontal lines indicate the average person measure for each of the five different categories listed in Table 10.6.

We saw the expected differences in performance. The high school students performed better on the post-test than the pre-test ($t = 2.82, p < .01$). The MTurk adults performed better than the high school students ($t = -3.65, p < .001$), but there was no significant difference in performance between the two groups of MTurk samples ($t = -1.08, p > 0.10$). The high school biology teachers performed the best with an average measure of 1.70 logits, which was significantly higher than the MTurk "Teacher" sample ($t = -6.22, p < .001$).

Table 10.6 Summary of person measures by sample

Category	Minimum measure	Maximum measure	Mean measure
1. Pre-test students	−2.59	1.10	−0.91
2. Post-test students	−4.31	1.99	−0.55
3. MTurk "students"	−3.01	2.71	0.20
4. MTurk "teachers"	−1.62	2.98	0.46
5. HS teachers	0.25	3.33	1.70

Ideally, we would have collected pilot test data from high school students and teachers in 2020. However, the COVID-19 pandemic made it nearly impossible to collect data from K-12 schools. We turned to using MTurk as a way to continue our assessment development efforts by collecting data from adults who we thought would be good proxies for our intended population. Our findings support our decision by demonstrating how the MTurk samples overlap both the student and teacher samples.

10.4.6 Cross-Plots

We used a cross-plot of the person measures to investigate the correlation between performance on the MBR tasks and CF items. The cross-plot in Fig. 10.8 plots the person measures based only on responses to the MBR tasks against the person measures based only on responses to the CF items. While there are some outliers, most of the points are found within the 95% confidence bands. A statistically significant ($p < 0.001$) Pearson correlation coefficient of 0.724 was found, which is considered strong. This finding supports the results of the principal component analysis and leads us to conclude that, overall, the MBR tasks and CF items form a unidimensional construct.

While the items that make up the MBR tasks and CF items measure the same construct, their locations on the Wright map suggest that they are measuring different aspects of the construct. The CF items span the mid-range of the difficulty scale from about −2 logits to 1 logit. The using models to predict or identify patterns items within the MBR tasks fill the lower end of the scale below −2 logits, and the model-based explanation items within the MBR tasks are located at the higher end of the scale above 1 logit.

10.5 Conclusions

This chapter addresses the need for new assessments that measure students' multidimensional science understanding as described in the NGSS and provides insights into how Rasch measurement supports the development of these

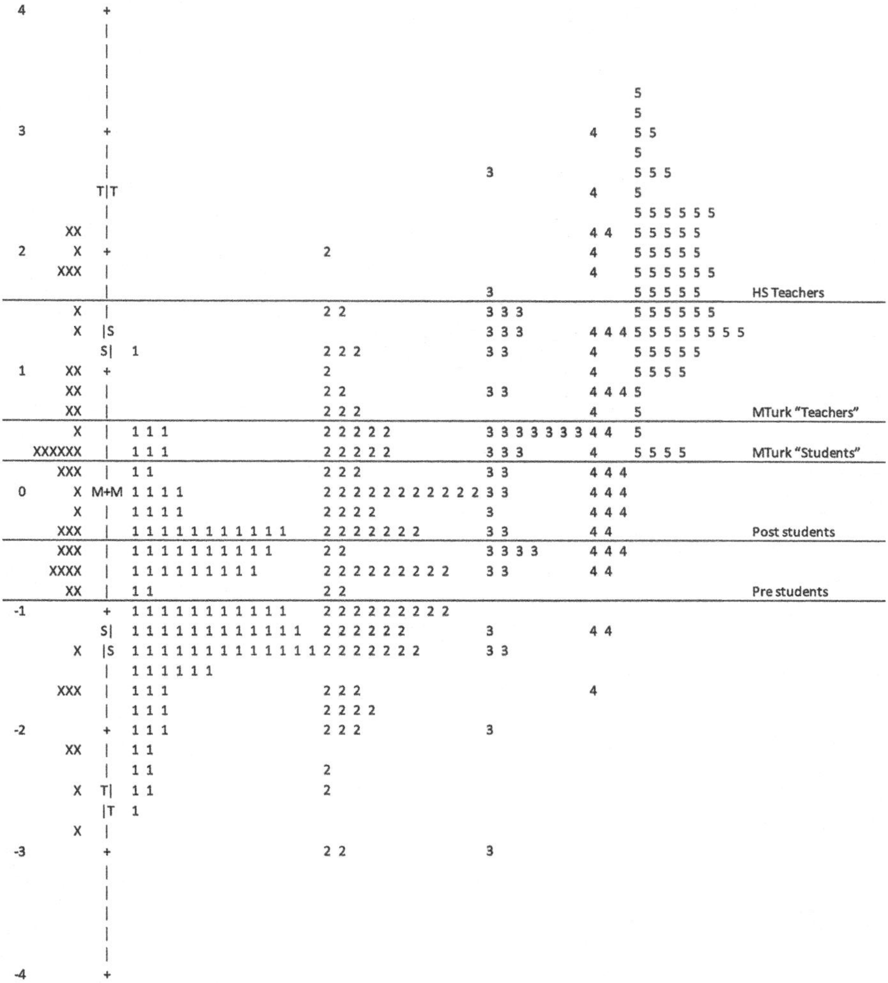

Fig. 10.7 Wright map showing the person measures (right side) and item difficulties (left side). Each "1" represents a pre-test student measure, "2" represents post-test student measures, "3" represent MTurk "student" measures, "4" represent MTurk "teacher" measures, and "5" represent high school teachers

assessments. The assessments we describe in this chapter are being used to measure the efficacy of model-based reasoning instructional materials and teacher resources in a large-scale quasi-experiment. Four versions of the instrument that combine content-focused items with model-based reasoning tasks were developed and tested with a variety of respondents including high school students, high school biology teachers, and crowd-sourced adults. Below we discuss how the findings help answer the research questions about (1) the extent to which an instrument made up of model-based reasoning tasks and content-focused items forms a unidimensional scale and

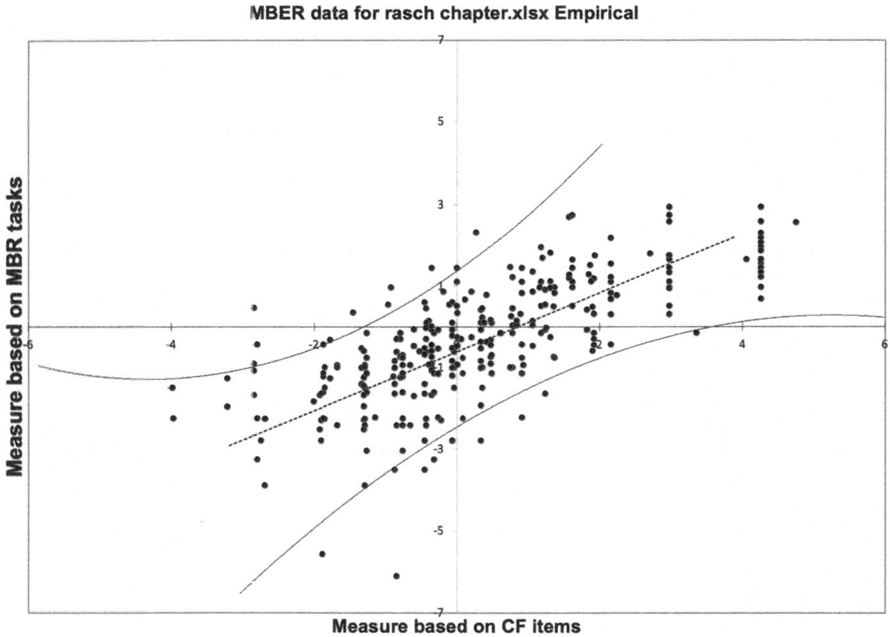

Fig. 10.8 Cross-plot of person measures based on MBR tasks and CF items

(2) the extent to which the scale formed by the instrument covers the full range of student and teacher performance.

RQ1. Unidimensional Scale
The principal component analysis of the Rasch residuals and the cross-plots of performance on the MBR tasks versus the CF items showed that the CF items and the items that make up the MBR tasks form a unidimensional construct and supported the idea that the MBR tasks and CF items measure different aspects of the construct. The results also showed that within the MBR tasks, items that required respondents to write model-based explanations of phenomena were more difficult than items that asked respondents to evaluate whether data support a given model. Additionally, multiple-choice items embedded within MBR tasks that asked students to predict or identify patterns were easier than the constructed-response items. These results provide support for using a combination of easily scored, content-focused items and more time intensive performance tasks to efficiently measure multidimensional science understanding during a large-scale efficacy study.

RQ2. Covering the Range of Student and Teacher Performance
The person Wright map showed the expected differences in the distributions of the categories of respondents from pre-test students to post-test students to crowd-sourced adults to high school biology teachers. The overlap between these distributions and the distribution of item difficulties confirmed that the items form a scale wide enough to measure both students' and teachers' performance levels. There were

a handful of teachers and crowd-sourced adults with measures above the highest item difficulty, however none of the respondents received all of the possible points on the items. Therefore, there is still room for even the best performing teachers to improve on a post-test.

As we navigate the context of the NGSS, assessment plays a key role in helping to build the evidentiary basis for the efficacy of new instructional materials. This chapter focuses on the development of new assessments that can be used to measure model-based reasoning skills in the context of high school biology and demonstrates the role Rasch modeling can play in the development of assessments that measure the complex science learning outcomes described in the NGSS. The results can inform best practices in how to measure students' multidimensional science understanding in large scale settings feasibly and efficiently.

Acknowledgements This material is based upon work supported by the National Science Foundation under DRL#1813538. Any opinions, findings, and conclusions or recommendations expressed in this material are those of the authors and do not necessarily reflect the views of the National Science Foundation.

Appendix

```
&INST
Title= "MBER data for Rasch chapter.xlsx"
; Excel file created or last modified: 5/23/2022 5:32:13 PM
; Excel Cases processed = 348
; Excel Variables processed = 50
ITEM1 = 1 ; Starting column of item responses
NI = 46 ; Number of items
NAME1 = 48 ; Starting column for person label in data record
NAMLEN = 29 ; Length of person label
XWIDE = 1 ; Matches the widest data value observed
GROUPS = 0 ; Partial Credit model: in case items have different rating
scales
CODES = "0123456" ; matches the data
TOTALSCORE = Yes ; Include extreme responses in reported scores
; Person Label variables: columns in label: columns in line
@Respondent-ID = 1E14 ; $C48W14
@group = 16E22 ; $C63W7
&END ; Item labels follow: columns in label
miceQ1 ; Item 1 : 1-1
miceQ2 ; Item 2 : 2-2
miceQ4Q5 ; Item 3 : 3-3
miceQ6Q7 ; Item 4 : 4-4
miceQ8Q9 ; Item 5 : 5-5
mice310 ; Item 6 : 6-6
miceQ11 ; Item 7 : 7-7
TaySachsQ1 ; Item 8 : 8-8
```

```
TaySachsQ2 ; Item 9 : 9-9
TaySachsQ3 ; Item 10 : 10-10
TaySachsQ4 ; Item 11 : 11-11
TaySachsQ5 ; Item 12 : 12-12
TaySachsQ6 ; Item 13 : 13-13
TaySachsQ7 ; Item 14 : 14-14
LactoseQ1 ; Item 15 : 15-15
LactoseQ2 ; Item 16 : 16-16
LactoseQ3 ; Item 17 : 17-17
LactoseQ4 ; Item 18 : 18-18
CatsQ2Q3 ; Item 19 : 19-19
CatsQ4 ; Item 20 : 20-20
CatsQ5 ; Item 21 : 21-21
Cats16 ; Item 22 : 22-22
GLAI30 ; Item 23 : 23-23
GLAI7 ; Item 24 : 24-24
GLAI16 ; Item 25 : 25-25
GLAI8 ; Item 26 : 26-26
GLAI4 ; Item 27 : 27-27
GLAI22 ; Item 28 : 28-28
GLAI1 ; Item 29 : 29-29
GLAI20 ; Item 30 : 30-30
GLAI25 ; Item 31 : 31-31
GLAI15 ; Item 32 : 32-32
GLAI6 ; Item 33 : 33-33
GLAI28 ; Item 34 : 34-34
Lizard1 ; Item 35 : 35-35
Lizard2 ; Item 36 : 36-36
Lizard3 ; Item 37 : 37-37
Lizard4 ; Item 38 : 38-38
Lizard5 ; Item 39 : 39-39
Lizard6 ; Item 40 : 40-40
Finch1 ; Item 41 : 41-41
Finch2 ; Item 42 : 42-42
Finch3 ; Item 43 : 43-43
Finch4 ; Item 44 : 44-44
Finch5 ; Item 45 : 45-45
Finch6 ; Item 46 : 46-46
END NAMES
```

References

Achieve. (2019). *Task annotation project in science*. Retrieved from https://www.achieve.org

Anderson, D. L., Fisher, K. M., & Norman, G. J. (2002). Development and evaluation of the conceptual inventory of natural selection. *Journal of Research in Science Teaching, 39*, 952–978.

Andrich, D., Marais, I., & Humphry, S. (2012). Using a theorem by Andersen and the dichotomous Rasch model to assess the presence of random guessing in multiple choice items. *Journal of Educational and Behavioral Statistics, 37*(3), 417–442.

Bond, T. G., & Fox, C. M. (2007). *Applying the Rasch model: Fundamental measurement in the human sciences* (2nd ed.). Lawrence Erlbaum Associates.

Boone, W. J., Staver, J. R., & Yale, M. S. (2014). *Rasch analysis in the human sciences*. Springer.

Bowling, B. V., Acra, E. E., Wang, L., Myers, M. F., Dean, G. E., Markle, G. C., Moskalik, C. L., & Huether, C. A. (2008). Development and evaluation of a genetics literacy assessment instrument for undergraduates. *Genetics, 178*(1), 15–22.

DeBoer, G. E., Herrmann-Abell, C. F., Gogos, A., Michiels, A., Regan, T., & Wilson, P. (2008). Assessment linked to science learning goals: Probing student thinking through assessment. In J. Coffey, R. Douglas, & C. Stearns (Eds.), *Assessing student learning: Perspectives from research and practice* (pp. 231–252). NSTA Press.

Gouvea, J., & Passmore, C. (2017). Models of versus models for. *Science & Education, 26*(1–2), 49–63.

Koo, T. K., & Li, M. Y. (2016). A guideline of selecting and reporting intraclass correlation coefficients for reliability research. *Journal of Chiropractic Medicine, 15*(2), 155–163.

Linacre, J. M. (2022). *Winsteps® Rasch measurement computer program (version 5.2.3)*. Winsteps.com.

Linacre, M. (n.d.). *DIF – DPF – bias – interactions concepts*. Winsteps.Com. https://www.winsteps.com/winman/difconcepts.htm

Litman, L., Robinson, J., & Abberbock, T. (2017). TurkPrime.com: A versatile crowdsourcing data acquisition platform for the behavioral sciences. *Behavior Research Methods, 49*(2), 433–442.

Masters, G. N. (1982). A Rasch model for partial credit scoring. *Psychometrika, 47*, 149–174.

Mislevy, R. J., Almond, R. G., & Lukas, J. F. (2003). *A brief introduction to evidence-centered design*. (Research Report 03–16) Educational Testing Service.

Mislevy, R. J., Haertel, G., Riconscente, M., Rutstein, D. W., & Ziker, C. (2017). *Assessing model-based reasoning using evidence-centered design: A suite of research-based design patterns*. Springer. https://doi.org/10.1007/978-3-319-52246-3

NGSS Lead States. (2013). *Next generation science standards: For states, by states*. The National Academies Press.

NRC. (2014). *Developing assessments for the next generation science standards*. The National Academies Press.

Passmore, C., Gouvea, J. S., & Giere, R. (2014). Models in science and in learning science: Focusing scientific practice on sense-making. In *International handbook of research in history, philosophy and science teaching* (pp. 1171–1202). Springer.

Penuel, W. R., Turner, M. L., Jacobs, J. K., Van Horne, K., & Sumner, T. (2019). Developing tasks to assess phenomenon-based science learning: Challenges and lessons learned from building proximal transfer tasks. *Science Education, 103*(6), 1367–1395.

Sadler, P. M., Sonner, G., Coyle, H. P., & Miller, K. A. (2016). Identifying promising items: The use of crowdsourcing in the development of assessment instruments. *Educational Assessment, 21*(3), 196–214.

Wilson, M. (1994). Objective measurement: Theory into practice. .

Wilson, M. (2005). *Constructing measures: An item response modeling approach*. Lawrence Erlbaum Associates.

Cari F. Herrmann-Abell is a senior research scientist at BSCS Science Learning. Her work focuses primarily on the research and development of assessment resources aligned to K-12 science

Molly A.M. Stuhlsatz is a senior research scientist at BSCS Science Learning. She has spent her career leading research and evaluation projects at BSCS and enjoys developing creative solutions to persistent challenges in the assessment of student and teacher outcomes.

Christopher D. Wilson is a senior research scientist and director of the research division at BSCS Science Learning. His current research focuses on the assessment of teacher and student learning in science education, the impact of lesson analysis based professional development on teacher and student outcomes, the application of automated scoring techniques to the measurement of teacher PCK and student argumentation, and the development and impact of three-dimensional curriculum materials.

Jeffery Snowden is a research scientist at BSCS Science Learning. His work largely focuses on the development and implementation of research and efficacy studies for K-12 science materials.

Brian M. Donovan is a senior research scientist at BSCS Science Learning. His research explores how genetics education interacts with social-cognitive biases to influence how students make sense of complex biological and social phenomena.

Chapter 11
Using Rasch Analysis to Assess Students' Learning Progression in Stability and Change across Middle School Grades

Shaohui Chi, Zuhao Wang, and Ya Zhu

Abstract This study developed a measurement instrument to assess students' learning progressions in the crosscutting concept of Stability and Change across middle school grades (from Grades 7 to 9). Based on existing concept development models, frameworks, and research on student learning progressions, this study's learning progression framework comprises four primary levels (i.e., Identifying, Understanding, Analyzing, and Designing) with three sub-levels (i.e., Static, Dynamic, and Cyclic), from basic to the most sophisticated. During the field test, three versions of the test comprising a total of 24 constructed-response items were administered to 136 seventh graders, 139 eighth graders, and 67 ninth graders. A partial credit Rasch model analysis was employed to inform instrument development and evaluation. Specifically, this study used step calibrations and item measures anchoring to express student performance across three grades on the same linear scale. Results provided evidence of reliability, content validity, construct validity, and predictive validity of measures of the instrument, suggesting the measurement instrument meets the quality benchmarks. The results illustrated that higher-grade students were more proficient than lower-grade students in Identifying, Understanding, Analyzing, and Designing regarding Stability and Change. None of the seventh graders and less than 5% of eighth graders were proficient at Cyclic level in Understanding, Analyzing, and Designing, whereas between 3% and 16.4% of ninth graders were proficient at level 3 in Understanding, Analyzing, and Designing.

11.1 Introduction

Crosscutting concepts are overarching scientific themes or concepts that link science knowledge across disciplines in standards, curriculum, instruction, and assessment (Fick et al., 2021), providing a tool and lens to analyze science phenomena from

S. Chi (✉) · Z. Wang · Y. Zhu
College of Teacher Education, Faculty of Education, East China Normal University, Shanghai, China
e-mail: charlcy@163.com

X. Liu, W. J. Boone (eds.), *Advances in Applications of Rasch Measurement in Science Education*, Contemporary Trends and Issues in Science Education 57,
https://doi.org/10.1007/978-3-031-28776-3_11

different perspectives (Rivet et al., 2016). Crosscutting concepts have played an essential role in students' science learning (Duschl, 2012), enhancing students' deep understandings of science core ideas and application of practices (NRC, 2012), and helping students develop comprehensive and coherent views of science (Goldston & Wilensky, 2008).

Developing students' coherent and increasingly sophisticated understandings of crosscutting concepts throughout the secondary school years has been one of the primary goals in science education (Duschl, 2012). Specifically, learning progressions clarify the pathway students follow as they move toward more sophisticated scientific understandings over time (Stevens et al., 2010) and monitor learning within and across grade levels (Liu, 2012). However, most learning progressions in science education start hypothetical without empirical data (Stevens et al., 2010). Empirical evidence from students thus is needed to validate these progressions (Wilson, 2009).

This chapter describes a study to develop a diagnostic assessment that can be used to evaluate students' learning progressions in Stability and Change across middle school grades. This study characterized the learning progressions in Stability and Change as developmental pathways that students are expected to progress over a particular instructional time (from Grades 7 to 9). This study addressed two research questions:

(1) What evidence supports the reliability and validity of measures of the instrument developed in this study for assessing students' progress on the learning progressions in Stability and Change across middle school grades?
(2) How do students develop an understanding of Stability and Change from Grades 7 to 9?

11.2 Literature Review

11.2.1 Stability and Change

Stability and Change is one of the seven crosscutting concepts identified in the National Research Council's *Framework for K-12 Science Education* (NRC, 2012) and the *Next Generation of Science Standards* (NRC, 2013). Stability and Change is essentially centered on understanding, creating, and controlling changes in systems (AAAS, 1993). Stability and Change plays a central role in disciplines such as biology (e.g., evolutionary change), geography (e.g., the motion of Earth), physics (e.g., conservation of energy), and chemistry (e.g., rates of chemical change). Many countries and regions, including Australia, China, Germany, Singapore, the U.K., and the U.S., have acknowledged the role of Stability and Change in standards, curriculum, instruction, and assessments. Stability and Change is also a means of thinking rather than a concept (AAAS, 1993). Understanding and enacting Stability and Change can help students make connections within systems and across disciplines, recognize patterns, understand phenomena, and see how systems influence each other (Riley & Biernat, 2019).

Even very young children learn how natural things change around them by observing, counting, and communicating with each other (AAAS, 1993). However, compared with disciplinary core ideas and scientific practices, crosscutting concepts are more challenging to discuss and develop (Rivet et al., 2016). Many researchers have found that students have difficulties understanding and applying knowledge related to Stability and Change (Fick et al., 2021). For instance, based on a review of past studies about students' understanding of chemical changes, Johnson (2000) concluded that secondary school students "have a very poor understanding of chemical change...the difficulties persist post 16" (p. 719). Stavy (1990) noted that complex changes, such as chemical reactions, are challenging for students. She found students developed their understanding of the reversibility beginning at age 12, yet only half of the 12 years old students could conserve matter with evaporation. According to Ben-zvi-Assarf and Orion (2005), middle school students encounter difficulties developing a dynamic, cyclic, and systemic perception of the system (e.g., water cycle). Testa et al. (2015) reported that many secondary school students still hold naïve understandings of the change of seasons, solar and lunar eclipses, and Moon phases.

11.2.2 Measuring Learning Progressions

According to Wilson (2005, 2009), a learning progression can be considered an assessment structure to identify students' learning development. He further outlined a four-principled approach, Construct Modeling, to measure learning progressions. The four principles are (1) a developmental perspective (construct maps), (2) a match between instruction and assessment (items design), (3) management by teachers (outcome space), and (4) evidence of high-quality assessment (Wright maps).

Many learning progression studies have successfully applied Wilson's Construct Modeling Approach to investigate students' learning progressions in some concepts. For instance, Jin and Anderson (2012) developed a learning progression framework focusing on K-12 students' performance using energy concepts in socio-ecological systems. Similarly, Lee and Liu (2010) measured middle school students' learning progressions of energy concepts across physical, life, and earth science contexts. Plummer and Maynard (2014) constructed learning progressions to explore students' reasoning about the seasons associated with the crosscutting concepts of energy and patterns. Kaldaras et al. (2021) presented a three-dimensional learning progression focusing on crosscutting concepts, disciplinary core ideas, and scientific and engineering practices.

Existing studies have made valuable contributions to investigating students' learning progressions in several concept areas. However, to our knowledge, the currently articulated learning progressions of crosscutting concepts are still hypothetical because of a lack of empirical evidence generated from research (Fick et al.,

2021). "Too little attention is given to how students' understanding of a topic can be supported and enhanced from grade to grade" (NRC, 2007, p. 213). Therefore, there is an urgent call for valid and reliable assessments to evaluate the hypothetical learning progressions in crosscutting concepts (Lee & Liu, 2010). In order to address this gap, the current study aims to validate a learning progression regarding Stability and Change across middle school grades (from Grades 7 to 9) by applying Wilson's Construct Modeling Approach. The findings of this study have the potential to support further integration of the crosscutting concepts into curriculum and instruction.

11.2.3 Learning Progression Framework

Following the *Framework for K-12 Science Education* (NRC, 2012), this study defined *Stability* as a stable condition where some aspects of a system do not change. Constancy, conservation, equilibrium, steady-state, and symmetry are alternative terms of *stability*. *Change* here denotes replacing an old state with a new one, and many aspects of science are related to change.

Various models and frameworks exist to generate the complexity levels of concept development, e.g., the model of complexity of tasks (Commons et al., 1998), Structure of the Observed Learning Outcome (SOLO) (Biggs, 1999), and the knowledge integration framework (Liu et al., 2008). Based on the above models and frameworks, as well as the *Framework for K-12 Science Education* (NRC, 2012), Bloom's (1956) Taxonomy of the Cognitive Domain, existing research about the learning progression of students' understanding of the crosscutting concepts (e.g., Jin & Anderson, 2012, Lee & Liu, 2010; Neumann et al., 2013; Songer et al., 2009), this study developed a learning progression framework to describe students' development of Stability and Change.

Specifically, three elements were considered regarding the learning progression framework development: (1) content domains: across four science disciplines (i.e., geography, physics, biology, chemistry), (2) range of grade levels: Grades 7 to 9, (3) learning progression levels: four primary hierarchical proficiency levels with three sub-levels.

Given that crosscutting concepts apply beyond disciplinary boundaries, students' learning progression in Stability and Change should be mapped out across disciplines (Park & Liu, 2016). This study identified eight science topics involving Stability and Change in four science disciplines based on a thorough review of science curriculum standards (i.e., MoE, 2011; NGSS Lead States, 2013) and science textbooks. The eight science topics are (1) Motions of the Earth, (2) Changes in Landscape, (3) Cycles of Matter, (4) Ecosystems, (5) Forces and Motion, (6) Energy Conservation, (7) Motion of Molecules, and (8) Chemical Change.

According to the National Research Council (NRC, 2007), a typical structure for measuring learning progressions should comprise qualitatively different levels that students are expected to progress through, including lower (as a starting point), intermediate, and upper levels (endpoint). The American Association for the Advancement of Science (AAAS) suggested that the first step in learning Stability and Change is recognizing and describing the changes (AAAS, 1993). After being able to identify concrete changes, students are then ready to "think about patterns of change in the abstract" (p. 272). According to the Next Generation Science Standards, students gradually move from the perspective of "a static equilibrium" (e.g., a stationary table sitting on the floor) to the more sophisticated view of "a dynamic equilibrium" (e.g., a reversible chemical reaction), and finally to "a repeating pattern of cyclic change" (e.g., the moon orbiting Earth) (NGSS Lead States, 2013, p. 10).

In light of the above suggestions, this study's learning progression framework comprises four primary levels (i.e., Identifying, Understanding, Analyzing, and Designing) with three sub-levels (i.e., Static, Dynamic, and Cyclic), from basic to the most sophisticated (shown in Table 11.1). This study assumed that middle school students (from Grades 7 to 9) would first be able to identify and describe a static

Table 11.1 A Learning Progression Framework of Stability and Change Development

Ability	Level	Performance examples
Identifying	Level 1–1: Static	Students are able to identify and describe a static equilibrium, e.g., a table staying on the floor.
	Level 1–2: Dynamic	Students are able to identify and describe a dynamic equilibrium, e.g., a reversible chemical reaction.
	Level 1–3: Cyclic	Students are able to identify and describe a repeating pattern of cyclic change, e.g., the moon orbiting earth.
Understanding	Level 2–1: Static	Students are able to understand that a static equilibrium is a form of stability.
	Level 2–2: Dynamic	Students are able to understand that a dynamic equilibrium is a form of stability.
	Level 2–3: Cyclic	Students are able to understand that a repeating pattern of cyclic change could be considered a stable situation, even though it is not static.
Analyzing	Level 3–1: Static	Students are able to analyze the phenomena of stability and change based on knowledge from daily life.
	Level 3–2: Dynamic	Students are able to analyze how things change and how they remain stable considering the external controls and internal feedback loops, e.g., systems in dynamic equilibrium are stable because of a balance of feedback mechanisms.
	Level 3–3: Cyclic	Students are able to analyze how things change and how they remain stable considering the systemic factors, e.g., gradual changes that accumulate over time would disturb stability.
Designing	Level 4–1: Static	In designing systems for stable operation, students are able to consider the factual knowledge of change and stability.
	Level 4–2: Dynamic	In designing systems for stable operation, students are able to consider the mechanisms of external controls and internal feedback loops.
	Level 4–3: Cyclic	In designing systems for stable operation, students are able to consider the systemic factors influencing the changes and stability, e.g., time scale or atomic scale.

equilibrium (Level 1–1: Static/Identifying). Then, they are able to identify and describe a dynamic equilibrium (Level 1–2: Dynamic/Identifying). Learning continues with students being able to identify and describe a repeating pattern of cyclic change (Level 1–3: Cyclic/Identifying). Eventually, students are expected to progress towards a more sophisticated ability to consider the systemic factors influencing the changes and stability in designing systems for stable operation (Level 4–3: Cyclic/Designing). However, note that there would be some overlaps among sub-levels of different primary levels. For instance, we would expect a student who is able to identify and describe a repeating pattern of cyclic change (Level 1–3) will sometimes consider the mechanisms of external controls and internal feedback loops in designing systems for stable operation (Level 4–2).

11.3 Methods

11.3.1 Instrument Development

This study applied the Construct Modeling Approach (Wilson, 2005) to guide the instrument development through an iterative process. In particular, the iterative process comprised six steps: (1) Developing a hypothetical construct map, (2) Generating items, (3) Developing scoring rubrics, (4) Administering a pilot test and revising items, (5) Conducting the field test, and (6) Rasch analysis.

Developing a Hypothetical Construct Map A construct map articulates a proposed continuum of knowledge that learners are expected to progress through for the target construct (Rittle-Johnson et al., 2011). As mentioned before (see Table 11.1), this study's construct map comprised twelve inter-related hierarchical levels mapped as likely trajectories of middle school students learning toward more sophisticated understanding and higher levels of performance complexity regarding Stability and Change (Morell et al., 2017).

Generating Items The initial item pool was generated based on a comprehensive review of related measurements, including the American Association for the Advancement of Science's assessment-item bank (AAAS Project 2061, 2013), the Next Generation Science Assessment task portal, and the National Assessment of Educational Progress.

The items in the initial pool were reviewed and revised using the following five tenets. The test items should (1) focus on Stability and Change, (2) represent one upper primary level (i.e., 1–3, or 2–3, 3–3, 4–3), (3) correspond to one of the eight science topics across middle school grades (from Grades 7 to 9), (4) allow accurate placement of students on the hypothetical construct map, and (5) be embedded in real-life contexts.

In order to ensure the content validity of the items, each item was reviewed multiple times by a group of experts (two science education professors and three experienced middle school science teachers). They checked the items for clarity, wording, and alignment to the science topics and learning progression levels. They also determined the ordering and spacing of items from easy to hard.

Developing Scoring Rubrics We authored scoring rubrics for each item as outcome spaces to evaluate student performance concerning the three sub-levels (i.e., Static, Dynamic, and Cyclic). An outcome space comprises a set of qualitatively distinct categories for scoring respondents' performance (Wilson, 2005). As shown in Table 11.2, students' responses to the items were categorized at one of four levels: level 3 (Cyclic level, with a score of 3), level 2 (Dynamic level, with a score of 2), level 1 (Static level, with a score of 1), and below level 1 (with a score of 0). Note that the item samples reported in this chapter have all been translated from the original Chinese version into English.

Administering a Pilot Test and Revising Pilot Test Items A total of 24 pilot items were completed by 49 seventh graders, 47 eighth graders, and 48 ninth graders recruited using convenience sampling from one junior high school in Jiangsu, China. The results of the pilot test were analyzed using Rasch analysis to check the quality of the items. Based on the pilot test results, we revised the items by analyzing each misfitting item through a focus group discussion with experts and six student interviews (two students per grade who took part in the pilot test). With consensus decisions made in the focus group, we revised three misfitting items (example shown in Table 11.3) and one mistyped item, and substituted two redundant items (i.e., landed at the same point on the Wright map) for two new items.

Conducting the Field Test The 40-minute field test was undertaken using convenience sampling with 136 seventh graders, 139 eighth graders, and 67 ninth graders enrolled in the other urban junior high school from Jiangsu during students' first mid-semester. These students were academically diverse, including low-, medium-, and high-performers.

Several cycles of item revisions and expert reviews resulted in three versions of the test, comprising a total of 24 constructed-response items. Table 11.4 presents the distribution of the items across the three grades. The first version was taken by 136 seventh graders and included 12 items covering four science topics. The second version, with 12 items covering four science topics, was administered to 139 eighth graders. Sixty-seven ninth graders responded to the third version with 12 items involving five science topics.

Although students in Grades 7 to 9 completed a unique combination of items, common items were present in the three versions to allow for the expression of student performance on the same equal interval logit scale (Boone & Scantlebury, 2006). Seventh graders shared three common items (Wzxh1, Wzxh2, and Wzxh 3) with eighth graders and another three common items (Stxt1, Stxt2, and Stxt 3) with ninth graders. The eighth and ninth graders had nine common items (Lyyd1, Lyyd2, Lyyd3, Lyyd4, Nlsh1, Nlsh2, Fzyd1, Fzyd2, Fzyd3).

Table 11.2 Sample Scoring Rubric of One Item (NLSH1)

(Nlsh1)

Smith places an ice pack in a small empty lunch box and then quickly closes the box. The below-left graph shows the thermal energy of the ice pack, the thermal energy of the air in the lunch box, and the total amount of thermal energy right after Smith closes the box. The below-right graph shows the thermal energy of the ice pack, the thermal energy of the air in the lunch box, and the total amount of thermal energy two hours later.

Please briefly describe what happened in the lunch box after two hours.

NOTE. Assuming that no energy is transferred into or out of the lunch box after it is closed.

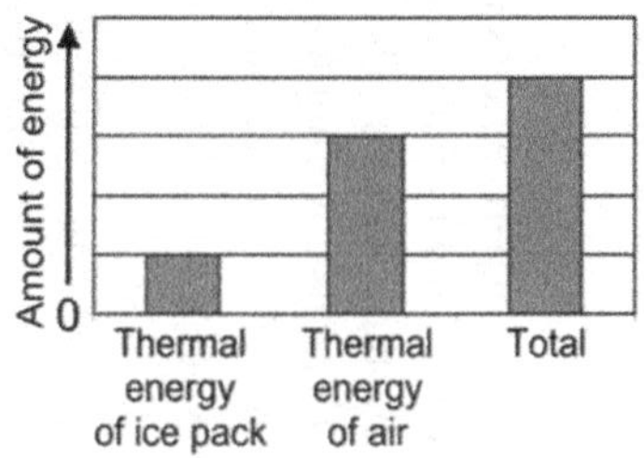

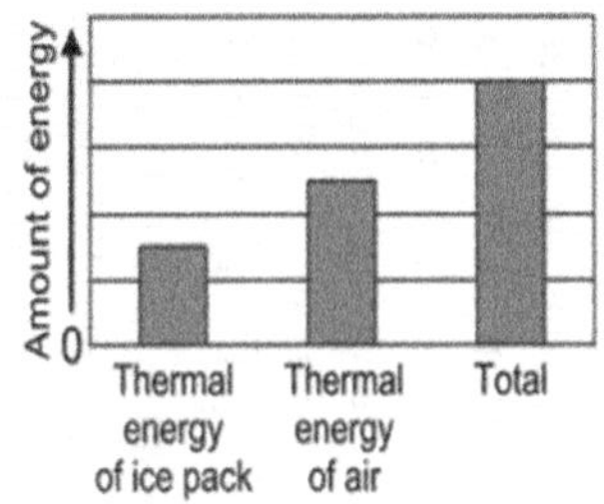

Primary level	Score	Sub-level	Response examples
Identifying	0	Below level 1	No response/wrong answer/response is unrelated to the item.
	1	Level 1	The ice is melting, making the temperature drop.
	2	Level 2	The increased thermal energy for the ice pack equals the loss in the amount of air's thermal energy.
	3	Level 3	The energy from high-temperature air transfers to the ice pack of low temperature because the temperature of ice (0 °C) is lower than the room temperature. In addition, there is no energy exchange between the lunch box and the outside world, so the total amount of thermal energy remains stable.

Table 11.3 Revising Item Example (Wzxh2)

Pilot test item	One day, Mary experimented on how a mouse could survive in a jar with a mint plant:  Mary found that the amount of oxygen (O_2) in this sealed jar with a mint plant kept stable for five minutes. Then, she took out the mint plant outside the jar and found that the amount of oxygen (O_2) in this sealed jar without a mint plant kept dropping, which caused a threat to the survival of the mouse. She finally stopped the experiment and opened the jar before the mouse died. ***Please analyze why and how the mouse was able to survive when placed in the jar with a mint plant.***
Comments	The question is too open-ended, and its meaning is not well-focused. In the pilot test, more than half of the participants only talked about the process of photo-synthesis and did not notice "the change of the amount of oxygen", which is the critical key to the response.
Modification	***Please analyze why and how the amount of oxygen was kept stable only with the mint plant so the mouse could survive.***

Table 11.4 Formation of the Measurement Instrument

Discipline	Science topic	Item	Identifying			Understanding			Analyzing			Designing			Grade		
			L1	L2	L3	L1	L2	L3	L1	L2	L3	L1	L2	L3	7	8	9
Geography	Motions of the Earth	Dqyd1	√	√	√										√		
		Dqyd2				√	√	√							√		
		Dqyd3				√	√	√							√		
	Changes in Landscape	Dxbh1							√	√	√				√		
		Dxbh2	√	√	√										√		
		Dxbh3				√	√	√							√		
Biology	Cycles of Matter	Wzxh1	√	√	√										√	√	
		Wzxh2							√	√	√				√	√	
		Wzxh3										√	√	√	√	√	
	Ecosystems	Stxt1							√	√	√				√		√
		Stxt2	√	√	√										√		√
		Stxt3				√	√	√							√		√
Physics	Forces and Motion	Lyyd1	√	√	√											√	√
		Lyyd2							√	√	√					√	√
		Lyyd3							√	√	√					√	√
		Lyyd4										√	√	√		√	√
	Energy Conservation	Nlsh1	√	√	√											√	√
		Nlsh2										√	√	√		√	√
Chemistry	Motion of Molecules	Fzyd1							√	√	√					√	√
		Fzyd2				√	√	√								√	√
		Fzyd3							√	√	√					√	√
	Chemical Change	Hxbh1				√	√	√									√
		Hxbh2							√	√	√						√
		Hxbh3				√	√	√									√

Rasch Analysis Rasch measurement has been continuously and routinely used for test development and analysis in educational settings. The Rasch model is "a supposition about what happens when a person responds to an item" (Wright, 1977, p. 114). As a probabilistic mathematical model, the Rasch model determines the probability of one person succeeding on a specific item with two parameters: item difficulty and person ability. Specifically, item difficulty and person ability measures are placed along the same interval logit scale, and do not depend on the sample used (Boone et al., 2014). The properties of the Rasch model make it particularly appropriate for measurement construction and instrument quality evaluation (Planinic et al., 2019). For instance, model-data fit statistics generated by the Rasch model can identify problematic items and idiosyncratic respondents and test the validity and reliability of the measurement scales (Andrich & Marais, 2019).

One of the Rasch model extensions is the partial credit Rasch model (Masters, 1982), which was developed for situations where instrument items do not share the same category structure (Fox, 1999). The partial credit Rasch model could take the following form: $\ln(P_{nij}/P_{ni(j\text{-}1)}) = B_n - D_i - F_{ij}$.
where:

$P_{nij}/P_{ni(j\text{-}1)}$ = the probability of a person n endorsing category j over the previous category $(j - 1)$;
B_n = person n's ability;
D_i = Item i's overall difficulty;
F_{ij} = the threshold between categories j and $j - 1$ on item i.

The development of this study's assessment instrument and subsequent analysis of pilot and field test results were all guided by the application of Rasch measurement. Given that all the 24 items in this study were polytomous and did not share the same scale steps, partial credit Rasch model analysis with Winsteps software (Linacre, 2014) version 4.5.2 was applied for measurement development and instrument evaluation. All raw data were converted into Rasch measures (in logits) to generate interval data for further statistical calculations (Bond & Fox, 2015).

In order to compare student performance obtained from the three versions of the test with partial credit items, we used step calibrations and item measures anchoring to express student performance across three grades on the same linear scale (Boone et al., 2014). First, the data from the seventh graders were analyzed, and the item measures and step values (Rasch-Andrich threshold values) of six common items (Wzxh1, Wzxh3, Wzxh3, Stxt1, Stxt2, Stxt3) were noted. The Rasch-Andrich threshold is the point at which there is a 50% likelihood of an individual being scored in either adjacent category. Then, the six common items' item measures and step values were anchored in the eighth-grade data analysis (Wzxh1, Wzxh3, Wzxh3) and ninth-grade data analysis (Stxt1, Stxt2, Stxt3), respectively. Similarly, in analyzing the data from the eighth graders, the item measures and step values of nine common items (Lyyd1, Lyyd2, Lyyd3, Lyyd4, Nlsh1, Nlsh2, Fzyd1, Fzyd2, Fzyd3) were recorded for anchoring those item measures and step values in ninth-grade data analysis.

In order to evaluate the psychometric quality of the instrument, we examined reliability by using the person separation index and person/item reliability indices, content validity by Lynn's Content Validity Index, construct validity by fit statistics, person-item distribution map (Wright map), unidimensionality, and absence of bias by gender differential item functioning (DIF). All the indices were generated by Winsteps software (Linacre, 2014), except Lynn's Content Validity Index.

11.4 Findings

Q1: What evidence supports the reliability and validity of measures of the instrument developed in this study for assessing students' progress on the learning progressions in Stability and Change across middle school grades?

Reliability Reliability was checked using person separation index and person/item reliability indices generated by Winsteps. In Rasch analysis, a person separation is an indicator of the number of strata or groups of person ability discriminated by the items (Gothwal et al., 2009). A person separation index of at least 2.0 represents a good level of separation and is considered the minimum preferable value (Boone & Staver, 2020). Person/item reliability indices are analogous to a Cronbach alpha (Brann et al., 2021). A high person reliability value (i.e., close to 1) represents a high internal consistency of a measure, implying that the test is sensitive enough to detect differences among participants. Item reliability means "reproducibility of relative measure location" (Linacre, 2014, p. 671), which can verify the item hierarchy. A high item reliability value shows that the sample is large enough to confirm the item difficulty hierarchy. This study referenced Fisher's (2007) rating scale instrument quality criteria as quality criteria of the person and item measurement reliability, i.e., reliability indices <.67 identified as "Poor", .67–.80 identified as "Fair", .81–.90 as "Good", .91–.94 as "Very good", and > .94 as "Excellent".

As shown in Table 11.5, all three scales showed adequate person separation, with person separation values greater than 2.0. In addition, "Fair" or "Good" person reliability (.77 to.83) and "Excellent" item reliability (.95 to.98) were observed. These values confirmed that items on the instrument could better differentiate between high-performing and low-performing participants, and the sample was sufficiently large to support the item difficulty hierarchy (Boone et al., 2014).

Content Validity Content validity refers to the extent to which the content of an instrument adequately reflects the target construct (Aiken, 2003). This study used Lynn's (1986) Content Validity Index (CVI) to determine content validity. Five experts reviewed and rated the degree to which each item represented the target construct. They rated each item through a four-point Likert scale, 1 = not relevant, 2 = somewhat targeting, 3 = quite targeting, and 4 = highly targeting. Then each item CVI was calculated as the proportion of experts giving a rating of 3 or 4. After

Table 11.5 Field test summary statistics from Rasch model analysis

Grade	Mean	N	Mean range	Mean measure (SE)	Infit		Outfit		Separation	Reliability
					MNSQ	ZSTD	MNSQ	ZSTD		
7	Person	136	−2.50 to 1.42	−0.18 (.41)	.99	0.0	.95	0.0	2.01	.77
	Item	12	−1.74 to 1.14	0.00 (.14)	.99	0.0	.95	−0.2	5.82	.97
8	Person	139	−1.14 to 2.84	0.49 (.39)	1.02	0.1	1.01	0.1	2.07	.81
	Item	12	−0.89 to 1.85	0.72 (.11)	1.04	0.3	1.00	0.1	6.42	.98
9	Person	67	−0.98 to 2.45	0.91 (.40)	1.18	0.3	1.20	0.2	2.30	.83
	Item	15	−0.89 to 1.85	0.58 (.20)	1.11	0.5	1.19	0.4	4.52	.95

several cycles of item revising, all the 24 field test items yielded an index of 1.0 (none of the items was rated 1 or 2 by any experts), verifying the content validity.

Construct Validity Construct validity refers to whether the items follow a predicted ordering to evaluate the latent construct (Boone & Staver, 2020). The validity of the internal construct could be verified by checking fit statistics, the person-item distribution map (Wright map), the existence of sub-dimensions irrelevant to the target construct (unidimensionality), and differential item functioning (DIF) (de Jong et al., 2012).

Winsteps produces a series of fit statistics to evaluate the extent to which each item matches the expectations of the Rasch model. In this study, we reviewed Outfit/Infit MNSQ and ZSTD, with MNSQ values between 0.7 and 1.3 and ZSTD values between −2 and +2 as fit criteria for quality assurance (Bond & Fox, 2015). Infit is an information-weighted misfit statistic, whereby the Outfit is an unweighted statistic and is relatively more sensitive to outliers. MNSQ and ZSTD denote "the differences between the Rasch model's theoretical expectation of item performance and the performance actually encountered for that item in the data matrix" (Bond & Fox, 2015, p. 55). As shown in Table 11.6, all the 24 field items had MNSQ and ZSTD values within the acceptance range.

The Wright map (person-item map) visually presents person ability and item difficulty on a single linear logit scale (Black et al., 2011). Wright Maps can help in the evaluation of construct validity by showing whether the item hierarchy is exhibited as expected (Boone & Staver, 2020). Figure 11.1 presents the Wright map showing the hierarchy of item difficulty and student ability. The 24 field items covered a range of item difficulty from −1.74 to 1.85 logits. Ordered as hypothesized, the higher the items plotted on the Wright map, the higher the complexity level, thus providing evidence of the construct validity of the scales.

Unidimensionality means that all the test items are expected to measure the same latent construct (Fan & Bond, 2019). This study applied the Principal Components Analysis (PCA) of the Rasch residuals to examine the unidimensionality of the scales. A small eigenvalue of the first contrast less than 2.0 with a large proportion explained variance greater than 40% provides a sign of unidimensionality (Linacre, 2014). Our results showed that the first latent variable explained between 43% and 45.2% of the total variance. The eigenvalue of the first contrast was between 1.7 and 1.9. Unidimensionality and construct validity were thus satisfied.

The existence of DIF is a threat to construct validity, as it is an indicator of construct irrelevant variance (Chan & Subramaniam, 2020). Gender DIF has been examined most frequently across DIF studies in achievement testing (Bond & Fox, 2015). The existence of gender DIF signifies the items are functioning differently between gender groups. In this study, the analysis of gender DIF was carried out using the DIF contrast values, where DIF with values less than 0.43 logits is negligible (Linacre, 2014). As shown in Table 11.7, none of the DIF was noticeable (less than 0.43), thus generating another piece of evidence of construct validity.

Table 11.6 Fit statistics of 28 field items from Rasch model analysis

Item	Grade	Measure	SE	Infit.MNSQ	Infit.ZSTD	Outfit.MNSQ	Outfit.ZSTD
Dqyd1	7	−1.74	0.15	0.89	−0.87	0.87	−1.02
Dqyd2		0.27	0.09	0.90	−0.99	0.91	−0.73
Dqyd3		−0.40	0.09	1.00	0.02	1.01	0.10
Dxbh1		0.19	0.11	1.09	0.83	1.11	0.95
Dxbh2		0.03	0.10	0.93	−0.68	0.91	−0.83
Dxbh3		0.03	0.16	0.87	−0.72	0.78	−1.12
Wzxh1	7	−0.08	0.11	0.89	−0.95	0.88	−1.03
	8		0.11	0.98	−0.13	0.97	−0.28
Wzxh2	7	0.45	0.10	0.99	−0.07	0.98	−0.11
	8		0.09	0.82	−1.70	0.75	−1.83
Wzxh3	7	1.14	0.13	0.99	−0.02	0.98	−0.17
	8		0.13	0.91	−0.81	0.91	−0.80
Stxt1	7	0.47	0.15	1.03	0.29	1.02	0.20
	9		0.19	1.28	1.98	1.30	1.99
Stxt2	7	−0.63	0.09	1.24	1.91	1.29	0.42
	9		0.18	1.29	1.97	1.28	1.99
Stxt3	7	0.28	0.14	1.21	1.22	1.30	1.73
	9		0.16	1.28	1.93	1.30	1.98
Lyyd1	8	−0.89	0.12	0.93	−0.57	0.94	−0.49
	9		0.25	1.29	1.84	1.30	1.98
Lyyd2	8	1.08	0.11	1.04	0.40	1.08	0.75
	9		0.15	1.03	0.22	1.04	0.30
Lyyd3	8	1.04	0.09	0.98	−0.15	0.96	−0.23
	9		0.12	1.17	1.18	1.25	1.45
Lyyd4	8	1.65	0.11	1.02	0.20	1.11	0.43
	9		0.13	1.15	0.98	1.24	1.58
Nlsh1	8	0.57	0.08	1.08	0.91	1.17	1.14
	9		0.12	1.01	0.13	1.05	0.28
Nlsh2	8	1.85	0.11	1.00	−0.02	0.96	−0.33
	9		0.15	1.28	1.92	1.30	1.98
Fzyd1	8	0.76	0.14	1.10	0.69	1.07	0.45
	9		0.16	1.25	1.56	1.26	1.78
Fzyd2	8	0.33	0.14	1.18	1.06	1.18	0.97
	9		0.21	0.87	−0.01	0.89	0.06
Fzyd3	8	0.74	0.10	1.01	0.13	0.96	−0.27
	9		0.15	0.73	−1.80	0.76	−1.45
Hxbh1	9	0.60	0.17	0.99	−0.04	0.99	−0.01
Hxbh2		0.61	0.15	0.95	−0.26	0.96	−0.17
Hxbh3		0.25	0.15	0.95	−0.28	0.90	−0.62

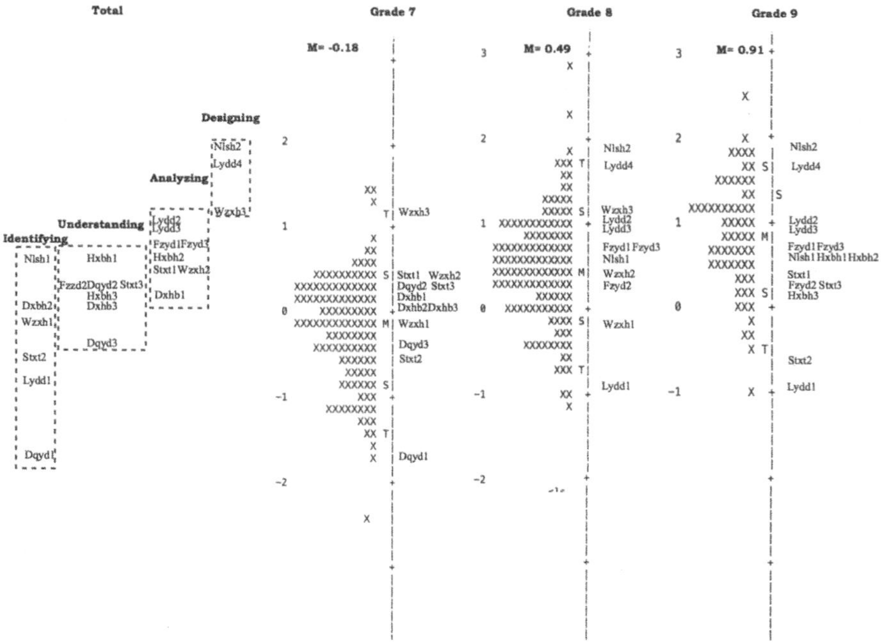

Fig. 11.1 Wright Map for the Distributions of Item Difficulty and Person Ability
Note. Each "X" represents one student

Q2: How do students develop an understanding of Stability and Change from Grades 7 to 9?

As aforementioned, we authored scoring rubrics for each item to evaluate student performance concerning the three sub-levels (i.e., Cyclic level, with a score of 3; Dynamic level, with a score of 2; Static level, with a score of 1). We also used Rasch-Andrich threshold values to determine each item's cutoff value of three sub-levels. For example, the Rasch-Andrich threshold values of item Dqyd1 ($M = -1.74$ logits) are -1.70 logits, -0.75 logits, and 2.46 logits, respectively. Accordingly, this item's cutoff values corresponding to the three sub-levels are -3.44 logits (level 1–1), $-$ 2.49 logits (level 1–2). and 0.72 logits (level 1–3). Then, each sub-level's cutoff value was calculated using the average measure of items' related Andrich threshold values.

As shown in Table 11.8, the cutoff value increased along with the proficiency levels, as hypothesized. On average, items of higher levels are more difficult than items of lower levels. Regarding the four primary levels, "Designing" ($M = 1.02$ logits) is the most difficult, followed by "Analyzing" ($M = 0.67$ logits), "Understanding" ($M = 0.23$ logits), and "Identifying" ($M = -0.46$ logits). The average item difficulty showed a linear function of complexity within the primary levels in terms of the sub-levels. However, there were overlapping sub-levels of different primary

Table 11.7 Gender DIF

Item	Male		Female		DIF Contrast	*p*
	DIF Measure	SE	DIF Measure	SE		
Dqyd1	−1.82	0.20	−1.66	0.23	−0.17	.59
Dqyd2	0.17	0.12	0.39	0.15	−0.22	.25
Dqyd3	−0.31	0.11	−0.55	0.13	0.24	.17
Dxbh1	0.35	0.15	−0.03	0.17	0.38	.10
Dxbh2	0.09	0.13	−0.08	0.15	0.16	.41
Dxbh3	0.19	0.20	−0.23	0.26	0.41	.21
Wzxh1	−0.04	0.10	−0.04	0.11	−0.01	.96
Wzxh2	0.51	0.10	0.57	0.11	−0.06	.70
Wzxh3	1.34	0.12	1.33	0.14	0.01	.97
Stxt1	0.27	0.15	0.41	0.17	−0.14	.54
Stxt2	−0.70	0.12	−0.29	0.12	−0.41	.01
Stxt3	0.18	0.13	0.19	0.15	−0.01	.96
Lyyd1	−0.69	0.14	−0.91	0.15	0.23	.26
Lyyd2	0.65	0.13	1.00	0.12	−0.35	.05
Lyyd3	1.13	0.10	1.16	0.11	−0.03	.82
Lyyd4	1.74	0.11	2.01	0.13	−0.28	.11
Nlsh1	0.67	0.09	0.61	0.09	0.05	.67
Nlsh2	1.36	0.11	1.35	0.12	0.00	.98
Fzyd1	0.45	0.15	0.16	0.16	0.29	.18
Fzyd2	0.53	0.19	0.42	0.20	0.11	.68
Fzyd3	0.74	0.11	0.68	0.11	0.06	.70
Hxbh1	1.02	0.24	0.51	0.24	0.51	.14
Hxbh2	0.97	0.20	0.55	0.22	0.42	.16
Hxbh3	0.33	0.21	0.49	0.21	−0.16	.59

levels, just as expected. The item difficulty level was presented in Fig. 11.2, ordered from the most difficult level (Level 4–3: Systems/Designing, $M = 2.26$ logits) to the easiest level (Level 1–1: Static/Identifying, $M = -1.70$ logits).

In Rasch analysis, respondents have a 50/50 chance of correctly answering an item if they have an identical ability measure with this item's difficulty measure (Boone et al., 2014). We thus classified a student as having attained a particular level when this student had an identical ability measure as the cutoff value of this level. For example, a student with 1.0 logits reached Level 1–3 but had not yet achieved Level 4–2.

As shown in Fig. 11.3, overall, more than half of the seventh graders in this sample attained level 2 or higher in Identifying (72.7%), level 1 or higher in Understanding (93.3%), Analyzing (80.8%), and Designing (59.5%). About one-fifth of the seventh graders performed at level 2 or higher in Understanding (21.3%). Less than 5% of them attained proficiency level 3 or higher in Identifying (2.9%) and level 2 or higher in Analyzing (4.4%) and Designing (2.2%). None of the

Table 11.8 Cutoff values of 12 levels

Primary level	Sub-level	Item cutoff value (logits)	Cutoff value (logits)	Mean cutoff (logits)
Identifying	1–1	Dqyd1(−3.44), Dxbh2(−0.94), Stxt2(−1.16), Wzxh1(−1.78), Lydd1(−2.35), Nlsh1(−0.54)	-1.70	−0.46
	1–2	Dqyd1(−2.49), Dxbh2(0.36), Stxt2(−0.88), Wzxh1(0.47), Lydd1(−0.66), Nlsh1(0.48)	−0.45	
	1–3	Dqyd1(0.72), Dxbh2(0.66), Stxt2(0.15), Wzxh1(1.06), Lydd1(0.33), Nlsh1(1.78)	0.78	
Understanding	2–1	Dqyd2(−0.23), Dqyd3(−0.74), Dxbh3(−1.65), Stxt3(−1.72), Fzyd2(−1.69), Hxbh1(−1.51), Hxbh3(−0.91)	−1.21	0.23
	2–2	Dqyd2(0.43), Dqyd3(−0.42), Dxbh3(0.01), Stxt3(0.34), Fzyd2(−0.30), Hxbh1(1.03), Hxbh3(0.65)	0.25	
	2–3	Dqyd2(0.61), Dqyd3(−0.04), Dxbh3(2.55), Stxt3(2.22), Fzyd2(2.97), Hxbh1(2.29), Hxbh3 (0.99)	1.66	
Analyzing	3–1	Dxbh1(−1.45), Stxt1(−2.04), Lydd2(−0.26), Fzyd1(−1.23), Fzyd3(−0.24), Lydd3(0.13), Wzxh2(−0.58), Hxbh2(−0.07)	−0.72	0.67
	3–2	Dxbh1(0.58), Stxt1(1.04), Lydd2(0.16), Wzxh2 (0.70), Fzyd1(1.09), Fzyd3(0.36), Lydd3(1.02), Hxbh2(0.14)	0.64	
	3–3	Dxbh1(1.45), Stxt1(2.40), Lydd2(3.34), Wzxh2 (1.18), Fzyd1(2.43), Fzyd3(2.10), Lydd3(1.97), Hxbh2(1.77)	2.08	
Designing	4–1	Wzxh3(−0.90), Lydd4(−0.52), Nlsh2(0.77)	−0.22	1.02
	4–2	Wzxh3(0.53), Lydd4(1.62), Nlsh2(0.94)	1.03	
	4–3	Wzxh3(2.27), Lydd4(2.15), Nlsh2(2.35)	2.26	

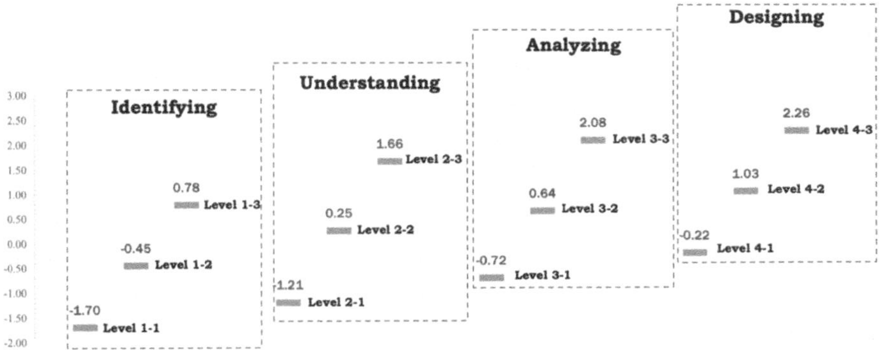

Fig. 11.2 Item difficulty level

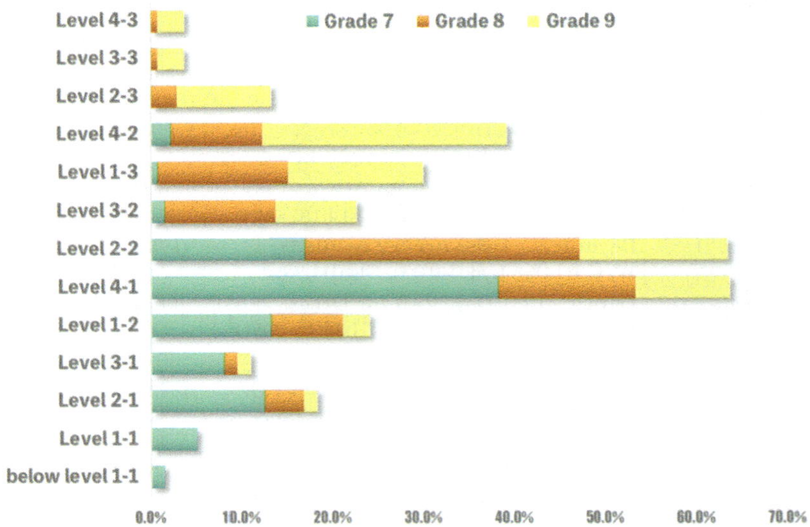

	below level 1-1	Level 1-1	Level 2-1	Level 3-1	Level 1-2	Level 4-1	Level 2-2	Level 3-2	Level 1-3	Level 4-2	Level 2-3	Level 3-3	Level 4-3
Grade 7	1.5%	5.1%	12.5%	8.1%	13.2%	38.2%	16.9%	1.5%	0.7%	2.2%	0.0%	0.0%	0.0%
Grade 8	0.0%	0.0%	4.3%	1.4%	7.9%	15.1%	30.2%	12.2%	14.4%	10.1%	2.9%	0.7%	0.7%
Grade 9	0.0%	0.0%	1.5%	1.5%	3.0%	10.4%	16.4%	9.0%	14.9%	26.9%	10.4%	3.0%	3.0%

Fig. 11.3 Students at different levels of proficiency, as a percentage of each grade
Note. Level 1–1: Static/Identifying; Level 1–2: Dynamic/Identifying; Level 1–3: Cyclic/Identifying; Level 2–1: Static/Understanding; Level 2–2: Dynamic/Understanding; Level 2–3: Cyclic/Understanding; Level 3–1: Static/Analyzing; Level 3–2: Dynamic/Analyzing; Level 3–3: Cyclic/Analyzing; Level 4–1: Static/Designing; Level 4–2: Dynamic/Designing; Level 4–3: Cyclic/Designing

seventh graders were proficient at level 3 in Understanding, Analyzing, and Designing.

Regarding eighth graders, over 70% of them were proficient at level 2 or higher in Identifying (94.2%) and Understanding (71.2%), level 1 or higher in Analyzing (95.2%) and Designing (86.3%). Between 14% and 42% of eighth graders performed at level 2 or higher in Analyzing (41.0%) and Designing (14.4%), and level 3 or higher in Identifying (28.8%). Less than 5% of students reached proficiency level 3 in Understanding (4.3%), Analyzing (1.4%), and Designing (0.7%).

More than half of ninth graders performed at level 2 or higher in Understanding (83.6%) and Analyzing (67.2%) and achieved level 3 or higher in Identifying (58.2%). About 40% of ninth graders attained level 2 or higher in Designing (43.3%), whereas 16.4% of students were proficient at level 3 or higher in Understanding. Only a few of ninth graders reached the highest level in Analyzing (6.0%) and Designing (3.0%).

11.5 Discussions and Implications

This study aimed to investigate middle school students' (from Grades 7 to 9) learning progression in Stability and Change. In order to achieve this aim, we developed three versions of an instrument with anchored items and steps to collect data on students' progression in developing this crosscutting concept. The instrument's functioning and students' progression in developing Stability and Change are discussed in the following section.

11.5.1 Instrument Functioning

Overall, the results of the Rasch analysis confirmed that the instrument satisfied the quality benchmarks. Specifically, this study checked various sources of reliability and validity evidence. Reliability was supported by the acceptable person separation values, "Fair" or "Good" person reliability, and "Excellent" item reliability. Regarding content validity, this study referenced 13 widely used middle school science textbooks and pertinent large-scale assessment items to develop a set of assessment items. All items were evaluated by a panel of experts, pilot tested with middle school students, and revised accordingly before being field tested. Analysis of item fit statistics showed that none of the items was identified as misfitting. Items plotting on the Wright map ordered as expected, lack of gender DIF, and absence of irrelevant sub-dimensions all supported the construct validity.

Predictive validity refers to how the measure has power in forecasting future outcomes (Baghaei, 2008). Boone and Staver (2020) proposed that researchers look at groups of students and individual students when evaluating predictive validity. That is, students with high measures should indeed be high achievers. As expected, this study's findings showed that students in higher grades outperformed peers in lower grades in terms of the mean Rasch measure ($M_{\text{ninth grader}} = 0.91$; $M_{\text{eighth grader}} = 0.49$; $M_{\text{seventh grader}} = -0.18$). In addition, on average, more students reached higher levels of the learning progression in the upper grades than in the lower grades. In these regards, the predictive validity of the scales could be supported by the above indicators.

Evaluating concurrent validity means examining whether "the person ability measures correlate well with other test instruments probing the same latent variable" (Baghaei, 2008, p. 1145). However, this study did not evaluate the concurrent validity because there were currently no alternative standard tests for measuring students' understandings of Stability and Change. Future studies are needed to develop more related measurement instruments and test concurrent validity.

Nevertheless, our instrument could be improved in several aspects. The person separation values were slightly lower than expected, partly because of the relatively small number of items administered to the students, i.e., 12–14 items per grade (Neumann et al., 2013). Moreover, analysis of the Wright map showed that two kinds of gaps should call attention to item addition or revision. First, along the continuum of item measures, there were notable gaps which are over 0.5 logits (Linacre, 2014) between two adjacent items, e.g., between item Wzxh 3 (1.14 logits) and item Lyyd 4 (1.65 logits). Second, gaps were above the most challenging item (Nlsh2, 1.85 logits). The ability measures of a few high performing students were greater (i.e., above 1.85 logits) than the most challenging item's difficulty measure. Although this study focused on assessing student learning progression rather than differentiating between different students, more items should be added to fill the gaps to target respondents accurately, and more complicated items are needed to estimate the most knowledgeable students.

11.5.2 Students' Progression in Developing Stability and Change

This study's results illustrate that higher-grade students were more proficient than lower-grade students regarding Stability and Change. In line with previous research (e.g., McGinnis, 2020; NGSS Lead States, 2013), the findings suggest students might first develop their understanding of stability, taking the forms of "a static equilibrium" and later "a dynamic equilibrium." Not until in Grade 8 are students gradually able to identify and describe a repeating pattern of cyclic change. Only a few of the ninth graders in this sample could understand that even though a repeating cyclic change pattern is not static, it could be considered a stable situation. In addition, the overlap between adjacent levels (i.e., level 2–2 is more difficult to achieve than Level 4–1) might imply that, for middle school students, understanding dynamic equilibrium is more challenging than considering factual knowledge of change and stability in designing systems for stable operation.

These results also echo previous studies' findings that students have difficulties understanding dynamic equilibrium, partly because dynamic equilibrium is usually not perceived in everyday life (Niebert et al., 2012). However, comprehension of dynamic equilibrium is the basis for understanding the stability and changes in the complex system (NGSS Lead States, 2013). Therefore, tremendous and specific efforts are required to help students understand dynamic equilibrium. In particular, classroom teachers could use the instrument to evaluate their students' position on the learning progression and then implement targeted instruction that moves students through the concept progression. For example, as McGinnis (2020) suggested, to facilitate students' learning of Stability and Change, teachers could make students predict how a system would be affected by sudden events or gradual changes and ask them to judge whether a system is stable using system-related vocabulary. In addition, teachers could engage students in investigations and discussions about linking the stability and changes to systems across disciplinary boundaries.

11.6 Limitations

This study has some limitations that point to future research. First, the data were from a single school, though this school did provide a diverse sample. Moreover, both pilot and field tests were based on convenience sampling methods to recruit the participants. Because of the invariance properties of Rasch measures (Liu & Lesniak, 2005), the measures are generalizable to samples with similar variations in student abilities as in this study's sample. Nevertheless, there is a need for further studies to validate the measures from more comprehensive geographical locations and academic backgrounds. Second, as this is a cross-sectional study, caution should be taken when interpreting the results. Future studies might conduct longitudinal analyses to investigate how an individual student advances on a learning progression in Stability and Change over time.

Statements and Declarations This work was supported by the Shanghai Pujiang Program (No. 2020PJC032) and the MOE Key Research Institute of Humanities and Social Sciences (No. 17JJD880007).

All authors have no relevant financial or non-financial interests to disclose.

Appendix

Rasch Analysis Codes (Winsteps Codes)

```
&INST
 TITLE = 789th stability and change
 PERSON = Person ; persons are ...
 ITEM = Item ; items are ...
 ITEM1 = 7 ; column of response to first item in data record
 NI = 24 ; number of items
 NAME1 = 1 ; column of first character of person identifying label
 NAMELEN = 6; length of person label
 XWIDE = 1 ; number of columns per item response
 CODES = "0123" ; valid codes in data file
 UIMEAN = 0 ; item mean for local origin
 USCALE = 1 ; user scaling for logits
 UDECIM = 2 ; reported decimal places for user scaling
 MISSCORE= -1
 T1P#=1
 @GENDER=6
 DIF=@GENDER
 ISGROUPS=0
 IAFILE= *
1 -1.74
 ........
 *
 SAFILE=*
 1 0 0
```

```
 1 1 -1.70
 1 2 -0.75
1 3 2.46
 ........
 *
 &END
 Dqyd1
 Dqyd2
 Dqyd3
 ........
 Hxbh1
 Hxbh2
 Hxbh3
 END LABELS
 740011333222331131XXXXXXXXXXXX
 740021332122121131XXXXXXXXXXXX
 740031213232110131XXXXXXXXXXXX
 740041230122031131XXXXXXXXXXXX
 740051201122112131XXXXXXXXXXXX
 ...........................................................
 811371XXXXXX332XXX300132223XXX
 811381XXXXXX322XXX311031223XXX
 811391XXXXXX121XXX311030220XXX
 811401XXXXXX332XXX213032223XXX
 811411XXXXXX212XXX100002221XXX
 ...........................................................
 933380XXXXXXXXX333123033223122
 933390XXXXXXXXX121230033222121
 933400XXXXXXXXX112320021222103
 933410XXXXXXXXX131222200221001
 933420XXXXXXXXX231320031321232
```

References

AAAS Project 2061. (2013). AAAS project 2061 science assessment website. Retrieved from assess.bscs.org/science/topics

Aiken, S. (2003). Estimation of item parameters. In G. H. Fischer & I. W. Molenaar (Eds.), *Rasch models, foundations, recent developments, and applications* (pp. 39–52). Springer.

American Association for the Advancement of Science. (1993). *Benchmarks for science literncy: Project 2061*. Oxford University Press.

Andrich, D., & Marais, I. (2019). *A course in Rasch measurement theory: Measuring in the educational, social and health sciences*. Springer Nature Singapore Pte Ltd..

Baghaei, P. (2008). The Rasch model as a construct validation tool. *Rasch Measurement Transactions, 22*(1), 1145.

Ben-zvi-Assarf, O., & Orion, N. (2005). A study of junior high students' perceptions of the water cycle. *Journal of Geoscience Education, 53*(4), 366–373.

Biggs, J. B. (1999). *Teaching for quality learning at university*. SRHE and Open University Press.

Black, P., Wilson, M., & Yao, S. Y. (2011). Road maps for learning: A guide to the navigation of learning progressions. *Measurement: Interdisciplinary Research and Perspective, 9*(2–3), 71–123.

Bloom, B. S., Engelhart, M. D., Furst, E. J., Hill, W. H., & Krathwohl, D. R. (1956). *Taxonomy of educational objetives: The classification of educational goals: Handbook I: Cognitive domain* (No. 373.19 C734t). D. Mckay.

Bond, T., & Fox, C. M. (2015). *Applying the Rasch model: Fundamental measurement in the human sciences* (3rd ed.). Taylor & Francis.

Boone, W. J., & Scantlebury, K. (2006). The role of Rasch analysis when conducting science education research utilizing multiple-choice tests. *Science Education, 90*(2), 253–269.

Boone, W., Staver, J., & Yale, M. (2014). *Rasch analysis in the human sciences*. Springer.

Boone, W. J., & Staver, J. R. (2020). *Advances in Rasch analyses in the human sciences*. Springer.

Brann, K. L., Boone, W. J., Splett, J. W., Clemons, C., & Bidwell, S. L. (2021). Development of the school mental Health self-efficacy teacher survey using Rasch analysis. *Journal of Psychoeducational Assessment, 39*(2), 197–211.

Chan, M., & Subramaniam, R. (2020). Validation of a science concept inventory by Rasch analysis. In M. Khine (Ed.), *Rasch measurement: Applications in quantitative educational research* (pp. 159–178). Springer. Singapore.

Commons, M. L., Trudeau, E. J., Stein, S. A., Richards, F. A., & Krause, S. R. (1998). Hierarchical complexity of tasks shows the existence of developmental stages. *Developmental Review, 18*(3), 237–278.

de Jong, A. E., Tuinebreijer, W. E., Bremer, M., van Komen, R., Middelkoop, E., & van Loey, N. (2012). Construct validity of two pain behavior observation measurement instruments for young children with burns by Rasch analysis. *Pain, 153*(11), 2260–2266.

Duschl, R. A. (2012). The second dimension crosscutting concepts: Understanding a framework for K–12 science education. *The Science Teacher, 79*(2), 34–38.

Fan, J., & Bond, T. (2019). Unidimensionality and local independence. In V. Aryadoust & M. Rachelle (Eds.), *Quantitative data analysis for language assessment (Volume I): Fundamental techniques* (pp. 83–102). Routledge.

Fick, S. J., McAlister, A. M., Chiu, J. L., & McElhaney, K. W. (2021). Using students' conceptual models to represent understanding of crosscutting concepts in an NGSS-aligned curriculum unit about urban water runoff. *Journal of Science Education and Technology, 30*(5), 678–691.

Fisher, W. P. (2007). Rating scale instrument quality criteria. *Rasch Measurement Transactions, 21*(1), 1095.

Fox, C. (1999). An introduction to the partial credit model for developing nursing assessments. *Journal of Nursing Education, 38*(8), 340–346.

Goldstone, R. L., & Wilensky, U. (2008). Promoting transfer by grounding complex systems principles. *Journal of the Learning Sciences, 17*(4), 465–516.

Gothwal, V. K., Wright, T. A., Lamoureux, E. L., & Pesudovs, K. (2009). Rasch analysis of visual function and quality of life questionnaires. *Optometry and Vision Science, 86*(10), 1160–1168.

Jin, H., & Anderson, C. W. (2012). A learning progression for energy in socio-ecological systems. *Journal of Research in Science Teaching, 49*(9), 1149–1180.

Johnson, P. (2000). Children's understanding of substances, part 1: Recognizing chemical change. *International Journal of Science Education, 22*(7), 719–737.

Kaldaras, L., Akaeze, H., & Krajcik, J. (2021). Developing and validating next generation science standards-aligned learning progression to track three-dimensional learning of electrical interactions in high school physical science. *Journal of Research in Science Teaching, 58*(4), 589–618.

Lee, H. S., & Liu, O. L. (2010). Assessing learning progression of energy concepts across middle school grades: The knowledge integration perspective. *Science Education, 94*(4), 665–688.

Linacre, J. M. (2014). *A user's guide to Winsteps Ministep Rasch-model computer programs: Program manual 3.81.0*. Winsteps.com.

Liu, O. L., Lee, H.-S., Hofstetter, C., & Linn, M. (2008). Assessing knowledge integration in science: Construct, measures, and evidence. *Educational Assessment, 13*(1), 33–55.

Liu, X., & Lesniak, K. M. (2005). Students' progression of understanding the matter concept from elementary to high school. *Science Education, 89*(3), 433–450.

Liu, X. (2012). Using learning progression to organize learning outcomes: Implications for assessments. In S. Bernholt, K. Neumann & Nentwig (Eds.), Making it tangible-learning outcomes in science education (pp. 285–301). : Waxmann.

Lynn, M. R. (1986). Determination and quantification of content validity. *Nursing Research, 35*(6), 382–386.

Masters, G. N. (1982). A Rasch model for partial credit scoring. *Psychometrika, 47*(2), 149–174.

McGinnis, P. (2020). Stability and change: Integral to systems. *Science Scope, 703*, 312–9273.

Ministry of Education (MoE). (2011). *Compulsory junior high school science curriculum standards*. Beijing Normal University Press. (in Chinese).

Morell, L., Collier, T., Black, P., & Wilson, M. (2017). A construct-modeling approach to develop a learning progression of how students understand the structure of matter. *Journal of Research in Science Teaching, 54*(8), 1024–1048.

National Research Council (NRC). (2007). *Taking science to school: Learning and teaching science in grade K-8*. The National Academies Press.

National Research Council. (2012). *A framework for K–12 science education: Practices, crosscutting concepts, and core ideas*. National Academies Press.

National Research Council. (2013). *Next generation science standards*. National Academies Press.

Neumann, K., Viering, T., Boone, W. J., & Fischer, H. E. (2013). Towards a learning progression of energy. *Journal of Research in Science Teaching, 50*(2), 162–188.

Niebert, K., Marsch, S., & Treagust, D. F. (2012). Understanding needs embodiment: A theory-guided reanalysis of the role of metaphors and analogies in understanding science. *Science Education, 96*(5), 849–877.

NGSS Lead States. (2013). *Next generation science standards: For states, by states*. National Academies Press.

Lead States, N. G. S. S. (2013). *Next generation science standards: For states, by states. Volume 2: Appendix G- crosscutting Concetps*. National Academies Press.

Park, M., & Liu, X. (2016). Assessing understanding of the energy concept in different science disciplines. *Science Education, 100*(3), 483–516.

Planinic, M., Boone, W. J., Susac, A., & Ivanjek, L. (2019). Rasch analysis in physics education research: Why measurement matters. *Physical Review Physics Education Research, 15*(2), 020111.

Plummer, J. D., & Maynard, L. (2014). Building a learning progression for celestial motion: An exploration of students' reasoning about the seasons. *Journal of Research in Science Teaching, 51*(7), 902–929.

Riley, L., & Biernat, K. (2019). Making sense of stability and change. *Science Scope, 42*(6), 32–35.

Rittle-Johnson, B., Matthews, P. G., Taylor, R. S., & McEldoon, K. L. (2011). Assessing knowledge of mathematical equivalence: A construct-modeling approach. *Journal of Educational Psychology, 103*(1), 85–104.

Rivet, A. E., Weiser, G., Lyu, X., Li, Y., & Rojas-Perilla, D. (2016). What are crosscutting concepts in science? Four metaphorical perspectives. In C. K. Looi, J. L. Polman, U. Cress, & P. Reimann (Eds.), *Transforming learning, empowering learners: The international conference of the learning sciences (ICLS)* (Vol. 2, pp. 970–973). International Society of the Learning Sciences.

Songer, N. B., Kelcey, B., & Gotwals, A. W. (2009). How and when does complex reasoning occur? Empirically driven development of a learning progression focused on complex reasoning about biodiversity. *Journal of Research in Science Teaching, 46*(6), 610–631.

Stavy, R. (1990). Children's conceptions of changes in the state of matter: From liquid (or solid) to gas. *Journal of Research in Science Teaching, 27*, 247–266.

Stevens, S., Delgado, C., & Krajcik, J. S. (2010). Developing a hypothetical multidimensional learning progression for the nature of matter. *Journal of Research in Science Teaching, 47*(6), 687–715.

Testa, I., Galano, S., Leccia, S., & Puddu, E. (2015). Development and validation of a learning progression for change of seasons, solar and lunar eclipses, and moon phases. *Physical Review Special Topics-Physics Education Research, 11*(2), 020102.
Wilson, M. (2005). *Constructing measures: An item response modeling approach.* Taylor & Francis Group.
Wilson, M. (2009). Measuring progressions: Assessment structures underlying a learning progression. *Journal of Research in Science Teaching, 46*(6), 716–730.
Wright, B. D. (1977). Solving measurement problems with the Rasch model. *Journal of Educational Measurement, 14*(2), 97–116.

Shaohui Chi She is a researcher in science education. Her research focuses on measuring and evaluating student outcomes in science. She is also interested in improving scientific literacy among 6–12 students.

Zuhao Wang He is known widely for his scholarship on chemistry learning and instruction. He is also an expert on chemistry curriculum development and evaluation and the editor-in-chief of a series of chemistry textbooks.

Ya Zhu She is currently a Ph.D. student focusing on chemistry learning and instruction.

Chapter 12
Using Rasch Measurement to Visualize How Clerkship and Extracurricular Experiences Impact Preparedness for Residency in an Undergraduate Medical Program

Amber Todd and William Romine

Abstract National survey data provided to institutions are often only used for reporting purposes and not typically used for secondary data analysis to provide actionable changes. We describe the use of Rasch analysis on data from the AAMC GQ from one allopathic undergraduate medical school to determine the impact of the clinical clerkship experience and participation in extracurricular activities on perception of preparedness for residency. We show how to use Rasch analysis to elucidate the types of activities that could be beneficial to students given their clerkship experience and how other institutions can do the same to help inform curricular changes.

12.1 Introduction/Purpose

As part of a mechanism to evaluate undergraduate medical programs, the Association of American Medical Colleges (AAMC) sends out the annual Medical School Graduation Questionnaire (GQ) to all students who are expected to graduate that academic year (Association of American Medical Colleges, 2022). Data from the GQ used by the Liaison Committee on Medical Education (LCME) and schools for accreditation purposes primarily focus on student satisfaction, perceived quality of educational experiences, and participation in elective opportunities and how these benchmarks compare to the other medical schools (*DCI | Standards, Publications, & Notification Forms | LCME*, 2022), which amounts to purely reporting the raw data,

A. Todd (✉)
Department of Medical Education, Boonshoft School of Medicine, Wright State University, Dayton, OH, USA
e-mail: amber.todd@wright.edu

W. Romine
Department of Biological Sciences, Wright State University, Dayton, OH, USA
e-mail: william.romine@wright.edu

X. Liu, W. J. Boone (eds.), *Advances in Applications of Rasch Measurement in Science Education*, Contemporary Trends and Issues in Science Education 57,
https://doi.org/10.1007/978-3-031-28776-3_12

not doing any analysis. While there have been several studies of late using data from the GQ for research, the analysis mostly stops at correlative (multivariate/logistic regression) studies or comparisons among groups of students (e.g., Dyrbye et al., 2021; Nguyen et al., 2021; O'Marr et al., 2022; Ryus et al., 2022; Shields & Dudley-Javoroski, 2021). Here, we provide a model of how schools can use Rasch analysis on their national survey data to not only determine how variables interact but also use these models to deliver a more sample-free, qualitative, descriptive interpretation. Specifically, we demonstrate how to use the Wright maps to facilitate the extraction of descriptive qualitative information from the correlational models. We argue that this can assist medical programs in strategic planning, including identifying curricular changes that can be implemented with subsequent classes. In this study, we use data from the AAMC GQ to determine the impact of the core clinical clerkship experience and participation in extracurricular activities on student perception of preparedness for residency in a Midwestern United States public allopathic (MD) medical school. Using these data, we provide evidence for the types of extracurricular activities that could be beneficial for these students, depending on their clerkship experience, to feel prepared to enter residency.

12.1.1 Background and Context

Ensuring that graduates are prepared for residency is critically important because we are entrusting these physicians to deliver safe and effective care for their patients. While there has been research on programs aimed to facilitate the transition from undergraduate (MD/DO) to graduate (residency) medical education (Kassam et al., 2021), many studies have noted the lack of requirements in fourth year could be an area of interest to ensure that students are prepared, especially for procedural residencies such as surgery and obstetrics & gynecology (Benson et al., 2015; Langdale et al., 2003; Minter et al., 2015).

The typical undergraduate (MD/DO) medical school curriculum consists of two phases: the pre-clinical/foundational phase and the clinical phase. During the pre-clinical phase (typically 1.5–2 years), students spend much of their time in coursework to understand the foundational science behind normal and disease states. Students then move to the clinical phase, typically consisting of a year of standard clerkships (multi-week rotations in each core specialty) followed by a year of electives/sub-internships/away rotations. In the spring of the fourth year, students "match" into a residency in their desired specialty, graduate with their degree, and then begin their resident training in graduate medical education in July. Throughout medical school, but especially within the fourth year, students can have a wide variety of experiences and electives. Using data from the GQ, we wanted to determine the impact of the core clinical clerkship experience and participation in extracurricular activities on student perception of preparedness for residency and how, for our population of students, these experiences can help students feel prepared for residency. We also provide evidence for the types of activities that could be beneficial for students in their fourth year who wish to feel more prepared for residency, based on their clerkship experience in the third year.

12.2 A Definition of Measurement

Early in the twentieth century, Louis Thurstone addressed the question: What if behavioral phenomena could be measured in the same way that we measure more directly observable qualities such as weight, blood pressure, or solution concentration, in a laboratory? What would an appropriate model of measurement entail? He concluded that what all scientific measures have in common is the property of invariance, and that this is indeed a property of how we define measurement itself (Thurstone, 1928). In order to obtain a useful measure for a patient's blood pressure, for example, we must have a fixed unit of measure that stays constant from patient-to-patient and visit-to-visit; and further, the location of and spacing between the tick marks on the scale should stay constant whether we are measuring a patient with high or low blood pressure. As another example, if we want a useful measure of a person's weight, the definition of the unit must be fixed and invariant no matter which patient we are measuring or the particular time we are taking the measurement. In sum, a useful measurement system is one which delivers measures that are non-arbitrary—their definition is independent of the person being measured or the time which measurement is taking place.

Using the same argument, in order to gather trustworthy and useful measures for behavioral constructs—in the context of this study, constructs like perceived preparation for residency, perceived quality of clerkship experiences, and participation in electives—we must employ a measurement framework which preserves this property of scale invariance. To this end, Rasch (1960) put forth a novel method for defining the law of invariance for measures constructed from categorical observations such as those derived from surveys and tests: the probability that a student selects a particular response on a question can be defined by the product of a difficulty parameter, which pertains only to the question's response category, and an ability parameter, which pertains only to the student completing the question (Wright & Stone, 1979a, b). Consequently, the log-odds of a student selecting a particular response to a question is constrained to be proportional only to the difference between the difficulty of the selected response and the ability of the student (Boone et al., 2013). This definition sets up a geometric measurement framework where the student and item estimates are derived independently; hence, the estimates of question difficulty are independent of which students took the survey and the estimates of student ability are independent of the particular questions on the survey. This gives us a definition of measurement that is constant across the domain of the scale and from sample to sample which sets the stage for scientific attitudinal measurement using measures like the GQ.

Rasch models can be defined mathematically within several modeling paradigms. For example, in the context of generalized mixed models, the Rasch model is often conceived as a mixed effects model with a log-odds (logit) link function where the target of measurement, such as students, is specified as a random factor and the observations of measurement, such as test questions, are specified as a fixed factor (Raudenbush & Bryk, 2002). The Rasch model can also be defined through a logistic confirmatory factor analytic/structural equation modeling framework where the loadings of the items with the underlying dimension of interest are fixed to unity and item estimation errors are constrained to be independent. However, the Rasch model has most often been defined within the paradigm of Item Response Theory (IRT) as a 1-parameter IRT model. Although all of these specifications result in an accurate mathematical definition, these approaches nonetheless ignore the measurement implications since factor analysis and IRT are confirmationist paradigms. In other words, they give a mathematical definition but lead us astray in terms of how this definition should be applied as a law of measurement. Specifically, in light of the falsificationist perspective on scientific inquiry, using the Rasch model as a law of measurement implies the need for a fixed philosophical specification which is held to be true *a priori*. This in turn facilitates a hypothetico-deductive approach to understanding the strengths and limitations of a measure and the framework that underlies it. New information is generated by holding the model as true and rationalizing anomalies that exist in the data (Popper, 1963). It follows that new information is generated not through fitting increasingly complex models as is often done in factor analysis and IRT traditions, but instead through the intellectual exercise of rationalizing the areas of misfit that exist in the measure with respect to a pre-specified model which we hold to be true and invariant (Popper, 1963).

Toward understanding the role of clerkship experiences and extracurricular activities in medical students' perceived preparation for residency, we use the Rasch rating scale model as a definition of measurement. Although this model, and the Rasch approach to measurement in general, may at first seem limiting in light of the common perception that simpler models tend to give poorer fit than complex models, we will see that this provides a sample-free definition of measurement which facilitates new understandings regarding the construct validity of the GQ through inspection of misfit. In addition, since the student ability and item difficulty measures are hypothesized to be related only through the single latent factor being measured, student measures can be mapped onto a hierarchy which is qualitatively defined by the questions on the assessment facilitating qualitative/descriptive interpretations for specific locations along the scale (Boone, 2016).

12.3 Study Design

12.3.1 Sample and Survey

An invitation to complete the AAMC GQ in the spring of 2021 was sent to 116 students graduating in the class of 2021 from a Midwestern United States public allopathic (MD) medical school. Eighty-eight students (76%) responded; 49 (56%)

were female and 39 (44%) were male. Deidentified individual-level data were provided by the AAMC for secondary data analysis. Some items were excluded to avoid identification. We were interested in three parts of the GQ: (1) students' perceptions of how prepared they are for residency, and how this is affected by (2) students' required clerkship experiences, mainly during Year 3, and (3) the extracurricular activities students choose to undertake during their time in medical school.

Preparedness for a residency program was measured by Question 12 on the GQ: "*Indicate whether you agree or disagree with the following statements about your preparedness for beginning a residency program.*" The responses took a 5-level Likert scale: 1 = Strongly Disagree, 2 = Disagree, 3 = Neutral, 4 = Agree, 5 = Strongly Agree. Six aspects of preparedness were provided: (1) adequate preparation to care for patients from different backgrounds, (2) understanding common conditions and their management, (3) communication skills to interact with patients and health professionals, (4) skills in clinical decision making and application of evidence-based information to medical practices, (5) understanding social science issues in medicine, and (6) understanding ethical and professional values. Three items on the GQ (clinical skills, skills to apply the principles of high value care in medical decision-making, and skills to address social determinants of health) were not included in the data provided by the AAMC.

Quality of clerkship experience was measured by Question 10 on the GQ: "*Rate the quality of your educational experiences in the following clerkships.*" Responses took a 4-level Likert scale: 1 = Poor, 2 = Fair, 3 = Good, and 4 = Excellent. The eight clerkships measured in this program were: (1) Emergency Medicine, (2) Family Medicine, (3) Internal Medicine, (4) Neurology, (5) Obstetrics and Gynecology, (6) Pediatrics, (7) Psychiatry, and (8) Surgery.

Participation in electives was measured by Question 14 on the GQ: "*Indicate the activities you will have participated in during medical school on an elective (for credit) or volunteer (not required) basis.*" Responses were scored dichotomously (1 = participated, 0 = did not participate). Data were provided for 15 different activities: (1) independent study project for credit, (2) research project with faculty member, (3) authorship [sole or joint] of a peer-reviewed paper submitted for publication, (4) global health experience, (5) educating elementary, high school or college students about careers in health professions or biological sciences, (6) providing health education [e.g., HIV/AIDS education, breast cancer awareness, smoking cessation, obesity], (7) field experience in providing health education in the community [e.g., adult/child protective services, family violence program, rape crisis hotline], (8) field experience in home care, (9) learned another language in order to improve communication with patients, (10) learned the proper use of the interpreter when needed, (11) experience related to health disparities, (12) experience related to cultural awareness and cultural competence, (13) community-based research project, (14) field experience in nursing home care, and (15) experience with a free clinic for the underserved population. Students were also able to indicate "other" and specify additional elective activities, however due to this being a "catch all" of activities, this was not included in our analysis. One item on the GQ (authorship [sole or joint] of a peer-reviewed oral or poster presentation) was not included in the data provided by the AAMC.

12.3.2 Application of Rasch Modeling

The goal of Rasch modeling was two-fold: (1) to provide evidence for unidimensional hierarchical structure and construct validity of the scales for perceived preparedness for residency, quality of clerkship experience, and participation in electives, and (2) to provide a qualitative predictive framework to help us better understand how formal clerkship experiences and the comparatively informal process of participating in electives work together to provide students the perception of being prepared for residency. Rasch models are often specified mathematically as a 1-parameter IRT model. However, the similarity ends there (Andrich, 2004). Unlike IRT, where increasingly complex models are selected to optimize fit with the data, Rasch models hold the philosophical ideal of scale invariance as the gold standard to which the data are expected to conform. In this case, we caution against the practice of choosing a "best-fitting validity model" and instead recommend choosing a validity model that makes sense given the structure of the survey and the purpose of the research. For this study, given our Likert scale and sample size of 88 students, it was conceivable that not all categories across the ordinal Likert responses would be well-represented on all items; hence we chose to use the rating scale model (Andrich, 1982) as a criterion for validity of the "*preparedness for residency*" and "*clerkship experience*" scales. Although the rating scale model holds the data to a more difficult-to-meet standard than partial credit or nominal response models, we will see that it pays dividends by improving our ability to interpolate the expected categories across all of the items in the scale, facilitating interpretation. The dichotomous model, which is sometimes referred to as the simple logistic model (Wright & Stone, 1979a, b), was used for *participation in electives*.

To explore the extent to which the hierarchy of competencies form a unidimensional scale meeting the above criteria, we used principal components analysis (PCA) on standardized residuals. If the competencies are measuring a single construct, then a first eigenvalue at or below 2, which indicates randomly distributed residuals, is expected (Linacre & Tennant, 2009). To evaluate the fit or misfit of individual items within each scale, we used mean squares infit and outfit measures. These have expected values of 1 (Wright & Stone, 1979a, b), but Wright et al. (1994) suggest that these indices can range between 0.5–1.5 for useful items and scenarios in low-stakes situations.

12.3.3 *Probable Response Modes and Regression Modeling*

Based on the step thresholds in the respective rating scales, we first inspected the individual scales to understand the types of responses that students at different locations on the scale were predicted to make. We then extended this to understanding how electives and clerkship experiences facilitate perception of preparedness for residency. We did this by first fitting a linear regression model with preparedness for residency (in logits) as the outcome and clerkship experience (in logits), electives (in logits), and the elective x clerkship interaction as predictors. Since the distribution of preparedness for residency was negatively skewed and leptokurtic, bootstrap standard errors (based on 10,000 resamples with replacement from the data) were obtained, and statistical significance was evaluated at the 95% confidence level using the percentile method. After a model was fitted, we overlaid the probable response modes obtained from the Rasch models over the regression lines and axes in order to facilitate qualitative diagnostics regarding the types of support different types of medical students need in order to feel best prepared for residency.

12.4 Construct Validity of the GQ

The six elements of preparedness for residency formed a reliable ($r_{person} = 0.84$, $r_{item} = 0.96$) and unidimensional scale (first eigenvalue = 1.78) with respect to the rating scale model. The eight clerkship experiences formed a unidimensional scale (first eigenvalue = 1.57) and displayed satisfactory reliability ($r_{person} = 0.79$, $r_{item} = 0.93$) with respect to the rating scale model. The 15 electives measured in this study also created a scale which was unidimensional (first eigenvalue = 1.94) and reliable ($r_{person} = 0.72$, $r_{item} = 0.95$). Given the diversity of the elements of residency preparedness, clerkships, and electives, we were initially surprised to find that the Rasch models supported the hypothesis of unidimensionality for the students in this medical program (Table 12.1). This tells us that although the competencies are qualitatively different, these differences can be accounted for by their hierarchy along their respective unidimensional scales. Infit and outfit indices also generally support the validity of the individual questions as a basis for their respective scale. Within preparedness for residency, the only misfitting competency was recognition of social science issues (outfit = 1.51); but with a point biserial correlation of 0.56, this could be categorized as an informative-noisy response pattern (Smith, 1996). The same could be said for the clerkship experience in Neurology (outfit = 1.45, point biserial = 0.34). All electives had near expected levels of fit with the dichotomous model (outfit = 0.46–1.26; infit = 0.75–1.23), which is exceptional given their considerable diversity. This means participation in the electives which are more difficult to engage within implies participation in those which are more accessible. In other words, students who participated in an elective with a low participation rate, such as learning a new language or practicing in another country, were likely to also

Table 12.1 Rasch item statistics for residency, clerkship, and electives based on the Rasch rating scale model. All models were unidimensional (first eigenvalues of 1.78, 1.57, and 1.94, respectively, based on PCA on standardized residuals)

Competency	Measure	SE	Infit	Outfit	PtBis
Preparation for Residency ($r_{person} = 0.84$, $r_{item} = 0.96$)					
Common Conditions	1.81	0.22	0.73	0.99	0.52
Clinical Decision Making	1.07	0.24	0.77	0.70	0.68
Social Issues	0.55	0.25	1.23	1.51	0.56
Diverse Patients	−0.29	0.27	1.34	1.19	0.37
Ethical Values	−1.48	0.29	0.78	0.70	0.53
Communication Skills	−1.66	0.29	0.97	0.61	0.65
Clerkship Experience ($r_{person} = 0.79$, $r_{item} = 0.93$)					
ObGyn	0.99	0.14	1.11	1.10	0.51
Surgery	0.88	0.15	0.99	1.06	0.41
Neurology	0.14	0.16	1.28	1.45	0.34
Internal Medicine	−0.04	0.16	0.71	0.76	0.63
Family Medicine	−0.18	0.17	1.12	1.24	0.50
Pediatrics	−0.21	0.17	0.73	0.78	0.66
Psychiatry	−0.52	0.18	1.09	0.95	0.50
Emergency Medicine	−1.06	0.21	0.93	0.83	0.51
Electives ($r_{person} = 0.72$, $r_{item} = 0.95$)					
Learn a New Language	1.72	0.25	1.04	0.90	0.20
International Study	1.65	0.25	1.17	1.04	0.09
Community Health	1.52	0.25	1.03	0.95	0.20
Community Project	1.10	0.23	1.12	1.19	0.09
Home Care	0.82	0.23	0.94	0.93	0.28
Nursing Home Care	0.65	0.22	1.02	0.99	0.22
Authorship on Paper	0.17	0.23	1.23	1.26	0.02
Educating Students	0.11	0.23	0.98	0.90	0.28
Health Education	0.05	0.23	0.91	1.11	0.34
Independent Study	−0.17	0.23	1.11	1.26	0.14
Free Clinic	−0.90	0.27	0.92	0.85	0.32
Interpreter	−1.15	0.29	0.91	0.73	0.36
Cultural Awareness	−1.43	0.31	0.79	0.53	0.48
Research with Faculty	−2.06	0.38	0.88	0.84	0.30
Health Disparities	−2.06	0.38	0.75	0.46	0.45

participate in the more accessible electives such as pursuing an authorship on a paper or working in the free clinic. Conversely, not participating in a more accessible elective such as doing research with a faculty member, was predictive of not participating in a more difficult-to-access electives such as working in health education.

12.5 Defining the Scales Qualitatively Using the Probable Response Modes

Satisfactory fit of the items with the Rasch model and their collective conformity to the unidimensionality assumption suggests that they conform to the definition of measurement put forth by the rating scale model and simple logistic model, respectively. This means that if a student's ability measure is at or above the difficulty measure for a particular threshold on a question, then it is more likely than not that they have met or exceeded that threshold. In this sense, we can predict which thresholds on particular questions a particular student has or has not met solely based on their location on the scale. This is the logic behind plotting a map of student measures and item difficulty measures along the common scale—the Wright map—as a diagnostic tool (Boone, 2016). Toward generating a descriptive qualitative understanding of each scale, we were interested in the most likely response to each question within a specific scale for different locations along the scale.

12.5.1 Preparedness for Residency

For perception of preparedness for residency, the Wright map (Fig. 12.1) shows that communication skills and ethical values were the easiest aspects of residency for which students felt prepared. Understanding how to work with diverse patients, the social science issues in medicine, and clinical decision making, were of moderate difficulty. Understanding of common conditions was the most difficult aspect for which students felt prepared. As we look at the hierarchy, we see a general progression from soft skills related to understanding and working with patients to the harder

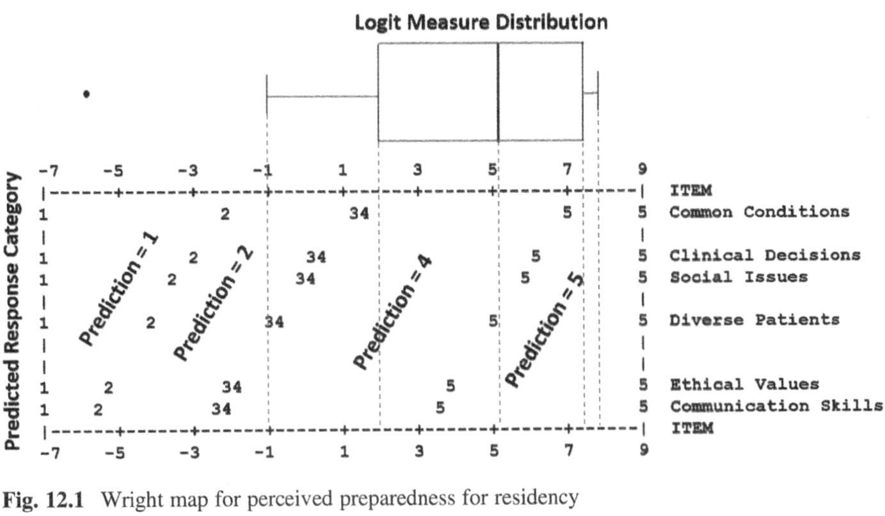

Fig. 12.1 Wright map for perceived preparedness for residency

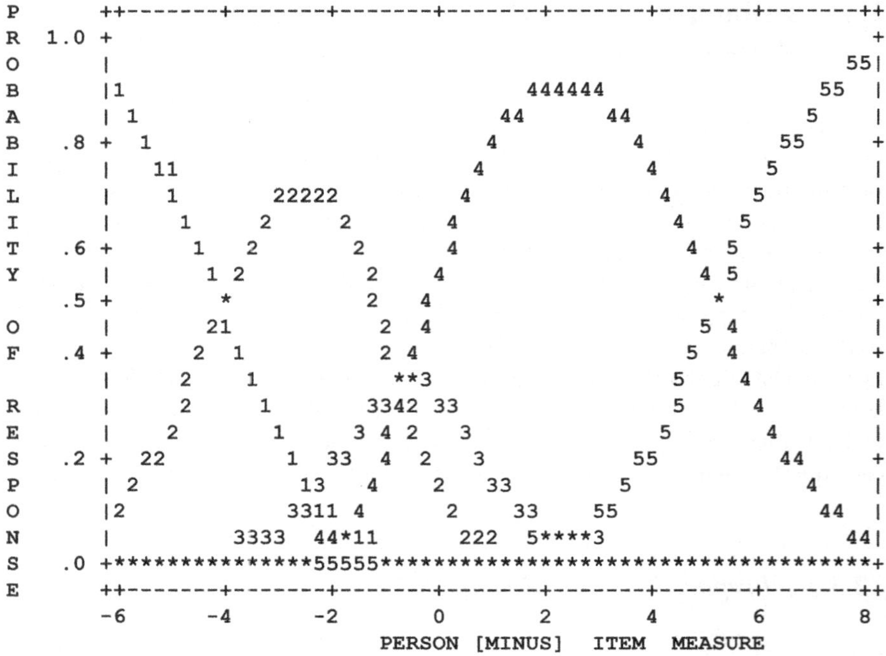

Fig. 12.2 Category curves for perceived preparedness for residency

diagnostic skills at the top of the scale. We can conclude from this that ideas related to working with and around patients instilled confidence fairly easily, whereas more hesitancy existed when students visualized themselves being put into a position where they needed to make specific medical diagnoses and decisions of consequence to a patient. Overall, we found that the student distribution was negatively skewed, meaning that the measures tended toward the top of the scale. A student in the second quartile (median) is predicted to select "agree" for common conditions, clinical decision making, and social science issues, and select "strongly agree" for working with diverse patients, ethical values, and communication skills. Students in the third quartile and above were predicted to select "strongly agree" for all competencies. At the other end of the spectrum, a student in the first quartile of the distribution was predicted select "agree" on all of the items. Only the students at the bottom of the scale were predicted to disagree with the more difficult decision-making aspects of residency. Students in our sample also showed little uncertainty about their perceived preparedness for residency, as shown by the lack of Prediction = 3 (uncertain) in Fig. 12.1. Category curves for perceived preparedness for residency are illustrated in Fig. 12.2 and also show the lack of response level three (uncertain).

12.5.2 Clerkship Experiences

We observed a large difficulty range in the various clerkship experiences. Emergency Medicine was the easiest clerkship experience for students to rate highly, followed by Psychiatry. Pediatrics, Family Medicine, Internal Medicine, and Neurology sat at the middle of the scale, indicating that these took a moderate level of difficulty. The most difficult clerkships for students to rate highly were Surgery and Obstetrics/Gynecology. In this particular MD program, the Emergency Medicine clerkship consists of eleven 9-hour shifts across four weeks with additional simulation sessions, didactics, and a national exam as the final exam. Students comment positively in the evaluation of the clerkship about the residents and attendings and clerkship director, express their enthusiasm for the simulation sessions, and appreciate the work/life balance of shift work. From the data, we can posit that this unique structure of the clerkship (also the only one to be a fourth-year clerkship as opposed to the others done in the third year) leads to the relative ease at which students highly rate this clerkship. On the other end of the spectrum, students found it difficult to highly rate Surgery and Obstetrics/Gynecology. These procedural specialties are typically rated more poorly by students across all schools (Association of American Medical Colleges, 2021), due in part to the long work hours, assisting with surgeries/operating room (OR) procedures, and the learning environment/specialty culture. In these data, we see the delineation between the clerkships that can provide a more positive work/life balance and learning environment and the clerkships that have a more difficult time achieving that balance and environment. It is notable that this cohort of students took the Emergency Medicine clerkship between July 2020 and May of 2021 during the height of the COVID-19 pandemic. Though these students found it easier to highly rate this clerkship than the others, it will be interesting to see how frontline physician burnout (Melnikow et al., 2022) may impact medical student clerkship experience.

Based on the Wright map (Fig. 12.3), a student in the second quartile (median) was expected to give an "excellent" rating for Emergency Medicine and "good" ratings for the rest of the clerkships. A third quartile student was predicted to give "excellent" ratings for all of the experiences except for Obstetrics and Gynecology, which students at this level tended to rate as "good" (Fig. 12.3). Indeed, only students at the top of the scale were predicted to give an "excellent" rating for Obstetrics/Gynecology. On the other end of the spectrum, students in the first quartile were predicted to give "good" ratings for Emergency Medicine, Psychiatry, Pediatrics, and Family Medicine and "fair" ratings for Internal Medicine, Neurology, Surgery, and Obstetrics/Gynecology. Students at the bottom of the scale were predicted to rate all of the clerkships as "poor" with the exception of Emergency Medicine, which they would rate as "fair." Category curves for quality of clerkship experiences are illustrated in Fig. 12.4.

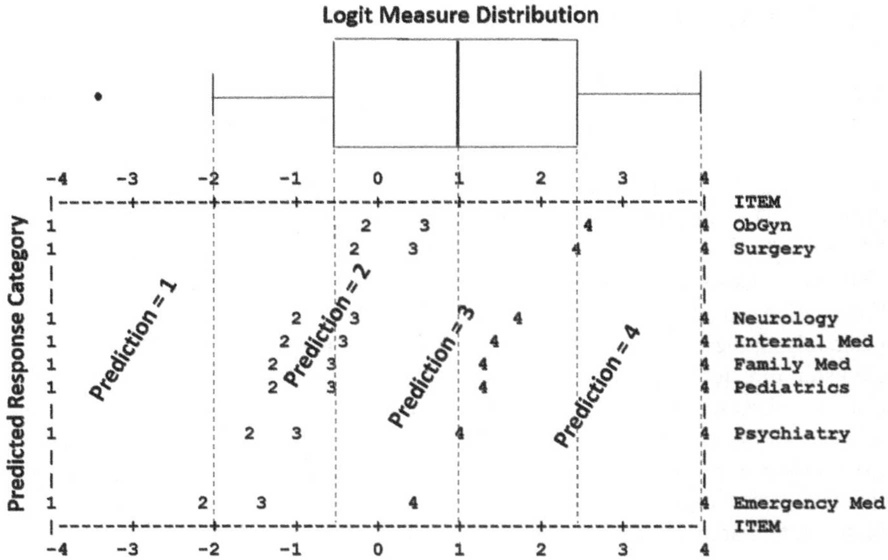

Fig. 12.3 Wright map for quality of clerkship experiences

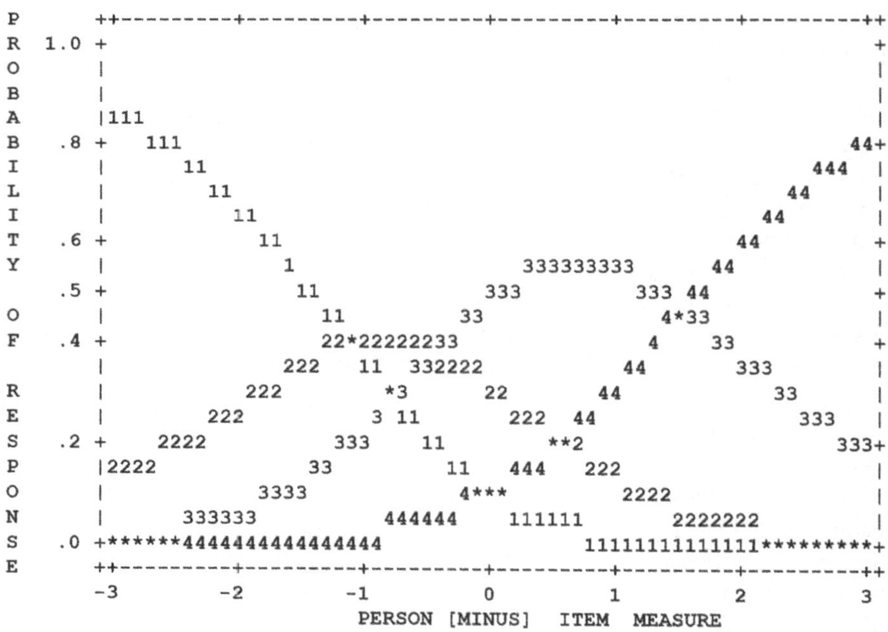

Fig. 12.4 Category curves for quality of clerkship experiences

12.5.3 Participation in Electives—A Measure of Self-Driven Initiative

Although the individual electives in which the medical students have the opportunity to participate are quite diverse, the data indicate that these form a unidimensional hierarchy. What this means is that there is a common underlying trait or cause that governs the extent to which medical students choose to participate in the various electives. We argue that a student's position along this scale is a *measure of their self-driven initiative*. Inspecting the bottom of the scale, we see that nearly all students are predicted to report working with people with health disparities and doing research with faculty. This is not surprising because the curriculum requires completing a scholarly research project and several courses and clerkships emphasize care for patients with health disparities, hence these opportunities would be difficult to avoid even for students with little initiative. As students move up the scale, we see more involved and independent activities such as working in the free clinic and learning about the role of an interpreter. This medical school has a robust relationship with a free clinic in the area, so it is not surprising that this is an easier activity for students to complete. Despite a high number of students who report learning about the role of an interpreter, the school has plans to introduce simulated patient scenarios involving interpreters; once included, this should further decrease the difficulty of this elective experience. As the students move up to the middle of the scale, they begin participating in electives that demand more self-driven initiative such as doing an independent study, working on community education initiatives, and co-authoring with faculty on research products. Students closer to the top of the scale report even more community involvement including work with eldercare and taking the lead on community health efforts. For this cohort of students, an experience in geriatric care was not required, however a Geriatrics clerkship has been introduced for subsequent classes; we anticipate that experience in eldercare will move toward the bottom of the scale in future iterations since engagement in this activity would require less initiative on the student's part. Only students with the highest levels of initiative are expected to report interacting with and leading efforts among international communities. The medical school has a robust Global Health Scholars program and though many students are interested in the program during the first two years of medical school, far fewer students participate in a global health experience abroad.

For participation in electives, a student in the 2nd quartile (median) will tend to report participating in an experience related to health disparities, pursuing research with faculty, increasing cultural awareness, learning about the role of an interpreter in medical practice, working in a free clinic, pursuit of independent study, engaging in health education in schools and the community, and contributing as an author on manuscripts. However, students with median initiative generally will not pursue electives which require more intensive involvement such as working with elderly populations, leading community projects, and working in international communities. Students in the 3rd quartile will additionally participate in activities related to home

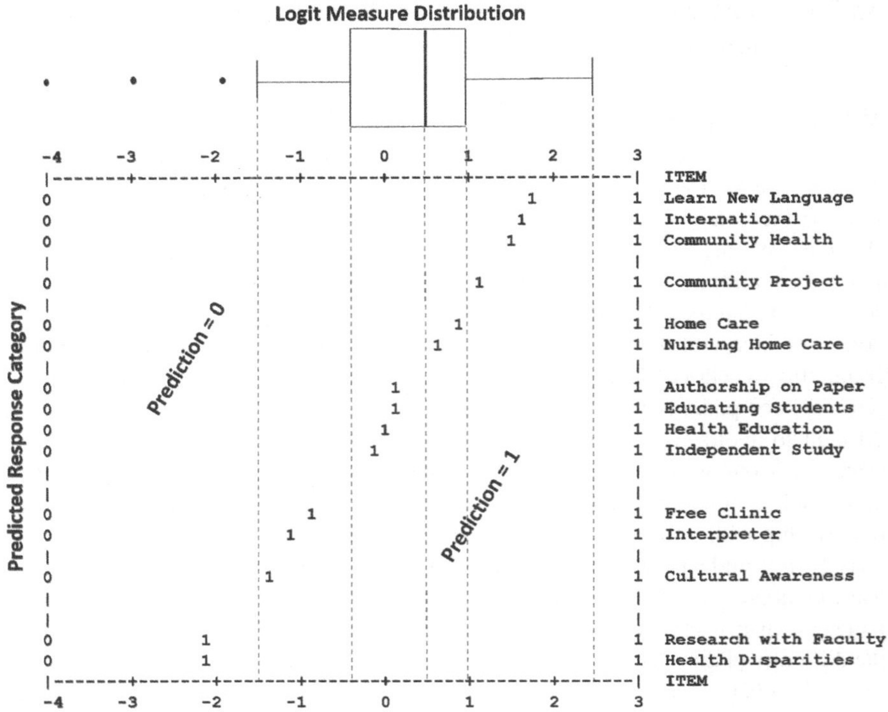

Fig. 12.5 Wright map for participation in elective opportunities

care and nursing home care. Only the most driven students will initiate community health projects, actively participate in global health initiatives, and learn a new language to better communicate with patients (Fig. 12.5). On the opposite end of the spectrum, even students in the 1st quartile will report working in the free clinic, doing research with faculty, and making efforts to expand their awareness of cultural differences and health disparities. The elective placement along the scale is certainly dependent on the context of the school, and holds strategic importance in that we can give consideration to measures we can take to make electives easier for students to access.

12.5.4 The Role of Clerkship Experiences and Self-Driven Initiative in Facilitating Preparedness for Residency

The multivariate regression model ($F_{3,82} = 24.96$, $p < < 0.001$, $r^2_{adj} = 0.46$) shows that clerkship experiences ($b = 0.83$, $SE = 0.17$, $p < < 0.001$, $r^2_{partial} = 0.25$) and self-driven initiative ($b = 0.78$, $SE = 0.19$, $p < < 0.001$, $r^2_{partial} = 0.17$) are strong predictors of perceived preparation for residency. A significant 2-way interaction

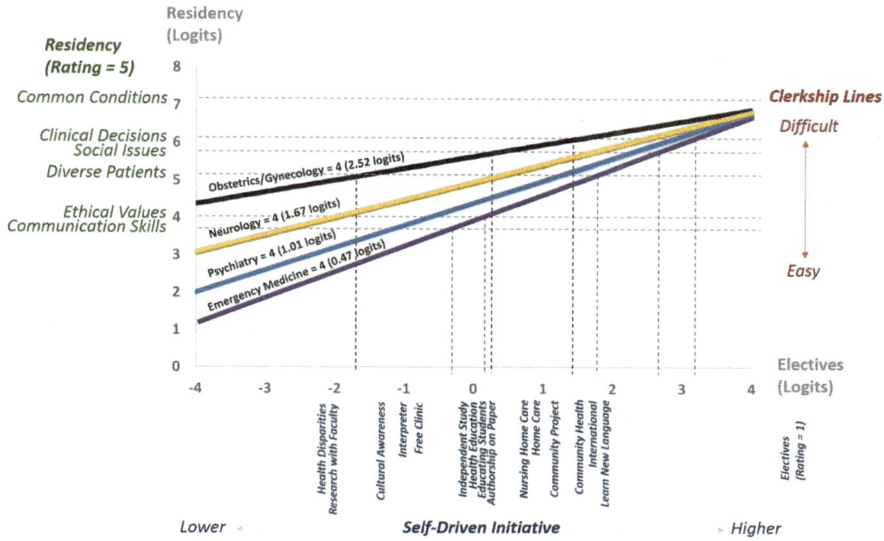

Fig. 12.6 Plot of the regression model with the axis values qualitatively defined by the Wright maps for the respective measures (Figs. 12.1, 12.3, and 12.5). The gray horizontal lines indicate the logit measure at which the student is predicted to give a "Strongly Agree" (5) rating to their perceived preparation for residency. The purple vertical lines show the locations at which the Emergency Medicine clerkship line (the easiest clerkship) intersects with the residencies. The black vertical lines show the locations at which the Obstetrics/Gynecology clerkship line (the most difficult clerkship) intersects with the residencies. These vertical lines collectively show the level of self-driven initiative (and associated electives [those to the left of each line]) necessary to achieve a "Strongly Agree" perception on preparation for each aspect of residency given a particular clerkship experience

shows that these do not act independently ($b = -0.18$, $SE = 0.093$, $p = 0.012$, $r^2_{partial} = 0.059$). The model indicates that effect of self-driven initiative on one's perceived preparation for residency decreases as the perceived quality of the clerkship experiences increases; or conversely, the effect of clerkship experiences becomes less important as the medical students become more self-driven (Fig. 12.6).

12.6 Discussion

This is the point where many Rasch studies which utilize regression models on logit measures conclude. Although use of logit measures in lieu of raw scores has many benefits in its own right, including proper accounting of linearity of the measures which reduces floor and ceiling effects and makes the measures more suitable for parametric hypothesis tests, the practice of equating students' measures and item difficulty on the same scale allows us assign qualitative meaning to different parts of the scale. In turn, Rasch modeling allows us to use our models to make qualitative predictions provided that the data fit the model adequately (i.e., the surveys have adequate construct validity).

Through consideration of the step thresholds in Figs. 12.1, 12.2, 12.3, 12.4 and 12.5, we were able to attach qualitative meaning to our regression model (Fig. 12.6). The inferential statistics indicated that quality clerkship experiences and self-driven initiative are important factors governing perceived preparation for residency, and that these negatively interact. But what does this tell us in terms of the types of needs that particular medical students have in light of the medical program's clerkship experiences?

We constructed Fig. 12.6 with the ideal that we would like for students to give a "strongly agree" (5) rating on their perception of preparedness for residency, and that sufficient quality in their clerkship experiences and participation in electives could facilitate this. Hence, we placed the step thresholds for the "5" rating for preparation for residency (Fig. 12.1) on the y-axis of Fig. 12.6. On the x-axis, in order to define qualitatively the types of participation in electives we expect to see at different levels of self-driven initiative, we plotted the item difficulty measures, which in terms of the simple logistic model were the step thresholds between the "0" and "1" category. Finally, given the expectation that would like for students to give a "4" (excellent) rating for their clerkship experiences, we plotted the linear functions derived from the step thresholds for the "4" level (Fig. 12.3). Since plotting the functions for all eight of the clerkship experiences would make the figure crowded and difficult to interpret, we instead chose four representative clerkships at key parts of the scale: (1) Emergency Medicine (bottom of the scale), (2) Psychiatry (the bottom of the middle cluster), (3) Neurology (the top of the middle cluster), and (4) Obstetrics/ Gynecology (the top of the scale). The idea is that the effects of the other clerkships not shown in the graph can be interpolated. For example, the "Family Medicine" line, which is not shown in Fig. 12.6, would lie between the "Psychiatry" line and the "Neurology" line.

Perhaps the most striking observation from the plot (Fig. 12.6) is that for students with very low levels of self-driven initiative, the clerkship experiences have a dramatic effect on the students' perceived preparation for residency. However, as self-driven initiative increases, the importance of the clerkship experiences diminishes in a linear fashion up the point where nearly all students with very high levels of self-driven initiative will consider themselves prepared for the various residencies in spite of how they felt about their clerkship experiences. Looking at the big picture, this suggests that, toward preparing undergraduate medical students for residency, we can capitalize on both formal and informal learning experiences. The type of learning experience on which to focus may depend very much on the personality of the particular student. Students who have high levels of independence and self-motivation will tend to actively pursue elective opportunities and learn from these. For the students who have the lowest levels of independence and self-motivation, the clerkship experiences become the primary mode of preparation for medical practice. But how do these specifically work together? And what does it take for a student to "strongly agree" that they are prepared for residency?

From Fig. 12.6, we glean that a student rating their Emergency Medicine clerkship as "excellent" (but rating all other clerkships less than "excellent"; Emergency Medicine = 4 regression line) would need to at least participate in a free clinic in order to give a rating of "strongly agree" on feeling prepared in the area of communication skills (vertical dashed line to the right of free clinic and horizontal dashed line at communication skills at the bottom Emergency Medicine line). This student would need to participate in public and school education activities in order to give a "strongly agree" rating on ethical values (following Emergency Medicine line up to the horizontal dashed line for ethical values). This student would also need to participate in international/global health in order to give a "strongly agree" rating to preparation to work with diverse patients (again following Emergency Medicine line up). As we continue to trace the Emergency Medicine line, we find that it is unlikely that elective participation would be sufficient to help such a student reach a level where they are selecting "strongly agree" for social science issues, clinical decision making, or dealing with common conditions (vertical lines are well to the right of any electives).

At the upper extreme, a student who had a very strong and positive clerkship experience sufficient to give an "excellent" rating on all of the clerkships (the Obstetrics/Gynecology regression line, Fig. 12.6) would be able to meet the "strongly agree" level of perceived preparation for all aspects of residency except for the most difficult competency, understanding common conditions and their management, through participation in electives (the Obstetrics/Gynecology line never reaches the top dashed line indicating the "5" threshold for fundamental understanding of common conditions and their management). Even without participation in electives, students giving "excellent" ratings to all of the clerkships are expected to "strongly agree" that they are prepared for residency with regards to communication skills and ethical values (the value of the Obstetrics/Gynecology line at the bottom of the electives scale). Research with faculty would be required to get this student to a level where they are giving their preparedness to work with diverse patients a "strongly agree" rating (black vertical line is to the right of these electives). Authorship on a paper would be required before this student would be expected to "strongly agree" with feeling prepared for understanding social science issues; engagement in community projects would be needed to get this student to a level where they would report strong perceived preparedness in terms of clinical decision making (black vertical lines to the right of these electives).

For students with lower levels of independence and self-motivation, inspecting the regression line values at the lower extreme of the elective participation scale (Fig. 12.6) can provide insight for curricular reform. Students who report no participation in electives are dependent upon the core medical school curriculum to provide preparation for residency. Understandably, there is a greater difference in perceived preparation for residency between students who report a more positive clerkship experience (OBGYN regression line) than students who report a more negative clerkship experience (Emergency Medicine regression line) when students do not participate in electives.

Though these data are specific to one graduating cohort at a Midwestern United States allopathic medical school and reflects institutional availability and culture around electives as well as curriculum and culture in clerkships, there does remain some information that could be generalizable across medical schools. As with this data set (Fig. 12.3), the procedural specialty clerkships (i.e., OBGYN and Surgery) have the lowest percentage of students rating them as good or excellent (Association of American Medical Colleges, 2021). Our data indicate that students who highly rate their experiences in these clerkships are less dependent upon participation in electives to provide them with the perception of being prepared for residency whether or not they have the independence and self-motivation to pursue a variety of elective experiences. Conversely, this independence and self-motivation to participate in electives is most important for students who rated their experiences in clerkships poorly. It is possible that these types of students feel fatigued by the more formal aspects of medical training. In order to ensure that these students are as prepared as possible for residency, it is essential to provide a variety of opportunities to interact with the medical community and engage in medical practice through less formal modalities. With the general lack of requirements for fourth year, this seems like the ideal place for providing a variety of less formal elective opportunities to help students increase their perceived preparedness for residency.

Given that considerable diversity exists in many medical programs, we expect that these trends and subsequent recommendations would be applicable to strategic planning for many medical programs. However, in order to use the scale for participation in electives for strategic planning, it is important to remember that the participation in electives which we observed in the data are actually a measure of our medical students' independence and self-motivation, which is something that medical programs do not necessarily have much control over. The implication here is that coercing a less independent or self-motivated student to engage in activities at the top of the scale such as leading community health efforts, working internationally, or learning a new language, is not likely to improve that student's preparation for residency. In fact, if participation in electives were coerced, this would likely show up through misfit of the coerced electives with the Rasch model since these would no longer provide a valid measure for self-driven initiative. In this sense, the Rasch model tells us that the electives we observe students undertaking should represent the outcome of their self-driven initiative, not the outcome of a coercive curriculum. Although students' self-driven initiative is something medical programs have comparatively little control over, this quality and its subsequent effect on elective participation and preparation for residency can be maximized by: (1) looking for qualities indicating self-motivation and independence when reviewing applications to medical school, and (2) looking for ways to provide self-motivated students a diverse array of elective opportunities.

One important limitation to using data from the GQ or other surveys for curriculum reform is the fact that student perceptions are relatively subjective and may not align with objective measurements of student performance or actual student experiences. This is why we overtly used the word "perceived" when discussing preparation for residency and quality of clerkship experiences. It is important for readers to remember that these entail the student reported perception which may or may not align with reality. Notably for this cohort (graduating class of 2021), students were unable to participate in away rotations during their fourth year due to the pandemic, which limited their interaction with medical students from other institutions. Anecdotally, we have heard that our students realize the strength of their clinical skills training when they complete away rotations and interact with students trained at other institutions. Without this experience, we have reason to believe that this cohort of students may have underestimated their perceived preparedness for residency. This is only one example of the problems with measuring perception on surveys and using it as a standard or marker for quality. There are many factors that influence student perceptions and the fact that student perception is being utilized heavily in decisions by accrediting bodies and school-level program evaluation is problematic from an assessment point of view because it very well may not be an accurate marker. Debating the merits of using perception data to drive curricular reform is beyond the scope of this chapter, but we would be remiss to not discuss this limitation. Regardless of this, we are able to illustrate how schools can utilize Rasch modeling in an innovative way to better use national survey data. Researchers interested in the validity of the GQ may consider seeing how well GQ data predict performance in residency based on observational measures collected during students' residency programs. Deploying expert observers is much more time and resource intensive than sending the student a survey for self-assessment; however, such an effort may pay dividends toward better understanding the validity of the GQ as a measure of effectiveness for undergraduate medical programs.

12.7 Conclusions

Undergraduate medical schools receive data each year from the AAMC, purportedly to help evaluate the quality of the medical program. While there has been some research using GQ data to answer correlative or comparison research questions, the GQ data are predominantly used in a reporting fashion for accreditation purposes. This chapter provides a model of how schools can use Rasch analysis on national survey data to assist with strategic planning—identifying curricular changes that can be implemented for subsequent classes to increase student perceived value of the program. Specifically, we illustrated how measures for (1) perceived preparedness for residency, (2) quality of clerkship experiences, and (3) participation in elective opportunities formed unidimensional scales and how these scales can be combined with a regression model to extract descriptive qualitative information related to how clerkship experiences and participation in electives impact perceived preparedness for residency. Our data illustrate that providing robust elective opportunities for students and increasing self-driven initiative to participate in electives, especially for those who had suboptimal clerkship experiences, can lead to increased perceptions of preparedness for residency.

Using this framework, schools could use GQ and Y2Q (AAMC Year Two Questionnaire) data to inform actionable changes instead of using it solely for reporting purposes to determine how well students were "satisfied" with certain aspects of the school. Nothing acts in a vacuum. Student satisfaction in one area bleeds into satisfaction in other areas. As schools are preparing for accreditation and undergoing continuous quality improvement in their programs, it is important to identify the root causes of any student dissatisfaction or suboptimal quality to enact meaningful change that reverberates through the program instead of enacting small isolated changes to several areas. Using Rasch modeling, we were able to demonstrate that the data show that the school should provide a diverse array of elective opportunities for students and also recruit students who exhibit qualities of self-motivation and independence if the school wants to produce graduates who enter graduate medical education confident in their abilities and preparation for the next level of training. The data used in this study was only a sliver of the data available in the GQ for medical schools. If we were able to extract these qualitative action items using Rasch from only three sets of questions, think about how much potential this method provides for other medical schools or institutions to enact curricular changes that will positively impact learners.

Appendix (Rasch Analysis Codes with Annotations)

BIGSTEPS Control File for "Residency"

```
&INST ;indicates beginning of the command syntax
TITLE='resid' ;title chosen by the researcher
NI=6 ;six items
ITEM1=1 ;first item begins on the first column
XWIDE=1 ;each item is one column wide
CODES=12345 ;these are the ordinal codes that are used. Any other code is
treated as missing.
NCOL=6 ;the data matrix is six columns wide
MODELS=R ;standard default option for dichotomous, rating scale, and
partial credit models
STBIAS=Y ;corrects for estimation bias
PRCOMP=S ;principal components analysis on standardized residuals
GROUPS=AAAAAA ;all six residencies share the same rating scale
TABLES=1111111111111111111111 ;all tables requested
&END ;indicates end of the command syntax
1=CommonCondit ;name for item 1
2=CommSkills ;name for item 2
3=ClinDecision ;name for item 3
4=SocSci ;name for item 4
5=Values ;name for item 5
6=Diversity ;name for item 6
END NAMES ;indicates the end of the naming syntax. Data are pasted below
in matrix form
444444
```

```
555555
454554
......
```

BIGSTEPS Control File for Clerkships

```
&INST ;indicates beginning of the command syntax
TITLE='clerkship' ;title chosen by the researcher
NI=8 ;eight items
ITEM1=1 ;first item begins on the first column
XWIDE=1 ;each item is one column wide
CODES=1234 ;these are the ordinal codes that are used. Any other code is
treated as missing.
NCOL=8 ;the data matrix is eight columns wide
MODELS=R ;standard default option for dichotomous, rating scale, and
partial credit models
STBIAS=Y ;corrects for estimation bias
PRCOMP=S ;principal components analysis on standardized residuals
GROUPS=AAAAAAAA ;all eight clerkships share the same rating scale
TABLES=1111111111111111111111 ;all tables requested
&END ;indicates end of the command syntax
1=EmergMed ;name of first item
2=FamilyMed ;name of second item
3=InternalMed ;name of third item
4=ObGyn ;name of fourth item
5=Neurology ;name of fifth item
6=Pediatrics ;name of sixth item
7=Psychiatry ;name of seventh item
8=Surgery ;name of eighth item
END NAMES ;indicates the end of the naming syntax. Data are pasted below
in matrix form. Note that "9" indicates missing since it is not specified
in the CODES command above.
93333333
43324343
34423433
......
```

BIGSTEPS Control File for Electives

```
&INST ;indicates beginning of the command syntax
TITLE='electives' ;title chosen by the researcher
NI=15 ;fifteen items
ITEM1=1 ;first item begins on the first column
XWIDE=1 ;each item is one column wide
CODES=01 ;these are the ordinal codes that are used. Any other code is
treated as missing.
NCOL=15 ;the data matrix is fifteen columns wide
MODELS=R ;standard default option for dichotomous, rating scale, and
```

```
partial credit models
STBIAS=Y ;corrects for estimation bias
PRCOMP=S ;principal components analysis on standardized residuals
GROUPS=0 :all fifteen items are allowed to have different scales
TABLES=1111111111111111111111 ;all tables requested
&END ;indicates end of the command syntax
1=IndepStudy ;name of first item
2=Research ;name of second item
3=Author ;name of third item
4=International ;name of fourth item
5=EducStudents ;name of fifth item
6=HlthEd ;name of sixth item
7=CommHlth ;name of seventh item
8=HomeCare ;name of eighth item
9=Nursing ;name of ninth item
10=NewLanguage ;name of tenth item
11=Interpreter ;name of eleventh item
12=HlthDisparities ;name of twelfth item
13=CultureAware ;name of thirteenth item
14=CommProject ;name of fourteenth item
15=FreeClinic ;name of fifteenth item
END NAMES ;indicates the end of the naming syntax. Data are pasted below
in matrix form
011001001001001
111010010011100
010111001011101
.....
```

References

Andrich, D. (1982). An Extension of the Rasch Model for Ratings Providing Both Location and Dispersion Parameters. *Psychometrika, 47*(1), 105–113. https://doi.org/10.1007/BF02293856

Andrich, D. (2004). Controversy and the Rasch Model: A Characteristic of Incompatible Paradigms? *Medical Care, 42*(1), I7–I16. http://www.jstor.org/stable/4640697

Association of American Medical Colleges. (2021). *Medical School Graduation Questionnaire 2021 Supplementary Benchmarking School Report.*

Association of American Medical Colleges. (2022). *AAMC Graduation Questionnaire*. https://www.aamc.org/data-reports/students-residents/report/graduation-questionnaire-gq

Benson, N. M., Stickle, T. R., & Raszka, W. v. (2015). Going "Fourth" From Medical School: Fourth-Year Medical Students' Perspectives on the Fourth Year of Medical School. *Academic Medicine, 90*(10), 1386–1393. https://doi.org/10.1097/ACM.0000000000000802

Boone, W. J. (2016). Rasch analysis for instrument development: Why, when, and how? *CBE Life Sciences Education, 15*(4). https://doi.org/10.1187/CBE.16-04-0148

Boone, W. J., Staver, J. R., & Yale, M. S. (2013). *Rasch analysis in the human sciences*. Springer.

DCI | *Standards, Publications, & Notification Forms | LCME.* (2022). https://lcme.org/publications/#DCI

Dyrbye, L. N., Satele, D., & West, C. P. (2021). Association of Characteristics of the Learning Environment and US Medical Student Burnout, Empathy, and Career Regret. *JAMA Network Open, 4*(8). https://doi.org/10.1001/JAMANETWORKOPEN.2021.19110

Kassam, A., Nickell, L., Pethrick, H., Mountjoy, M., Topps, M., & Lorenzetti, D. L. (2021). Facilitating Learner-Centered Transition to Residency: A Scoping Review of Programs Aimed at Intrinsic Competencies. *Teaching and Learning in Medicine, 33*(1), 10–20. https://doi.org/10.1080/10401334.2020.1789466

Langdale, L. A., Schaad, D., Wipf, J., Marshall, S., Vontver, L., & Scott, C. S. (2003). Preparing Graduates for the First Year of Residency: Are Medical Schools Meeting the Need? *Academic Medicine, 78*, 39–44. http://journals.lww.com/academicmedicine

Linacre, J. M., & Tennant, A. (2009). More about critical eigenvalue sizes (variances) in standardized residual principal components analysis (PCA). *Rasch Measurement Transactions, 23*(3), 1228.

Melnikow, J., Padovani, A., & Miller, M. (2022). Frontline physician burnout during the COVID-19 pandemic: national survey findings. *BMC Health Services Research, 22*(1), 1–8.

Minter, R. M., Amos, K. D., Bentz, M. L., Blair, P. G., Brandt, C., D'Cunha, J., Davis, E., Delman, K. A., Deutsch, E. S., Divino, C., Kingsley, D., Klingensmith, M., Meterissian, S., Sachdeva, A. K., Terhune, K., Termuhlen, P. M., & Mullan, P. B. (2015). Transition to surgical residency: A multi-institutional study of perceived intern preparedness and the effect of a formal residency preparatory course in the fourth year of medical school. *Academic Medicine, 90*(8), 1116–1124. https://doi.org/10.1097/ACM.0000000000000680

Nguyen, M., Cerasani, M., Dinka, L. A., Rodriguez, J. A., Omoruan, M., Acosta, E., Alder, K. D., Brutus, N. N., Termuhlen, P. M., Dardik, A., Mason, H. R. C., & Stain, S. C. (2021). Association of Demographic Factors and Medical School Experiences With Students' Intention to Pursue a Surgical Specialty and Practice in Underserved Areas. *JAMA Surgery, 156*(12), e214898. https://doi.org/10.1001/JAMASURG.2021.4898

O'Marr, J. M., Chan, S. M., Crawford, L., Wong, A. H., Samuels, E., & Boatright, D. (2022). Perceptions on Burnout and the Medical School Learning Environment of Medical Students Who Are Underrepresented in Medicine. *JAMA Network Open, 5*(2), E220115. https://doi.org/10.1001/JAMANETWORKOPEN.2022.0115

Popper, K. (1963). Science: Conjectures and refutations. *Conjectures and Refutations*, 33–39.

Rasch, G. (1960). *Studies in mathematical psychology: I. Probabilistic models for some intelligence and attainment tests*. Nielsen & Lydiche. https://psycnet.apa.org/record/1962-07791-000

Raudenbush, S. W., & Bryk, A. S. (2002). *Hierarchical linear models: Applications and data analysis methods* (Vol. 1). Sage.

Ryus, C. R., Samuels, E. A., Wong, A. H., Hill, K. A., Huot, S., & Boatright, D. (2022). Burnout and Perception of Medical School Learning Environments Among Gay, Lesbian, and Bisexual Medical Students. *JAMA Network Open, 5*(4), e229596. https://doi.org/10.1001/JAMANETWORKOPEN.2022.9596

Shields, R. K., & Dudley-Javoroski, S. (2021). Benchmarking in Academic Physical Therapy: A Multicenter Trial Using the PT-GQ Survey. *Physical Therapy, 101*(12). https://doi.org/10.1093/PTJ/PZAB229

Smith, R. M. (1996). Polytomous mean-square fit statistics. *Rasch Measurement Transactions, 10*(3), 516–517.

Thurstone, L. L. (1928). Attitudes Can Be Measured. *American Journal of Sociology, 33*(4), 529–554. https://doi.org/10.1086/214483

Wright, B. (1994). Reasonable mean-square fit values. *Rasch Measurement Transactions, 8*(3), 370.

Wright, B. D., & Stone, M. H. (1979a). *Best test design*.

Wright, B., & Stone, M. (1979b). Best test design. *Measurement and Statistics*. https://research.acer.edu.au/measurement/1

Amber Todd, Ph.D. is an Assistant Professor and Assistant Dean of Medical Education & Accreditation in the Department of Medical Education at the Wright State University Boonshoft School of Medicine.

William Romine, Ph.D. is an Associate Professor of Biological Sciences at Wright State University. He serves as the director of the *Data Science for Education Laboratory* at Wright State University.

Chapter 13
Applying Rasch Measurement to Assess Knowledge-in-Use in Science Education

Peng He, Xiaoming Zhai, Namsoo Shin, and Joseph Krajcik

Abstract This study applied the many-facet Rasch measurement (MFRM) to assess students' knowledge-in-use in middle school physical science. 240 students completed three knowledge-in-use classroom assessment tasks on an online platform. We developed transformable scoring rubrics to score students' responses, including a task-generic polytomous rubric (applicable to the three tasks), a task-specific polytomous rubric (for each task), and a task-specific dichotomous rubric (for each task). Three qualified raters scored 240 students' responses to the three tasks. MFRM reported student ability, item difficulty, rater severity, and interaction effects, which helped improve the assessment tasks and rubrics.

13.1 Introduction

The rapidly changing world calls for developing knowledge-in-use rather than memorizing facts without understanding. Knowledge-in-use requires individuals to apply knowledge to solve real-world problems in uncertain situations (Pellegrino & Hilton, 2012). In science education, there is a consensus among contemporary standards (Chinese Ministry of Education, 2017; Finnish National Board of Education, 2016; Next Generation Science Standards [NGSS] Lead States, 2013) and international assessment projects (e.g., [PISA], OECD, 2019) that knowledge-in-use requires students using and applying disciplinary and interdisciplinary core ideas with scientific practices to make sense of real-world phenomena or solve complex problems. While there is a tremendous interest in this new direction, assessing

P. He (✉) · N. Shin · J. Krajcik
CREATE for STEM Institute, Michigan State University, East Lansing, MI, USA
e-mail: hepeng1@msu.edu

X. Zhai
Department of Mathematics, Science, and Social Studies Education, University of Georgia, Athens, GA, USA
e-mail: Xiaoming.Zhai@uga.edu

X. Liu, W. J. Boone (eds.), *Advances in Applications of Rasch Measurement in Science Education*, Contemporary Trends and Issues in Science Education 57,
https://doi.org/10.1007/978-3-031-28776-3_13

students' knowledge-in-use has been recognized as a tangible goal and brought new challenges to science education. Using such assessments in science classrooms, teachers could adjust teaching and plan lessons to better support students in developing the competence to use scientific knowledge. As a complex construct, knowledge-in-use presents challenges to assess using traditional multiple-choice items. Researchers focus on developing novel assessments to examine students' scientific knowledge while engaging them in scientific practices (e.g., Harris et al., 2019; He et al., 2023). These assessments are usually performance-based, which support valid assessment conclusions only if accompanied with evidence-based scoring rubrics that are validated by psychometric evidence. Developing scoring rubrics that could assess students' performance on the standard-aligned learning goals of knowledge-in-use assessments is critical. However, we have limited knowledge about the features and psychometric properties of scoring rubrics (e.g., dichotomous, polytomous). Moreover, scores from these performance-based assessment tasks are usually provided by teachers or multiple raters, creating additional construct irrelevant variance when scoring (Zhai et al., 2020). These aspects could potentially impact the validity of the conclusions made on the scores.

Using the assessment tasks developed in the Next Generation Science Assessment (NGSA) project (Harris et al., 2019; NGSA, 2022), this study demonstrates the application of many-facet Rasch measurement (MFRM; Linacre, 1989) to assess students' knowledge-in-use in science education. Specifically, this study discusses the design of transformable rubrics for scoring students' responses to three assessment tasks, the training of multiple raters to use the scoring rubrics, and the psychometric properties of assessment tasks when using different scoring rubrics.

13.2 Theoretical Background

Knowledge-In-Use

Knowledge-in-use is grounded in situated cognition theory (Greeno et al., 1996), which contends that knowledge is situated, being part of a product of activities, and context and culture-dependent, based on the environment in which it is developed and used (Brown et al., 1989). Situated cognition emphasizes real-world situations that are meaningful for learners to make sense of phenomena and motivate them to figure out solutions for problems in new contexts actively. While situated in diverse contexts, students may engage in the sensemaking process with different scientific practices. In the *Framework for K-12 Science Education* (NRC, 2012), eight scientific practices have been recognized as the fundamental ways scientists build scientific knowledge. The *Framework* (NRC, 2012) emphasizes the unique merits of each practice for students' science learning by doing science in real-world contexts. The Framework specifies three dimensions that need to work together to support knowledge-in-use. These three dimensions include disciplinary core ideas (DCIs, Duncan et al., 2017), scientific and engineering practices (SEPs, Schwartz et al., 2017), and crosscutting concepts (CCCs, Nordine & Lee, 2021). For a specific situation, students may apply the same DCI by either analyzing and interpreting data or

Table 13.1 Three NGSA tasks

Task name	Learning performance	NGSS performance expectation
T1. Gas filled balloons	Students analyze and interpret data to determine whether substances are the same based upon characteristic properties. (LP1)	MS-PS1-2. Analyze and interpret data on the properties of substances before and after the substances interact to determine if a chemical reaction has occurred. MS-PS1-5. Develop and use a model to describe how the total number of atoms does not change in a chemical reaction and thus mass is conserved
T2. Layers in a test tube	Students construct a scientific explanation about whether a chemical reaction has occurred by using patterns in data on the properties of substances before and after the substances interact. (LP2)	
T6. Battery under water	Students develop a model of a chemical reaction that explains the regrouping of atoms forms new substances, and that mass is conserved. (LP6)	

developing and using models to make sense of phenomena or solve problems. Knowledge-in-use also requires the use of CCCs such as cause and effect, patterns, etc. (Nordine & Lee, 2021) in conjunction with DCIs and SEPs. Although SEPs are unique to knowledge-in-use, they cannot be separated from DCIs and CCCs. The NGSS performance expectations (PEs) are demonstrated as knowledge-in-use statements with multiple representations of the integrations of SEPs, DCIs, and CCCs. For instance, in the NGSS chemical reactions, MS-PS1–2 is a PE of students' knowledge-in-use statement using the SEP of analyzing and interpreting data and the CCC of patterns, whereas MS-PS1–5 is another knowledge-in-use PE using the SEP of developing and using models and the CCC of energy and matter (See the example PEs in Table 13.1).

Moreover, applying scientific knowledge in scientific practices to figure out real-world phenomena or solve problems is often regarded as the process of sensemaking in science education (Kapon, 2017; Krist et al., 2019; NRC, 2012; Odden & Russ, 2019; Schwarz et al., 2017). Sensemaking involves explaining observed phenomena through coordination of theory and evidence (Crowder, 1996), making connections to the real-world or lived experience, and coordinating multiple representations to solve problems (Chen et al., 2013). Schwarz and colleagues addressed sensemaking as four following aspects: "what are we trying to figure out? How will we figure this out? How can we keep track of ideas? How does it all fit together?" (Schwarz et al., 2017, p. 18). Based on Schwarz et al.'s (2017) four questions of sensemaking and existing frameworks, we identify the common attributes of knowledge-in-use across different representations: claim building, evidence noticing, principle utilizing, and mechanistic reasoning. Claim building involves students in making conclusions using texts or drawn models to explain the anchored phenomena or problems. Evidence noticing attribute refers to the data patterns to support students' claims. Principle utilizing requires students to connect their claims with evidence using scientific principles explicitly. Finally, mechanistic reasoning emphasizes the reasoning process of using evidence, making claims, and employing scientific principles to address the anchored phenomena or problems.

13.2.1 Assessment

Assessments are premised on assumptions about how to interpret observations to make inferences (Mislevy et al., 2003). Eliciting observations of students' performance could infer their knowledge-in-use only if they are presented with authentic tasks. However, assessing knowledge-in-use has been recognized as a critical but challenging task in science education. Knowledge-in-use assessment tasks that integrate multiple dimensions (i.e., DCIs, SEPs, and CCCs) require students to experience complex cognitive processes in real-world situations that are both cognitively and behaviorally demanding. Such assessment tasks are usually accompanied by "authentic" scenarios which are innovative but also cognitively demanding.

Mislevy et al. (2003) developed an evidence-centered design (ECD) approach that has been widely adopted in knowledge-in-use assessment development projects (e.g., Harris et al., 2019; NRC, 2014). For instance, the NGSA project adapted the ECD approach (Harris et al., 2019) to develop hundreds of online, interactive assessment tasks in middle school physical science and life science that teachers and students widely use across the U.S. and other countries. Although traditional assessment tasks (e.g., multiple-choice or true/false) have advantages related to cost, time efficiency, and psychometric properties, they are less satisfactory for gathering reliable and valid information about complex constructs such as knowledge-in-use. Many researchers have adopted the construct-response format for the knowledge-in-use assessment tasks (e.g., He et al., 2023; Kaldaras et al., 2021; Penuel et al., 2019). Students must write a paragraph or draw a pictorial model/diagram to show their understanding in responding to such construct-response tasks.

13.2.2 Scoring Rubric

When using construct-response tasks, students' written or drawn responses demonstrate the heterogeneity of students' thinking. Therefore, designing rubrics to score a wide range of students' responses for eliciting students' achievement of the targeted learning performance goals is challenging. Several strategies are widely adopted to address this challenge. First, analytic rubrics with dichotomous codes (e.g., Liu et al., 2016) are commonly used to diagnose the presence of the given criteria in students' constructed responses. The criteria in the type of rubrics serve as a checklist for assessing items in that "1" represents the presence of the given criteria, and "0" means incorrect or missing criteria. Another type is a polytomous multi-leveled coding scheme to characterize the quality of complete student responses (e.g., Liu et al., 2016; Mao et al., 2018). Instead of coding students' responses with correctness or not, such coding schemes identify students' progression on several sub-constructs (or elements) of a complex construct, such as the qualities of claim, evidence, and reasoning for scientific explanations (Gotwals & Songer, 2013; Gotwals et al., 2012). Dichotomous checklist-type rubrics and polytomous multi-leveled rubrics (Harris et al., 2006) are equally valuable in providing student information for teacher instructional and evaluation purposes. However, it remains another challenge to create a rubric system that allows the transformation of coding schemes (Jescovitch et al., 2021; Wang et al., 2021) for diagnostic and evaluation purposes across multiple types of complex assessment tasks. This study discusses the design of transformable rubrics to score students' constructed responses on three types of NGSA assessment tasks.

13.2.3 Rater

The heterogeneity of students' written and drawn responses on constructed response tasks also challenges human scoring. Research has shown that human scoring is regarded as one of the largest sources of variation in assessment studies (Leckie & Baird, 2011; Styck et al., 2021). We usually recruit multiple raters to score the same task performance to offset the score variations. However, it is difficult to identify the ideal number of raters to balance the scoring variability. In addition, training multiple raters consumes time and finances. The minimum number of raters that can bring stability across their scores leads to an effort-efficient process. Increasing the number of raters reduces scoring time and efforts but brings more uncertainty in scoring due to the variations in rater background, understanding of the rubrics, and fatigue. Therefore, minimizing raters'effects in the scoring process is critical, especially in scoring students' constructed responses on low-stakes assessment tasks. Existing studies (e.g., He et al., 2023; Jescovitch et al., 2021) have revealed that the research team needs to recruit qualified raters, train them with the rationale and process of using rubrics, and constantly check the agreement of their scores on a small sample of common cases. For judging raters' consistency of scoring, inter-rater reliability (IRR) is typically used by calculating the correlations among rater scores, such as Cohen's kappa (Cohen, 1960) for two raters and Fleiss's kappa (Fleiss & Cohen, 1973) for more than three raters. It may lead to unintended interpretations of a scoring rubric if the IRR is not at a high rate (Eckes, 2008), meeting the cut-off value of 0.70. After the IRR meets the cut-off, raters assigned scores to student responses independently. However, raters' scoring may fluctuate over time due to their familiarity with the rubrics and rater fatigue, etc. Therefore, strategies such as assigning overlapping cases to individual raters are needed so that we can check the IRR. If the IRRs are below the benchmark, a retrain procedure with raters should be conducted. That requires extra cost and time for scoring. Moreover, skilled raters' variability may still exist even with sufficient training. Therefore, there is a need to consider raters' reliability in estimating students' performances on construct-response assessment tasks.

13.2.4 Analytic Approach

While novel construct-response assessments can elicit students' knowledge-in-use, raters' variability must be considered in the analytic models. Researchers usually calculate the average scores of multiple raters before entering the data in their final analytic models (e.g., He et al., 2016; Zhai et al., 2022). In addition, psychometric approaches, such as classic test theory (CTT) and Rasch modeling, have been widely adopted to examine the reliability of assessment tasks, regardless of rater effects. Numerous studies (e.g., Boone & Staver, 2020; Liu, 2020; Yang et al., 2018) have preferred Rasch modeling rather than CTT approaches because the Rasch approach avoids the threat of sample dependence and estimates measures with true intervals. However, most existing studies only apply Rasch modeling and include two essential facets: 1) examinee ability and 2) item difficulty. This two-facet Rasch modeling ignores the variability among raters, especially in scoring individual cases by each rater, which may create significant construct irrelevant variance in student scores (Zhai et al., 2020). Moreover, several items are typically nested in a construct-response task, leading to task dependence on student performance. Such a structure of constructed response tasks violates the assumption of Rasch modeling that all items need to be independent. The above issues could hardly be dealt with using classical testing theories or classic two-facet Rasch modeling, but Many-facet Rasch Measurement (MFRM) could well handle these problems.

The MFRM extends the classical Rasch modeling by handling more than two facets (Eckes, 2015). First, MFRM allows test scores to be clustered based on shared features (Boone et al., 2016). Second, MFRM has been used to investigate some critical research agendas in science education, such as the judging severity of using an observation protocol (Boone et al., 2016), the comparison of items with different features (Chi et al., 2021), and machine scoring of science teachers' pedagogical content knowledge (Zhai et al., 2020). This study employed the MFRM to explore our research agenda.

13.3 Research Purposes and Questions

This study selected three NGSA tasks (see Table 13.1) to assess middle school students' knowledge-in-use in physical science. We then developed scoring rubrics, consisting of a task-generic polytomous rubric (for comparison use across multiple types of tasks), a task-specific polytomous rubric (for evaluation use in a specific task), and a task-specific dichotomous rubric (for diagnostic use in a specific task) (see Figs. 13.1, 13.2, and 13.3). Three raters with subject content knowledge were recruited and trained to score students' responses on the three tasks. We then applied MFRMs to estimate the measures of each facet for providing empirical evidence to improve the scoring rubrics (i.e., criteria, specific elements, and common attributes) for eliciting students' knowledge-in-use performances. Next, we explore the

Gas filled balloons (ID#: 034.02-c01)

Alice did an experiment that caused four balloons to fill with gas, as shown in the figure to the right. Alice tested the flammability of each gas. She also measured the volume and mass of each gas to calculate the density. The tests and measures all occurred under the same conditions. The data is in Table 1.

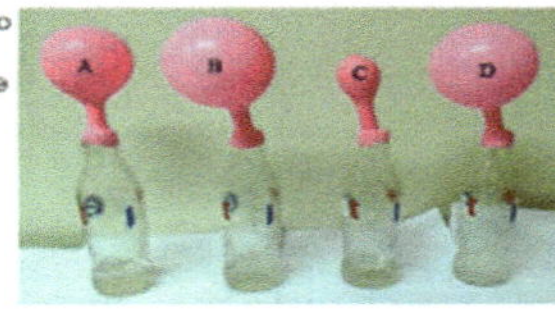

Table 1. Data of four gases in the balloons.

Sample	Flammability	Density	Volume
Gas A	Yes	0.089 g/L	180 cm^3
Gas B	No	1.422 g/L	270 cm^3
Gas C	No	1.981 g/L	35 cm^3
Gas D	Yes	0.089 g/L	269 cm^3

Question #1

Which, if any, of the gases listed in the data table could be the same? Using information from the table, explain your answer.

Type answer here

Next >

Coding rubric for "T1. Gas filled balloon" task

Common attribute	Element	Criteria	Score
Com1. Claim building	**AID1.** Identification of substances	**T1_c1.** Student states that Gases A & D could be the same substance.	1/0
Com2. Evidence noticing	**AID2.** Patterns of Data	**T1_c2.** Student supports a claim by referring to a data pattern that the flammability of Gas A & D is the same in the table. (Comparisons of data in different columns).	1/0
		T1_c3. Student supports a claim by referring to a pattern of data that the density of Gas A & D is the same in the table. (Comparisons of data in different columns).	1/0
Com3. Principle utilizing	**AID3.** Characteristic properties of Substances	**T1_c4.** Student indicates flammability is one of the characteristic properties to identify substances.	1/0
		T1_c5. Student indicates density is one of the characteristic properties to identify substances.	1/0

Fig. 13.1 "T1. Gas filled balloon" task and rubric

following research questions (RQs): RQ1. What empirical evidence do MFRMs provide to improve a rubric system (i.e., criteria, specific elements, and common attributes) for scoring students' responses to knowledge-in-use tasks? RQ2. After considering rater severity, what general patterns can be observed in the measures of

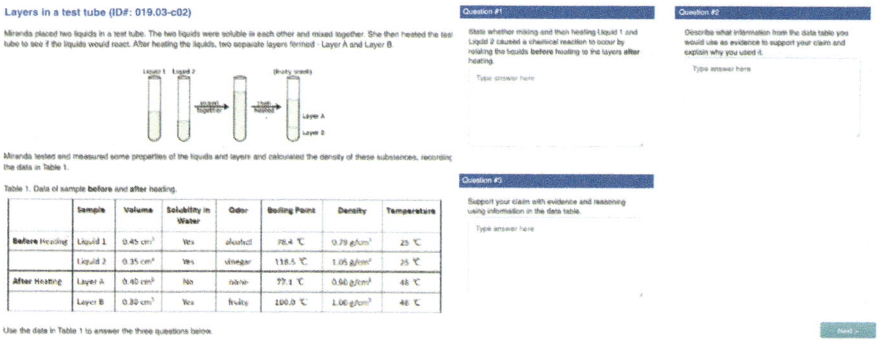

Coding rubric for "T2. Layers in a test tube" task

Common attribute	Element	Criteria	Score
Com 1. Claim building	**CSE1.** Claim	**T2_e1.** Student states that a chemical reaction occurs (when Liquid 1 and Liquid 2 were mixed and heated).	1/0
Com 2. Evidence noticing	**CSE2.** Pattern of data as evidence	**T2_e2.** Student describes the pattern (comparing before and after) in solubility data as evidence to support the claim.	1/0
		T2_e3. Student describes the pattern (comparing before and after) in odor data as evidence to support the claim.	1/0
		T2_e4. Student describes the pattern (comparing before and after) in boiling point data as evidence to support the claim.	1/0
		T2_e5. Student describes the pattern (comparing before and after) in density data as evidence to support the claim.	1/0
Com 3. Principle utilizing	**CSE3.** Characteristic properties of Substances	**T2_e6.** Student indicates solubility is the characteristic property to identify substances. (Similar words for "characteristic" such as unique, important).	1/0
		T2_e7. Student indicates odor is the characteristic property to identify substances. (Similar words for "characteristic" such as unique, important).	1/0
		T2_e8. Student indicates boiling point is the characteristic property to identify substances. (Similar words for "charecteristic" such as unique, important).	1/0
		T2_e9. Student indicates density is the characteristic property to identify substances. (Similar words for "characteristic" such as unique, important).	1/0
Com4. Mechanistic reasoning	**CSE4.** Reasoning	**T2_e10.** Student states layer A and B are new substances by comparing the differences of characteristic properties (before and after Liquid 1 and 2 were mixed).	1/0
		T2_e11. Student states that the new substances formed means that a chemical reaction occurs.	1/0

Fig. 13.2 "T2. Layers in a test tube" task and rubric

student performances on criteria, specific elements, and common attributes across multiple assessment tasks?

13.4 Method

13.4.1 Data Sources

All NGSA tasks have been placed in a free web portal accessible to teachers and students. More than 69,285 students, 9808 teachers, and 6453 classes registered for user accounts on the portal. We downloaded students' responses to three selected assessment tasks based on the research purposes. To protect students' privacy, we had no access to students' demographic information. Instead, we anchored student responses to three tasks (T1, T2, and T6, see Table 13.1) using the schools' and students' IDs. Then, we selected students (N = 240) who responded to all three tasks using data from our online data portal.

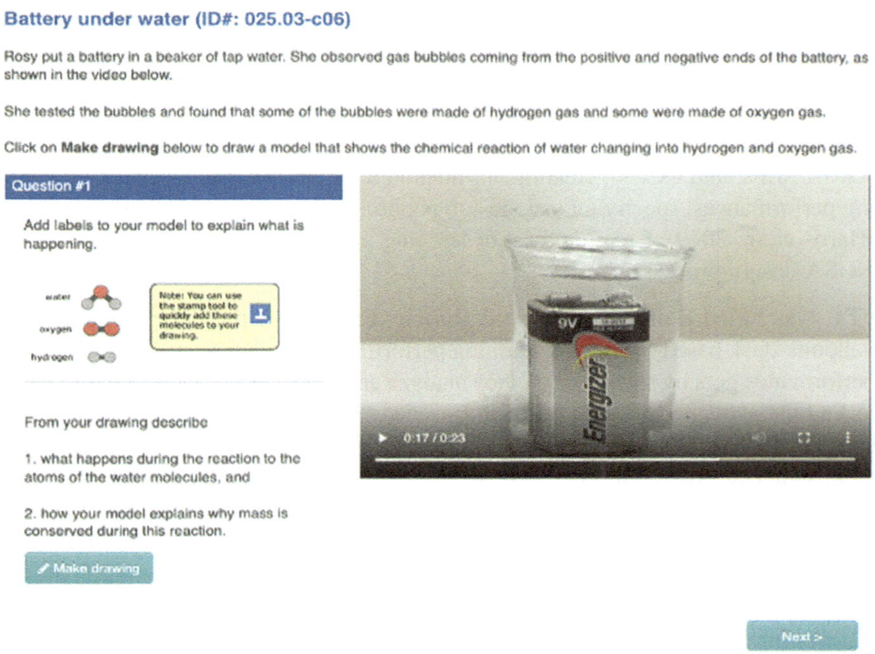

Coding rubric for "T6. Battery under water" task

Common attribute	Element	Criteria	Score
Com 1. Claim building	**DUM1.** Components in a model	**T6_c1.** Student clearly draws the water, oxygen, and hydrogen molecules.	1/0
Com 2. Evidence noticing	**DUM2.** Relationships between components in a model	**T6_c2.** Student correctly draws water molecule as the reactant and oxygen and hydrogen molecules as the products.	1/0
		T6_c3. Student correctly draw the ratio of water, oxygen, and hydrogen molecules during the reaction: 2, 1, and 2.	1/0
Com 4. Mechanistic reasoning	**DUM3.** Causal explanation for using a model	**T6_c4.** Student correctly state that the atoms of the water molecules are regrouped (or rearranged/broken down/became) to form new molecules- oxygen and hydrogen molecules.	1/0
		T6_c5. Student correctly state that before and after the chemical reaction, the number and type of atoms are the same, (so the mass is conserved).	1/0

Fig. 13.3 "T6. Battery under water" task and rubric

13.4.2 NGSA Assessment Tasks

This study used three tasks (T1, T2, and T6, see Table 13.1) in middle school physical science. Table 13.1 shows the three tasks, associated with knowledge-in-use learning performance goals. Learning performances are knowledge-in-use statements incorporating aspects of DCIs, SEPs, and CCCs but are smaller in scope than the PE. A series of LPs build toward meeting a PE (Harris et al., 2019). When constructing assessment tasks, we bundled two PEs that work together (See Table 13.1). All NGSA assessment tasks were designed using the ECD-based design process, consisting of identifying target NGSS PEs, unpacking three dimensions (DCIs, SEPs, and CCCs), creating an integrated dimension map, articulating learning performances, specifying task design patterns, and developing tasks and rubrics (Harris et al., 2019). Creating a set of learning performances is the keystone of the NGSA design process.

"T1. Gas Filled Balloon" Task and Rubric Figure 13.1 presents "T1. Gas filled balloons" task based on the LP1 learning performance. Students achieve the learning performance goal by showing that they analyze and interpret the data from four gases to find the same gases based on their knowledge of the characteristic properties of substances. The task includes a meaningful scenario with an image, data sources in a table, and a question to prompt students to figure out the given problem described in the front scenario. Finally, students respond to the task by typing the answer in the blank box. Based on the modified ECD-based evidence statements (Harris et al., 2019), we created a coding rubric of five criteria to score responses for each task. The five criteria are categorized into three elements of analyzing and interpreting data (AID) practice (Chen & Terada, 2021; Kim et al., 2020; Schwarz et al., 2017): Identifications of substances (T1_c1, standing for Task 1 and criteria 1), Pattern of data (T1_c2 and T1_c3), and Characteristic properties of substances (T1_c4 and T1_c5). Using the criteria in the rubric, content experts score 1 or 0 on students' responses to indicate achieved or not achieved, respectively.

"T2. Layers in a Test Tube" Task and Rubric Figure 13.2 shows the "T2. Layers in a test tube" task based on the LP2 learning performance. The task sets up a scenario in the lab where two liquids are mixed and heated, resulting in two layers with changes in properties. The three prompts are designed based on the three elements of a scientific explanation (CSE): claim, evidence, and reasoning (McNeill et al., 2006; Gotwals & Songer, 2013). Students achieve the learning goal by constructing a scientific explanation with evidence and ideas to claim that mixing the two liquids caused a chemical reaction. In the rubric design, we explicitly derived the element of scientific ideas from the reasoning part. We articulated four elements of scientific explanations to categorize the 11 criteria: claim (T2_c1), pattern of data as evidence (T2_c2, c3, c4, and c5), characteristic properties of substances (T2_c6, c7, c8, and c9), and reasoning (T2_c10 and c11). A score of 1 indicates a student achieves the statement in each criterion, and a score of 0 means the student does not achieve the criterion.

"T6. Battery Under Water" Task and Rubric Figure 13.3 presents "T6. Battery under water" task based on the LP6 learning performance. The task describes a real-world scenario and shows a scientific phenomenon in a video of a battery under water. Next, students play the video to observe the phenomenon before constructing a model to explain it. In the drawing process, the task provides drawing tools and has two prompts to assist students in drawing their models. Finally, students achieve the learning performance goal by drawing a model to explain the reaction of water and the conservation of matter during the reaction. According to previous work (He et al., 2023; Mayer & Krajcik, 2015; Zangori et al., 2017), this study employed three elements of a developing and using model (DUM) to categorize the five criteria: components in a model (T6_c1), relationships between components in a model (T6_c2 and c3), and causal explanations for using a model (T6_c4 and c5). Each criterion is coded as 1 (achieved) or 0 (not achieved) based on the scores of students' responses.

13.4.3 Scoring Design

This study recruited three graduate students (Rater A, B, and C) with bachelor's degrees in chemistry from a teacher preparation program. Before the formal scoring work, we provided professional learning for them to use the rubrics. Then, we developed a training workbook with selected sampling responses and discussed the discrepancies of their scores. The 40 common cases were scored by the three raters independently. The remaining 200 cases were divided and assigned to the three raters to score respectively. Table 13.2 shows the design of the scoring matrix across the three raters.

13.4.4 Inter-Rater Reliability

We used Fleiss's kappa (Fleiss & Cohen, 1973) and the percentage agreement to show the inter-rater reliability of the three raters' scores. The cut-off value of Fleiss's kappa is 0.70 (Fleiss et al., 2003). Table 13.3 presents the values of Fleiss kappa for all the items in the three tasks, which are higher than 0.70. Meanwhile, the agreement

Table 13.2 Scoring matrix design

	Common cases	Individual cases		
	ID1–40	ID41–106	ID107–172	ID173–240
Rater A	√	√		
Rater B	√		√	
Rater C	√			√

Table 13.3 Inter-rater reliability

Task	Fleiss Kappa (per item)	Percentage (%, per item)
T1	[0.742, 1.000]	[85.7, 100]
T2	[0.740, 1.000]	[91.4, 100]
T6	[0.708, 0.835]	[82.9, 94.3]

percentage for each item is higher than 80%. The high kappa value and the percentage agreement show that the three raters had high reliability in scoring students' responses using our rubrics.

13.4.5 Many-Facet Rasch Measurement

Classical Rasch measurement models estimate two facets of parameters (person abilities and item difficulty) and place them on the same logit scale. The probability of correctly responding to a question depends on item difficulty and the examinee's ability in such a model. However, if other factors (e.g., rater severity in scoring performance-based assessments) substantially contribute to the variance of student scores, two-facet Rasch modeling may generate errors in estimations. Therefore, Linacre (1989) developed the many-facet Rasch measurement (MFRM) that can estimate additional facets. MFRM includes the rater effect (e.g., Boone et al., 2016), examinee gender effect (Wu et al., 2007), shared features of assessment tasks (Chi et al., 2021), and students' sub-competence that is part of a complex construct (e.g., Zhai, 2022). Because multiple raters scored NGSA tasks, this study employed MFRM to analyze the student scores with different rubrics. This study employed the Conquest 2.0 software (Wu et al., 2007) to conduct the MFRM analysis.

Using the MFRM, each of the following three sections presents results, including (a) the reliability of measures in each facet; (b) the psychometric properties of the items and raters; and (c) the Wright map to investigate the variability of raters' severity and item difficulties. To examine the reliability, we reported three measures: Chi-square test index, separation reliability, and EAP (i.e., the expected student posterior latent ability)/PV (i.e., the plausible student ability value). The Chi-square test (Wu et al., 2007) indicates the equality of the items (or common attributes) in each facet. The separation reliability, ranging from 0 to 1, shows the degree to which the item parameters are separated (Wright & Stone, 1979). The larger the value, the better the item parameters are separated. In addition, comparable to KR-20 and Cronbach's alpha reliability, the EAP/PV reliability indicates the ratio of true variance to observed variance for each facet. Finally, this study examined the psychometric properties of the items in each facet using two typical parameters: unweighted and weighted mean square item fit (MNSQ). Both unweighted and weighted MNSQ are used to detect the misfit of items, but the outfit statistics are more sensitive to outliers. The ranges of the two indices are 0 to infinity, with an ideal value of 1.0. An MNSQ value lower than 1.0 indicates the score has minor variation than expected, meaning that the Rasch model may be overly predictable than expected, and an over 1.0 MNSQ indicates that the Rasch model may be inadequately predictable. This study adopts the cut-off points for MNSQs ranging from 0.5 to 1.5, as Linacre et. (1989) suggested for being productive for measurement. This study preferred the weighted MNSQ over the unweighted MNSQ because of the potential number of outliers in the dataset. We also present Wright maps to visually show the variability of parameters in each facet, such as person ability, item and task difficulty, and rater severity.

13.5 Analytic Strategies and Results

13.5.1 Applying Dichotomous MFRMs to Analyze Scores Using Task-Specific Dichotomous Rubrics

Analytic rubrics provide fine-grained information about students' performance and thus could serve diagnostic purposes. The criteria in each task are used to examine what students have already achieved or not yet achieved regarding a target learning performance goal. Thus, the estimates of those criteria are needed for improving the task-specific dichotomous rubrics.

Data Framing Students' responses to the three tasks were coded into 21 dichotomous scores (T1_c1-c5; T2_c1-c11; T6_c1-c5). Students in the common case group (ID1–40) had three rows of 21 dichotomous scores because three raters coded all those responses. Students in the individual case group (ID41–240) were only coded once by one rater. Therefore, a variable named "Rater" was created in the dataset with three categories (Rater A, Rater B, and Rater C). Rater variability could potentially create local dependency as the scores assigned by the same rater may reflect the rater's scoring severity if different from other raters. To examine the rater effects, we checked the local independence of the criteria using the residual correlation analyses. A residual correlation below 0.7 indicates the independence of those criteria in our rubrics. We found that the correlation coefficients of residuals for some pairs of criteria are higher than 0.70, including pairs of T1_c4 and T1_c5 ($r = 0.90$), T2_c2 and T2_c6 ($r = 0.77$), and T6_c2 and T6_c3 ($r = 0.70$). Those criteria are highly correlated with each other, which indicates a violation of the assumption of the classical Rasch measurement (Zhai, 2022). Based on the preliminary analysis and expert review, we deleted T1_c4, T2_c06, and T6_c03 and then applied MFRM to analyze the data.

Analytic Model Given the data structure, this section employed a three-facet dichotomous Rasch model, including (1) rater facet (rater severity), (2) rubric criteria facet (item difficulty), and (3) examinee's ability facet. The formula of this MFRM is,

$$\log\left(\frac{P_{nji}}{1-P_{nji}}\right) = B_n + L_j - D_i + \varepsilon$$

Where,

Table 13.4 Fit statistics for rater and item estimates

	Estimate	Error	Unweighted Fit			Weighted Fit		
			MNSQ	CI	T	MNSQ	CI	T
Rater								
Rater A	−0.005	0.086	0.89	(0.73, 1.27)	−0.8	0.87	(0.71, 1.29)	−0.8
Rater B	−0.062	0.089	0.81	(0.73, 1.27)	−1.5	0.82	(0.71, 1.29)	−1.3
Rater C	0.067*		0.89	(0.73, 1.27)	−0.8	0.88	(0.71, 1.29)	−0.8
Criteria								
T1_c1	−2.595	0.225	2.64	(0.82, 1.18)	12.6	1.80	(0.70, 1.30)	4.3
T1_c2	−1.713	0.192	1.21	(0.82, 1.18)	2.2	1.57	(0.74, 1.26)	3.7
T1_c3	−2.268	0.211	1.33	(0.82, 1.18)	3.3	1.56	(0.72, 1.28)	3.4
T1_c5	−0.333	0.159	1.26	(0.82, 1.18)	2.6	1.86	(0.79, 1.21)	6.4
T2_c1	−3.337	0.262	0.25	(0.82, 1.18)	−12.0	0.76	(0.66, 1.34)	−1.4
T2_c2	−0.788	0.168	0.82	(0.82, 1.18)	−2.1	0.92	(0.77, 1.23)	−0.6
T2_c3	0.730	0.147	2.46	(0.82, 1.18)	11.5	2.69	(0.80, 1.20)	11.7
T2_c4	−1.342	0.181	0.81	(0.82, 1.18)	−2.2	0.95	(0.75, 1.25)	−0.4
T2_c5	−1.343	0.181	0.98	(0.82, 1.18)	−0.2	1.27	(0.75, 1.25)	2.0
T2_c7	1.585	0.148	1.28	(0.82, 1.18)	2.8	1.85	(0.79, 1.21)	6.5
T2_c8	−0.580	0.164	0.68	(0.82, 1.18)	−54.0	0.99	(0.78, 1.22)	−0.0
T2_c9	−0.555	0.163	0.74	(0.82, 1.18)	−3.0	1.15	(0.78, 1.22)	1.3
T2_c10	6.069	0.497	9.52	(0.79, 1.21)	31.3	1.78	(0.04, 1.96)	1.5
T2_c11	6.140	0.773	4.24	(0.73, 1.27)	13.5	2.05	(0.00, 2.62)	1.3
T6_c1	−0.602	0.166	1.13	(0.82, 1.18)	1.4	1.33	(0.78, 1.22)	2.7
T6_c2	−0.015	0.156	1.14	(0.82, 1.18)	1.5	1.30	(0.79, 1.20)	2.7
T6_c4	0.359	0.160	1.03	(0.81, 1.19)	0.3	1.29	(0.78, 1.22)	2.4
T6_c5	0.586*		1.35	(0.81, 1.19)	3.3	1.42	(0.79, 1.21)	3.4

P_{nji} = probability of person n receiving the score of 1 assigned by rater j on item i.
B_n = the ability of person n; L_j = the severity of rater j; D_i = the difficulty of item i; ε is the error.

Main Findings Table 13.4 shows the model-data fit statistics for rater and item estimates. First, regarding the fit statistics for rater severity, all MNSQs are in the range of 0.81–0.89, meeting the cut-off criteria of 0.5–1.5. The result shows that the scores rated by the three raters were well fitted with the MFRM. Result also shows that Rater C (0.067) is more severe than Rater A (−0.005) and Rater B (−0.062). However, the Chi-square test result is not significant ($\chi^2(2) = 0.48$, $p = 0.785$), and the separation reliability is 0. The result indicates that the differences in scoring severity between the three raters are not statistically significant. Second, controlling for the raters' severity, we find that the infit MNSQs for seven of the total 18 criteria is over 1.5, indicating those criteria may not be adequately used for estimating students' abilities due to the misfit statistics. The results indicate that those misfit criteria might have the disorder estimations of students' abilities. Students with overall high performances may have trouble in some of those criteria, whereas those with low performances may respond appropriately. Overall, we find that the separation reliability value for item parameters is 0.988, with $\chi^2(20) = 1020.51$, $p < 0.001$. The EAP/PV value is 0.856. The results show that the facet of item (criteria) parameters is sufficiently and reliably separated.

The Wright map in Fig. 13.4 shows student ability, rater severity, and item difficulty estimates in three columns separated by the three vertical dash lines, respectively. The vertical scale ranging from −6 to 6 logits is shared by the estimates of students' abilities, rater severities, and item difficulties. Each "x" represents 2.0 students on the leftmost column, and its location corresponding to the vertical scale indicates student ability. Figure 13.4 shows that students' abilities are typically distributed from the bottom to the top, with a large proportion of students in the middle with ability values around −1 to 2 logit. The higher the ability value on the map, the greater probability of success that students' responses to the assessment tasks meet the rubric criteria. Looking at the second column, we further confirm that the three raters had similar severity in using the rubrics to score students' responses to the assessment tasks. Regarding the third vertical dash line, we determine that the difficulties of the criteria items in the T1 task are under 0 logit, whereas the difficulties of the five criteria in the T6 task are above −1 logit. We also find that the range of item difficulties for both T1 and T6 tasks are around 3 logits, which is insignificant compared to the T2 task. T2 task has a large variability of item difficulties ranging from −4 logit to 6 logits. We conclude that most item difficulties are between −4 logit and 2 logits, except for T2_c11 and T2_c10. The two items are the most challenging; most students could not achieve the two criteria. Overall, all items across the three tasks have a good range of difficulties in measuring students' abilities.

13.5.2 *Applying Partial Credit MFRMs to Analyze Scores Using Task-Specific Polytomous Rubrics*

In addition to the diagnostic purpose of the rubrics, it would be helpful to know the extent of students' performances at the element level that may include several items (i.e., criteria) in each task. Therefore, the original scores were transformed into a composite score on the element of the three tasks. This section estimated the measure of rater severity and all elements in the three tasks to improve the task-specific polytomous rubrics of the assessment tasks.

Data Framing Based on our design approach, elements have a variety of criteria (ranging from 1 to 4), depending on the task features and the complexity of the scenarios (see Table 13.5). For instance, the first element in the T1 task (AID1 – Identification of substances) only has one item (T1_c1) with the original code of 0 and 1. So, we kept the original scores for criterion T1_c1 for the element AID1. The same strategy was used for T2_c1 and T6_c1. For elements with four items (CSE2 and CSE3), we calculated the original scores (ranging from 0 to 4) and

```
=====================================================================================
MAP OF LATENT DISTRIBUTIONS AND RESPONSE MODEL PARAMETER ESTIMATES
=====================================================================================
                    Terms in the Model (excl Step terms)
                           +rater                          +criteria
-------------------------------------------------------------------------------------
                 |                          |.        T2_c11                    |
  6              |                          |.        T2_c10                    |
                 |                          |                                   |
                 |                          |                                   |
  5              |                          |                                   |
                X|                          |                                   |
                X|                          |                                   |
                X|                          |                                   |
  4            XX|                          |                                   |
               XX|                          |                                   |
               XX|                          |                                   |
  3            XX|                          |                                   |
              XXX|                          |                                   |
              XXX|                          |                                   |
  2         XXXXX|                          |                                   |
           XXXXXX|                          |                                   |
         XXXXXXXX|                          |.          T2_c7                   |
       XXXXXXXXXX|                          |                                   |
  1       XXXXXXX|                          |                                   |
       XXXXXXXXXX|                          |           T2_c3          T6_c5    |
         XXXXXXXX|                          |.                         T6_c4    |
  0        XXXXXX|R1 R2 R3                  |.                         T6_c2    |
            XXXXX|                          |T1_c5                              |
            XXXXX|                          |.          T2_c8 T2_c9    T6_c1    |
 -1         XXXXX|                          |.          T2_c2                   |
            XXXXX|                          |                                   |
              XXX|                          |           T2_c4 T2_c5             |
              XXX|                          |T1_c2                              |
 -2             X|                          |                                   |
               XX|                          |T1_c3                              |
               XX|                          |T1_c1                              |
 -3            XX|                          |                                   |
               XX|                          |.          T2_c1                   |
               XX|                          |                                   |
                X|                          |                                   |
 -4             X|                          |                                   |
                X|                          |                                   |
                X|                          |                                   |
 -5             X|                          |                                   |
                 |                          |                                   |
                 |                          |                                   |
 -6             X|                          |                                   |
                 |                          |                                   |
                X|                          |                                   |
=====================================================================================
Each 'X' represents    2.0 cases
=====================================================================================
```

Fig. 13.4 The Wright map with the three facets of student ability, rater severity, and criteria difficulty

Table 13.5 The overview of data structure in the three tasks

Task	Element	Item ID	New item ID	Scores	Recode
T1	AID1. Conclusion of identifying substances	T1_c1	T1_c1	0,1	Remain the same
	AID2. Patterns of data	T1_c2, T1_c3	T1_c2c3	0,1,2	Composite the original score
	AID3. Characteristic properties of substances	T1_c4, T1_c5	T1_c4c5	0,1,2	Composite the original score
T2	CSE1. Claim	T2_c1	T2_c1	0,1	Remain the same
	CSE2. Pattern of data as evidence	T2_c2, T2_c3, T2_c4, T2_c5	T2_c2345	0,1,2	Recode the composite scores
	CSE3. Characteristic properties of substances	T2_c6, T2_c7, T2_c8, T2_c9	T2_c6789	0,1,2	Recode the composite scores
	CSE4. Reasoning	T2_c10, T2_c11	T2_c10c11	0,1,2	Composite the original score
T6	DUM1. Drawn of components in a model	T6_c1	T6_c1	0,1	Remain the same
	DUM2. Relationships between components in a model	T6_c2 T6_c3	T6_c2c3	0,1,2	Composite the original score
	DUM3. Causal explanation for using a model	T6_c4 T6_c5	T6_c4c5	0,1,2	Composite the original score

transformed them into 0, 1, and 2. For instance, the CSE2 is about the evidence to support the claim that a chemical reaction occurs based on the changes in characteristic properties. We believe that two of the four items (T2_c2, c3, c4, and c5) have sufficient evidence to support the claim. Thus, we record the composite scores as follows: 2 (the original 2, 3, and 4), 1 (the original 1), and 0 (the original 0). For elements with two items (AID2, AID3, CSE4, DUM2, and DUM3), we calculated the original scores as the scores of those elements. Thus, we organized the data of 10 elements (with new item IDs) with the partial credit scores (ranging from 0 to 2) in the three tasks scored by three raters.

Analytic Model For analyzing the data above, this section applied four-facet partial credit Rasch models, including (1) rater facet (rater severity), (2) element facet (difficulty), (3) examinee's ability facet, and (4) two interaction facets of rater and element, and rater, element, and step. The formula of this MFRM is,

$$\log\left(\frac{P_{njik}}{P_{nji(k-1)}}\right) = B_n + L_j - D_i - S_{ijk}$$

Where:

Table 13.6 Fit statistics for rater and element estimates

	Estimate	Error	Unweighted Fit			Weighted Fit		
			MNSQ	CI	T	MNSQ	CI	T
Rater								
Rater A	−0.027	0.127	0.78	(0.73, 1.27)	−1.7	0.78	(0.70, 1.30)	−1.5
Rater B	0.295	0.127	0.93	(0.73, 1.27)	−0.5	0.91	(0.72, 1.28)	−0.6
Rater C	−0.268*		0.92	(0.73, 1.27)	−0.6	0.89	(0.72, 1.28)	−0.8
Element								
T1_c1 (AID1)	−2.405	0.237	1.23	(0.82, 1.18)	2.4	1.70	(0.70, 1.30)	3.8
T1_c2c3 (AID2)	−1.446	0.157	1.06	(0.82, 1.18)	0.7	1.79	(0.71, 1.29)	4.3
T1_c4c5 (AID3)	0.050	0.113	2.14	(0.82, 1.18)	9.5	2.54	(0.77, 1.23)	9.2
T2_c1 (CSE1)	−3.511	0.419	0.27	(0.82, 1.18)	−11.6	0.76	(0.67, 1.33)	−1.5
T2_c2345 (CSE2)	0.332	0.136	1.10	(0.82, 1.18)	1.1	1.49	(0.80, 1.20)	4.3
T2_c6789 (CSE3)	0.608	0.124	1.03	(0.82, 1.18)	0.3	1.56	(0.79, 1.21)	4.4
T2_c1011 (CSE4)	5.547	0.403	3.84	(0.82, 1.18)	18.6	1.64	(0.30, 1.70)	1.6
T6_c1 (DUM1)	−0.489	0.159	0.78	(0.82, 1.18)	−2.5	1.04	(0.80, 1.20)	0.5
T6_c2c3 (DUM2)	0.679	0.111	1.23	(0.82, 1.18)	2.4	1.55	(0.78, 1.22)	4.3
T6_c4c5 (DUM3)	0.635*		0.85	(0.81, 1.19)	−1.6	1.26	(0.76, 1.24)	2.0

P_{njik} = probability of person n getting scores of k by rater j to element i
$P_{nji(k-1)}$ = probability of person n getting scores of $k - 1$ by rater j to element i
B_n = the ability of person n; L_j = the leniency of rater j; D_i = the difficulty of element i
S_{ijk} = the interaction estimates of rater, element, and step.

Main Findings Table 13.6 presents the model-data fit statistics for rater and element estimates. The unweighted and weighted MNSQs for all raters range from 0.78 to 0.93, indicating that the scores of elements rated by the three raters fit well with the Rasch model. No irrelevant variances exist at the element level. The separation reliability value for rater parameter is 0.688, with $\chi^2(2) = 5.43$, $p < 0.066$. The result also suggests that the scoring severity among the three raters cannot be separated.

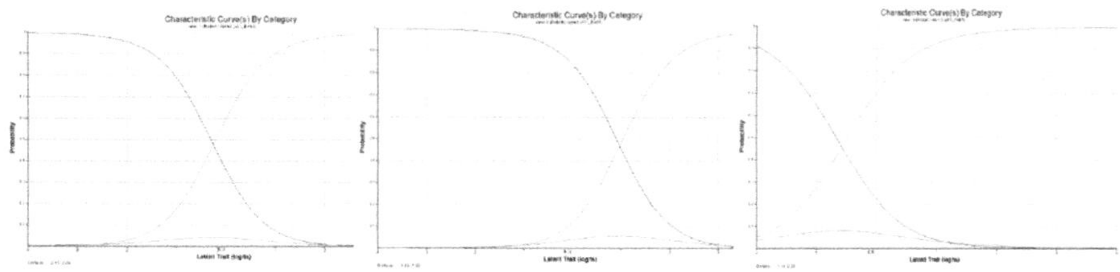

Fig. 13.5 Characteristic curves by categories for element AID3

Six of the ten elements had MNSQ values over 1.50, indicating they may not be appropriate for estimating students' overall ability, especially the element AID3 (i.e., T1_c4c5; Weight MNSQ = 2.54). We further examined the estimates of steps in those elements and found that the score of 1 is underestimated, which may not be needed to score students' achievement on those elements. Figure 13.5 shows that the three raters scored element AID3 either 0 or 2 with very few scores of 1. The results suggest the future revision of scoring those elements to improve the accuracy of reporting students' overall ability. According to the final analytic model, the separation reliability value for element parameters is 0.991, with $\chi^2(9) = 524.53$, $p < 0.001$. The EAP/PV value is 0.789. The results indicate that element parameters' facets are sufficiently and reliably separated.

Figure 13.6 shows the four facet Wright map with estimates of student ability, rater severity, element difficulty, and the estimate of rater and element interaction on the same scale (the four vertical dash lines, respectively). The vertical scale ranges from −5.0 to 5.0 logits for the four facets. Each "x" represents 2.4 students on the map. There is also a normal distribution of students' abilities, with the average at 1.0 logit. From the second vertical dash line, we find that the three raters' severity estimates are placed around the middle of the scale, and they are close to each other, indicating that their scores on the elements across the three tasks are consistent with each other. The element column (the third vertical dash line) shows that most elements are located within −2 to 1. Two elements (T1_c1 and T2_c1) are the least challenging elements for students at the bottom. In contrast, the element CSE (i.e., T2_c10c11) is placed at the top of the map. Most students' responses on T2 tasks could not meet the element's two criteria (T2_c10 and T2_c11). The interaction between raters and elements was added to model variation to account for how the raters' severity affects the difficulties of the elements. The fourth vertical dash line shows that rater and element interaction estimates range from −1.5 to 1.0, indicating relative variability. Besides, the Chi-square test of parameter equality for the interaction is $\chi^2(18) = 75.82$, $p < 0.001$, indicating that significant interaction between the raters and the elements exists so that the rater effect cannot be ignored in the Rasch model.

```
================================================================================================================
MAP OF LATENT DISTRIBUTIONS AND RESPONSE MODEL PARAMETER ESTIMATES
================================================================================================================
                                 Terms in the Model (excl Step terms)
                           +rater                  +element                  +rater*element
----------------------------------------------------------------------------------------------------------------
                   |        |                              |                                                    |
                   |        |T2_c10c11                     |                                                    |
                   |        |                              |                                                    |
  5                |        |                              |                                                    |
                   |        |                              |                                                    |
                   |        |                              |                                                    |
                   |        |                              |                                                    |
                  X|        |                              |                                                    |
  4               X|        |                              |                                                    |
                  X|        |                              |                                                    |
                  X|        |                              |                                                    |
                  X|        |                              |                                                    |
  3              XX|        |                              |                                                    |
                 XX|        |                              |                                                    |
                 XX|        |                              |                                                    |
               XXXX|        |                              |                                                    |
  2             XXX|        |                              |                                                    |
              XXXXX|        |                              |                                                    |
              XXXXX|        |                              |                                                    |
           XXXXXXXX|        |                              |                                                    |
  1         XXXXXXX|        |                              |R2.T2_c1 R3.T6_c1 R3.T6_c4c5                        |
         XXXXXXXXXX|        |T2_c6789 T6_c2c3 T6_c4c5      |R2.T1_c4c5 R3.T2_c2345 R2.T2_c6789 R1.T2_c10c11     |
             XXXXXX|        |                              |R1.T1_c1 R1.T2_c1 R3.T6_c2c3                        |
              XXXXX|R2      |T2_c2345                      |R2.T1_c1 R2.T1_c2c3                                 |
  0            XXXX|R1      |T1_c4c5                       |R3.T1_c2c3 R1.T2_c6789 R3.T2_c10c11 R1.T6_c2c3      |
              XXXXX|R3      |                              |R1.T1_c2c3 R1.T1_c4c5 R3.T1_c4c5 R1.T2_c2345        |
               XXXX|        |T6_c1                         |R3.T1_c1 R2.T2_c2345 R2.T2_c10c11 R1.T6_c2c3        |
                XXX|        |                              |R3.T2_c6789 R2.T6_c1 R2.T6_c2c3                     |
 -1             XXX|        |                              |                                                    |
                 XX|        |                              |                                                    |
                  X|        |T1_c2c3                       |R3.T2_c1                                            |
                 XX|        |                              |                                                    |
 -2              XX|        |                              |                                                    |
                  X|        |                              |                                                    |
                 XX|        |T1_c1                         |                                                    |
                  X|        |                              |                                                    |
 -3               X|        |                              |                                                    |
                  X|        |                              |                                                    |
                  X|        |T2_c1                         |                                                    |
                  X|        |                              |                                                    |
 -4               X|        |                              |                                                    |
                   |        |                              |                                                    |
                   |        |                              |                                                    |
                  X|        |                              |                                                    |
                   |        |                              |                                                    |
 -5               X|        |                              |                                                    |
================================================================================================================
Each 'X' represents   2.4 cases
Some parameters could not be fitted on the display
================================================================================================================
```

Fig. 13.6 The Wright map with the four facets of student ability, rater severity, element difficulty, and rater-item interaction estimates

13.5.3 Applying Partial Credit MFRMs to Analyze Scores Using Task-Generic Polytomous Rubrics

As many elements are nested within one task (Zhai, 2022), it would be necessary to estimate the three tasks' difficulties with common attributes of knowledge-in-use. We created four common attributes to restructure the elements across the three tasks based on the literature (Chen et al., 2013; Kapon, 2017; Krist et al., 2019; Schwarz et al., 2017). The MFRM models can estimate the difficulties of the three tasks and the four common attributes that would capture students' performance at the task and attribute levels.

Data Framing According to our design approach, we categorized the elements in each task into four common attributes (See Table 13.7). For instance, three elements (AID1, 2, and 3) were categorized into claim building (Com1), evidence noticing (Com2), and principle utilizing (Com3). The Mechanistic reasoning (Com4) was not included in the T1 task, so it was coded as missing in the dataset. Partial credit scores (0, 1, and 2) were also used in this dataset. Thus, the dataset in this section has the following variables: Rater (Rater A, Rater B, and Rater C), Task (T1, T2, and T6), and four common attributes (Com1, Com2, Com3, and Com4).

Analytic Model We used many-facet Rasch measurement (MFRM, Linacre, 1989) to analyze the dataset, including multiple raters' scores on the three NGSA tasks. Rasch measurement converts the nonlinear raw scores to the linear measures of person ability and item difficulty using the transformation of log odds (Bond & Fox, 2015; Liu, 2020). Classical Rasch measurement models estimate two facets of person abilities, item difficulty, and position them at the same logit scale. However, the MFRM incorporates more facets than the typical Rasch measurement model (Boone et al., 2016).

Given the data structure above, this section employed the five-facet partial credit Rasch models, including (1) examinee's ability facet, (2) rater facet (rater severity), (3) task facet (difficulty), (4) common attribute facet (difficulty), and (5) three interaction estimates of rater and task, rater and common attribute, and rater, common attribute, and step. The formula of this MFRM is,

$$\log\left(\frac{P_{njmik}}{P_{njmi(k-1)}}\right) = B_n + L_j - T_m - C_i - S_{jmik}$$

Where:

P_{njmik} = probability of person n getting scores of k by rater j to common attribute i in task m.

$P_{njmi(k-1)}$ = probability of person n getting scores of k-1 by rater j to common attribute i in task m.

B_n = the ability of person n; L_j = the leniency of rater j; C_i = the difficulty of common attribute i

T_m = the difficulty of task m; S_{jmik} = the interaction estimates of rater, task, common attribute and step.

Main Findings We conducted the five-facet partial credit Rasch analysis and examined the reliability of each facet and model-data fit statistics. Table 13.8 shows the model-data fit statistics for rater, task, and common attribute estimates. The unweighted and weighted MNSQs for the three raters range from 0.63 to 0.91, meaning that the rater facet fits well with the Rasch model. Scores at the common attribute level with different raters did not increase the irrelevant variances. The

Table 13.7 The overview of data structure in the three tasks

	Com1: Claim Building		Com2: Evidence Noticing		Com3: Principle Utilizing		Com4: Mechanistic Reasoning	
Task	Element	Code	Element	Code	Element	Code	Element	Code
T1	AID1. Conclusion of identifying substances	0,1	AID2. Patterns of data	0,1,2	AID3. Characteristic properties of substances	0,1,2	N/A	N/A
T2	CSE1. Claim	0,1	CSE2. Pattern of data as evidence	0,1,2	CSE3. Characteristic properties of substances	0,1,2	CSE4. Reasoning	0,1,2
T6	DUM1. Drawn of components in a model	0,1	DUM2. Relationships between components in a model	0,1,2	N/A		DUM3. Causal explanation for using a model	0,1,2

Table 13.8 Fit statistics for raters, tasks, and common attributes' estimates

			Unweighted Fit			Weighted Fit		
	Estimate	Error	MNSQ	CI	T	MNSQ	CI	T
Rater								
Rater A	0.275	0.118	0.63	(0.84, 1.16)	−5.4	0.81	(0.80, 1.20)	−2.0
Rater B	−0.029	0.114	0.81	(0.84, 1.16)	−2.6	0.84	(0.81, 1.19)	−1.6
Rater C	−0.247*		0.76	(0.85, 1.15)	−3.3	0.91	(0.81, 1.19)	−0.9
Task								
Gas balloon (T1)	−1.997	0.140	0.52	(0.82, 1.18)	−6.5	0.94	(0.70, 1.30)	−0.3
Test tube (T2)	1.804	0.085	1.25	(0.82, 1.18)	2.6	0.92	(0.77, 1.23)	−0.6
Battery under water (T3)	0.193*		0.87	(0.82, 1.18)	−1.4	1.07	(0.71, 0.29)	0.5
Common attribute								
Com1	−3.272	0.138	2.46	(0.90, 1.10)	19.8	1.36	(0.80, 1.20)	3.1
Com2	−0.355	0.092	0.86	(0.90, 1.10)	−2.8	1.47	(0.86, 1.14)	5.6
Com3	1.025	0.088	0.68	(0.87, 1.13)	−5.6	1.22	(0.81, 1.19)	2.1
Com4	2.602*		21.93	(0.87, 1.13)	80.5	2.00	(0.76, 1.24)	6.3

separation reliability value for rater parameter is 0.708, with $\chi^2(2) = 5.48$, $p = 0.065 > 0.05$. The result also suggests that the scoring severity among the three raters cannot be separated.

Moreover, the MNSQ values for the three tasks are within the range of 0.5–1.5, indicating that the three tasks produce a good model that fits the data. In addition, the MNSQs for the four common attributes have a wide range, especially the unweighted fit values for Com4 (21.93) and Com1 (2.46) and the weighted fit value for Com4 (2.00). The results indicate that the Com4 has a misfit estimate due to the outliers and the design of scores. In addition, Table 13.9 further shows the model-data fit statistics for rater-task and rater-common attribute interaction estimates. The weighted MNSQ values for the interaction terms of rater-task and rater-common attributes are within the range of 0.5 to 1.5. The separation reliability values for task facet and common attribute facet are 0.998, with $\chi^2(2) = 656.77$, $p < 0.001$ and $\chi^2(3) = 714.57$, $p < 0.001$, respectively. The EAP/PV value is 0.749. The results indicate that the facets of the task and common attributes parameters are both sufficiently and reliably separated.

Figure 13.7 presents the five facet Wright map with student ability estimates, rater severity, task difficulty, common attribute difficulty, and two interaction estimates as

Table 13.9 Fit statistics of rater-task and rater-common attribute interaction estimates

		Estimate	Error	Unweighted Fit			Weighted Fit		
				MNSQ	CI	T	MNSQ	CI	T
Rater*task									
Rater A	Gas balloon (T1)	0.150	0.165	0.39	(0.73, 1.27)	−5.8	0.68	(0.60, 1.40)	−1.7
Rater B	Gas balloon (T1)	0.076	0.154	0.60	(0.73, 1.27)	−3.4	0.91	(0.66, 1.34)	−0.5
Rater C	Gas balloon (T1)	−0.227*		0.34	(0.73, 1.27)	−6.7	1.12	(0.54, 1.46)	0.6
Rater A	Test tube (T2)	0.216	0.112	0.81	(0.73, 1.27)	−1.4	0.76	(0.72, 1.28)	−1.7
Rater B	Test tube (T2)	−0.093	0.112	1.00	(0.73, 1.27)	0.1	0.81	(0.70, 1.30)	−1.3
Rater C	Test tube (T2)	−0.123*		1.23	(0.73, 1.27)	1.6	0.89	(0.74, 1.26)	−0.8
Rater A	Battery under water (T3)	−0.366*		0.68	(0.73, 1.27)	−2.6	1.01	(0.65, 1.35)	0.1
Rater B	Battery under water (T3)	0.017*		0.83	(0.73, 1.27)	−1.3	0.84	(0.66, 1.34)	−0.9
Rater C	Battery under water (T3)	0.350*		0.72	(0.73, 1.27)	−2.2	0.82	(0.66, 1.34)	−1.0
Rater*common attribute									
Rater A	Com1	0.433	0.225	4.15	(0.84, 1.16)	22.9	0.96	(0.68, 1.32)	−0.2
Rater B	Com1	−0.108	0.204	1.02	(0.84, 1.16)	0.3	1.02	(0.71, 1.29)	0.2
Rater C	Com1	−0.325*		1.42	(0.85, 1.15)	4.7	0.97	(0.60, 1.40)	−0.1
Rater A	Com2	0.297	0.125	0.59	(0.84, 1.16)	−6.0	0.84	(0.81, 1.19)	−1.7
Rater B	Com2	−0.344	0.122	0.85	(0.84, 1.16)	−1.9	1.07	(0.83, 1.17)	0.8
Rater C	Com2	0.047*		0.62	(0.85, 1.15)	−5.5	0.91	(0.81, 1.19)	−0.9
Rater A	Com3	−0.365	0.131	0.55	(0.81, 1.19)	−5.5	0.74	(0.79, 1.21)	−2.6
Rater B	Com3	0.539	0.125	0.66	(0.81, 1.19)	−4.0	0.98	(0.76, 1.24)	−0.2
Rater C	Com3	−0.174*		0.52	(0.81, 1.19)	−6.1	0.90	(0.77, 1.23)	−0.8
Rater A	Com4	−0.365*		2.07	(0.81, 1.19)	8.4	0.95	(0.71, 1.29)	−0.3
Rater B	Com4	−0.087*		50.19	(0.81, 1.19)	81.3	1.06	(0.73, 1.27)	0.4
Rater C	Com4	0.452*		1.54	(0.80, 1.20)	4.6	1.21	(0.75, 1.25)	1.6

```
=====================================================================================================================
MAP OF LATENT DISTRIBUTIONS AND RESPONSE MODEL PARAMETER ESTIMATES
=====================================================================================================================
                                        Terms in the Model (excl Step terms)
                 +rater        +task       +component        +rater*task              +rater*component
---------------------------------------------------------------------------------------------------------------------
              |             |             |             |                          |                                  |
 10          X|             |             |             |                          |                                  |
  9          X|             |             |             |                          |                                  |
             X|             |             |             |                          |                                  |
  8          X|             |             |             |                          |                                  |
            XX|             |             |             |                          |                                  |
  7          X|             |             |             |                          |                                  |
  6         XX|             |             |             |                          |                                  |
           XXX|             |             |             |                          |                                  |
  5        XXX|             |             |             |                          |                                  |
        XXXXXX|             |             |             |                          |                                  |
  4    XXXXXXX|             |             |             |                          |                                  |
  3 XXXXXXXXXX|             |             |             |                          |                                  |
     XXXXXXXXX|             |             |Com4         |                          |                                  |
  2  XXXXXXXXX|             |C2           |             |                          |                                  |
        XXXXXX|             |             |             |                          |                                  |
  1       XXXX|             |             |Com3         |R3.C2 R2.C3               |R2.Com3 R2.Com4 R3.Com4           |
  0        XXX|R1 R2        |C6           |             |R1.C1 R2.C1 R1.C2 R2.C2   |R1.Com1 R2.Com1 R1.Com2 R3.Com2   |
          XXXX|R3           |             |Com2         |R3.C1 R1.C3 R3.C3         |R3.Com1 R2.Com2 R1.Com3 R3.Com3   |
 -1       XXXX|             |             |             |                          |R1.Com4                           |
          XXXX|             |             |             |                          |                                  |
 -2       XXXX|             |C1           |             |                          |                                  |
 -3        XXX|             |             |             |                          |                                  |
            XX|             |             |Com1         |                          |                                  |
 -4        XXX|             |             |             |                          |                                  |
           XXX|             |             |             |                          |                                  |
 -5         XX|             |             |             |                          |                                  |
 -6         XX|             |             |             |                          |                                  |
            XX|             |             |             |                          |                                  |
 -7          X|             |             |             |                          |                                  |
             X|             |             |             |                          |                                  |
 -8          X|             |             |             |                          |                                  |
 -9          X|             |             |             |                          |                                  |
              |             |             |             |                          |                                  |
-10          X|             |             |             |                          |                                  |
              |             |             |             |                          |                                  |
=====================================================================================================================
Each 'X' represents   6.5 cases.
=====================================================================================================================
```

Fig. 13.7 The Wright map with the five facets of student ability, rater severity, task, and common attribute estimates, and the estimates of rater-task interaction and rater-common attribute interaction

rater-task and rater-common attributes (the six vertical dash lines, respectively). The vertical scale for all facets ranges from -10 to 10, representing a large variability of students' abilities. Each "x" represents 6.5 students on the map. A skewed normal distribution of students' abilities with the average at 3.0 logit indicates that many students could achieve the learning goals at the task and common attribute levels. The second vertical dash line shows that the three raters are placed closely on the map. The result confirms that the three raters consistently scored students' responses at task and common attribute levels. However, the third vertical dash line presents significant gaps among the estimates of the three tasks, indicating that the T2 task is the most challenging, followed by the T6 task and T1 task. The common attribute column (the fourth vertical dash line) indicates the challenge of the four common attributes from complex to easy: Com4 (mechanistic reasoning), Com3 (principle utilizing), Com2 (evidence noticing), and Com1 (claim building). In the fifth vertical dash line, the range of estimates for the rater-task interaction is between -1.0 and 1.0, which is narrow on the map. The Chi-square test of parameter equality for the rater-task interaction is $\chi^2(4) = 5.47$, $p = 0.242 > 0.05$, indicating there is no interaction between rater and task with a small effect that could not be considered.

However, the interaction between the rater and common attribute is significant as the Chi-square test of parameter equality for the interaction is $\chi^2(6) = 43.94$, $p < 0.001$. The right column (the sixth vertical dash line) also shows a wide range of interaction estimates, indicating rater facet affects the variability of the estimates of the four common attributes.

13.6 Discussion

13.6.1 The Functionality of Scoring Rubrics for Eliciting Knowledge-in-Use

As many previous studies have discussed the challenges of scoring students' responses to knowledge-in-use assessment tasks (e.g., Harris et al., 2006; He et al., 2023; Krajcik, 2021), this study provides a potential solution for designing scoring rubrics for three tasks (AID, CSE, and DUM). With the framework of sensemaking (e.g., Chen et al., 2013; Kapon, 2017; Krist et al., 2019; Schwarz et al., 2017) and the elements of scientific practices (e.g., Chen & Terada, 2021; Gotwals et al., 2013; Zangori et al., 2017), we constructed the rubrics with three-level hierarchical structure: criteria, elements, and common attributes, respectively. Compared to other types of rubrics, our rubric system takes multiple perspectives to understand students' performance on knowledge-in-use assessment tasks comprehensively. In addition, our rubric design approach has addressed Jescovitch et al. (2021) concerns about the transformation of scores among the three hierarchical levels. The strategy we used makes it possible that the original scores at the criteria level could be transformed into the element level scores to evaluate knowledge-in-use performance. We believe our rubric system could be helpful for teachers and researchers for many purposes, including diagnosis of students with insignificant weaknesses, evaluation of student quality of usable knowledge with practices, and comparison of student's performance on multiple types of knowledge-in-use assessment tasks.

13.6.2 The Usability of MFRM for Measuring Knowledge-in-Use

As recommended by previous research (e.g., Boone et al., 2016; Chi et al., 2021; Zhai, 2022), this study demonstrates the usability of MFRM to handle the estimations of students' performances on knowledge-in-use assessment tasks using NGSA as an example. To reduce raters' workloads on scoring, we assigned three raters with coding books containing unique cases for a specific rater and overlapping cases shared by the three raters. Our study suggests that the MFRM makes it possible to examine raters' scoring effects and compare raters' severity. Moreover, construct irrelevant variances resulting from rater severity were eliminated when estimating students' abilities so that the variability of raters' scoring on complex assessment tasks could be accounted. The scoring process would be eased due to the time and cost savings because it is not necessary to constantly check the IRRs among raters and retrain them, as long as they have a solid and comprehensive understanding of scoring rubrics. Moreover, this study extends the existing studies of scoring rubrics (e.g., Jescovitch et al., 2021; Wang et al., 2021) that used MFRMs to meet the purpose of the estimations regarding students' performances on criteria, elements, and common attributes. As for the scores on criteria with multiple raters, this study employed many-facet dichotomous Rasch models. With the composite scores on elements, this study conducted many-facet partial credit Rasch models, which could estimate the appropriateness of multiple score points for evaluating knowledge-in-use quality. Most importantly, the MFRM provides sufficient information about the psychometric properties of knowledge-in-use assessment tasks from different perspectives, such as fit statistics of rater effect, criteria, elements, and common attributes, and the Wright maps consisting of all facets. The information could help revise the scoring rubrics for the knowledge-in-use tasks.

13.6.3 The Variability of Essential Facets for Demonstrating Knowledge-in-Use

The main findings of using MFRM provide helpful information about students' knowledge-in-use performances on NGSS tasks. First, interaction estimations show that rater severity matters in estimating students' abilities, indicating that rater variability cannot be ignored even if the raters' severity is consistent. Second, we found that the T2 task (constructing scientific explanations) is the most challenging task for students in our current sample, followed by the T6 task (developing and using models) and the T1 task (analyzing and interpreting data). Although generalizable conclusions need more evidence, we speculate that the T2 task provides much more information than the other two tasks, which use a greater cognitive load to accomplish T2. Besides, the difficulties of common attributes across the three NGSA tasks (see Table 13.8) might also interest the field. Students had the slightest difficulty building a claim and less difficulty finding patterns of data as evidence; however, they felt challenged in linking their claims and evidence using scientific principles and even more challenged in providing a chain of mechanistic reasoning processes to explain a given phenomenon or problem. When contextualizing those common attributes in each of the three NGSA tasks, the trends of the elements'estimations (see Table 13.5) are consistent with previous studies (analyzing and interpreting data, Chen & Terada, 2021; constructing scientific explanations, Gotwals et al., 2013; developing and using models, He et al., 2023). These trends can also be explained by the cognitive load theory (Chandler & Sweller, 1991; Paas et al., 2003; Sweller, 1994) in that the more information added to the common attributes/elements, the working memory's intrinsic cognitive load increases to make them even more challenging.

13.7 Limitations

Several limitations need further addressing. First, our sample was selected from a large online dataset without any information about school districts, the description of classrooms, the orientation of NGSS teaching, and student background. Besides, the sample was not selected randomly. By merging the datasets, we only found 240 students who competed in the three tasks on the online portal. The constraints of sample size and sampling method may impede the results of the MFRM analyses. Future work should ensure that students in the sample completed the NGSA tasks in NGSS-aligned classrooms (i.e., students had the opportunity to learn). In addition, a robust sampling technique should be applied to accurately generate the psychometric analysis of the assessment tasks and generalize the estimations of students' knowledge-in-use to a large population. Second, some criteria and elements with poor fit statistics should be reviewed and modified. Besides, when the original criteria scores were transformed into the composite scores for elements, the partial scores (1 point) for several elements were usually not necessary for evaluating the quality of elements in each task. Thus, the rubrics at the element level should be examined further.

13.8 Implications and Conclusions

Our study draws attention to researchers who are interested in developing rubrics for scoring assessment tasks with complex constructs, such as knowledge-in-use. Our findings suggest that a system of scoring rubrics with hierarchical structure can address different assessment purposes. The task-specific dichotomous rubrics with detailed criteria are particularly appropriate for diagnosing student learning need. However, some of the criteria do not help examine student proficiency. Thus, the criteria of task-specific polytomous rubrics should be scrutinized to meet the evaluation purposes. Moreover, task-generic polytomous rubrics make it possible to compare student performance across different types of knowledge-in-use tasks. Finally, we suggest selecting appropriate measurement models based on the rubrics and students' responses to constructed response assessment tasks. In this study, we employed the MFRM because we had more than two facets, item dependence, and interactions among facets.

In conclusion, this study contributes to the fields of science education and Rasch measurement by (a) designing a rubric system for scoring complex constructed response assessment tasks with multiple purposes; (b) demonstrating an exemplar case to employ MFRM analysis to assess complex constructed response assessments, such as knowledge-in-use assessment tasks; and c) providing psychometric evidence for improving the rubric system for eliciting students' knowledge-in-use performance.

Acknowledgements This study is supported by the National Science Foundation (Grant Numbers 2101104, 2100964, 2201068; DRL-1903103, DRL-1316874, DRL-1316903, DRL-1316908). Any opinions, findings, and conclusions or recommendations expressed in this material are those of the authors and do not necessarily reflect the views of the National Science Foundation.

Appendix: ConQuest Many-Facet Rasch Modeling Codes

MFRM analysis code (by ConQuest)	Annotation
Datafile Fulldata01.Dat; Format rater 5 responses 7–27 / Rater 5 responses 7–27 / Rater 5 responses 7–27! Criteria(21); Label << NGSA240.Nam; Set update = yes,warning = no; Model rater + criteria; Export parameters >> NGSA240.Prm; Export reg >> NGSA240.Reg; Export cov >> NGSA240.Cov; Estimate! Nodes = 10,stderr = full; Show parameters!Estimates = latent, tables = 1:2:4> > NGSA240.Shw;	The final model in the section of applying dichotomous MFRMs to analyze scores using task-specific dichotomous rubrics. See the main findings in Table 13.4 and Fig. 13.4.
Datafile Fulldata Recode012.Dat; Format rater 5 responses 7–16 / Rater 5 responses 7–16 / Rater 5 responses 7–16! Element(10); Label << NGSA240r.Nam; Set update = yes,warning = no; Model rater + element + rater*element + rater*element*step; Export parameters >> NGSA240r.Prm; Export reg >> NGSA240r.Reg; Export cov >> NGSA240r.Cov; Estimate! Nodes = 10,stderr = full; Show parameters!Estimates = latent, tables = 1:2:4> > NGSA240r.Shw;	The final model (model 4) in the section of applying partial credit MFRMs to analyze scores using task-specific Polytomous rubric. See the main findings in Table 13.5 and Fig. 13.6.
Datafile Fulldata Recode012Com.Dat; Format rater 5 task 7 responses 9–12 / Rater 5 task 7 responses 9–12 / Rater 5 task 7 responses 9–12! Com (4); Label << NGSA240r2.Nam; Set update = yes,warning = no;	The final model (model 2) in the section of applying partial credit MFRMs to analyze scores using task-generic Polytomous rubrics. See the main findings in Tables 13.8 and 13.9 and Fig. 13.7.

(continued)

MFRM analysis code (by ConQuest)	Annotation
Model rater + task + com + rater*task + rater*com + rater*com*step; Export parameters >> NGSA240r2.Prm; Export reg >> NGSA240r2.Reg; Export cov >> NGSA240r2.Cov; Estimate! Nodes = 10,stderr = full; Show parameters!Estimates = latent, tables = 1:2:4> > NGSA240r2.Shw;	

References

Bond, T. G., & Fox, C. M. (2015). *Applying the Rasch models: Fundamental measurement in the human sciences*. Lawrence Erlbaum Associate.

Boone, W. J., & Staver, J. R. (2020). *Advances in Rasch analyses in the human sciences*. Springer.

Boone, W. J., Townsend, J. S., & Staver, J. R. (2016). Utilizing multifaceted Rasch measurement through FACETS to evaluate science education data sets composed of judges, respondents, and rating scale items: An exemplar utilizing the elementary science teaching analysis matrix instrument. *Science Education, 100*(2), 221–238.

Brown, J. S., Collins, A., & Duguid, P. (1989). Situated cognition and the culture of learning. *Educational Researcher, 18*(1), 32–42.

Chandler, P., & Sweller, J. (1991). Cognitive load theory and the format of instruction. *Cognition and Instruction, 8*(4), 293–332.

Chen, Y.-C., & Terada, T. (2021). Development and validation of an observation-based protocol to measure the eight scientific practices of the next generation science standards in K-12 science classrooms. *Journal of Research in Science Teaching, 58*(10), 1489–1526.

Chen, Y., Irving, P. W., & Sayre, E. C. (2013). Epistemic game for answer making in learning about hydrostatics. *Physical Review Special Topics—Physics Education Research, 9*(1), 1–7. https://doi.org/10.1103/PhysRevSTPER.9.010108

Chi, S., Liu, X., & Wang, Z. (2021). Comparing student science performance between hands-on and traditional item types: A many-facet Rasch analysis. *Studies in Educational Evaluation, 70*, 100998.

Cohen, J. E. (1960). A coefficient of agreement for nominal scales. *Educational and Psychological Measurement, 20*(1), 37–46.

Crowder, E. M. (1996). Gestures at work in sensemaking science talk. *Journal of the Learning Sciences, 5*(3), 173–208.

Duncan, R. G., Krajcik, J. S., & Rivet, A. E. (Eds.). (2017). *Disciplinary core ideas: Reshaping teaching and learning*. NTSA Press.

Eckes, T. (2008). Rater types in writing performance assessments: A classification approach to rater variability. *Language Testing, 25*(2), 155–185.

Eckes, T. (2015). *Introduction to many-facet Rasch measurement: Analyzing and evaluating rater-mediated assessments* (2nd ed.). Peter Lang.

Finnish National Board of Education. (2016). *National core curriculum for basic education 2014*. Finnish National Board of Education.

Fleiss, J. L., & Cohen, J. (1973). The equivalence of weighted kappa and the intraclass correlation coefficient as measures of agreement. *Educational and Psychological Measurement., 33*, 613–619.

Fleiss, J., Levin, B., & Paik, M. (2003). *Statistical methods for rates & proportions* (3rd ed.). Wiley & Sons.

Gotwals, A. W., & Songer, N. B. (2013). Validity evidence for learning progression-based assessment items that fuse core disciplinary ideas and science practices. *Journal of Research in Science Teaching, 50*(5), 597–626.

Gotwals, A. W., Songer, N. B., & Bullard, L. (2012). Assessing students' progressing abilities to construct scientific explanations. In *Learning progressions in science* (pp. 183–210). Brill Sense.

Greeno, J. G., Collins, A. M., & Resnick, L. B. (1996). Cognition and learning. *Handbook of Educational Psychology, 77*, 15–46.

Harris, C. J., McNeill, K. L., Lizotte, D. L., Marx, R. W., & Krajcik, J. (2006). Usable assessments for teaching science content and inquiry standards. In M. McMahon, P. Simmons, R. Sommers, D. DeBaets, & F. Crowley (Eds.), *Assessment in science: Practical experiences and education research* (pp. 67–88). National Science Teachers Association Press.

Harris, C. J., Krajcik, J. S., Pellegrino, J. W., & DeBarger, A. H. (2019). Designing knowledge-in-use assessments to promote deeper learning. *Educational Measurement: Issues and Practice., 38*(2), 53–67.

He, P., Liu, X., Zheng, C., & Jia, M. (2016). Using Rasch measurement to validate an instrument for measuring the quality of classroom teaching in secondary chemistry lessons. *Chemistry Education Research and Practice, 17*(2), 381–393.

He, P., Chen, I., Touitou, I., Bartz, K., Schneider, B., & Krajcik, J. (2023). Predicting student science achievement using post-unit assessment performances in a coherent high school chemistry project-based learning system. *Journal of Research in Science Teaching, 60*(4), 724–760.

Jescovitch, L. N., Scott, E. E., Cerchiara, J. A., Merrill, J., Urban-Lurain, M., Doherty, J. H., & Haudek, K. C. (2021). Comparison of machine learning performance using analytic and holistic coding approaches across constructed response assessments aligned to a science learning progression. *Journal of Science Education and Technology, 30*(2), 150–167.

Kaldaras, L., Akaeze, H., & Krajcik, J. (2021). Developing and validating next generation science standards-aligned learning progression to track three-dimensional learning of electrical interactions in high school physical science. *Journal of Research in Science Teaching, 58*(4), 589–618.

Kapon, S. (2017). Unpacking sensemaking. *Science Education, 101*(1), 165–198.

Kim, E. M., Nabors Oláh, L., & Peters, S. (2020). A learning progression for constructing and interpreting data display. *ETS Research Report Series, 2020*(1), 1–27.

Krajcik, J. S. (2021). Commentary—Applying machine learning in science assessment: Opportunity and challenges. *Journal of Science Education and Technology, 30*(2), 313–318.

Krist, C., Schwarz, C. V., & Reiser, B. J. (2019). Identifying essential epistemic heuristics for guiding mechanistic reasoning in science learning. *Journal of the Learning Sciences, 28*(2), 160–205.

Leckie, G., & Baird, J. A. (2011). Rater effects on essay scoring: A multilevel analysis of severity drift, central tendency, and rater experience. *Journal of Educational Measurement, 48*(4), 399–418.

Linacre, J. M. (1989). *Many-faceted Rasch measurement*. MESA Press.

Liu, X. (2020). *Using and developing measurement instruments in science education: A Rasch modeling approach* (2nd ed.). IAP.

Liu, O. L., Rios, J. A., Heilman, M., Gerard, L., & Linn, M. C. (2016). Validation of automated scoring science assessments. *Journal of Research in Science Teaching, 53*(2), 215–233.

Mao, L., Liu, O. L., Roohr, K., Belur, V., Mulholland, M., Lee, H.-S., & Pallant, A. (2018). Validation of automated scoring for a formative assessment that employs scientific argumentation. *Educational Assessment, 23*(2), 121–138.

Mayer, K., & Krajcik, J. (2015). Designing and assessing scientific modeling tasks. In *Encyclopedia of science education* (pp. 291–297). Springer.

McNeill, K. L., Lizotte, D. J., Krajcik, J., & Marx, R. W. (2006). Supporting students' construction of scientific explanations by fading scaffolds in instructional materials. *Journal of the Learning Sciences, 15*(2), 153–191.
Ministry of Education, P. R. China. (2017). Chemistry curriculum standards for senior high school [普通高中化学课程标准]. People's Education Press.
Mislevy, R. J., Almond, R. G., & Lukas, J. F. (2003). A brief introduction to evidence-centered design. *ETS Research Report Series, 2003*(1), i–29.
National Research Council. (2012). *A framework for K-12 science education: Practices, crosscutting concepts, and core ideas*. The National Academies Press.
NGSA. (2022). *Next generation science assessment*. https://ngss-assessment.portal.concord.org
NGSS Lead States. (2013). *Next generation science standards: For states, by states*. The National Academies Press.
Nordine, J., & Lee, O. (Eds.). (2021). *Crosscutting concepts: Strengthening science and engineering learning*. National Science Teaching Association.
Odden, T. O. B., & Russ, R. S. (2019). Defining sensemaking: Bringing clarity to a fragmented theoretical construct. *Science Education, 103*(1), 187–205.
Organization for Economic Cooperation and Development. (2019). *PISA 2018 assessment and analytical framework*. OECD Publishing. https://doi.org/10.1787/b25efab8-en
Paas, F., Renkl, A., & Sweller, J. (2003). Cognitive load theory and instructional design: Recent developments. *Educational Psychologist, 38*(1), 1–4.
Pellegrino, J. W., & Hilton, M. L. (2012). *Education for life and work: Developing transferable knowledge and skills in the 21st century*. National Academies Press.
Penuel, W. R., Turner, M. L., Jacobs, J. K., Van Horne, K., & Sumner, T. (2019). Developing tasks to assess phenomenon-based science learning: Challenges and lessons learned from building proximal transfer tasks. *Science Education, 103*(6), 1367–1395.
Schwarz, C., Passmore, C., & Reiser, B. J. (2017). *Helping students make sense of the world using next generation science and engineering practices*. National Science Teachers Association.
Styck, K. M., Anthony, C. J., Sandilos, L. E., & DiPerna, J. C. (2021). Examining rater effects on the classroom assessment scoring system. *Child Development, 92*(3), 976–993.
Sweller, J. (1994). Cognitive load theory, learning difficulty, and instructional design. *Learning and Instruction, 4*(4), 295–312.
Wang, C., Liu, X., Wang, L., Sun, Y., & Zhang, H. (2021). Automated scoring of Chinese grades 7–9 students' competence in interpreting and arguing from evidence. *Journal of Science Education and Technology, 30*(2), 269–282.
Wright, B. D., & Stone, M. H. (1979). *Best test design*. MESA press.
Wu, M. L., Adams, R. J., Wilson, M. R., & Haldane, S. A. (2007). *ACER ConQuest version 2.0: Generalised item response modelling software*. ACER Press.
Yang, Y., He, P., & Liu, X. (2018). Validation of an instrument for measuring students' understanding of interdisciplinary science in grades 4-8 over multiple semesters: A Rasch measurement study. *International Journal of Science and Mathematics Education, 16*(4), 639–654.
Zangori, L., Peel, A., Kinslow, A., Friedrichsen, P., & Sadler, T. D. (2017). Student development of model-based reasoning about carbon cycling and climate change in a socio-scientific issues unit. *Journal of Research in Science Teaching, 54*(10), 1249–1273.
Zhai, X. (2022). Assessing high-school students' modeling performance on newtonian mechanics. *Journal of Research in Science Teaching., 59*, 1–41.
Zhai, X., Haudek, K. C., Stuhlsatz, M. A., & Wilson, C. (2020). Evaluation of construct-irrelevant variance yielded by machine and human scoring of a science teacher PCK constructed response assessment. *Studies in Educational Evaluation, 67*, 100916.
Zhai, X., He, P., & Krajcik, J. (2022). Applying machine learning to automatically assess scientific models. *Journal of Research in Science Teaching., 59 (10)*, 1765–1794.

Peng He, research assistant professor of science education in the College of Education at Michigan State University. His research interests include designing and testing learning environment with innovative technologies, classroom assessment, teacher professional learning, science measurement.

Xiaoming Zhai, assistant professor of science education and affiliated assistant professor of Artificial Intelligence, Director of AI4STEM Education Center at the University of Georgia. He is interested in developing innovative assessments and technologies to support science teaching and learning.

Namsoo Shin, associate research professor in the College of Education at Michigan State University with expertise in the research and design of individualized, customized learning environments and assessments to support student learning of all students. She is participating as a senior researcher in multi-institution, collaborative projects on assessment design and feedback, and 3-D learning integrated system and computational thinking.

Joseph Krajcik, the director of the CREATE for STEM Institute and the Lappan-Phillips Professor of Science Education and a University Distinguished Professor at Michigan State University. Joe served as president of the NARST, from which he received the Distinguished Contributions to Science Education through Research in 2010. In 2020, he was elected to the National Academy of Education and received the McGraw Prize for Innovation in Pre-K-12 Education and, in 2021, the International Society for Design and Development in Education Prize for Excellence in Educational Design. His research focused on designing and testing project-based learning environments to reform instruction and research student learning.

Chapter 14
Measuring Student Readiness to Learn College Chemistry: A Rasch Modeling Approach

Dennis L. Danipog, Nona Marlene B. Ferido, Rachel Patricia B. Ramirez, Maria Michelle V. Junio, and Joel I. Ballesteros

Abstract Despite the wealth of research on persistence and completion in science, technology, engineering, and mathematics (STEM) fields, little emphasis has been given to measuring student readiness to learn in college science courses (Wang X, Am Educ Res J 50(5):1081–1121, 2013). This chapter presents the results of a research project implemented in the Philippines that discusses an innovative approach to assessing the senior high school STEM students' understanding of chemistry concepts and skills using Rasch measurement. This approach determines student readiness in engaging with the general college chemistry course to inform teaching and learning.

Readiness level is considered as among the entrance behaviors of learners to a new teaching process (Yuksel & Yuksel, 2012). Measuring student readiness could be useful in evaluating the structure of the intended curriculum. Determining the alignment between high school offerings and academic expectations at the college level represents a promising step towards promoting greater student interest and entrance into postsecondary STEM courses (Wang, 2013). Thus, taking the step to

D. L. Danipog (✉) · M. M. V. Junio
National Institute for Science and Mathematics Education Development (NISMED), University of the Philippines Diliman, Quezon City, Philippines
e-mail: dldanipog@up.edu.ph

N. M. B. Ferido
Assessment, Curriculum and Technology, & Research Centre (ACTRC), Quezon City, Philippines

R. P. B. Ramirez
College of Education, University of the Philippines Diliman, Quezon City, Philippines

J. I. Ballesteros
Institute of Chemistry, University of the Philippines Diliman, Quezon City, Philippines

X. Liu, W. J. Boone (eds.), *Advances in Applications of Rasch Measurement in Science Education*, Contemporary Trends and Issues in Science Education 57,
https://doi.org/10.1007/978-3-031-28776-3_14

measure student readiness to learn college science courses could inform curriculum design. In this study, the students' readiness to learn college chemistry was assessed by looking at their understanding of prerequisite chemistry topics introduced in high school. Mastery of basic laboratory skills, students' emotion, attitude, self-efficacy, and some other factors were not included in the assessment of students' readiness.

In the Philippines, despite the implementation of the K to 12 basic education curriculum, national assessments show poor performance of high school students in science. This was evident from the previous results of the National Achievement Test (NAT), an annual national test given by the Department of Education (DepEd) for Grades 3, 6 and 10 (Benito, 2010; DepEd, 2020). In 2015, 1.5 million Grade 10 students obtained 46.56% mean percentage score (MPS) in science, a 1.76% increase from the previous score in 2014 (44.80%), but still way below the passing MPS set by DepEd which is 75% (Espinosa et al., 2013; Ocampo, 2016). The overall performance of Filipino students in science is also lagging behind the international standards. Based on the results of the PISA 2018, Filipino students did not perform well in science. Among the 79 participating countries, the Philippines ranked 78th in science (Schleicher, 2018). These alarming results may suggest that the preparation of students in pre-tertiary science education could be problematic, which may affect their performance later on when they take up college STEM courses. With the ongoing science curriculum refinements being done in the Philippines, an examination of the readiness of students to engage with the college science curriculum is essential in order to evaluate the degree to which the science curriculum in the K to 12 Program equips students with the necessary prerequisite knowledge and skills.

Considered as the central science, chemistry is linked to other science subjects. It serves as a foundational science, which is often a required course for STEM-related undergraduate courses (Cohen & Kelly, 2019). Despite its known significance, chemistry has been considered as one of the difficult subjects for high school students (Danipog & Ferido, 2011) and perceived as a difficult course by college students (Cracolice & Busby, 2015). Research reports have even highlighted that general chemistry is a challenging course in which college students receive a failing grade rate higher than other STEM courses (Cohen & Kelly, 2019).

Danipog and Ferido (2011) described chemistry as a very complex subject for students. While a great deal of its concepts can be observed at the macroscopic level, these concepts can only be explained at the sub-microscopic (particulate) level. They further explained that students' difficulty in learning chemistry arises from the failure of many students to make connections between the macroscopic and the sub-microscopic (particulate) levels of understanding matter.

High school chemistry in the Philippines plays an essential role because it introduces the students to both macroscopic and sub-microscopic representations of matter. According to Deters (2003), high school chemistry is important because it is a preparatory subject to college chemistry, and at the same time a general education course that provides another path for students to view the world and further interest them in science. With these purposes, Deters aimed to identify the knowledge and skills college professors perceive their students should have mastered before taking a freshman-level college chemistry course.

In a case study on a STEM intervention program that enhances college readiness among students, Lane et al. (2017) described college readiness as one of the most critical issues in modern higher education. The authors emphasized the importance of understanding readiness in the context of STEM disciplines as it is one of the most predictive factors to student success in college.

A study conducted by Cracolice and Busby (2015) reported a meta-analysis of 225 published and unpublished research studies which revealed that the average failure rate of students in general chemistry course was over 50% versus 33.4% for other STEM courses. According to the authors, students enter the general chemistry course lacking the necessary general chemistry content knowledge and procedural knowledge. Further, results of this study show that students' prior knowledge and their scientific reasoning ability at the onset of the course remain positively correlated to their performance on the first term General Chemistry Examination. In terms of scientific reasoning ability, this study found out that those students at risk of not completing the general chemistry course can be predicted by the scientific reasoning ability pretest. This implies that the development of students' scientific reasoning abilities should be a curricular goal for preparatory courses to college chemistry.

To measure the readiness of students to learn higher grade level concepts and skills, some studies have focused on measuring the prerequisite skills, concepts, and competencies of learners (ACTRC, 2017; Means et al., 2016; Wang, 2013).

Although the literature stresses the significance of general chemistry in college (Cohen & Kelly, 2019), there is scarcity of studies on investigating the readiness of students under the STEM-track curriculum to learn general chemistry at the collegiate level (Wang, 2013). This information justifies the need to measure student readiness in the context of chemistry. The following research question was answered by this study:

> Do students under the STEM-track curriculum enter the university with the level of conceptual knowledge and skills required in order to engage with the general college chemistry curriculum?

In this study, the readiness test was developed to determine whether the entry conceptual knowledge and skills of students are sufficient for them to engage in the general college chemistry curriculum of the University of the Philippines Diliman (UPD). The readiness test was a multiple-choice test used to assess students' knowledge of prerequisite concepts on selected topics essential in studying general college chemistry. The details of test development are discussed in the Methods section of this chapter.

14.1 Background on K to 12 Education Reform in the Philippines

Science education is perceived to have great significance in promoting national goals, and plays a vital role for the future of societies as it is linked to technology and industry. Many nations allot substantial resources to continuously find ways to improve the quality of science education.

In the Philippines, seven years before this research project was initiated, efforts had been directed towards the improvement of pre-tertiary education. From 1945 to 2010, pre-tertiary education covered Grades 1–6 for elementary level and first year to fourth year for secondary level. In 2011, the Department of Education initiated this pre-tertiary education reform known as the *Enhanced Basic Education Program* or the K to 12 Program. This program covered Kindergarten and 12 years of pre-tertiary education: six years of elementary education, four years of junior high school, and additional two years of senior high school. The rationale for this reform was to provide students with sufficient time for mastery of concepts and skills; develop lifelong learners; and prepare students for tertiary education, middle-level skills development, employment, and entrepreneurship (Official Gazette of the Philippine Government, n.d.; Southeast Asian Ministers of Education Organization—Regional Center for Education Innovation and Technology [SEAMEO—INNOTECH], 2015).

As part of the K to 12 program, new curricula in core subject areas were developed and implemented in schools across the country (Montebon, 2014). In science, the curriculum was designed to be student-centered and inquiry-based. The organization of the curriculum had also changed in that concepts and skills were revisited at each grade level with increasing depth, creating a 'spiral' progression. Through this new curriculum, DepEd aims to develop scientific literacy among students that will enable them to make judgments and decisions on the applications of scientific knowledge that may have significant impact in everyday life (DepEd, 2016a). Furthermore, the introduction of senior high school curriculum in the Philippines aims to prepare students for either further education or employment. In senior high school, students undertake a standard core curriculum and can choose from four different track subjects: academic; technical-vocational and livelihood (TVL); sports; and arts and design. In the academic track, students can choose from four strands: accountancy, business, and management (ABM); humanities and social sciences (HUMSS); science, technology, engineering, and mathematics (STEM); and general academic (Official Gazette, n.d.). In the STEM strand, senior high school students are expected to take advanced science and mathematics subjects to acquire knowledge and skills which will prepare them for college STEM courses.

Under this STEM strand are the General Chemistry 1 and 2 specialized subjects which focus on the composition, structure, and properties of matter; quantitative principles, kinetics, and energetics of transformations of matter; and fundamental concepts of organic chemistry (DepEd, 2016b). The contents are divided into four quarters in which two quarters are distributed to each of the general chemistry subjects (i.e., General Chemistry 1 covers Quarters 1 and 2 and General

Chemistry 2 covers Quarters 3 and 4). The major contents of these specialized subjects include—.

Quarter 1 – General Chemistry 1: matter and its properties; measurements; atoms, molecules, and ions; stoichiometry; gases;
Quarter 2 – General Chemistry 1: electronic structure of atoms; electronic structure and periodicity; chemical bonding (ionic and covalent bonds); organic compounds;
Quarter 3 – General Chemistry 2: intermolecular forces and liquids and solids; physical properties of solutions; thermochemistry; chemical kinetics; and
Quarter 4 – General Chemistry 2: chemical thermodynamics; chemical equilibrium; acid-base equilibria and salt equilibria; electrochemistry.

The General Chemistry 1 and 2 curriculum also features content and performance standards; and a total of 183 specific learning competencies pertaining to the subject contents.

Details presented in the following section generate information on the prerequisite concepts and skills necessary for the general college chemistry curriculum.

14.2 The Rasch Model and Measurement of Readiness

The use of Rasch (1960) modeling has been considered by many to be one of the best approaches for measurement (Boone et al., 2010; Liu, 2010; Sondergeld & Johnson, 2014). Today, many researchers have used Rasch modeling in developing tests and in obtaining student ability and item difficulty estimates (e.g., Areepattamannil & Khine, 2017; Boone & Scantlebury, 2006; Karim et al., 2014; Phillipson & Tse, 2007). This is because the Rasch model can address some measurement issues that the Classical Test Theory (CTT), which was ever prevalent in measurement history, fails to deal with. Data analysis using the Rasch modeling approach provides researchers with many techniques to evaluate assessment reliability and validity. It provides a deeper understanding of the instrument's strengths and weaknesses than the CTT's approach to measurement. It has tight standards for the data to be analyzed. It is not designed to fit the data, but instead is derived to define measurement (Wright, 1992). Therefore, it is an ideal model that addresses the requirements of scientific measurement (Boone et al., 2014).

In the Rasch model, the probability of a person's successful response in an item is expressed mathematically as follows (Wright & Stone, 1979):

$$P(X_{vi} = 1|\beta_v; \delta_i) = \frac{\exp(\beta_v - \delta_i)}{1 + \exp(\beta_v - \delta_i)}$$

In this equation, X_{vi} indicates the success or failure of person v on the item i. When $X_{vi} = 1$, it means success (or a correct response) on the item, and if $X_{vi} = 0$ this means failure (or an incorrect response) on the item. β_v is a person-parameter, representing the ability level of person v on the latent variable scale, and δ_i is the difficulty of the item which is on the same measurement scale as β_v.

Through this equation, the Rasch model explains that a student whose ability is equal to the item difficulty has 50% chance of answering the item correctly. If an item's difficulty is lower than a student's ability, then the student has greater than 50% chance of answering the item correctly. If an item's difficulty is higher than a student's ability, then the student has less than 50% chance of answering the item correctly. Further, the above equation representing the Rasch model defines a unit called "logits" for student ability and item difficulty estimates (Wright, 1977). This equation shows that the Rasch model enables comparison of the students' ability and the difficulty of the skills assessed by the test. Because each item assesses a specific skill, the interpretation of the results may also be extended to include the conclusion that a student whose ability is equal to an item difficulty is ready to learn the skill, which the item is assessing (Griffin, 2007). For this reason, it can be argued that the Rasch modeling approach to analysis is beneficial for determining the students' readiness in a particular learning area.

In recent studies, the Rasch model was used to investigate students' readiness for e-learning during the COVID-19 pandemic. A group of researchers in South East Asia used the Rasch model to assess any significant differences between students' gender, age, ethnicity, level of education, field of study and their readiness for an e-learning environment (Adams et al., 2021). They found that most students in higher education are ready for an e-learning mode of instruction, and that there were differences in students' readiness for e-learning based on their demographic profiles. Researchers from the Philippines also used the Rasch model to assess the Filipino higher education students' readiness for e-learning during the pandemic (Reyes et al., 2021). They argued that the Rasch model could give the expected response to items based on the assumption that the probability that a person will agree to an item is a logistic function of the difference between a person's level of e-learning readiness and the level of readiness express by the item in the questionnaire. They examined their data against these expectations using a variety of fit statistics.

Furthermore, the Rasch model was also used to investigate readiness in a different context. In a study that investigated the readiness of Internet users, Salman and Aziz (2015) applied the Rasch measurement model to analyze how ready the respondents are to be participants in a digital society. In this study, the variable map clearly summarizes the correlations between the prescribed tasks and the ability of the respondents involved in the study. This enabled a comparison of the respondents and items to better understand how prepared the respondents are to participate in a digital society. This study also showed that the Rasch model could offer a value-added platform in analyzing data that deals with the behavior of people.

These studies have reported the suitability of the Rasch measurement approach in analyzing data to examine readiness in different contexts. However, there is a lack of deeper investigation of how the findings on readiness from the Rasch analysis could be used to evaluate the structure of curriculum, which could inform teaching and learning. This chapter addresses this issue drawn from the findings of the Rasch measurement approach in determining student readiness to engage in college chemistry. Consequently, these findings were used to evaluate the structure of the chemistry curriculum.

14.3 Methods

14.3.1 Selection of Research Samples

The data for this study were collected from eight public mainstream secondary schools located in two cities of the National Capital Region (NCR) in the Philippines. These schools were selected based on the following criteria: (1) availability of the STEM program for senior high school students, (2) large population of senior high school STEM students, and (3) better location of and easy access to schools. The schools were invited to participate in this study upon permission from the Department of Education's (DepEd) Schools Division Offices and following formal approval for their participation in the research. Consenting school principals provided contact details of senior high school STEM classes, which enabled the researchers to invite them through their respective teachers. The consenting schools had classes, whose students consented to participate in the research.

To determine the readiness of students to learn general college chemistry, test data were collected from 459 Grade 11 students in 13 STEM classes from eight sample schools. The test was administered during the second semester of the school year 2018–2019. The average number of students in a class is 33. The sample students from these schools consisted of more males (54.8%) than females (45.2%). Some of these students obtained their junior high school (Grade 7 to Grade 10) education from other schools. Because of this, students may have had different learning experiences in chemistry. The sample students belong to a cohort that was exposed to the enhanced K to 12 science curriculum where spiral progression of concepts and skills was emphasized. This curriculum included General Chemistry subjects for the senior high school STEM program. In the spiral progression approach, the concepts and skills in the curriculum were presented with increasing levels of complexity from one grade level to another, with the intention of paving the way to a deeper understanding of core concepts. In this approach, the scope and sequence of the content of the curriculum are developed such that concepts and skills are revisited at each grade level with increasing depth. As more facts and principles on each topic are encountered, learning of knowledge, skills, values, and attitudes increases in depth and breadth, creating a metaphorical spiral. This approach is a significant change in the new curriculum since previously, science was taught by discipline (discipline-based) per grade level in secondary schools. For example, in Grade 8, students studied biology for the whole school year, in Grade 9, students took chemistry, and physics was taught at Grade 10. The intention of this change is to better equip school leavers to make a contribution as "scientifically, technologically, and environmentally literate and productive members of society" (DepEd, 2016a, p. 2).

14.3.2 Development of the Readiness Test in General College Chemistry

The following are the processes performed in the development of the test.

1. ***Review of general college chemistry curriculum and senior high school general chemistry curricula.*** Both the current general college chemistry 1 curriculum developed by the UPD Institute of Chemistry and the general chemistry 1 and 2 curricula for senior high school STEM program developed by the Philippine Department of Education (DepEd) were examined to identify key topics in chemistry that have prerequisite concepts and skills on essential topics in general college chemistry. After curricular examination, eleven key topics were identified—*matter, quantum theory, chemical periodicity, chemical bonding, molecular geometry, stoichiometry, thermochemistry, liquids, solids, and phase changes, gases, solutions, and acids and bases.* In each of these topics, learning competencies were identified using the general chemistry curriculum for senior high school STEM program. These topics and their competencies were used in the creation of the readiness test blueprint.
2. ***Development of a test blueprint.*** To structure the development of items for the readiness test, a test blueprint was created using the information from the curriculum review. A workshop was conducted to design the test blueprint. The workshop was attended by a number of experts in chemistry education—the researchers of this study, an associate professor in chemistry education, a chemistry education specialist, and a senior high school chemistry teacher to collaboratively determine the prerequisite concepts and skills of essential topics in general college chemistry. The final blueprint for the test was composed of identified prerequisite concepts and skills necessary to learn the topics in general college chemistry. After the development of a test blueprint, the number of multiple-choice items for each of the prerequisite concepts and skills was also identified. Sixty multiple-choice items were intended to be included in the test.
3. ***Construction of test items.*** The final test blueprint served as the basis for the construction of items. Multiple-choice test items were developed by the chemistry education experts using the prerequisite concepts and skills identified in the blueprint. The items were constructed to cover a range of difficulties.
4. ***Review of constructed test items.*** A series of meetings was conducted by the research team to review the test items. Once each item was written based on prerequisite concepts and skills, it underwent a review process considering the chemistry content of the items and their face validity. This process ensured that test items adhered to guidelines for best practice in objective item writing and these items incorporated appropriate language and accurate concepts. During the

review process, the research team discussed important points such as the knowledge and skills required to answer the item correctly, the difficulty, and the quality of the item. The necessity of general consensus on each of the items to be retained or removed from the test was always observed during the meetings.

5. ***Pilot test of items.*** After a series of review meetings, 60 items were piloted in 11 general chemistry laboratory classes at the UPD Institute of Chemistry. The purpose of the pilot test was to evaluate the performance of each item and to obtain item characteristics in order to develop a psychometrically sound test. Two hundred and thirty-three (233) University freshmen students who were newly enrolled in general chemistry classes took the test for pilot testing. A majority of these students took the STEM strand in senior high school.
6. ***Item calibration.*** Pilot test data were analyzed using the Rasch one parameter simple logistic model (Rasch, 1960) via *ConQuest* generalized item response modeling software (Wu et al., 2007) to determine its psychometric properties. Item characteristics such as fit indices, item characteristic curves, and difficulty indices were examined in the analysis. Based on the results, 12 items in the test were recommended for deletion since some of the items were too easy and some did not discriminate students well according to their ability.
7. ***Finalization of the test.*** The output of the Rasch analysis was used to finalize the test. After all the necessary revisions were done, the research team came up with the final readiness test in chemistry. The calibrated test was composed of 48 multiple-choice questions. This version of the test was used in the actual research data collection in schools. The content of the test with samples of key competencies and skills tested is shown in Table 14.1.

Some examples of the types of items in the test are shown in Fig. 14.1. It is important to note that due to the need for test security, only three items are presented here.

14.3.3 Administration of the Test in Sample Schools

The readiness test was administered to 459 Grade 11 STEM students in eight public secondary schools located in the National Capital Region of the Philippines. The test was administered during the second semester of the school year 2018–2019. Test administrators (an examiner and a proctor) gave the test in each Grade 11 STEM class for 1 h. Before the test started, the examiner explained briefly to the students the protocol and the purpose of administering the test. Each student received a test booklet, a scannable answer sheet, and a pencil during the test administration.

Table 14.1 Content of the readiness test in general college chemistry

Topic	Number of items	Key competencies/skills
1. Matter: Properties, composition, and changes	5	Differentiate pure substances from mixtures, compare the attractive forces of solid, liquid, and gas; recognize the formula of substances
2. Quantum theory and atomic structure	6	Recognize the atomic number and mass, atoms, molecules, and ions; determine the number of electrons from atomic number and mass; perform criss-cross method for writing ionic formula; use prefixes/suffixes for writing molecular formula; apply law of conservation of mass in chemical reactions
3. Chemical periodicity	2	Compare the atomic size and electronegativity of elements
4. Chemical bonding	3	Distinguish through the Lewis structure ionic and molecular compounds; determine if a compound is ionic or covalent; recognize that atoms which satisfy the octet rule have noble gas configuration
5. Molecular geometry	3	Use vector addition to determine polarity of a molecule; use electronegativity values to determine bond polarity; determine the number of valence electrons
6. Stoichiometry	14	Write correct chemical formula based on the fundamental laws of matter; balance chemical equations; use ratio and proportion to: (1) calculate number of moles or mass of reactants or products in chemical reactions and (2) determine the mass of a given number of moles of elements/compounds, and the number of moles given the mass of elements/compounds; explain the meaning of coefficients and subscripts in terms of the law of conservation of mass; determine the molar mass of an element or a compound; identify stoichiometric relationships among reactants and products
7. Thermochemistry	7	Identify the system and the surrounding in a diagram; interpret meaning of (+) or (−) sign of ΔH values; describe the energy involved in a reaction; recognize heat as energy in transit; differentiate heat from temperature; interpret potential energy diagram to differentiate endothermic from exothermic reaction
8. Liquids, solids, phase changes	4	Relate the polarity of water with its molecular structure; explain why molecules of liquids exhibit surface tension; relate boiling point of liquids with intermolecular forces of attraction
9. Gases	5	Predict and calculate the pressure, volume, or temperature of a gas under certain conditions of change; explain the assumptions of the kinetic molecular theory; explain gas laws using kinetic molecular theory
10. Solutions	7	Predict the change in boiling point or freezing point given concentration of solutions; calculate molarity of solutions; relate vapor pressure or osmotic pressure

(continued)

Table 14.1 (continued)

Topic	Number of items	Key competencies/skills
		with concentration of solutions; differentiate molality from molarity; express the mole fraction given molality of solutions
11. Acids and bases	4	Compare pH of solutions; describe dissociation of water molecules; distinguish acids from bases; relate pH of solutions with their hydronium and/or hydroxide ion concentrations
Total items	**48**	

14.3.4 Examination of Dimensionality of the Readiness Test

Three steps were used to examine the unidimensionality of the data: (1) checking for items with negative point-biserial correlations and (2) using Rasch fit indicators (Linacre, 1998), and (3) modified parallel analysis (Drasgow & Lissak, 1983).

Items with negative point-biserial correlations were deemed to be not behaving as expected and are therefore removed from the test. The unexpected behavior of these items could be explained by some underlying trait apart from that which is measured by the test. The Rasch fit indicators may also provide evidence on possible multidimensionality of the data. The outfit and infit statistics were used for this purpose. Outfit is based on the unweighted squared residual between observed and expected. Infit is based on the weighted squared residual between observed and expected. The modified parallel analysis involved the following steps: (1) parameter estimates were obtained from a real data, (2) item parameter estimates were used to simulate a truly unidimensional response data, (3) tetrachoric correlations were computed, the largest off-diagonal correlations were used as communalities and the eigenvalues of the reduced correlation matrix were obtained, and (4) corresponding eigenvalues from the real and simulated data were compared. The data is considered multidimensional if the second eigenvalue from the real data is significantly larger than the simulated data.

14.3.5 Determination of Student Ability and Item Difficulty Estimates

To determine the student ability estimates in the test, test data were analyzed using the Rasch model. This analysis also gave the item difficulty estimates. When student raw scores are converted into the logits scale, the student ability estimate is produced. Based on the Rasch model, this estimate represents the students' true ability. The measure of how difficult the items are for the students is represented by the item difficulty estimate. This estimate is also expressed in logits scale and is derived from

26. The combustion reaction of butane is

$$2\ C_4H_{10}(g) + 13\ O_2(g) \rightarrow 8\ CO_2(g) + 10\ H_2O\ (l)$$

How many grams of CO_2 is produced from the combustion of 1.00 mole C_4H_{10}? (Molar mass: CO_2 = 44 g/mol)

A. 11.0 g CO_2
B. 44.0 g CO_2
C. 176 g CO_2
D. 352 g CO_2

27. The setup below shows a coffee-cup calorimeter to measure heat change in a solution. The components of the calorimeter are labelled A, B, C, and D. In the setup, which letter refers to the reaction system?

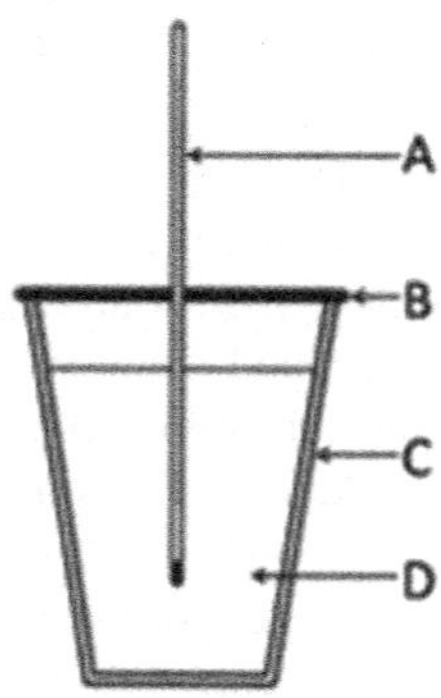

28. Which of the following is an example of a molecule?

A. Cu
B. CO
C. Cl^-
D. C

8

Fig. 14.1 Sample items from the readiness test

the proportion of students who answered each of the items correctly. These two estimates were compared in the Rasch model.

Student test data were cleaned prior to analysis. Test items were calibrated to examine the psychometric properties of the test and to identify which items work well, in other words, which items have good fit to the Rasch model. These items were identified by examining the results of calibration particularly the item fit indices (weighted mean square [MNSQ] fit), item difficulty estimates, and item discrimination indices. After item calibration, 12 items were identified that did not fit well to the Rasch model. These items had low discrimination indices (below the recommended index of 0.2) and high fit indices (greater than the recommended of 1.0) and were then removed from the 60-item dataset. In addition, the stems and distractors of each of these 12 items were analyzed to gain insight on why the items did not fit well to the model. One item was too easy, others too difficult, and still others were found to be confusing for the students. The distractor analysis revealed that these items involved some other factors that might have prevented students from accurately manifesting their level of ability when they responded to the items. By examining the stems and the distractors, it was revealed that some items required several prerequisite concepts and skills. Items requiring calculation involved at least two steps, making them too difficult and taxing. For example, one item required students to solve for the number of moles of a reactant first, then use this value to calculate the amount of the substance in grams. In another, the number of moles had to be solved first, before the molarity of the solution was computed. Hence, instead of including items which tested these skills in combination, it was decided to just include those items which tested these skills individually. Aside from considering the similarity of these items with those which were retained, the decision to remove such items also depended on whether or not these nonfunctioning items involved several prerequisite concepts. For example, one item asked students to identify the compound having the highest boiling point based on their structural formula. For students to get the correct answer, they should have a clear understanding on how intermolecular forces of attraction operate in different kinds of molecules, as well as on how these forces differ in terms of strength, which would then give clues to the boiling point of the substances. It was important that only those items which would clearly show how students differ in terms of their ability to understand and correctly respond to test items for specific chemistry concepts and skills were included in the final data analysis. With this, there were 48 calibrated items in the readiness test used for data analysis. This analysis generated the student ability and item difficulty estimates in the test.

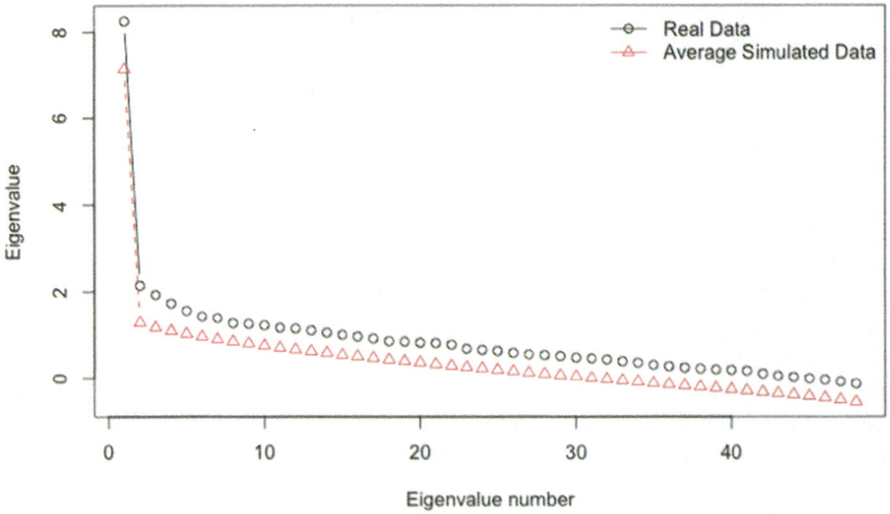

Fig. 14.2 Modified parallel analysis eigenvalue plot derived from actual data and simulated data based on the assumed model (Rasch)

14.3.6 Determination of Difficulty Order in the Readiness Test

The benefit of using the Rasch model in the analysis is that the students and items can be placed on the same scale. Because of this, the Rasch model enables the comparison of students with the concepts and skills assessed by the test items, which is shown in the variable map. The variable map (Fig. 14.2) is presented and discussed on page 15. In the variable map, when the item numbers are linked with the concept or skill assessed by the item, the horizontal alignment identifies which concepts or skills the aligned students are ready to learn.

The difficulty order was determined using the variable map, which was generated after item calibration. The two estimates (student ability and item difficulty) were used to group the readiness test items that assess sets of skills into a similar level of difficulty. After grouping the items, four different groups were identified in the map and each group contained similar skills. Each group was assigned a level (Levels A, B, C, and D). In each level, descriptions of the skills being assessed by the items were provided to encapsulate the main ideas that students in each level are ready to learn. The difference in the difficulty of the items was shown by the cognitive demand of knowledge and skills being assessed by the items. The categories of cognitive demand for science (Blank et al. 2001; Porter & Smithson, 2001) were used as guide to show this difference in difficulty of the items: (1) memorize facts/definitions/formulas, (2) communicate understanding of science concepts, (3) analyze information and advance scientific arguments, and (4) apply concepts/make connections.

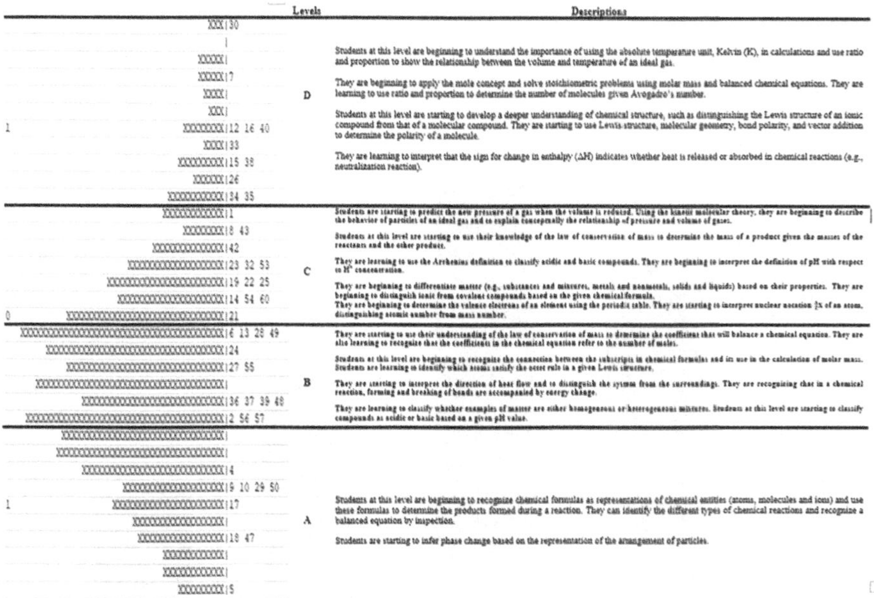

Fig. 14.3 The variable map of student and item distribution with skill level descriptions

The levels were defined by identifying the cut-off points. The cut-off points were determined based on two main considerations: (1) similarity between clusters of items and (2) the resulting width of the levels. For the first consideration, the underlying skills/abilities were inferred from the items in the cluster. Items near the boundary between clusters are grouped under a level that is more consistent with the item in terms of the underlying skill. This step is highly dependent on expert judgement (Griffin, 2007). The second consideration was to ensure that the resulting width of the levels were enough to accommodate fluctuations in the estimates due to error. This was done by seeing that the widths are at least twice the average amount of error associated with the difficulty of the items in the level. The whole process of determining the cut-off points was done iteratively until the best combination of levels and descriptions as well as the width were achieved. The items and the students whose ability fall within a level's cut-off point were assigned to that level. The students assigned to a level were considered ready to learn the concepts and skills in that level.

Using these levels, the percentages of students who were ready and not ready to take general college chemistry were then determined. The levels with descriptions (Fig. 14.3) and the percentage of students in each level (Fig. 14.4) are presented on pages 18 and 19 in the results section of this paper.

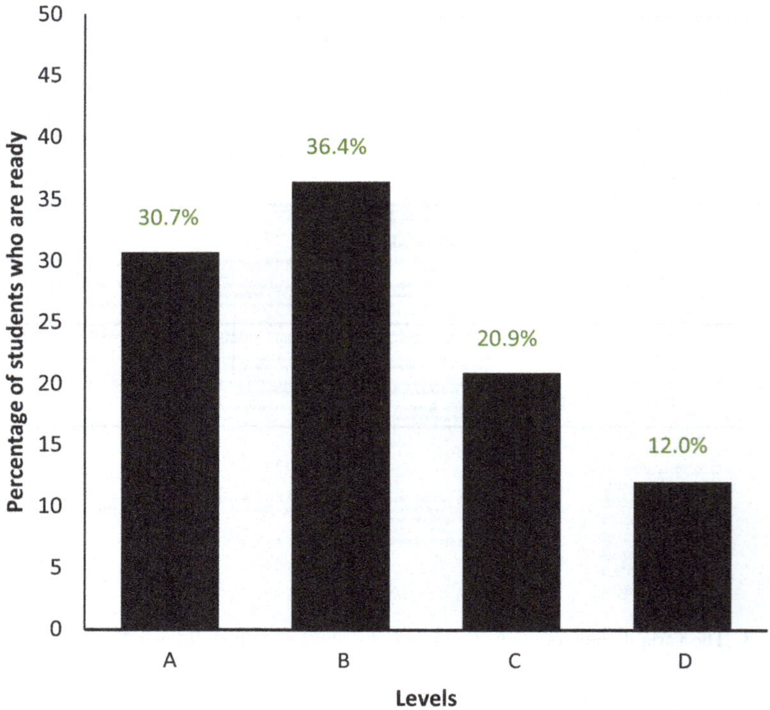

Fig. 14.4 Distribution of Grade 11 STEM students across levels

14.4 Results

14.4.1 Results of the Rasch Analysis of the Readiness Test in General College Chemistry

The item statistics for the readiness test in general college chemistry is presented in Table 14.2. Test items were examined for item difficulty (item delta) and item fit (weighted mean squared [MNSQ] fit). The item fit indices for all items were within the acceptable range of 0.8–1.2 logits (Wright & Linacre, 1994). This means that the items were found to fit the model, showing that they measure the same construct. The Rasch analysis codes is located in Appendix A.

The expected a posteriori/plausible value (EAP/PV) item separation and the weighted maximum likelihood estimation (WLE) person separation reliabilities were acceptable for the test as shown in Table 14.3. Values greater than 0.70 are considered acceptable (Grigg & Manderson, 2016; Yang, He, & Liu, 2017). This means that the test performed well in having items with different levels of difficulty to discriminate between students belonging to different levels, thus providing support for the reliability of the test.

Table 14.2 Item statistics for the readiness test in general college chemistry

Item	Item difficulty (Item delta)	Error	Item fit (Weighted MNSQ)
1 (1)	0.506	0.104	1.02
2 (2)	−0.529	0.098	1.03
3 (4)	−0.836	0.099	0.94
4 (5)	−1.473	0.109	0.93
5 (6)	−0.059	0.098	1.02
6 (7)	1.217	0.120	0.98
7 (8)	0.485	0.104	1.04
8 (9)	−0.929	0.100	0.93
9 (10)	−0.868	0.100	0.96
10 (12)	1.033	0.115	0.95
11 (13)	−0.092	0.098	1.00
12 (14)	0.081	0.099	0.99
13 (15)	0.831	0.110	1.06
14 (16)	1.033	0.115	0.93
15 (17)	−0.949	0.101	0.97
16 (18)	−1.118	0.103	0.91
17 (19)	0.170	0.100	1.09
18 (21)	0.017	0.098	1.00
19 (22)	0.211	0.100	1.00
20 (23)	0.313	0.101	1.05
21 (24)	−0.209	0.097	1.02
22 (25)	0.201	0.100	0.94
23 (26)	0.679	0.108	1.03
24 (27)	−0.288	0.097	1.06
25 (28)	−0.095	0.098	0.96
26 (29)	−0.927	0.100	0.94
27 (30)	1.558	0.133	1.03
28 (32)	0.272	0.101	0.99
29 (33)	0.897	0.112	0.96
30 (34)	0.601	0.105	1.10
31 (35)	0.593	0.105	1.04
32 (36)	−0.419	0.097	0.91
33 (37)	−0.476	0.097	0.85
34 (38)	0.770	0.109	1.01
35 (39)	−0.479	0.098	0.92
36 (40)	0.952	0.113	1.04
37 (42)	0.369	0.102	1.05
38 (43)	0.420	0.103	1.03
39 (47)	−1.125	0.103	1.04
40 (48)	−0.473	0.097	0.96
41 (49)	−0.104	0.098	1.06
42 (50)	0.898	0.101	1.03

(continued)

Table 14.2 (continued)

Item	Item difficulty (Item delta)	Error	Item fit (Weighted MNSQ)
43 (53)	0.291	0.101	0.99
44 (54)	0.094	0.099	1.06
45 (55)	−0.278	0.097	0.99
46 (56)	−0.514	0.098	0.96
47 (57)	−0.573	0.098	1.03
48 (60)	0.117	0.099	1.07

Table 14.3 Item and person separation reliabilities for the readiness test in general college chemistry

EAP/PV item separation reliability	WLE person separation reliability
0.800	0.809

14.4.2 Results of the Examination of Dimensionality

The final set of items used in the analysis did not include items that have negative point-biserial correlation (see Appendix C). Items that did not perform well were already removed in the earlier iterations of the analysis. Based on point-biserial correlation, the items in the final analysis did not include any that indicated possible multidimensionality in the data.

The outfit and infit statistics of the items (see Appendix D) indicate acceptable fit of the data to the Rasch model (range of 0.8–1.2). Adjusting for the sample size (Smith et al., 1998), the range of acceptable fit statistics are 0.72–1.28 and 0.91–1.09 for outfit and infit, respectively. Using this adjusted range, items 34 and 38 are flagged as problematic. Item 34 has an infit statistic that is above the acceptable range while item 38 is below that range. Despite these, the overall MNSQ for both outfit and infit are within the acceptable range. Thus, there is sufficient reason to proceed with the analysis using this data within the Rasch framework.

Visual inspection of the modified parallel analysis eigenplot value (see Fig. 14.2) shows a single dominant factor in the data. The pattern is similar to what was shown in Dragow and Lissak (1983) as indicative of a sufficiently unidimensional data that allows for the use of Rasch model. Modified parallel analysis, however, requires that, in order to rule out multidimensionality, the second eigenvalue for the real data must be shown to be not substantially larger than the second eigenvalue of the simulated data. Testing the difference between the second eigenvalues shows that there was a statistically significant difference between real data and simulated data based on the assumed model (see Table 14.4). This means that although there is a clearly dominant factor that could account for a unidimensional data, a second factor may well be present in the data and could warrant further investigation.

Although the point-biserial correlations and the fit statistics provided evidence that the data fit well with the Rasch model, thus demonstrating that the requirements

Table 14.4 Modified parallel analysis test statistics

Statistic	Value
Second eigenvalue in the observed data	2.1437
Average of second eigenvalues in Monte Carlo samples	1.2951
Monte Carlo samples	100
p-value	0.0099

for the use of Rasch model were met satisfactorily, the modified parallel analysis test indicated that examination of multidimensionality in the data is warranted. Given this result, it may still be said that a meaningful analysis may still be done within the Rasch framework. It has already been shown that unidimensional IRT models are quite robust to violations on the unidimensionality assumption (Anderson et al., 2017), thus the estimates may still be used for drawing inferences from the difficulties of the items and abilities of the students.

14.4.3 Results of the Student Readiness Test in General College Chemistry

Figure 14.3 shows the variable map generated from the Rasch analysis of the student test in general college chemistry. The left side of the map shows the distribution of Grade 11 STEM students where they are represented by 'X' (where X = 0.7 cases). The right side of the map shows the item number based on the test order. This map shows that item 30 is the most difficult item and item 5 is the easiest item. It is important to note that the item numbers are linked with the chemistry concept or skill assessed by the item. When an item is positioned next to a student, the student has a 50/50 chance of answering that particular item correctly. Thus, the horizontal alignment identifies which concepts or skills the aligned students (as represented by 'X's) are ready to learn. Furthermore, a description for each level (A, B, C, and D) was written to encapsulate the main ideas in chemistry that students in each level are ready to learn. Refer to Appendix B for the descriptions for each level.

There are four levels—A, B, C, and D that conceptually and empirically resulted on the basis of the Rasch analysis. These levels are arranged in an order based on the results of student test, from the easiest level (level A) to the most difficult level (level D). The levels contain descriptions of the prerequisite concepts and skills for learning general college chemistry. Some of these concepts and skills were taken by the students in junior high school chemistry (Grade 7 to Grade 10). Based on the analysis, the students assigned to each level are considered "ready to learn" the concepts and skills associated with the corresponding level.

Distribution of Students Across Levels

To identify the percentage of students who are well prepared to engage with the conceptual knowledge and skills required by the general college chemistry curriculum, the distribution of students across levels is presented in Fig. 14.4. This figure shows that only 12% of sample students are ready to learn the conceptual knowledge and skills required by the general college chemistry curriculum, which is shown at level D. This indicates that a large number of Grade 11 STEM students involved in this study, comprising 88% of the sample population, have struggled with the concepts and skills necessary to learning general college chemistry. It is important to note that students need to have the mastery of all knowledge and skills assessed at the four levels for them to be considered ready for college chemistry. It can then be argued that most students may not have sufficient prior knowledge and skills for understanding general college chemistry.

Levels A, B, and C indicate the prerequisite concepts and skills for general college chemistry. Figure 14.3 shows that majority of the students are located in these levels. Since students in levels A, B, and C are "ready to learn" these concepts and skills, they have not yet mastered the prerequisites for general college chemistry. Based on the map, these students are distributed across the levels, so some have mastered more of the assessed concepts and skills than the others in these levels. Based on the concepts and skills identified by the research team and chemistry education experts, the concepts and skills in level D are beyond those identified as being prerequisite knowledge. These are topics that are within the curriculum of general college chemistry, but not pre-requisite for the latter. This implies that students in level D (12.0%) are adequately prepared to engage with the general college chemistry curriculum. They already know the prerequisite concepts and skills in levels A, B, and C.

The most difficult level D contains topics that are more complex than those in levels A, B, and C. The complexity of the topics can be attributed to the number of basic concepts needed prior to understanding these topics. For instance, stoichiometry can be easily understood if one has prior knowledge of the concept of mole, balancing chemical equations, chemical formulas, and molar mass. Gas problems can be solved easily if one has prior knowledge of gas laws, as well as the concept of pressure, temperature, mole, and volume. The polarity of a compound can be easily determined if one knows how to draw a Lewis structure, determine molecular geometry, identify bond polarity, and apply vector addition. Based on the results of this study, some students are still in the process of mastering the prerequisite concepts, which makes it difficult for them to understand the level D topics. Poor understanding even of only one of the prerequisite concepts can make the understanding of the level D topics difficult for students.

Students in level A (30.7%) are ready to learn the following concepts and skills: recognizing chemical formulas as representations of chemical entities and using formula to determine the products formed during a reaction, recognizing a balanced chemical equation by inspection, and inferring phase change based on the representation of the arrangement of particles.

Those in level B (36.4%) are beginning to learn concepts, such as classifying matter as homogeneous or heterogeneous mixtures, acidic or basic based on pH values, using the law of conservation of mass to determine the coefficient that will balance a chemical equation, and recognizing that these coefficients refer to the number of moles, using subscripts in chemical formulas to calculate molar mass, identifying atoms that satisfy the octet rule in a Lewis structure, interpreting heat flow in a system, and recognizing that forming and breaking of bonds are accompanied by energy change. It should be noted that students in Level B have mastered the concepts and skills in level A. They already know the prerequisite concepts and skills identified in level A.

Students in level C (20.9%) are just starting to learn concepts, such as determining the valence electrons of an element using the periodic table, interpreting nuclear notation *ZAX* of an atom, distinguishing atomic number from mass number, differentiating matter based on their properties, distinguishing ionic from covalent compounds based on the given chemical formula, using the Arrhenius definition to classify acidic and basic compounds, interpreting the definition of pH with respect to H^+ concentration, using the law of conservation of mass to determine the mass of a product, and using the kinetic molecular theory to describe the behavior of an ideal gas, and to explain conceptually the relationship of pressure and volume of gases. Students in level C have mastered the concepts and skills in levels A and B. This cohort of students knows the prerequisite concepts skills identified in levels A and B.

In summary, these results indicate that a vast majority of Grade 11 STEM students involved in this study are not yet well prepared to engage with the conceptual knowledge and skills required by the general college chemistry curriculum. Without the prerequisite skills, they will not be able to fully understand the lessons presented to them as part of their general college chemistry experience.

14.5 Discussion

A college-ready student is one who "can cope with the content knowledge presented in a college course and can take away from the course the key intellectual lessons and dispositions the course was designed to convey and develop" (Conley, 2007, pp. 5–6). The dismal performance of a considerable number of senior high school STEM students involved in this study in the readiness test for general college chemistry is a clear indication of the difficulty they are going to experience in coping with the content knowledge and skills required of them later.

Figures 14.3 and 14.4 show that the 12% of sample students at Level D are those who are ready to engage with the conceptual knowledge and skills required by the general college chemistry curriculum. These students can understand the chemistry concepts, which require some skills in mathematical problem solving and in interpreting and applying structural representations (like the Lewis structure).

Students at this level possess both knowledge of chemistry content and essential cognitive skills. According to Conley (2007), these two (key cognitive strategies and content knowledge), are important dimensions from which successful academic preparation for college is rooted. In an earlier study, Tai et al. (2005) suggested that an emphasis on understanding in high school chemistry courses may contribute to future success of students in introductory college chemistry.

Consequently, at this level, 88% of the sample population can be considered not ready and are then expected to face difficulty in coping with the essential content knowledge and skills of the general college chemistry course. The poor performance demonstrated by this significant number of students in the readiness test may have stemmed from their lack of understanding of some basic chemistry concepts like chemical formulas, balanced chemical equations, mole concept and molar mass. This deficiency translated to their inability to interpret symbolic representations and to solve problems involving gases and stoichiometry. Conceptual understanding involves being able to represent and translate chemical problems using three forms of representation: macroscopic, particulate, and symbolic (Bowen & Bunce, 1997). This lack of understanding of basic concepts resulting to learning difficulties in more complex chemistry content knowledge and skills have been reported in the literature (Nurrenbern & Pickering, 1987; Gabel & Bunce, 1994; Herron, 1996; Tai et al., 2005). As Herron (1996) emphasizes, "If students are having trouble solving problems, the first thing to check is their understanding of the concepts of the problem" and their level of understanding should extend beyond the simple ability to use words to describe the concept (54, p. 103).

Problem solving skills are specifically important for students to engage with the quantitative problems of chemistry. However, problem solving is the most difficult part for many chemistry students (Bowen & Bunce, 1997). In this study, three topics (stoichiometry, gases, and solutions) have items involving problem solving that require students to do some calculation. The first two (stoichiometry and gases) will be used as examples to illustrate how the research findings were utilized to gain insight on the alignment of the content knowledge and skills possessed by students and those expected of them when they enter college.

Chemistry 16 (General Chemistry I), an introductory college chemistry course involves the study of the basic principles of chemistry and its applications. Focusing only on the two topics involving problem solving mentioned earlier, the Course Guide specifies the following learning outcomes: (1) apply the basic laws of chemistry to determine mass relationships in chemical formulas and chemical reactions; and (2) solve basic problems involving gases.

Specific problem solving skills expected of students relevant to stoichiometry and gases are the following: (1) perform conversions between number of molecules/ions/formula units/atoms, number of moles, and mass; (2) convert between moles and mass; (3) use molar ratios to convert between amounts of products or reactants; (4) determine changes in pressure, volume, temperature, and number of moles according to different gas laws. These expectations from the college chemistry curriculum assume that students have prior knowledge on essential prerequisite concepts such as the mole concept, molar mass, chemical formulas, and balanced chemical equations. When solving stoichiometry problems, students are expected to have already developed problem solving strategies, in addition to sound understanding of these more basic concepts.

The misalignment of these expectations with the content knowledge and skills of students entering tertiary education is evident in the number of students who did not make it to Level D. Figure 14.3 shows the level A and B descriptions that summarize what students are ready to learn about chemical formulas, an essential prerequisite concept for determining molar mass, for balancing chemical equations and for solving stoichiometric problems. It can be observed that at Level B, students are beginning to recognize the connection between the subscripts in the chemical formulas and its use in the calculation of molar mass. It should be noted that at this level B, only 36.4% are ready to learn these concepts and skills. Consequently, about two-thirds of the sample involved in the study are not ready to engage in classroom activities involving these topics. In addition, Level A description on chemical formulas states that at this level, students are beginning to recognize chemical formulas as representations of chemical entities (atoms, molecules, and ions), and use these formulas to determine the products formed during a reaction. These concepts and skills are required for students to make meaningful interpretations of chemical equations. In terms of the number of students found to be at this level, it is alarming that compared to those in Level B, an even smaller percentage make up this group of students, only 30.7%. Hence, about 70% are not ready to engage with the topic on chemical formulas.

It should be noted that in order to engage with the general college chemistry curriculum, students are expected to be able to know and do the following in relation to the topic, Chemical Formulas: (1) determine the empirical formula of a compound; and (2) determine the molecular formula of a compound. Hence, it is assumed that when students tackle these topics at the tertiary level, students are already expected to have a clear understanding of a chemical formula as an expression showing the elements that make up a compound, as well as the relative proportions of those elements. Considering the number of students at Levels A and B, a college instructor teaching the basics of stoichiometry would then be faced with a significant number of students with very limited understanding about the prerequisite concepts, and who are not ready to engage with the topic at hand. In dealing with problems on stoichiometry, students might be able to use algorithmic problem solving even when their understanding of the relevant concepts are lacking (BouJaoude & Barakat, 2003). It is interesting to note that chemistry is often taught in senior high school using the lecture format. Teachers who lecture emphasize problem solving and algorithmic calculations that do not address conceptual understanding (Salame & Casino, 2021).

Aside from knowledge of chemical formulas as a prerequisite for solving problems in stoichiometry, the mole concept is another essential topic that students find difficult to understand. As Kolb (1978) argued, "*there is probably no concept in the entire first-year chemistry course more important for students to understand than the mole and one of the main reasons the mole concept is so essential in the study of chemistry is stoichiometry*" (p.728). The difficulty of students in applying the mole concept in solving stoichiometric problems stems from their lack of a scientific conception of mole (Gabel & Bunce, 1994). In a later study, the difficulty experienced by students in handling this concept was reported by Furio et al. (2002). They conducted a review of the literature on the learning and teaching of the concepts "amount of substance" and "mole". They reported that a "*clear discrepancy exists between what is assumed as correct by the scientific community and the thinking of educators*" (p. 287). Hence, in a general college chemistry classroom, an instructor presenting a problem on stoichiometry, like one requiring students to convert from number of moles to mass, will encounter more than three-fourths of the class confused and possibly clueless about what mole is and how it can be used to solve the problem presented. This group of students make up the 88% of the sample who are not ready to learn the concepts and skills described in Level D (Fig. 14.3).

The topic of gases and their related properties (pressure, volume, amount of substance, and temperature) also require problem solving skills and basic mathematical concepts of ratio and proportion. Students at Levels C are beginning to describe the behavior of particles of an ideal gas and to explain conceptually the relationship of pressure and volume of gases. In addition, students at this level are starting to predict the new pressure of a gas when volume is reduced. Figure 14.2 shows that 20.9% of the sample involved in this study are at Level C. Similarly, students at Level D are beginning to understand the importance of using the absolute temperature unit, Kelvin (K) in calculations and use ratio and proportion to show the relationship between volume and temperature of an ideal gas. This group of students make up 12% of the sample in this study. The possible reason why the concept of volume – temperature relationship is at a higher level is due to the additional step of converting the unit of temperature to kelvin. In addition, problems on pressure – volume relationship are usually introduced first. As such, students might have been used to solving problems on gases by directly substituting the given quantities to the equations they have identified. Thus, when presented with a problem on volume – temperature relationship, they stick to their usual strategy. This is a clear indication of their faulty understanding of the basic concepts about the properties of gases, such as pressure, volume, temperature, and the mole. Indeed, as Nurrenbern and Pickering (1987) put it, "*Students can solve problems about gases without knowing anything much about the nature of a gas* (p. 510). Similar to the previous discussion on Stoichiometry, a college instructor expecting students to be able to determine changes in pressure, volume, temperature, and number of moles according to different gas laws, will encounter a considerable number of students who would struggle to answer such problems. About 80% will have difficulty engaging with the concept of pressure and volume relationship, and a larger number (88%) having trouble with the concept of volume and temperature relationship of gases. These examples discussed demonstrate that most students enter college unprepared for the intellectual demands and expectations of tertiary education. Clearly, a change in the way teachers handle high school chemistry classes should be done to better prepare their students for the rigors of higher level chemistry curriculum.

It is important to note that the prerequisite concepts for topics at Level D are introduced in Grade 7 until Grade 10 Chemistry classes. To better prepare students for Grade 11, chemistry teaching at the junior high school level should be strengthened and should address the lack of understanding of basic chemistry concepts. This connection between how chemistry is taught and how students understand chemistry concepts was the focus of the study conducted by Furioet al. (2002). They argued that the problem of lack of understanding of the concepts 'amount of substance' and 'mole' manifested by students is strongly connected to teachers' ideas and to the methodologies used in the teaching of chemistry. Hence, a conceptual approach in teaching could address the inability of students to form connections between the observable macroscopic properties of substances to their characteristics at the sub-microscopic atomic level. The sequence of content instruction could be arranged based on the relative difficulties of understanding the levels of representation as perceived by students. For example, in introducing the concept of quantity 'amount of substance' and its unit, the 'mole', Furio, Azcone and Guisasola (2002) recommended differentiating this macroscopic quantity from 'mass' and 'volume' first, then relate it to the microscopic world of substances, meaning, as a way to count atoms and molecules. They also suggested the use of familiar analogies to facilitate the learning of the mole concept. Review sessions at the start of topics that students find difficult like stoichiometry and gases may also enable struggling students to catch up and improve their understanding of basic concepts, before they are presented with more advanced topics. The use of formative assessment during instruction could also facilitate learning as this can provide immediate feedback for students having difficulty understanding the prerequisite concepts.

At the national level, a curriculum review is underway to address the need to decongest the curriculum, to ensure a 'horizontal alignment' of the competencies and to adapt to new approaches in teaching and learning (Hernando-Malipot, 2021). At the classroom level, teachers can use simulations and other computer-assisted strategies to teach abstract concepts like atoms, molecules, and moles. To address the inadequacy of the mechanical use of algorithms to solve problems in stoichiometry, teachers can employ faded worked examples as a problem solving tool. This approach can help students who are identified as mathematically underprepared in a college chemistry course, as the fading out of steps in the problem solving process allowed for better understanding (Hesser & Gregory, 2015). They reported that the approach enhanced student learning and problem solving abilities. The use of formative assessment could also facilitate learning chemistry concepts and skills. However, it is important for teachers to use both conceptual and problem solving goals to assess student understanding, particularly with items that require computation (Sandlin et al., 2015). In this way, teachers can check more accurately if students understood a particular concept based on their responses to formative assessment items requiring explanation and computation.

As mentioned earlier, the difficulties in learning the skills and concepts found in level D may have occurred due to lack of mastery of prior knowledge and skills from earlier grades. This claim can be supported by the results of earlier assessments on students' readiness to learn chemistry in junior high school.

In comparison with the previous studies (Table 14.5), there was no significant change in the percentages of students who were ready to take higher level of

Table 14.5 Comparison of student readiness from pre-grade 7 to pre-college

	Percent ready	Percent not ready	References
Pre-grade 7	39%	61%	(Ferido et al., 2015)
Pre-grade 8	27%	73%	(Ferido et al., 2016)
Pre-grade 9	13%	87%	(Ferido et al., 2017a)
Pre-grade 10	11%	89%	(Ferido et al., 2017b)
Pre-college	12%	88%	This study

chemistry from Grades 9 through 11. This result indicates that only around 11–13% of the regular high school students were capable of adapting to the transition from one grade level to the next. This small group of students was successful in mastering the needed concepts for them to advance to higher chemistry education. The percentage of ready students is critically low considering that these students chose the STEM track and will most likely take general chemistry courses in college. The result obtained by this study is an indication that the Philippine Department of Education and the educators handling classes in the STEM strand of the senior high school curriculum should continually address issues and concerns regarding the curriculum and instruction of chemistry subjects.

It should be noted that in this study, the sample is part of the cohort of senior high school students who have completed the junior high school K to 12 curriculum from Grade 7 to 10. The goal is for these students to enter college with the level of skills and conceptual knowledge they need in order to engage with the general college chemistry course. However, the shortcomings of these students in terms of prerequisite skills and conceptual knowledge could have accumulated from one lower grade level to higher grade levels. This is consistent with the results of Rasch analysis (as represented in the variable map), which show that concepts in levels A to D are interlinked, and foundational skills and concepts are mostly found in levels A and B. However, the percentage of students in level A who are ready to learn college chemistry already fell behind at 30.7%, followed by 36.4% in B, 20.9% in C, and only 12.0% in D.

These results and the data of ACTRC (2017) provide a clear trend of decrease in the levels of skills and conceptual knowledge that students have prior to learning college chemistry. These percentages (%) are already low and even significantly decrease from a lower grade level to a higher one. If students have insufficient prior knowledge and skills in chemistry (possibly starting at the elementary level) in the junior high school yet they move to senior high school, this only indicates that students' lack of readiness to learn college chemistry.

Chemistry is already described as a difficult subject both in high school (Danipog & Ferido, 2011) and college (Cracolice & Busby, 2015; Cohen & Kelly, 2019), despite serving as "gatekeeper for future study in many sciences" (Tai, Sadler & Loehr, 2005). In the Philippines, this is evident in the poor performance of students in chemistry in the achievement tests (Espinosa et al., 2013). However, this study presents other factors that could have led to students' decrease in their readiness to learn college chemistry. Results of interviews with teachers of sample students also revealed that there are persisting problems in teaching senior high school general chemistry. These include insufficient time to teach the subject, inadequacy of teaching resources, and lack of teachers who specialize in chemistry. In terms of their preparation for college coursework, success is said to be influenced by their high school chemistry teachers and the decisions these teachers make in designing the segments of their chemistry courses (Tai et al., 2005).

Thus, insufficient prior knowledge and skills in chemistry in junior high school, and the shortcomings in teaching general chemistry to senior high school STEM students could all be contributory to the poor performance of students in the readiness test to engage the general college chemistry curriculum.

14.6 Conclusion

The study demonstrated that Rasch measurement could be used to develop tests and obtain student ability in a progression. Based on the variable map of student and item distribution with skill level descriptions, as the difficulty level of the skills and concepts in chemistry progresses from A (least difficult) to D (most difficult), the number of students ready to learn college chemistry decreases.

Thus, most of the Filipino senior high school STEM students in the sample have low readiness to learn general college chemistry. This is due to their lack of prior knowledge and skills needed for learning general chemistry.

Among the conceptual knowledge students are having difficulty with are related to using the absolute temperature unit, Kelvin (K) in calculations; applying the mole concept; solving stoichiometric problems; determining the number of molecules given Avogadro's number; understanding chemical structure, such as distinguishing the Lewis structures of compound; using Lewis structure, molecular geometry, bond polarity, and vector addition to determine the polarity of a molecule; and interpreting the sign for change in enthalpy (ΔH).

Tracing the connections of these difficult topics to related concepts, one implication is that the students' difficulty in learning these skills and conceptual knowledge is attributed to their having insufficient prior knowledge and skills for these chemistry topics as they went through Grades 7 to 10, since these concepts progress in complexity through the spiral structure of the K to 12 curriculum. Other related factors are lack of time to teach, insufficient number of teaching and learning resources, and few teachers who specialize in chemistry.

Acknowledgement The authors acknowledge the Office of the Chancellor of the University of the Philippines Diliman, through the Office of the Vice Chancellor for Research and Development, for funding support through the Outright Research Grant.

Appendices

Appendix A: Rasch Analysis Codes (ConQuest)

```
LET X=nismed_v3;

let path=C:\Users\ACTRC\Documents\NISMED;

datafile %path%\NISMED Data.sav!filetype=spss, responses=Item_1 to
Item_60;

codes 1,2,3,4;

key 231321431222442344411332434214334132423114323142442144323411!
1;

delete !items(3,11,20,31,41,44,45,46,51,52,58,59);

model item;
set warnings=no, update=yes;
estimate !stderr=empirical, nodes=60, iterations=5000;

show cases !pfit=yes, estimates=wle>>%path%\WLE%X%.case;
Show !estimates=latent>>%path%\SHOW%X%.shw;
itanal !estimates=latent>>%path%\Itanal%X%.itn;

show !estimates=latent, FILETYPE=excel,itemlabels=yes >>%path%\SHOW
%X%.XLS;

itanal! FILETYPE=excel >> %path%\Olditanal%X%.XLS;

itanal! FILETYPE=excel,format=summary >> %path%\Itanal%X%.XLS;

show cases !estimates=wle, FILETYPE=SPSS >> %path%\WLE%X%.sav;
```

Appendix B: Level Descriptions

Level	Description	No. of students ready to learn at each level (N = 459)
A	Students at this level are beginning to recognize chemical formulas as representations of chemical entities (atoms, molecules and ions) and use these formulas to determine the products formed during a reaction. They can identify the different types of chemical reactions and recognize a balanced equation by inspection.	141 (30.7%)

(continued)

Level	Description	No. of students ready to learn at each level (N = 459)
	Students are starting to infer phase change based on the representation of the arrangement of particles.	
B	They are starting to use their understanding of the law of conservation of mass to determine the coefficient that will balance a chemical equation. They are also learning to recognize that the coefficients in the chemical equation refer to the number of moles. Students at this level are beginning to recognize the connection between the subscripts in chemical formulas and its use in the calculation of molar mass. Students are learning to identify which atoms satisfy the octet rule in a given Lewis structure. They are starting to interpret the direction of heat flow and to distinguish the system from the surroundings. They are recognizing that in a chemical reaction, forming and breaking of bonds are accompanied by energy change. They are learning to classify whether examples of matter are either homogeneous or heterogeneous mixtures. Students at this level are starting to classify compounds as acidic or basic based on a given pH value.	167 (36.4%)
C	Students are starting to predict the new pressure of a gas when the volume is reduced. Using the kinetic molecular theory, they are beginning to describe the behavior of particles of an ideal gas and to explain conceptually the relationship of pressure and volume of gases. Students at this level are starting to use their knowledge of the law of conservation of mass to determine the mass of a product given the masses of the reactants and the other product. They are learning to use the Arrhenius definition to classify acidic and basic compounds. They are beginning to interpret the definition of pH with respect to H^+ concentration. They are beginning to differentiate matter (e.g., substances and mixtures, metals and nonmetals, solids and liquids) based on their properties. They are beginning to distinguish ionic from covalent compounds based on the given chemical formula. They are beginning to determine the valence electrons of an element using the periodic table. They are starting to interpret nuclear notation $^{A}_{Z}X$ of an atom, distinguishing atomic number from mass number.	96 (20.9%)
D	Students at this level are beginning to understand the importance of using the absolute temperature unit, kelvin (K), in calculations and use ratio and proportion to show the relationship between the volume and temperature of an ideal gas. They are beginning to apply the mole concept and solve stoichiometric problems using molar mass and balanced chemical equations. They are learning to use ratio and	55 (12.0%)

(continued)

Level	Description	No. of students ready to learn at each level (N = 459)
	proportion to determine the number of molecules given Avogadro's number. Students at this level are starting to develop a deeper understanding of chemical structure, such as distinguishing the Lewis structure of an ionic compound from that of a molecular compound. They are starting to use Lewis structure, molecular geometry, bond polarity, and vector addition to determine the polarity of a molecule. They are learning to interpret that the sign for change in enthalpy (ΔH) indicates whether heat is released or absorbed in chemical reactions (e.g., neutralization reaction).	

Appendix C: Point-Biserial Correlations

Item number	Item-rest corr	Item-total corr	Item number	Item-rest corr	Item-total corr
1	0.23	0.29	28	0.34	0.40
2	0.21	0.27	29	0.37	0.42
4	0.36	0.42	30	*0.16*	0.20
5	0.34	0.39	32	0.27	0.33
6	0.24	0.30	33	0.32	0.38
7	0.28	0.33	34	*0.09*	*0.15*
8	0.20	0.26	35	0.22	0.28
9	0.37	0.42	36	0.41	0.47
10	0.34	0.39	37	0.52	0.57
12	0.32	0.37	38	0.23	0.29
13	0.26	0.32	39	0.42	0.47
14	0.31	0.37	40	*0.19*	0.24
15	*0.15*	0.21	42	*0.18*	0.24
16	0.36	0.41	43	0.24	0.30
17	0.32	0.38	47	*0.17*	0.23
18	0.40	0.45	48	0.33	0.39
19	*0.12*	*0.18*	49	*0.15*	0.22
21	0.27	0.33	50	*0.19*	0.25
22	0.27	0.33	53	0.29	0.35
23	*0.17*	0.23	54	*0.17*	0.23
24	0.23	0.29	55	0.29	0.35
25	0.38	0.44	56	0.34	0.40
26	0.20	0.26	57	0.23	0.30
27	*0.16*	0.23	60	*0.14*	0.21

Appendix D: Outfit and Infit Item Statistics

	Outfit			Infit		
Item number	MNSQ	Confidence interval		MNSQ	Confidence interval	
1	1.05	0.87	1.13	1.02	0.91	1.09
2	1.04	0.87	1.13	1.03	0.95	1.05
4	0.92	0.87	1.13	0.94	0.94	1.06
5	0.87	0.87	1.13	0.93	0.90	1.10
6	1.04	0.87	1.13	1.02	0.94	1.06
7	0.99	0.87	1.13	0.98	0.86	1.14
8	1.08	0.87	1.13	1.04	0.92	1.08
9	0.90	0.87	1.13	0.93	0.93	1.07
10	0.92	0.87	1.13	0.96	0.94	1.06
12	0.95	0.87	1.13	0.95	0.88	1.12
13	1.00	0.87	1.13	1.00	0.94	1.06
14	0.99	0.87	1.13	0.99	0.94	1.06
15	1.16	0.87	1.13	1.06	0.89	1.11
16	0.96	0.87	1.13	0.93	0.88	1.12
17	0.94	0.87	1.13	0.97	0.93	1.07
18	0.86	0.87	1.13	0.91	0.92	1.08
19	1.12	0.87	1.13	1.09	0.93	1.07
21	1.00	0.87	1.13	1.00	0.94	1.06
22	0.99	0.87	1.13	1.00	0.93	1.07
23	1.08	0.87	1.13	1.05	0.93	1.07
24	1.01	0.87	1.13	1.02	0.95	1.05
25	0.94	0.87	1.13	0.94	0.93	1.07
26	1.08	0.87	1.13	1.03	0.90	1.10
27	1.08	0.87	1.13	1.06	0.95	1.05
28	0.97	0.87	1.13	0.96	0.94	1.06
29	0.91	0.87	1.13	0.94	0.93	1.07
30	1.10	0.87	1.13	1.03	0.83	1.17
32	1.01	0.87	1.13	0.99	0.93	1.07
33	0.95	0.87	1.13	0.96	0.89	1.11
34	1.16	0.87	1.13	1.10	0.91	1.09
35	1.07	0.87	1.13	1.04	0.91	1.09
36	0.91	0.87	1.13	0.91	0.95	1.05
37	0.83	0.87	1.13	0.85	0.95	1.05
38	1.08	0.87	1.13	1.01	0.90	1.10
39	0.89	0.87	1.13	0.92	0.95	1.05
40	1.11	0.87	1.13	1.04	0.88	1.12
42	1.06	0.87	1.13	1.05	0.92	1.08
43	1.04	0.87	1.13	1.03	0.92	1.08
47	1.06	0.87	1.13	1.04	0.92	1.08
48	0.96	0.87	1.13	0.96	0.95	1.05

(continued)

Item number	Outfit			Infit		
	MNSQ	Confidence interval		MNSQ	Confidence interval	
49	1.07	0.87	1.13	1.06	0.94	1.06
50	1.07	0.87	1.13	1.03	0.93	1.07
53	0.98	0.87	1.13	0.99	0.93	1.07
54	1.08	0.87	1.13	1.06	0.94	1.06
55	0.98	0.87	1.13	0.99	0.95	1.05
56	0.94	0.87	1.13	0.96	0.95	1.05
57	1.01	0.87	1.13	1.03	0.95	1.05
60	1.10	0.87	1.13	1.07	0.94	1.06

References

Adams, D., Chuah, K. M., Sumintono, B., & Mohamed, A. (2021). Students' readiness for e-learning during the COVID-19 pandemic in a South East Asian university: A Rasch analysis. *Asian Education and Development Studies, 11*(2), 324–339.

Anderson, D., Kahn, J., & Tindal, G. (2017). Exploring the robustness of a unidimensional item response theory model with empirically multidimensional data. *Applied Measurement in Education, 30*(2), 163–177. https://doi.org/10.1080/08957347.2017.131627

Areepattamannil, S., & Khine, M. S. (2017). Evaluating the psychometric properties of the original grit scale using Rasch analysis in an Arab adolescent sample. *Journal of Psychoeducational Assessment, 36*, 856–862. https://doi.org/10.1177/0734282917719976

Assessment, Curriculum and Technology Research Centre (ACTRC). (2017, January–February). The challenge of achieving student readiness for the 7–10 science curriculum, *ACTRC Newsletter*, Issue 6.

Benito, N. V. (2010). *National achievement test: An overview*. Retrieved from http://www.cfo-pso.org.ph/pdf/9thconferencepresentation/day2/National_Achievement_Test_Dr%20Benito.pdf on 10 May 2022.

Blank, R. K., Porter, A. C., & Smithson, J. (2001). *New tools for analyzing teaching, curriculum and standards in Mathematics & Science. Results from survey of enacted curriculum project. Final report* (978-1-884037-73-3). Retrieved from Washington D. C. https://curriculumanalysis.org/Reference/SECNewToolsreport01.pdf

Boone, W. J., & Scantlebury, K. (2006). The role of Rasch analysis when conducting science education research utilizing multiple-choice tests. *Science Education, 90*, 253–269.

Boone, W. J., Staver, J. R., & Yale, M. S. (2014). *Rasch analysis in the human sciences*. Springer.

Boone, W. J., Townsend, J. S., & Starver, J. (2010). Using Rasch theory to guide the practice of survey development and survey data analysis in science education and to inform science reform efforts: An exemplar utilizing STEBI self-efficacy data. *Science Education, 95*(2), 258–280.

Boujaoude, S., & Barakat, H. (2003). Students' problem solving strategies in stoichiometry and their relationships to conceptual understanding and learning approaches. *Electronic Journal for Research in Science and Mathematics, Education, 7*(3) Available online https://ejrsme.icrsme.com/article/view/7709

Bowen, C. W., & Bunce, D. M. (1997). Testing for conceptual understanding in general chemistry. *Chemical Educator, 2*, 1–17.

Cohen, R., & Kelly, A. (2019). Community college chemistry coursetaking and STEM academic persistence. *Journal of Chemical Education, 96*, 3–11.

Conley, D. (2007). *Redefining college readiness Eugene*. Educational Policy Improvement Center.

Cracolice, M., & Busby, B. (2015). Preparation for college general chemistry: More than just a matter of content knowledge acquisition. *Journal of Chemical Education, 92*, 1790–1797.

Danipog, D. L., & Ferido, M. B. (2011). Using art-based chemistry activities to improve students' conceptual understanding in chemistry. *Journal of Chemical Education, 88*, 1610–1615.

Department of Education. (2016a). *K to 12 curriculum guide: Science grade 3 to grade 10*. Author.

Department of Education. (2016b). *K to 12 basic education curriculum senior high school – Science, Technology, engineering, and mathematics (STEM) specialized subject*. Author.

Department of Education. (2020). *DepEd memorandum No. 033, s. 2020*. Retrieved from https://www.deped.gov.ph/wp-content/uploads/2020/03/DM_s2020_033.pdf on 24 May 2022.

Deters, K. M. (2003). What should we teach in high school chemistry? *Journal of Chemical Education, 8*(10), 1153–1155.

Drasgow, F., & Lissak, R. (1983). Modified parallel analysis: A procedure for examining the latent dimensionality of dichotomously scored item responses. *Journal of Applied Psychology, 68*, 363–373.

Espinosa, A., Monterola, S., & Punzalan, A. (2013). Career-oriented performance tasks in chemistry: Effects on students' integrated science process skills. *Cypriot Journal of Educational Sciences, 8*(2), 211–226.

Ferido, M., Robertson, P., Care, E., Pavlovic, M., & Harding, S. (2015). *Progress of students through the science curriculum: Focus on chemistry* (Stage 1 Pre-Grade 7 report). Report prepared for the Australian Department of Foreign Affairs and Trade and the Philippine Department of Education. Assessment Curriculum and Technology Research Centre (ACTRC).

Ferido, M., Robertson, P., Pavlovic, M., & Care, E. (2016). *Progress of students through the science curriculum: Focus on chemistry* (Stage 2 Pre-Grade 8 report). Report prepared for the Australian Department of Foreign Affairs and Trade and the Philippine Department of Education. Assessment Curriculum and Technology Research Centre (ACTRC).

Ferido, M., Marabe, J., Pavlovic, M., Bagui, L., Robertson, P., Ramirez, R., Bustos, T., & Care, E. (2017a). *Progress of students through the science curriculum: A focus on matter (chemistry)* (Stage 3 Pre-Grade 9 report). Report prepared for the Australian Department of Foreign Affairs and Trade and the Philippine Department of Education. Assessment Curriculum and Technology Research Centre (ACTRC).

Ferido, M., Robertson, P., Pavlovic, M., & Care, E. (2017b). *Progress of students through the science curriculum: A focus on matter (chemistry)* (Stage 4 Pre-Grade 10 report). Report prepared for the Australian Department of Foreign Affairs and Trade and the Philippine Department of Education. Assessment Curriculum and Technology Research Centre (ACTRC).

Furio, C., Azcona, R., & Guisasola, J. (2002). The learning and teaching of the concepts "amount of substance" and "mole": A review of the literature. *Chemical Education Research and Practice in Europe, 3*(3), 277–292.

Gabel, D. L., & Bunce, D. M. (1994). Research on problem solving: Chemistry. In D. L. Gabel (Ed.), *Handbook of research on science teaching and learning* (pp. 301–326). Macmillan.

Griffin, P. (2007). The comfort of competence and the uncertainty of assessment. *Studies in Educational Evaluation, 33*(1), 87–99.

Grigg, K., & Manderson, L. (2016). The Australian racism, acceptance, and cultural-ethnocentrism scale (RACES): Item response theory findings. *International Journal for Equity in Health, 15*(1), 49. https://doi.org/10.1186/s12939-016-0338-4

Hernando-Malipot, M. (2021, June 23). DepEd reviews 'congested' K to 12 curriculum. *Manila Bulletin*. https://mb.com.ph/2021/06/23/deped-reviews-congested-k-to-12-curriculum/

Herron, J. D. (1996). *The chemistry classroom: Formulas for successful teaching*. American Chemical Society.

Hesser, T. L., & Gregory, J. L. (2015). Exploring the use of faded worked examples as a problem solving approach for underprepared students. *Higher Education Studies, 5*(6), 36–46. https://files.eric.ed.gov/fulltext/EJ1086007.pdf

Karim, A. A., Shah, P. M., Din, R., Ahmad, M., & Lubis, M. A. (2014). Developing information skills test for Malaysian youth students using Rasch analysis. *International Education Studies, 7*(13), 112–122.

Kolb, D. (1978). The mole. *Journal of Chemical Education, 55*, 728–732.

Lane, T., Morgan, K., & Lopez, M. (2017). "A bridge between high school and college:" a case study of a STEM intervention program enhancing college readiness among underserved students. *Journal of College Student Retention: Research, Theory & Practice, 0*(0), 1–25.

Linacre, J. M. (1998). Detecting multidimensionality: Which residual data-type works best? *Journal of Outcome Measurement, 2*(3), 266–283.

Liu, X. (2010). *Using and developing measurement instruments in science education: A Rasch modeling approach*. Information Age.

Means, B., Wang, H., Young, V., Peters, V. L., & Lynch, S. J. (2016). STEM- focused high schools as a strategy for enhancing readiness for postsecondary STEM programs. *Journal of Research in Science Teaching, 53*(5), 709–736.

Montebon, D. R. T. (2014). K to 12 science program in The Philippines: Student perception on its implementation. *International Journal of Education and Research., 2*(12), 153–164.

Nurrenbern, S. C., & Pickering, M. (1987). Concept learning versus problem solving: Is there a difference? *Journal of Chemical Education, 64*(6), 508–510.

Ocampo, D.J.S. (2016, September 20). *Updates in the K to 12 basic education program*. Powerpoint presentation in the Department of Education.

Official Gazette of the Philippine Government. (n.d.). *The K to 12 basic education program*. Retrieved from https://www.officialgazette.gov.ph/k-12/ on 9 May 2022.

Phillipson, S. N., & Tse, A. K. (2007). Discovering patterns of achievement in Hong Kong students: An application of the Rasch measurement model. *High Ability Studies, 18*(2), 173–190.

Porter, A. C., & Smithson, J. L. (2001). *Defining, developing, ang using curriculum indicators*. Consortium for Policy Research in Education.

Rasch, G. (1960). *Probabilistic models for some intelligence and attainment tests*. Danish Institute for Educational Research.

Reyes, J. R. S., Grajo, J. D. L., Comia, L. N., Talento, M. S. D. P., Ebal, L. P. A., & Mendoza, J. J. O. (2021). Assessment of Filipino higher education students' readiness for e-learning during a pandemic: A Rasch technique application. *Philippine Journal of Science, 150*(3), 1007–1018.

Salame, I., & Casino, P (2021). Using chemistry concepts inventory to identify alternative conceptions and their persistence in general chemistry courses. *International Journal of Instruction, 14*(3), 787–806. https://doi.org/10.29333/iji.2021.14346a

Salman, A., & Aziz, A. A. (2015). Evaluating user readiness towards digital society: A Rasch measurement model analysis. *Procedia Computer Science, 65*, 1154–1159.

Sandlin, B., Harshman, J., & Yezierski, E. (2015). Formative assessment in high school chemistry teaching: Investigating the alignment of teachers' goals with their items. *Journal of Chemical Education, 92*(10), 1619–1625.

Schleicher, A. (2018). *PISA 2018 insights and interpretations*. OECD Publishing.

Southeast Asian Ministers of Education Organization—Regional Center for Educational Innovation and Technology. (2015). *Philippine education for all 2015 review report*. Retrieved from https://www.seameo-innotech.org/wp-content/uploads/2020/04/EFA-Part-2-V12.pdf on 26 April 2022.

Smith, R. M., Schumacker, R. E., & Bush, M. J. (1998). Using item mean squares to evaluate fit to the Rasch model. *Journal of Outcome Measurement, 2*, 66–78.

Sondergeld, T. A., & Johnson, C. C. (2014). Using Rasch measurement for the development and use of affective assessments in science education research. *Science Education, 98*, 581–613.

Tai, R., Sadler, P., & Loehr, J. (2005). Factors influencing success in introductory college chemistry. *Journal of Research in Science Teaching, 42*(9), 987–1012. https://doi.org/10.1002/tea.20082

Wang, X. (2013). Why students choose STEM majors: Motivation, high school learning, and postsecondary context of support. *American Educational Research Journal, 50*(5), 1081–1121.

Wright, B. D. (1977). Solving measurement problems with the Rasch model. *Journal of Educational Measurement, 14*(2), 97–116.
Wright, B. D. (1992). IRT in the 1990s: Which models work best? *Rasch Measurement Transactions, 6*(1), 196–200.
Wright, B. D., & Linacre, J. M. (1994). Reasonable mean-square fit values. *Rasch Measurement Transactions, 8*(3), 370.
Wright, B. D., & Stone, M. H. (1979). *Best test design Rasch measurement.* MESA Press.
Wu, M. L., Adams, R. J., & Wilson, M. R. (2007). *ACER ConQuest version 2.0: Generalized item response modelling software.* ACER Press.
Yang, Y., He, P., & Liu, X. (2017). Validation of an instrument for measuring students' understanding of interdisciplinary science in grades 4-8 over multiple semesters: A Rasch measurement study. *International Journal of Sciences and Mathematics Education, 16*(4), 639–654.
Yuksel, I., & Yuksel, I. (2012). The importance of students' readiness levels in program evaluation studies. *Elsevier, Procedia Social and Behavioral Sciences, 31*, 908–912.

Dennis L. Danipog is a Science Education Specialist at National Institute for Science and Mathematics Education Development (NISMED), University of the Philippines Diliman. He served as Project Leader in science education research projects at NISMED. He is also involved in developing science curriculum materials and in planning and conducting professional development programs for chemistry teachers. His current research interests focus on assessment of twenty-first century skills in STEM education, digital and AI-based assessments, and inquiry-based chemistry education.

Nona Marlene B. Ferido is presently the Philippine Deputy Director of the Assessment, Curriculum, and Technology Research Centre (ACTRC), a partnership between the University of Melbourne and the University of the Philippines, supported by the Australian government. Some of her research involvements in the Centre were: progress of students through the science curriculum, curriculum review of the Philippine K to 12 curriculum, feasibility study on micro-certification for the alternative learning system (ALS) in the Philippines, and the development and conduct of the professional development (PD) program for science, mathematics and reading teachers on assessment and emerging literacies.

Rachel Patricia B. Ramirez is a faculty member at the College of Education, University of the Philippines Diliman. Her areas of interest in research include Science curriculum design and evaluation, inquiry-based approaches in teaching Science, learning progression in Chemistry and creative teaching in Chemistry.

Maria Michelle V. Junio is a Science Education Specialist at the University of the Philippines National Institute for Science and Mathematics Education Development (UP NISMED). Her research interests include chemistry education, science curriculum, and medical tourism.

Joel I. Ballesteros is currently a University Research Associate of the Institute of Chemistry, University of the Philippines Diliman. His research interests include analytical chemistry and chemical education.

Chapter 15
Investigating Differential Severity Across Linguistic Subgroups in Automated Scoring of Student Argumentation

Zoë Buck Bracey, Molly A. M. Stuhlsatz, Christopher D. Wilson, Tina Cheuk, Marisol M. Santiago, Jonathan Osborne, Kevin Haudek, and Brian M. Donovan

Abstract This study uses Many-Facet Rasch Measurement (MFRM) to examine the extent to which computer scoring models for assessing students' argumentation in science might be more or less severe when scoring students who have been designated as English Learner (EL) students than humans scoring the same data. We found that while no one machine scoring approach produced significant bias, performance on certain items demonstrated that one machine model had significant potential to widen performance gaps.

In this chapter, we describe the results of a study that uses Many-Facet Rasch Measurement (MFRM) to examine the extent to which computer scoring models for assessing students' written argumentation from evidence in science might be more or less severe when scoring (English Learner) EL students than humans scoring the same data. As automated scoring of open-ended responses becomes more common in K-12 education, it is essential to consider the impact of this shift on science education outcomes for students from non-dominant linguistic backgrounds. Science education researchers agree that testing students designated as ELs is complex (Hutchinson & Hadjioannou, 2017), with the result that these students are systematically denied the opportunity to demonstrate their science competence on assessments (Abedi, 2006; Solano-Flores, 2008; Shaw et al., 2010). MFRM is a

Z. B. Bracey (✉) · M. A. M. Stuhlsatz · C. D. Wilson · B. M. Donovan
BSCS Science Learning, Colorado Springs, CO, USA
e-mail: zbuck@bscs.org

T. Cheuk
CalPoly San Luis Obispo, San Luis Obispo, CA, USA

M. M. Santiago
Michigan State University, East Lansing, MI, USA

J. Osborne · K. Haudek
Stanford University, Stanford, CA, USA

X. Liu, W. J. Boone (eds.), *Advances in Applications of Rasch Measurement in Science Education*, Contemporary Trends and Issues in Science Education 57,
https://doi.org/10.1007/978-3-031-28776-3_15

powerful tool for thinking critically about assessment and linguistic equity because it applies the principles of the Rasch measurement model to reveal linguistic biases that might not be clear otherwise.

Bias in MFRM is defined as differential rater severity (negative bias) and differential rater leniency (positive bias). Differential rater severity/leniency measures the tendency of a rater to assign scores to a particular group of students that are, on average, either lower or higher than the measurement model would expect for that group (Myford & Wolfe, 2004). Thus, an MFRM bias analysis helps us identify interactions between facets that suggest a consistent deviation from what we would expect given all the responses (Myford & Wolfe, 2003). This type of Rasch analysis is useful to determine if automated scoring of assessments puts EL students at an even greater risk of being misjudged or pushed out (Bradley & Renzulli, 2011) by the K-12 system, and how we might focus our efforts during assessment development to prevent this occurring.

This study is part of the Argulex project (*NSF #1561150, #1561159, #1561155*), which brings together an interdisciplinary team of science educators, cognitive scientists, and computational learning experts to develop automated scoring models for a set of assessment items aligned with a learning progression for scientific argumentation (Osborne et al., 2016). The goal of the project was to develop automated scoring models that can accurately and reliably score written responses at levels equal to human expert scorers. By developing computer scoring models that can reliably replace the time-consuming process of expert human scoring of students' writing on this task, the resulting instruments can be made available online, and provide rapid scores of student competency with argumentation for research and evaluation purposes, as well as formative feedback for teachers. But research suggests that these advantages may come with risks to EL students (Caliskan et al., 2017). We seek to determine if automated scoring of assessments puts EL students at an even greater risk of being misjudged or pushed out (Bradley & Renzulli, 2011) by the K-12 system, and how we might focus our efforts during assessment development to prevent this from occurring.

15.1 Background and Rationale

15.1.1 Human Capacity to Mitigate Bias

Written assessments have the capacity to expand the ways in which students can express their competences (Warren et al., 2001), but only when the scorer can read "past English fluency" to evaluate students' science ideas (Moschkovich, 2007). In other words, while content and language are inextricably linked, there are language invariant ideas and practices that teachers can extricate from students' written responses. Research across disciplines shows that humans have this ability to counteract their own biases consciously (Jackson et al., 2014; Devine et al., 2012; Monteith et al., 1993; Caliskan et al., 2017). For example, in a 12-week longitudinal study by Devine et al. (2012), people who received an intervention designed to increase their awareness of implicit bias showed dramatic reductions in implicit race bias. Similarly, Jackson et al. (2014) evaluated changes in implicit bias against women in STEM after explicit anti-bias training and found that personal implicit associations about women in STEM improved for the men in the study. The results of these studies suggest that humans can be trained to overcome implicit bias. But is this capacity lost when we task a computer with developing scoring models? If so, problematic biases inherent in human decision making could be magnified by the switch to machine learning algorithms.

15.1.2 Bias in Machine Learning Algorithms

Machine learning is a branch of the larger computer science domain known as artificial intelligence (AI), which describes the capacity of a machine to emulate human cognitive processes (Tyagi, 2020). Emerging research in AI outside of the educational context indicates that computers tend to acquire the inherent biases of the humans who train them, but lack the metacognitive ability that humans can develop to counteract these biases consciously. In 2017, researchers at Princeton University trained a machine learning algorithm to replicate human responses to the Implicit Association Test, a commonly used psychological tool designed to reveal and measure implicit bias (Greenwald et al., 1998). They found that the machine learning algorithm acquired the same biases as the humans used to train it, "whether these are morally neutral as towards insects or flowers, [or] problematic as towards race or gender" (Caliskan et al., 2017). This bias becomes particularly problematic when we realize how much human-trained AI is built into our modern lives (Buolamwini, 2020). Studies of facial recognition software inspired researcher Joy Buolamwini to consider how bias in AI can contribute to the dehumanization of certain populations in "AI, Ain't I A Woman?" (Buolamwini, 2019), and form the Algorithmic Justice League to raise awareness about the impacts of AI, and the dangers of allowing bias to be integrated into this technology. Similarly, researchers at Stanford University published results in 2020 indicating that there are large racial disparities in commercial speech recognition systems such as Alexa or Siri (Koenecke et al., 2020). The authors stress the need to audit emerging machine-learning systems to ensure they are broadly inclusive. Such calls need to be taken seriously, as the implication of these biased AI technologies can mean life or death. For example, researchers found that an algorithm to detect melanoma that was trained primarily on white skin struggles to detect cancerous moles in darker skin, which could mean an increase in risk of death for Black and other melanated people with skin cancer (Adamson & Smith, 2018). While the stakes of classroom assessment are not immediately this high, assessments determine educational outcomes, and it is well established that educational outcomes are correlated with any number of career and life outcomes, including mortality (Cutler & Lleras-Muney, 2006; Hummer & Hernandez, 2013). Thus, we would argue, the field should take bias in classroom assessments very seriously.

15.1.3 Machine Learning in Classrooms

Machine learning technology is already in classrooms. Researchers predict that artificial intelligence (AI) in U.S. education will have grown by 47.5% between the years of 2017–2021 (Technavio, 2018). Moreover, people have started to pay attention, calling for "due diligence and intellectual exploration before we deploy the technology to schools," and "intentionally build[ing] AI systems [deployed in schools] through a lens of racial equity" (Perry & Lee, 2019, para. 4). Likewise, we argue that assessment developers should intentionally build computerized assessments through the lens of linguistic equity if we are to avoid constructing automated assessments that are unfairly biased against students from non-dominant linguistic backgrounds.

But AI may not be all bad news for EL students. If we apply the technology fairly, it has the potential to be a powerful resource for supporting the success of EL students in demonstrating their competence in science classrooms going beyond superficial linguistic performativity. Limited prior research indicates that humans may in fact have a harder time consistently scoring responses with non-dominant linguistic patterns than machines (Ha & Nehm, 2016). By using computers to analyze student writing, we can analyze text at the lexical level and thereby ignore the grammar and syntax, avoiding the implicit bias and inference that students with good grammar and syntax have better knowledge of the construct being measured (Shermis, 2015). This suggests that automated scoring might have the potential to be a powerful tool for the valid assessment of EL students in science. We currently lack empirical evidence either way.

As the population of students classified as EL students grows (NCES, 2020), developers have an ethical duty to monitor the differential performance of assessments between this group, and their non-EL peers. This includes establishing the assessments' sensitivity to the distinction between linguistic fluency and competency in the construct(s) being measured. We do not yet know if automated analysis has this capacity, but as we develop more assessments that can be scored by a machine, it is important to explore this possibility, with the goal of building scoring models that capture the wide range in which our increasingly linguistically diverse student population expresses their knowledge and capabilities.

15.1.4 EL Designation

We recognize that the language that we use to describe marginalized groups is important. We have chosen to study students who have been classified by the state of California as English Learners (EL), based on their performance on the English Language Proficiency Assessments for California (ELPAC) exam. Throughout this study, we refer to these students as EL students. We do not equate this designation to a universal experience, or homogeneous linguistic background. We acknowledge that the designation of "EL" does not capture the multi-dimensional diversity of students' linguistic backgrounds in the United States. Millions of students in U.-S. schools are emergent multilingual learners, with varying levels of expertise across modalities in one or more of thousands of languages other than English (NCES, 2020). In addition, some students speak only English using a non-dominant register, dialect, or level of exposure to language (Eckert & Rickford, 2001). While this rich heterogeneity is not well captured by the EL student designation, we do not have access to additional linguistic information about the students in our dataset beyond their ELPAC scores.

15.2 Research Questions

We seek to determine if automated scoring of assessments puts EL students at an even greater risk of being misjudged and/or pushed out by the K-12 system, and how we might focus our efforts during assessment development to prevent this. We assume that there is already inherent bias built into human scoring of EL students' responses. According to Caliskan et al. (2017), "Bias is identical to meaning, and it is impossible to employ language meaningfully without incorporating human bias." That is, what we are looking for is differential scoring across demographic groups for students who might otherwise score the same. While this may or may not reflect an inherent bias associated with a rater, it can reveal patterns that affect students from certain groups in disparate ways, leading to unjust outcomes. In this study we explore the possibility that computer algorithms *could* widen the existing gap between assessment outcomes for EL designated students and their peers. We refer to this as *differential rater severity*. Thus, our first research question seeks to investigate:

1. Is machine scoring of student writing in response to an argumentation task differentially more severe than human scoring for students who have EL status, compared to students who do not have EL status?

We are also interested in how this differential severity might vary across different machine learning approaches, and how it might be affected by item difficulty. The raters across which we are comparing include the human consensus, and two different machine learning models, described in greater detail in the Methods section. Item difficulty is determined through Rasch modeling, and an established learning progression for scientific argumentation (Osborne et al., 2016), which is described in more detail in the Theoretical Framework. Thus our second research question seeks to investigate:

2. Do differences in severity between the automated scoring and the humans vary across:

    (a) Item difficulty?
    (b) Machine learning approach?

15.3 Theoretical Framework

This work comes from a sociocultural theoretical perspective, which frames learning as a complex, situated, social process (Lave & Wenger, 1991; Vygotsky, 1978). From the perspective of sociocultural theory, we must consider that both the experience of taking an assessment, and the very construct of scientific argumentation are mediated by artifacts, language, context, and culture.

15.3.1 Assessment, Context, and Consequences

Our model for meaningful assessment is the Assessment Triangle pictured in Fig. 15.1 (Pellegrino et al., 2001). The cognition corner of the triangle represents what is really being assessed: models of knowledge, skills and learning within the domain. In the project described here, what is being assessed is an established learning progression for argumentation described in the next section. The observation corner of the triangle represents what learners have to do: the tasks learners are asked to perform to demonstrate their knowledge and/or skills, such as answering assessment items or demonstrating proficiency via performance tasks. In this project, what learners have to do are complete a series of validated argumentation prompts aligned to the learning progression. The interpretation corner of the triangle represents how the assessment is scored and used: tools to make sense of the observations, such as scoring guides, rubrics and measurement models. This project makes

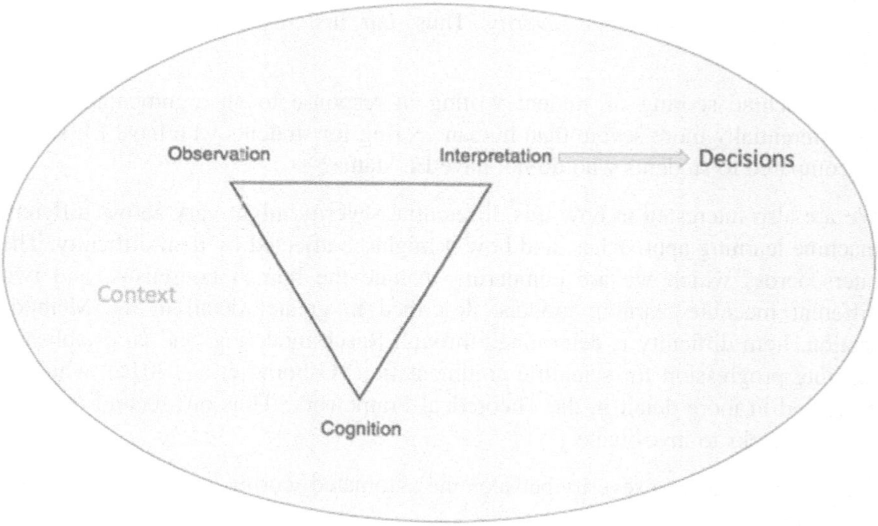

Fig. 15.1 The assessment triangle (Pellegrino et al., 2001), modified to include context and decisions

interpretation its focus. The assessment is scored in two ways: by human scorers using a rubric based on the argumentation learning progression, and by automated analysis of student responses.

The framework in Fig. 15.1 also includes two pieces that are often left out of models for assessment: (i) the context in which the assessment is embedded (McCullough, 2004), and (ii) the consequences: decisions that are made based on interpretation of the assessment (Shaw, 2005). The inclusion of context acknowledges that both assessment and argumentation are heavily mediated by artifacts, language, context, and culture. We cannot extricate cognition, observation or interpretation from who is taking the test, who is giving the test, where, when, and in what language (Solano-Flores & Nelson-Barber, 2001; Solano-Flores, 2008). Neither can we extricate the consequences of the assessment, or the perception of those consequences from the perspective of the test-taker. Although the assessment described here was designed to be used for research purposes only, the majority of assessments that students encounter carry higher stakes. Students who are tracked incorrectly or held back because of performance on an assessment may be increasingly susceptible to school failure or dropping out (Kaufman et al., 2000; Hakuta & Beatty, 2000). In addition, low aggregate performance by language subgroups contributes to larger Discourses of deficiency (Hakuta & Beatty, 2000; Moschkovich, 2007; Gee, 2015) that establish and maintain institutional racism and anti-immigrant sentiment. If computers are more severe on EL students than humans are, we are not just at risk of compromising validity, we are at risk of pushing students out of school and perpetuating racist Discourse.

15.3.2 Argumentation Learning Progression

The cognition corner of the assessment triangle is a model of the construct that we are trying to measure, in this case argumentation. Educators agree that argumentation is fundamental to the practice of science (Osborne, 2010; Erduran et al., 2004), and thus it is emphasized in standards documents adopted by a majority of the states in the U.S., like the Next Generation Science Standards, the Common Core State Standards for English Language Arts and Literacy in History/Social Studies, Science, and Technical Subjects, and the Common Core State Standards for Mathematics (Stage et al., 2013).

Scientific argumentation practices are often theorized using Toulmin's (1958) argument structure for analysis of student discourse (e.g., Cavagnetto, 2010; Lee et al., 2014; Osborne, 2010; Buck et al., 2014). Toulmin (1958) identified six structural elements that can be found universally across disciplines: claim, data, warrant, backing, qualifier, and rebuttal. In addition, we include the practice of formulating rebuttals and engage in the act of critique. Being able to explain why an idea is flawed is as important as being able to explain why it is right (Allchin, 2013).

Despite the fact that Toulmin's framework is used across disciplines and languages, we do not contend that scientific argumentation is linguistically neutral. Constructing an argument from evidence requires students to use scientific discourse to make reasonable claims, read and understand evidence, communicate their reasoning, and persuade other people. Constructing a text-based response in science always involves reading, comprehending and writing, but the subtleties of language make argumentation a particularly complex linguistic task. For example, scientific argument often communicates likelihood or uncertainty (Buck et al., 2014; Lee et al., 2014), and the words and phrases that convey this uncertainty could be easier to miss for students who are learning English (Symons et al., 2017). Scientific argumentation – including both construction and critique – is a competency that draws on diverse knowledge and practices (Henard & Roseaveare, 2012), including specific domain content (Osborne, 2010), rhetorical knowledge about the conventions of argumentation (Kelly & Takao, 2002), and requires an epistemic understanding of the use of evidence as the basis of belief (Sandoval, 2003). All of these features make the development of assessments for argumentation particularly complex.

A learning progression is a tool that can be used to guide the construction of such assessments, an empirical map of "successively more sophisticated ways of learning about a topic that can follow and build on one another as children learn and investigate a topic over a broad span of time" (National Research Council, 2007, p. 230). The construct map in Table 15.1 provides a summary of the learning progression used as the foundation of our assessment. Research has shown the construct to be psychometrically uni-dimensional, and supports the distinction between Constructing Arguments and Critiquing Arguments (Osborne et al., 2016).

Level 0 performances simply require the knowledge of the elements of an argument and the ability to provide such elements when asked or identify them in the arguments of others. Level 1 items require not only the provision of level 0 elements but the ability to construct relationships between them – that is, for instance, to provide the warrant connecting a piece of evidence to a claim. Level 2 items are more demanding as they require the construction of a rebuttal. This performance requires the student to be able to identify the elements of another's argument that might be flawed and then construct a related argument which justifies this views. Thus, the student must cognitively consider one argument and compare and contrast it with another.

15.3.3 Methodology

While we (the authors) locate this research within a critical theoretical perspective that sees knowledge as situated (Lave & Wenger, 1991; Vygotsky, 1978), and acknowledge that mainstream research practices generally reproduce class, race, and gender oppression (Kincheloe & McLaren, 1994), we do not see this stance as invalidating the usefulness of quantitative means to confront injustice by

Table 15.1 Scientific argumentation construct map from Osborne et al., 2016

Level	Constructing	Critiquing	Description
0a	Constructing a claim		Student states a relevant claim.
0b		Identifying a claim	Student identifies another person's claim.
0c	Providing evidence		Student supports a claim with a piece of evidence.
0d		Identifying evidence	
1a	Constructing a warrant		Student constructs an explicit warrant that links their claim to evidence.
1b		Identifying a warrant	Student identifies the warrant provided by another person.
1c	Constructing a complete argument		Student constructs a synthesis between the claim and the warrant.
1d	Providing an alternative counter argument		Student offers a counterargument as a way of rebutting another person's claim.
2a	Providing a counter-critique		Student critiques another's argument.
2b	Constructing a one-sided comparative argument		Student makes an evaluative judgment about the merits of two competing arguments
2c	Providing a two-sided comparative argument		Student provides an evaluative judgement about two competing arguments
2d	Constructing a counter claim with justification		Student explicitly compares and contrasts two competing arguments, and an argument as to why it is superior to each of the previous arguments.

Note: Levels 1b, 2a, 2c, and 2d are highlighted to illustrate the elements that were the most difficult. The four most difficult items were aligned with these levels, and manifested problematic patterns across machine learning models

demonstrating oppression using numbers (Stage & Wells, 2014). In the tradition of pragmatism, "We all try out things to see what works, what solves problems, and what helps us to survive," (Johnson & Onwuegbuzie, 2004, p. 18). Our research questions address the degree to which a new way of scoring assessments might put students classified as EL students at even greater risk of being left behind by the K-12 system due to biased scoring outcomes from automated scoring models. Answering these questions requires a large corpus of student assessment data. These data were analyzed quantitatively to reveal psychometric bias so that we can compare across human scored responses, and responses scored by two different machine-learning approaches.

In order to determine relative linguistic bias between the consensus human score and the two computer scoring models based on student EL status, the scored data were analyzed using Many-Facet Rasch Measurement (MFRM), which is grounded in generalizability-theory, or G theory (Cronbach et al., 1963; Briggs & Wilson, 2007). G theory reframes the conversation around validity and fairness on assessments as a question of measurement error, rather than as an examination of score gaps between student groups. According to Solano-Flores & Li, 2006, "An approach based on viewing language as a source of measurement error addresses the fact that bilingual individuals do not typically replicate their capacities across languages" (p. 2359). G theory distinguishes between two kinds of sources of score variation. One is student (s), the object of measurement; the other comprises facets—or sources of measurement error (Solano-Flores & Li, 2006). A facet could be, for example rater, demographic, item set, item, or EL status. MFRM analyses estimate score variance due to (1) the main effect of student (s); (2) the main effect of the facets; and (3) the interaction effect of all sources of score variation.

15.4 Methods

15.4.1 The Question Development Cycle

Our approach to developing assessments draws on the Question Development Cycle (QDC) for automated analysis (Urban-Lurain et al., 2013, Fig. 15.2). The first step of the QDC is to develop assessment questions (New Question Design). This is followed by Data Collection, Human Rubric Development, and Human Coding. During Confirmatory Analysis, lexical resources are used as dependent variables in

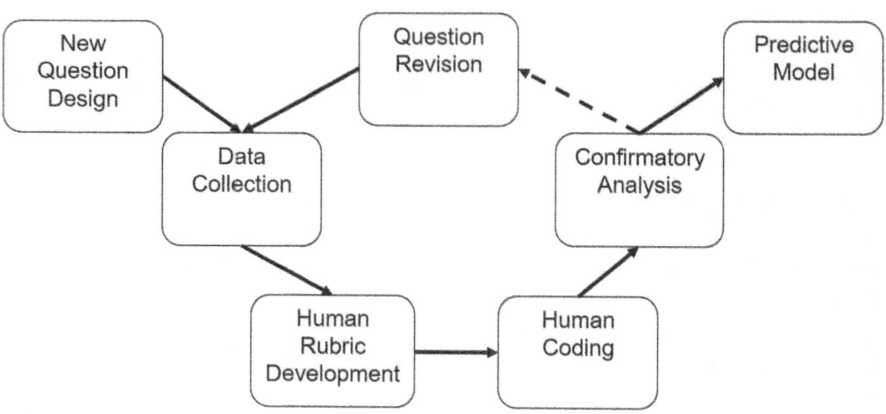

Fig. 15.2 Question Development Cycle (QDC). (Adapted from Urban-Lurain et al., 2013)

statistical and machine classification techniques to predict expert human coding, resulting in a Predictive Model that can be used to automate the scoring of new responses. This cycle may repeat several times. Our process, described in detail below, diverged from this cycle in two ways. First, we began with a validated set of questions aligned to a learning progression (Osborne et al., 2016), which we used as the basis for assessment development. Second, the result of our process was two predictive models, rather than one, as described below.

15.4.2 Item and Rubric Development

To date, three sets of questions in the contexts of sugar dissolving in water (Sugar), kinetic theory of gases (Gases) and spread of bacteria (Bacteria) have been developed with their scoring guides, each the product of an extensive process of development consisting of cognitive think alouds, and pilot testing with Amazon MTurk (Crowston, 2012). See (Wilson et al., in revision). Our rubric development approach defined the argument structures of an appropriate answer for a particular level of the learning progression. We examined the cognitive and linguistic demands for each question by unpacking what receptive processing is necessary, and what the student is required to produce. This helped us refine the questions, and identify key argument structures (e.g., claim, evidence) for each question. What emerged from this process was a set of analytic elements corresponding to these argument structures, which form the basis of an analytic rubric that was used by humans to code each student response (see Table 15.2). These analytic elements were then combined into a single holistic score using logic operators, which represents an holistic measure of student ability at the targeted level (see Table 15.3). Table 15.4 illustrates the correspondence of each question with the argumentation learning progression shown in Table 15.1.

Table 15.2 Alignment of assessment items with the argumentation learning progression

Gases items	Sugar and water items	Bacteria
G1 -> 0b	S1 -> 1c	B1 -> 0b
G2 -> 0b	S2 -> 0d	B2 -> 0d
G3 -> 1a	S3 -> 0d	B3 -> 1c
G4 -> 1a	S4 -> 2a	B4 -> 2d
G5 -> 1a	S5 -> 2c	B5 -> 2b
G6 -> 1c		B6 -> 1b
G7 -> 2a		
G8 -> 2c		

Table 15.3 Analytic coding rubric for item Sugar 1

Argumentation element	Component to be identified	Student example with component present
Claim	The sugar dissolved	The sugar was dissolved in the water. Because the grains of sugar are so small, they blend with the water and become unnoticeable to the naked eye.
Evidence	Cannot see the sugar or the sugar disappeared	The sugar dissolved into the water, forming a mixture, so that the individual sugar grains are no longer visible.
Evidence	Color change of image after mixing	The sugar dissolved into syrup, you can notice it has a darker shade but you still can't see the solid sugar. This is because when you mix sugar with water it turns into a liquid
Reasoning	The sugar broke into pieces	The sugar gets dissolved and the particles break up until they are spread out throughout the liquid and you no longer can see them with the naked eye.
Reasoning	The sugar bonded with the water molecules or the sugar mixed with water	The sugar grains dissolved in the liquid, essentially bonding with the liquid molecules.
Reasoning	The physical act of stirring	I think that the sugar in the cup dissolved in the water because after they stirred it, the sugar was no longer visible
Reasoning	Matter is conserved	I think that the sugar has dissolved because I have put sugar in water and the water appears clear. In science class we have learned matter cannot be created or destroyed. The sugar is therefore still in the cup.

If a component was present, the student received a 1 for that component. If the component was not present, the student received a 0 for that component. These components were then combined as described in Table 15.4 to produce a holistic score

Table 15.4 Holistic scores for Sugar 1 by combination of components

Holistic score	Combination	Student example
3	Claim **and** at least one Evidence **and** at least one Reasoning component present	The sugar dissolves after stirring it in the water for a few minutes, that's the reason they can't see the sugar particles anymore.
2	Claim **and** at least one Reasoning component present; **or** Claim **and** at least one Evidence components present	The grains of sugar dissolved after stirring the glass with a spoon.
1	Contains only one Claim **or** Evidence **or** Reasoning component present; **or** At least one Evidence **and** at least one Reasoning component present	The sugar disappeared and the sugar never turned back to its original shape
0	No claim, Evidence or Reasoning component present. Out of context response	What had happened to the grains of sugar?

15.4.3 Data Collection

Data for the Argulex project was collected in school districts across Northern California. After a short lesson on argumentation in science, students answered a set of argumentation-related science items on an online platform. The assessment was written in English, and students were asked to respond in English.

Sample Description The sample size of this study was 804 students, of whom 143 were EL students (18%), as determined by the state's English language proficiency scores. The data we present here were drawn from two sources, a private independent school (grades 5–8), and a set of five middle school (grades 6–8) science classrooms. Students in these classrooms came from a broad diversity of linguistic backgrounds, but the majority of students classified as ELs reported that they spoke Spanish at home.

15.4.4 Scoring

Human Coding Using Analytic Rubrics

The data were scored by eight human scorers across three institutions. Each institution scored data from one item set, focused on a single content area: either gases, sugar and water, or bacteria (see description of the items and rubrics above). At each institution, two human coders were trained on a set of items and associated rubrics. While MFRM analysis does not require high reliability between raters, allowing for rater-level parameters, training the machine to complete automated analysis requires a reliable data set. Thus the two coders went through multiple rounds of training using a random subset of 150 student responses to each item. Training rounds were iterated until interrater reliability (Cohen's kappa) between the coders was at least 0.6 on each rubric component. Then, the remaining data set was split into two subsets, including an overlapping set of 150 responses, and each coder scored one subset independently. The final Cohen's kappa value was calculated using the overlapping set of 150 responses coded independently; any scoring disagreements in this subset of responses was resolved by a third scorer and reported as the consensus score. To ensure the highest reliability possible, humans coded each response for the presence or absence (boolean logic) of several analytic elements as dictated by the rubrics. These analytic elements were then combined using the logic laid out in the rubrics. This process led to two data sets on which we could train the machines, one with the analytic codes assigned by the humans, and another with holistic scores calculated from the analytic codes.

Development of Automated Scoring Models

Both sets (analytic and holistic) of human-scored responses were used to develop automated text scoring models using a machine learning approach. Briefly, we used supervised machine learning to implement a text classification approach to assign each student written responses a classification or score (see Aggarwal & Zhai, 2012). During our machine learning process, each individual student response was treated as a document and the bins in the scoring rubric are treated as classes. The computerized scoring system then generated predictions on whether each given document is a member of each class. To generate these predictions we used an ensemble of eight individual machine learning algorithms (Jurka et al., 2013). Each algorithm votes independently on the categorization of a particular document, but these individual votes were combined to make a final categorization prediction of each response. Each individual algorithm was trained using a corpus of human-scored student responses. The computer model was generated using a ten-fold cross-validation approach, using consensus scores, when possible, resulting in the production of a set of predictive scoring models. We built a set of web-based applications that perform the necessary text parsing and supervised machine learning algorithms employed in this study. The R code is available at https://github.com/BeyondMultipleChoice/AACRAutoReport.

The machine was trained on both the raw analytic codes, and the holistic scores produced by those codes, producing two different models for predicting student scores. Figure 15.3 illustrates these two different approaches.

For the first approach, the machine was trained using the analytic codes, and holistic machine scores were calculated by combining the machine predicted

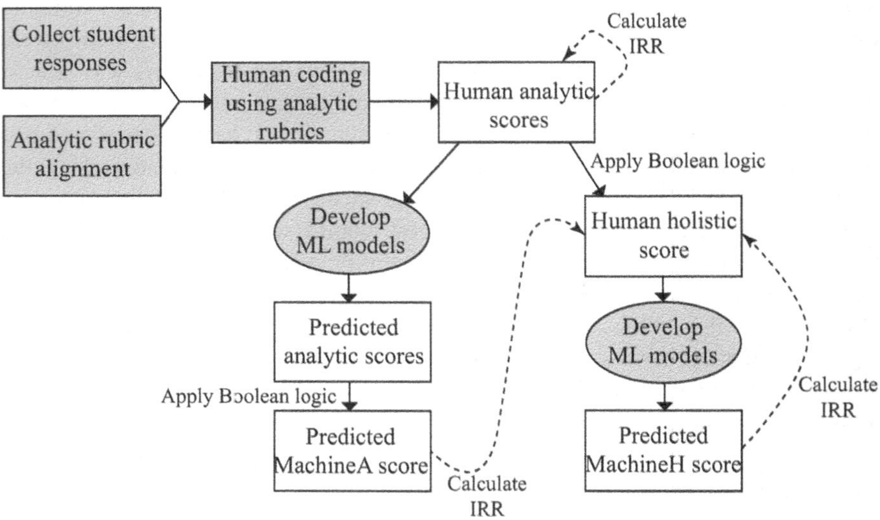

Fig. 15.3 Development of the MachineA and MachineH scoring models

Table 15.5 Interrater reliabilities and accuracy between human consensus and computer predicted scores

Item	Cohen's kappa HH-MA	Accuracy (HH-MA)	Cohen's kappa HH-MH	Accuracy (HH-MH)
S1	0.75	84.0	0.64	78.2
S2	0.82	91.2	0.77	88.6
S3	0.81	90.7	0.82	90.8
S4	0.73	82.5	0.59	74.3
S5	0.68	77.8	0.62	73.7
B1	0.93	96.3	0.90	94.4
B2	0.87	92.9	0.83	91.3
B3	0.69	78.9	0.64	76.3
B4	0.71	83.3	0.63	79.1
B5	0.74	92.7	0.72	73.7
B6	0.88	96.6	0.94	98.3
G1	0.63	91.3	0.63	91.0
G2	0.74	91.5	0.73	90.8
G3	0.60	73.0	0.59	72.8
G4	0.71	81.2	0.63	76.6
G5	0.71	87.0	0.67	85.1
G6	0.65	78.0	0.54	72.1
G7	0.58	78.1	0.42	73.2
G8	0.70	85.2	0.37	75.9

Note: HH refers to human holistic scores. MA refers to machine analytic scores. MH refers to machine holistic scores

analytic codes using the logic dictated by the rubrics. We call the scores predicted by this model the machine analytic score (MachineA). For the second approach, the machine was trained only on the holistic scores that emerged from the human coding, after combining the analytic elements. We call the scores predicted by this model the machine holistic score (MachineH). Thus, each response in our data set was assigned three scores; the human-assigned score, the MachineH predicted score and the MachineA predicted score. We have developed a set of machine analytic, and machine holistic scoring models for each of the 19 items on the assessment. Table 15.5 shows interrater reliabilities and accuracy between human consensus scores and computer predicted scores generated by each approach.

15.5 Data Analysis

15.5.1 Many-Facet Rasch Measurement (MFRM)

The purpose of Rasch analysis is to test whether the data fit a measurement scale built to be invariant across test-takers. The probability of a student receiving a correct response on an assessment is modeled as a function of parameters related to the test-taker (performance, often referred to as "ability"), and the items themselves (difficulty). The basic Rasch model cannot be applied to answer questions about parameters related to the rater, such as severity.

MFRM analysis extends the Rasch measurement model to include investigation of the impact of facets in addition to person ability and item difficulty on the rating process (Linacre, 1989; Myford & Wolf, 2003). These facets are re-mapped onto a linear, equal-interval logit scale, creating a single frame of reference for comparisons within and between the facets.

In the illustration in Fig. 15.4, we see how each facet can be conceptualized as a dimension of the data. Each small cube in the image represents a score given to a student by a rater on an item. In the case of this analysis, an additional fourth

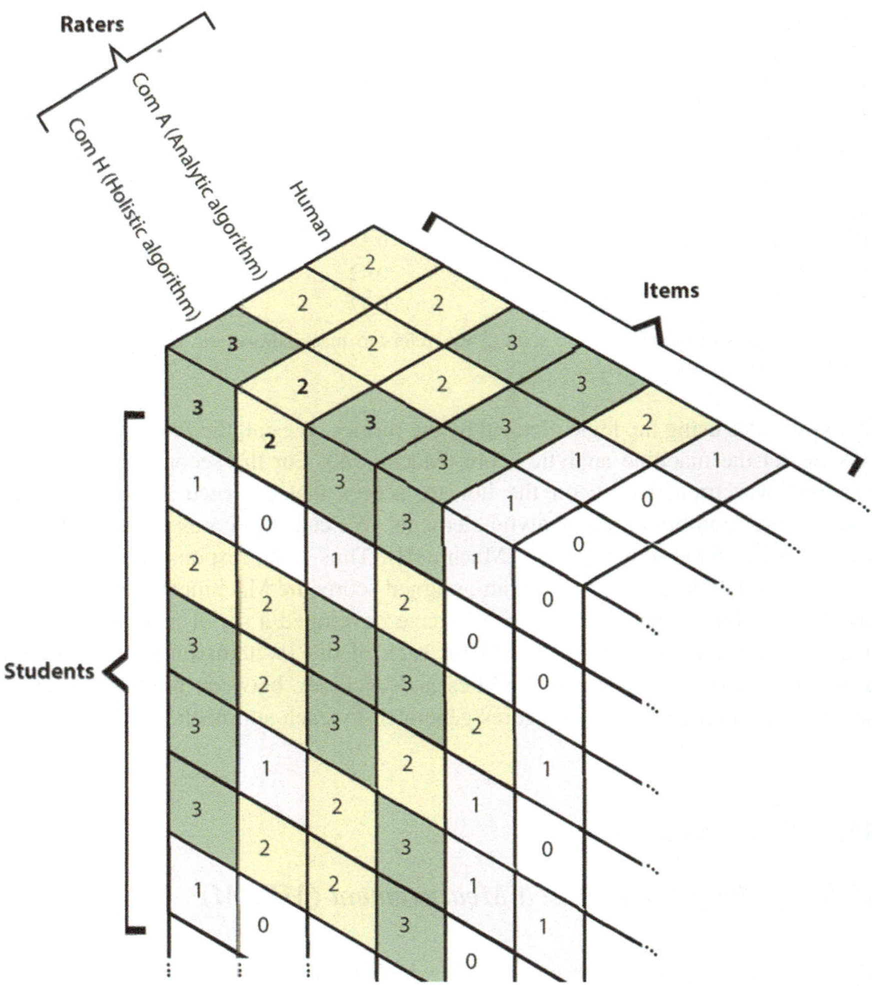

Fig. 15.4 Model for data facets in MFRM

dimension of the cube was added – the EL status of the students being rated. The simplest form of this model is as follows:

$$\mathrm{Log}(\mathrm{Pnijkl}/\mathrm{Pnij}(\mathrm{k}-1)) = \mathrm{Bn} - \mathrm{ELm}\text{-}\mathrm{Di} - \mathrm{Cj} - \mathrm{Fk} \tag{15.1}$$

In Eq. 15.1 above, the logit is the equivalent of the dependent variable, and each facet functions as an independent variable that impacts the log odds. The mathematical model shown in Eq. 15.1 specifies that that the probability of a student n with EL status m receiving a rating of k rather than k-1 on item i by rater j depends on the additive effects of difficulty introduced by EL status (Elm), difficulty of the trait (Di), the severity of the rater (Cj), the level of performance of the ratee (Bn), and the relative difficulty of scale category k when compared to scale category k-1 across all traits and all raters (Fk). This last parameter represents the point at which the probability is 50% of a ratee being rated in one or the other of these two adjacent categories, given that the student is in one of them.

In some ways, MFRM is analogous to an ANOVA approach, in that it allows us to investigate group-level rater effects. But with ANOVA, possible interaction effects contaminate possible main effects, making the interpretation more difficult. This is not true with MFRM. With an MFRM approach, we can look at group-level main effects for the rater (human, computer holistic, computer analytic), for the ratees (the students), and for a "trait" assigned to the ratees (EL status). These are the facets. The contribution of each facet can be separated out and examined independently of the other facets to determine to what extent they are functioning as intended, or for interactions to reveal possible bias. This is particularly useful for researchers who are interested in questions about differential levels of severity (or leniency) of particular raters (or in our case types of raters) scoring student responses. And more importantly for our research question, the output from an MFRM analysis provides rater effect interactions to answer questions about whether or not rater severity is invariant across sub-groups (EL status).

Measuring Bias with MFRM

Bias in MFRM is defined as differential rater severity (negative bias) and differential rater leniency (positive bias). Differential rater severity/leniency measures the tendency of a rater to assign scores to a particular group of students that are, on average, either lower or higher than the measurement model would expect for that group (Myford & Wolfe, 2004). Thus, an MFRM bias analysis helps us to identify interactions between facets that suggest a consistent deviation from what we would expect given all the responses (Myford & Wolfe, 2003). In order to answer our research question, we compared bias in the automated scoring (both holistic and analytic) to bias in the human scoring on which the machines were trained.

The Argulex project utilizes the Facets computer program to analyze the data using MFRM (Linacre, 2020). Parameter interactions or facet interactions (raters and EL status, items and EL status, items and raters, etc.) are permitted in Facets, which has the flexibility to define a wide range of models. For example, to generate rater bias estimates for EL status, a five-facet model with rater-EL status interaction can be used:

$$\mathrm{Log}(\mathrm{Pnijkl}/\mathrm{Pnij}(\mathrm{k}-1)) = \mathrm{Bn} - \mathrm{Emi} - \mathrm{Lim} - \mathrm{Di} - \mathrm{Cj} - \mathrm{Fk} \tag{15.2}$$

Thus the facets were: (1) student, (2) item, (3) rater type, and (4) EL status. This model is different from Eq. 15.1 because it includes interactions between a student facet and a rater facet: Emi is the severity of each rater i for each student with EL status m; and Lim is the ability of each group of students with EL status m as rated by each rater i. Computation of bias in *Facets* includes: (i) The program calculates estimates for the measures of each element, and the structure of the rating scale, and holds these constant (or fixed). (ii) The program calculated residuals by subtracting expected values of the observations from the observed values of the observations. (iii) The residuals for each interaction term (e.g., EL students rated by the human) are summed. (If this sum is not zero, this indicates an interaction). (iv) Finally, the program estimates the size and statistical significance of the interaction.

A second analysis was run to investigate the impact of item difficulty, as described in the Findings Section under the heading: "Difficulty Analysis." This analysis included five facets, because item difficulty was coded as an additional dummy facet.

15.6 Findings

15.6.1 Summary of Findings

Our first research question was: *Is machine scoring of an argumentation task differentially more severe than human scoring for students who have EL status, compared to students who do not have EL status?*

Our results indicate that across the full study population, the machine models are more severe than the human scorers. In addition, across both of the two machine models and the human consensus score, responses from EL students received lower scores. Machine H (the holistic model) is the most severe rater regardless of EL status, and human scoring is the most lenient.

Our second research question was: *Do differences in severity between the automated scoring and the humans vary across item difficulty and machine learning model?*

A closer analysis using MFRM reveals that Machine H, while the most severe, is less differentially severe across EL status groups than Machine A, particularly for the most difficult items. In other words, the model trained on the holistic scores is harsher than the human on every student equally, while the model trained on the analytic scores appears to be differentially harsher on EL students, thus actually widening existing gaps between linguistic groups.

15.6.2 Overview: The Facets Ruler

Figure 15.5 is a ruler, or the Facets equivalent of the Rasch Wright Map. It maps the elements of the four facets on their common log odds scale. The facets were: (1) student, (2) item, (3) rater type, and (4) EL status. Because all facets are on a common scale, we can compare elements within and between facets.

- The first column (Measr) of the ruler in Fig. 15.5 represents the linear, equal-interval logit scale upon which all facets in the analysis are positioned, creating a single frame of reference for comparisons within and between the facets.
- The second column (+Student) of the ruler in Fig. 15.5 represents the students' performance measures. These measures are produced as a result of including the B parameter in the mathematical model statement shown in Eq. 15.1. Performance measures are single number summaries on the logit scale of each ratee's tendency to receive high or low ratings across raters on the four traits. The ratees are ordered from higher performing (at the top of the column) to lower performing (at the bottom of the column). An asterisk denotes three ratees, while a dot denotes a single ratee.
- The third column (-Item) of the ruler in Fig. 15.5 represents the item measures. These measures are produced as a result of including the D parameter in Eq. 15.1. Items appearing higher in the column were more difficult for ratees to receive high ratings on than items appearing lower in the column.
- The fourth column (-EL) of the ruler in Fig. 15.5 represents a ratee trait: EL status. These measures are produced as a result of including the E and L parameters in Eq. 15.2. Traits appearing higher in the column were more difficult for ratees to receive high ratings on than traits appearing lower in the column.
- The fifth column (-Rater) displays the three rater severity measures. These measures are produced as a result of including the C parameter in Eq. 15.1. The raters are ordered in terms of the level of leniency/severity each rater exercised when evaluating the ratees. More severe raters appear at the top of the column, while more lenient raters appear at the bottom of the column.
- The sixth (S.2) and seventh (S.3) columns display the rating scales. A horizontal line across a column indicates the point at which the likelihood of a ratee receiving the next higher rating begins to exceed the likelihood of that ratee receiving the next lower rating (i.e., a rating scale category "threshold"). This is the F parameter in Eq. 15.1.

Rater Severity

Our results indicate that across the full study population, the machine is more severe than the human. In the ruler in Fig. 15.5, we see that the machine that was trained on the holistic scores (MachineH) is the most severe, followed by the machine trained

```
+-----------------------------------------------------------------------------+
|Measr|+Student     |-Item               |-EL     |-Rater     | S.2 | S.3 |
|-----+-------------+--------------------+--------+-----------+-----+-----|
|   2 + .           +                    +        +           + (3) + (2) |
|     |             |                    |        |           |     |     |
|     |             |                    |        |           |     |     |
|     | .           |                    |        |           |     |     |
|     |             |                    |        |           |     |     |
|     | .           | B6_1b  G8_2c       |        |           | --- |     |
|     | .           |                    |        |           |     |     |
|     | .           | B4_2d              |        |           |     |     |
|   1 + .           + G7_2a              +        +           +     +     |
|     | *.          |                    |        |           |     |     |
|     | .           |                    |        |           |     | --- |
|     | **.         |                    |        |           |     |     |
|     | **.         | G3_1a  G5_1a       |        |           |  2  |     |
|     | ****.       | G6_1c              |        |           |     |     |
|     | *****.      | G4_1a  S4_2a       |        |           |     |     |
|     | ******.     |                    | EL     | MachineH  |     |     |
*   0 * *****.      * B3_1c  S5_2c       *        * MachineA  *     *  1  *
|     | ********.   | S3_0d              | not EL | Human     | --- |     |
|     | *********.  | B1_0b  S1_1c  S2_0d |       |           |     |     |
|     | *******.    |                    |        |           |     |     |
|     | ********.   | B5_2b              |        |           |     |     |
|     | ******.     |                    |        |           |  1  |     |
|     | ******.     |                    |        |           |     | --- |
|     | *****.      |                    |        |           |     |     |
|  -1 + ****.       +                    +        +           +     +     |
|     | ***         |                    |        |           |     |     |
|     | **.         | B2_0d              |        |           |     |     |
|     | *.          |                    |        |           | --- |     |
|     | *.          |                    |        |           |     |     |
|     | *.          |                    |        |           |     |     |
|     | .           | G2_0b              |        |           |     |     |
|     | .           |                    |        |           |     |     |
|  -2 + .           +                    +        +           +     +     |
|     | .           |                    |        |           |     |     |
|     | .           | G1_0b              |        |           |     |     |
|     | .           |                    |        |           |     |     |
|     | .           |                    |        |           |     |     |
|     | .           |                    |        |           |     |     |
|     |             |                    |        |           |     |     |
|     | .           |                    |        |           |     |     |
|  -3 + .           +                    +        +           +     +     |
|     | .           |                    |        |           |     |     |
|     |             |                    |        |           |     |     |
|     |             |                    |        |           |     |     |
|     |             |                    |        |           |     |     |
|     | .           |                    |        |           |     |     |
|     | .           |                    |        |           |     |     |
|     | .           |                    |        |           |     |     |
|  -4 + .           +                    +        +           + (0) + (0) |
|-----+-------------+--------------------+--------+-----------+-----+-----|
|Measr| * = 9       |-Item               |-EL     |-Rater     | S.2 | S.3 |
+-----------------------------------------------------------------------------+
```

Fig. 15.5 The facets ruler maps the elements of the facets on a common log odds scale

on the analytic scores (MachineA). The human is the most lenient scorer (Human). In other words, for all students, regardless of EL status, the computer will be more likely to assign a lower score than the human.

Figure 15.6 shows the mean scores for EL students and non-EL students across all three raters. Note that MachineH is the most severe rater regardless of EL status, and the human is the most lenient. Also notice that it appears that the difference between the human and MachineA is more pronounced for EL students than for non-EL students. This was something we intended to explore further with Facets.

Table 15.6, the rater measurement report output from Facets, quantifies this overall difference in leniency. The column with the heading "Measure" indicates the logits at which the rater falls on the Rasch scale. The Human is at −.07 logits. The machines trained on the analytic and holistic scores are at −.02 and .09 logits respectively. A higher logit value corresponds to a more severe rater. In assigning

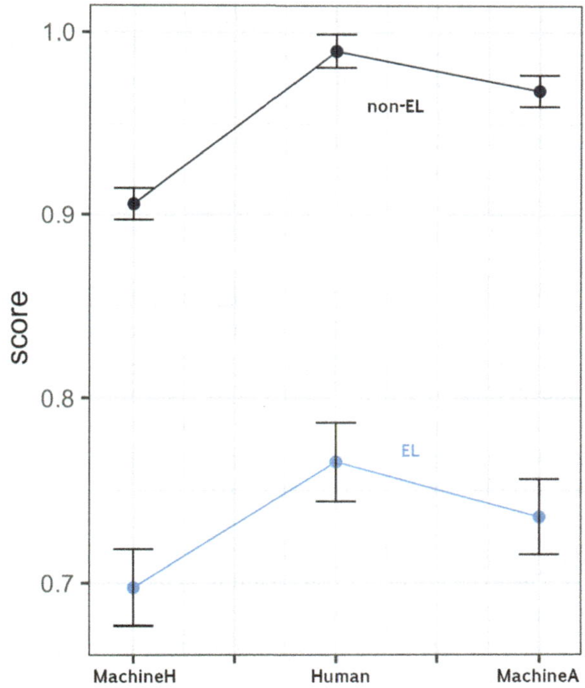

Fig. 15.6 Means for EL students and Non-EL students across all three raters

Table 15.6 Rater measurement report

Rater	Total Count	Fair Average	Measure	Model S.E.	Infit MnSq	Outfit MnSq
Human	12,281	.99	−.07	.01	1.04	1.05
MachineA	12,281	.96	−.02	.01	.97	.97
MachineH	12,281	.88	.09	.01	.98	.97

Sample separation: 6.52
Reliability (not inter-rater): .98 model
Model, fixed (all same) chi-square: 86.6, d.f.: 1, significance (p): .00
Model, random (normal) chi-square: 2.0, d.f.: 1, significance (p): .16
Inter-rater agreement = 84.4%

this measure, the software has normalized all facets for the same Rasch scale. The "Fair Average" takes the measure and translates it back into the rating scale of the average item on the assessment. So we can think of it as the average we would expect an average student to perform on an average item, scored by that rater on average. In other words, the model predicts that the average student would receive a score of about 0.99 on the average item when scored by a human (most items were scored on a scale from 0–2, with some 0–1, and others 0–3). But when that same student is scored by the machine algorithm trained on the raw analytic scores (MachineA), the model predicts that they will receive a score of .96. And when that same student is scored by the machine algorithm trained on the combined holistic scores (MachineH), the model predicts that they will receive a score of .88.

At the bottom of this table is a measure of the significance of this separation: the reliability of the separation index. A reliability of 1.00 indicates that the raters are very different. The reliability for the sample is quite high, at .98, indicating that the human, the machine trained on analytic scores, and the machine trained on holistic scores are scoring with different levels of severity. The fixed effect chi-square test for the raters' facet tests the hypothesis that this is not true, and that both machine approaches and the human raters exercised the same level of severity when evaluating students, after accounting for measurement error. The probability of this is .00, as shown in the bottom row of the table, again indicating differential severity across the three raters.

We can do the same analysis for the student status facet. Doing so indicates that students classified as EL students produce lower scoring performances than their peers. The table below quantifies the overall difference in performance of EL students and non-EL students across the whole assessment. The column with the heading "Measure" indicates the mean logits assigned by the Rasch model for that group of students. The fair average tells us that the model predicts that the average EL student would score about 0.76 on the average item (most items were scored on a scale from 0–2, with some 0–1, and others 0–3), while their non-EL peers would score about 0.98 (Table 15.7).

The fixed effect chi-square test here tests the hypothesis that students who are EL students and students who are not EL students experience the same level of difficulty, after accounting for measurement error. The probability of this is .00, as shown in the table.

This story becomes more complex when we look at how this difficulty is distributed across items. In Fig. 15.7, we see two Facets rulers overlaid – one with

Table 15.7 EL status measurement report

Student Status	Total Count	Fair Average	Measure	Model S. E.	Infit MnSq	Outfit MnSq
Not EL	32,109	.98	−16	.01	1.00	1.01
EL	4734	.76	.16	.02	.95	.95

Sample separation: 13.70
Reliability: .99
Model, fixed (all same) chi-square: 188.7, d.f.: 1, significance (p): .00

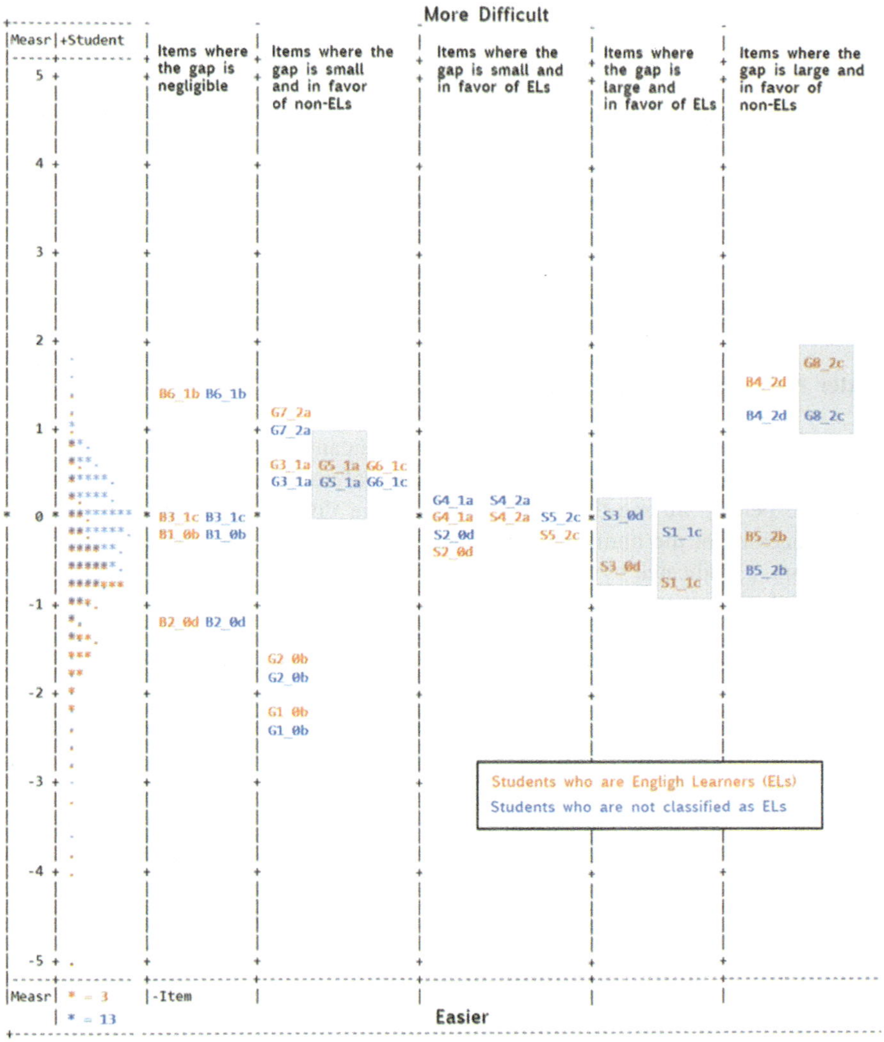

Fig. 15.7 Two facets rulers overlaid – EL students (in orange) and Non-EL students (in blue)

just EL students (in orange), and one with just non-EL students (in blue). We see that in column 2, EL students are distributed lower on the logit scale than non-EL students. In addition, we see that for certain items the gap between EL students and their non-EL peers appears to be almost negligible (column 3), while for other items the gap appears to be very large (column 7). The gaps that are statistically significant are: B5, G8, G5, S1, and S3 (indicated in grey). For all of the sugar items, and one of the gas items (6 out of 19 items in total) this gap is reversed, where EL students are scoring higher than non-EL students on the items across all scorers. The sugar items are anomalous in general, as they tend not to align well with the distribution predicted by the learning progression. This is a fascinating pattern that we would like to explore in future work. For the purposes of this study, we are interested in looking more closely at questions like G8 and B5, where the gap is large and significant ($p = .0069$, $.0003$ respectively), and reflects the pattern of unequal outcomes that we see across the entire assessment (EL students scoring lower). We want to compare those gaps across human and machine scoring. In order to do this, our next step was to run a bias/interaction analysis in Facets.

15.6.3 Findings of the Bias/Interaction Analysis

Rater × EL Status Interactions

Overall, it appears that neither machine was significantly more severe or lenient than the humans in grading students across a selection of items, regardless of EL status. In the table below, we see the bias size, indicating differential severity/leniency of either machine or the human, is small. More importantly, none of these biases are significant, as indicated by the p-values associated with each t-statistic. This statistic tests the hypothesis that the mean observed-expected residuals for EL students across the assessment is the same as the observed-expected residuals for non-EL students. Because the p-values are high, we cannot rule out the possibility that any apparent difference in these means is simply chance. However, it is interesting to note here that MachineA does appear to manifest a small bias (−.02) against EL students, while MachineH, the most severe rater, appears to manifest a small bias (.01) for EL students (Table 15.8).

In the pairwise table below, which again does not show any significant interactions, we see how this pattern might potentially manifest in scoring decisions. For students who are not EL students, found in the bottom three rows of the table, the target measure across all three algorithms is the same. In other words, all of these students should expect to score the same on the assessment, regardless of if they are being scored by the human, the analytic algorithm, or the holistic algorithm. But for EL students this may not be the case. For all three raters, there appears to be a slight difference in the way the EL students would be scored, although this difference is not

Table 15.8 Rater × EL status Bias/Interaction report

Rater	Student status	Obs-Exp Avg.	Bias size	t	p	Infit MnSq	Outfit MnSq
MachineA	EL	−.01	−.02	−.47	.65	.9	.9
MachineH	Not EL	.00	.00	−.13	.90	1.0	1.0
Human	Not EL	.00	.00	−.04	.97	1.0	1.1
MachineA	Not EL	.00	.00	.17	.87	1.0	1.0
Human	EL	.00	.00	.11	.92	1.0	1.0
MachineH	EL	.01	.01	.36	.72	1.0	.9

Table 15.9 Rater × EL status pairwise contrast report

Status	Meas.	Obs-Exp Avg.	Rater	Meas.	Obs-Exp Avg.	Rater	Contrast	t	p
EL	.14	.01	MachineH	.18	−.01	MachineA	−.03	−.58	.56
EL	.15	.00	Human	.18	−.01	MachineA	−.02	−.41	.68
EL	.14	.01	MachineH	.15	.00	Human	−.01	−.18	.86
Not EL	−.16	.00	MachineH	−.16	.00	Human	.00	.06	.95
Not EL	−.16	.00	Human	−.16	.00	MachineA	.00	.15	.88
Not EL	−.16	.00	MachineH	−.16	.00	MachineA	.00	.21	.84

statistically significant. MachineA is slightly more severe than the human, placing EL students an average of .02 logits below the human (row 3). MachineH, while the most severe rater overall, appears if anything, differentially lenient toward EL students' responses, placing EL students' responses an average of .01 logits above the where the human placed them (row 2). Comparing MachineA and MachineH (row 1), we see that using the former algorithm could be placing EL students' responses .03 logits lower than they would have been if they had been scored by MachineH (Table 15.9).

Note that while these relationships are not significant at the assessment level, our EL sample was small compared to the larger population, and our test form was modified by experts in the field of EL students and assessment, and in the field of argumentation, to ensure lower linguistic complexity while maintaining content validity and rigor. In addition, the interaction between linguistic subgroup status and rater did become significant when we looked at individual items, as described in the next section.

Item × Rater × EL Status Interactions

Even though there is not significant bias apparent across the humans and machines over the entire assessment, there are several individual items where we observe differential leniency/severity between the computer and the human based on EL status, with significance ($p < .05$). For example, in the table below are listed the interactions between rater, item and EL status that resulted in a calculated bias term over .30 (the dark grey column). In other words, the students with EL status as indicated in column 2 of this table, scored by a rater indicated in column 3, on an item indicated in column 1, received an average observed score that was at least .3 logits lower (or higher) than predicted by the model if there were no interactions. The biases where the corresponding probability are below .05 are considered significant (light grey column) (Table 15.10).

Table 15.10 Rater × EL status × Item Bias/Interaction report

Item	Status	Rater	Obs-Exp Avg.	Bias Size	S.E.	t	d.f.	p	Infit MnSq	Outfit MnSq
G8_2c	EL	MachineH	–.19	–0.95	0.37	-2.58	60	0.01*	1.7	0.8
B5_2b	EL	Human	–.41	–0.52	0.2	-2.64	33	0.01*	0.6	0.6
G8_2c	Not EL	MachineH	–.18	–0.48	0.08	-5.79	485	0.00*	1.6	1.3
G8_2c	EL	MachineA	–.13	–0.48	0.27	-1.74	60	0.09	1.2	0.7
B5_2b	EL	MachineA	–.34	–0.44	0.2	-2.23	33	0.03*	0.6	0.7
G1_0b	EL	Human	–.06	–0.4	0.23	-1.72	119	0.09	0.9	0.8
G5_1a	EL	MachineH	–.19	–0.39	0.16	-2.4	87	0.02*	1.8	1.1
B4_2d	EL	MachineA	–.12	–0.37	0.3	-1.22	37	0.23	1	1.1
G2_0b	EL	Human	–.06	–0.35	0.21	-1.65	118	0.10	0.8	0.8
B4_2d	EL	Human	–.12	–0.32	0.29	-1.11	37	0.27	0.9	1
S3_0d	EL	MachineH	.06	0.3	0.2	1.5	114	0.14	1.1	1.1
S1_1c	EL	MachineH	.24	0.32	0.11	3.05	118	0.00*	0.6	0.9
S3_0d	EL	MachineA	.07	0.34	0.2	1.72	114	0.09	1.1	1.1
B6_1b	EL	Human	.08	0.36	0.4	0.9	23	0.38	1.5	1.4
S3_0d	EL	Human	.11	0.49	0.2	2.49	114	0.01*	1.1	1.1

Fixed (all = 0) chi-square: 224.0 d.f.: 114 significance (probability): .00

Let's use item G8 as an example, as it has the highest significant negative bias listed in the table. MachineH was biased significantly when assigning scores to both EL students (bias = – .95, $p < .05$) and non-EL students (bias = – .48, $p < .0005$). MachineA was biased (although this is not significant) when assigning scores to EL students on G8 (bias = – .48, $.05 < p < 0.1$) as well. But MachineA did not demonstrate a similar negative bias toward non-EL students. In fact, MachineA was significantly more lenient on non-EL students on G8, scoring them on average .23 logits ($p < .0005$) higher than the model would predict if there were not interactions. Note that the human also manifested a significant positive bias toward non-EL students on G8, but that this bias manifested in only .18 logits ($p < .005$). The human did not manifest any measurable significant bias against EL students on this item.

Thus, MachineH was significantly more severe on all students on this question, and this bias impacted EL students about twice as much as non-EL students. MachineA, on the other hand, was only half as severe on EL students as MachineH, which at first glance may seem like MachineA is less biased. But what we are really interested in is differential severity of the two machine scores across these groups of students. In addition to being more severe against EL students (bias = – .48, $.05 < p < 0.1$), MachineA was also more lenient on non-EL students (bias = .23, $p < .0005$). That difference sums to .71 logits. The result is that even though an EL student scored by MachineH will receive a lower score on average than an EL student scored by MachineA, the gap in performance between EL students and non-EL students scored by MachineA will be wider than that of students scored by MachineH.

The table below shows pairwise EL-status comparisons of G8 for each rater to illustrate this widened gap (Table 15.11).

This table shows:

- the logit score for EL students scored by the humans on G8 were about .22 logits lower on average than the logit scores of non-EL students (not significant)

Table 15.11 Rater × EL status × item pairwise contrast report for item G8

		Not EL		EL					
Item	**Rater**	**Meas.**	**Obs-Exp Avg.**	**Meas.**	**Obs-Exp Avg.**	**Contrast (Not EL-EL)**	**t**	**d.f.**	**p**
G8_2c	MachineA	1.07	.12	1.78	.13	–.7	-2.52	65	0.014*
G8_2c	MachineH	1.90	–.18	2.37	–.19	–.47	-1.25	66	0.22
G8_2c	Human	1.08	–.10	1.30	–.01	–.22	-0.96	68	0.34

Table 15.12 Rater × EL status × item pairwise contrast report for item B5

		Not EL		EL					
Item	**Rater**	**Meas.**	**Obs-Exp Avg.**	**Meas.**	**Obs-Exp Avg.**	**Contrast (Not EL - EL)**	**t**	**d.f.**	**p**
B5_2b	Human	−.63	.00	−.11	−.41	−.52	- 2.53	39	0.02*
B5_2b	MachineA	−.59	−.00	−.14	−.34	−.44	- 2.15	39	0.04*
B5_2b	MachineH	−.58	−.08	−.22	−.19	−.36	- 1.74	39	09

- the logit score for EL students scored by MachineH were about .47 logits lower than the logit scores of non-EL students (not significant)
- the logit score for EL students scored by MachineA (trained on the human analytic scores) were about .70 logits lower on average than the logit scores of non-EL students (p < .05)

The increase in bias from the human to the more severe MachineH was .23 logits, representing almost 100% increase. But the increase in bias from the human to MachineA for G8 was .48 logits, representing an over 200% increase.

B5 is another interesting item to look more closely at. For this item, the human appears to manifest a significant bias against EL students (bias = − .52, p < .05), and no significant bias for or against non-EL students. The table below shows pairwise EL-status comparisons of B5 (Table 15.12).

This table shows:

- the logit score for EL students scored by the humans on B5 were about .52 logits lower on average than the logit scores of non-EL students (p < .05)
- the logit score for EL students scored by MachineH were about .36 logits lower than the logit scores of non-EL students (not significant)
- the logit score for EL students scored by MachineA (trained on the human analytic scores) were about .44 logits lower on average than the logit scores of non-EL students (p < .05)

Here we see a slightly different pattern than we did with G8. In this case, the human manifested the most bias with a gap of .52 logits between EL students and non-EL students. However, the increase in bias between MachineA and the human was relatively small, representing only .08 logits, or about 20%. For comparison, the increase in bias from the human to MachineA for G8 was .48 logits (from .22 to .70), representing an over 200% increase. And once again, we see MachineH, which is generally the most severe scorer, does not manifest a significant gap between EL students and non-EL students on B5.

It is interesting to note that these items were more difficult for EL students overall. EL students were scoring .4 logits lower on G8 and B5 than predicted by the model without interactions. But the patterns of differential severity between the humans and the machines, and between the two machine algorithms on these questions suggest that the story is more complex.

Difficulty Analysis

One possible explanation for the difference we see in differential severity of raters across sub-groups between questions like B5 and G8 are their difficulty. We were curious about if there was a consistent difference in the patterns of bias across items with different difficulty levels. So we took items over 1 logits to be difficult, items between −1 and 1 logits to be medium, and items below −1 logits to be easy. We then assigned dummy variables to these items in order to analyze whether the bias effect was dependent on the difficulty of the item (Table 15.13).

We found that MachineH is significantly more severe on hard items for all students, regardless of EL status, by an average of .21 logits. The human and the machine analytic scoring, on the other hand, were both significantly just a bit more lenient on all students on the hard items, by an average of only .08 and .09 logits respectively. On the easy and medium items, there was no significant bias (Table 15.14).

In general, the hard items on the assessment were more difficult for EL students. EL students scored on average .26 logits lower than their non-EL peers on these items (Table 15.15).

Table 15.13 Assessment items sorted by difficulty

Easy	Medium		Hard
G1_0b	G3_1a	B5_2b	B4_2d
G2_0b	G4_1a	S1_1c	B6_1b
B2_0d	G5_1a	S2_2c	G7_2a
	G6_1c	S3_0d	G8_2c
	B1_0b	S4_2a	
	B3_1c	S5_2c	

Table 15.14 Rater × difficulty Bias/Interaction report

Rater	Item difficulty	Obs-Exp Avg.	Bias size	t	p	Infit MnSq	Outfit MnSq
MachineH	Hard	–.08	–.21	-5.34	.00*	1.2	1.3
Human	Easy	–.02	–.09	-1.79	.07	.8	.9
MachineA	Medium	–.01	–.01	-1.05	.29	1.0	1.0
Human	Medium	.00	–.01	-0.51	.61	1.0	1.0
MachineA	Easy	.00	.00	-0.09	.93	.8	.9
MachineH	Medium	.01	.02	1.56	.12	1.0	1.0
Human	Hard	.04	.08	2.46	.01*	1.1	1.2
MachineH	Easy	.02	.09	1.79	.07	.8	.9
MachineA	Hard	.04	.09	2.62	.01*	1.1	1.1

Table 15.15 EL Status × difficulty pairwise contrast report

	Not EL		EL					
Difficulty	Meas.	Obs-Exp Avg.	Meas.	Obs-Exp Avg.	Contrast (Not EL - EL)	t	d.f.	p
Hard	–.02	.01	.24	–.07	–.26	-3.00	633	.00*
Easy	–.02	.00	.11	–.03	–.13	-1.73	1225	.08
Medium	.00	.01	–.03	.02	.04	1.54	4203	.12

These patterns may be a reflection of biases built into human coding. The human raters indeed tended to be a little more severe on EL students on the hard items, but this was not significant, and translated into human raters scoring EL students an average of about .10 logits lower on those items. But this bias appears to have been magnified by the machine scoring. For the machine analytic score, the bias was not significant, but MachineH tended to score EL students' responses an average of .29 logits lower than predicted by the model without interactions ($.05 < p < .1$), magnifying the bias of the humans by almost 3 times. MachineH was more severe than predicted by the model on EL students' responses an average of .38 logits, magnifying the potential bias of the humans by almost 4 times ($p < .01$) (Table 15.16).

But again, we need to dig closer to look for differential severity. According to the pairwise table below, the human scored EL students .19 logits lower than non-EL students on average, a difference that was not significant. Despite the drastic increase in severity manifested against EL students and non-EL students in the table above, MachineH did not appear to widen that gap (.19 logits, also not significant). But MachineA widened this gap on the harder items by 100% (.40 logits, $p < .05$). There was no significant gap between EL students and non-EL students for any scorer on the medium or easy questions (Table 15.17).

15.7 Discussion

15.7.1 Approaches to Training the ML Models

The rapidly expanding literature on machine learning indicates that human bias can be magnified by computer. And indeed we see this effect in our results. While both machine models (the model trained on the analytic scores, and the model trained on the holistic scores) were more severe than the human, MachineH was significantly more severe than the other raters across all items and students. But MachineA was more differentially severe than MachineH (or the human consensus scores). While this was not significant across the entire assessment, it became significant when we looked at the items that were the most difficult, at the top of the argumentation learning progression. This pattern illustrates why an analysis such as this one can be so powerful. Had we not included the EL-status facet in our modeling, for example with a basic Rasch model, we would have noticed the increase in severity of the machine trained on the holistic scores (MachineH), and the relatively small increase in severity of the machine trained on the analytic scores (MachineA). But we would not have noticed that MachineA, while more closely matched in overall severity to the human, was more differentially severe on EL students than either the human, or MachineH.

Table 15.16 Rater × EL status × difficulty bias/interaction report

Rater	Difficulty	EL status	Obs-Exp Avg.	Bias size	t	p	Infit MnSq	Outfit MnSq
MachineH	Hard	EL	–.09	–0.38	-2.37	0.02*	1.2	0.9
MachineA	Hard	EL	–.08	–0.29	-1.95	0.05*	1.1	0.9
Human	Easy	EL	–.05	–0.21	-1.74	0.08	0.9	0.9
MachineH	Hard	Not EL	–.07	–0.2	-4.89	0.00*	1.3	1.3
Human	Hard	EL	–.03	–0.1	-0.73	0.47	1.2	1.1
MachineA	Easy	EL	–.02	–0.08	-0.63	0.53	0.8	0.8
Human	Easy	Not EL	–.01	–0.06	-1.21	0.23	0.8	0.9
MachineH	Easy	EL	–.01	–0.05	-0.41	0.68	0.9	0.9
MachineA	Medium	Not EL	–.01	–0.02	-1.23	0.22	1.0	1.0
Human	Medium	Not EL	–.01	–0.01	-0.88	0.38	1.0	1.0
MachineA	Easy	Not EL	.00	0.01	0.16	0.87	0.8	0.9
MachineA	Medium	EL	.01	0.01	0.32	0.75	0.9	0.9
MachineH	Medium	Not EL	.01	0.02	1.21	0.23	1.0	1.0
Human	Medium	EL	.02	0.04	0.92	0.36	1.0	1.0
MachineH	Medium	EL	.03	0.05	1.27	0.20	1.0	1.0
Human	Hard	Not EL	.04	0.09	2.75	0.01*	1.1	1.2
MachineH	Easy	Not EL	.02	0.11	2.1	0.04*	0.8	0.9
MachineA	Hard	Not EL	.05	0.11	3.26	0.00*	1.1	1.1

Table 15.17 Rater × EL status × difficulty pairwise contrast report

		Not EL		**EL**					
Difficulty	**Rater**	**Meas.**	**Obs-Exp Avg.**	**Meas.**	**Obs-Exp Avg.**	**Contrast (Not EL-EL)**	**t**	**d.f.**	**p**
Hard	MachineA	–.14	.05	.26	–0.08	–.40	- 2.65	207	0.01
Hard	Human	–.16	.04	.03	–0.03	–.19	- 1.39	212	0.17
Hard	MachineH	.29	–.07	.48	–0.09	–.19	- 1.11	211	0.27
Easy	MachineH	–.02	.02	14	–0.01	–.16	- 1.21	408	0.23
Easy	Human	–.01	–.01	.14	–0.05	–.15	- 1.12	407	0.26
Easy	MachineA	–.03	.00	.05	–0.02	–.09	- 0.64	406	0.52
Medium	MachineA	–.01	–.01	–.04	0.01	.03	0.72	1400	0.47
Medium	MachineH	.07	.01	.04	0.03	.03	0.79	1394	0.43
Medium	Human	–.06	–.01	–.11	0.02	.05	1.17	1405	0.24

15.7.2 Argumentation Learning Progression

The questions that were the most difficult, for which the differential severity across raters became significant, included the following:

- B6 (a level 1b item, identifying a warrant)
- G7 (a level 2a item, providing a counter-critique)
- G8 (a level 2c item, providing a two-sided comparative argument.)
- B4 (a level 2d item, constructing a counter claim with justification)

These items all required students to engage in the practice of critique at a high level.

Future research and next steps should include a closer, qualitative look at these items to identify patterns that might explain their impact on differential severity between the human and the machine algorithms. Such an analysis is beyond the scope of this paper.

15.7.3 Limitations

This study relies heavily on probability-values (p-values) to identify interesting or significant patterns. P-values have been identified as a problematic measure of statistical significance for measuring the effect on marginalized groups. Because p-values depend on sample sizes, they could lead to selective reporting and selective attention (Amrhein & Greenland, 2018) and thus the further marginalization of injustices borne by already marginalized groups of students who represent a smaller proportion of the sample. In this study, some patterns that were clearly significant for the non-EL students became insignificant for EL students because the sample was smaller. A larger percentage of EL students in the sample would lessen the impact of this limitation on drawing our conclusions.

15.8 Implications

15.8.1 Ethical Implications of an "Unbiased" Assessment

Our primary finding as part of this study was that over the entire assessment, the automated scoring models were not significantly more severe than humans when assigning scores to responses from EL students versus non-EL students. At first glance, this seems like a mark of approval. We have achieved what we have set out to do with this project. But this is only true because the set of questions that were biased against EL students were almost balanced by a set of questions that were slightly balanced for EL students. Had the assessment been longer, or if the distribution of question difficulty had been different, the slight off-balance might have become significant. In addition, we need to consider the ethical implications of the fact that we don't know how the machine learning models are making decisions about score assignments. This is a conversation we should be having more often in the science education assessment community.

15.8.2 Making Choices About Which Machine Models to Use

The process of producing machine models that can score assessments reliably usually requires the comparison of several models. The primary factor used in this comparison is typically an inter-rater statistic that describes how well the machine mimics the human scorers. At first glance one might conclude that using the models trained on the analytic components will be the best way to move forward with scoring (see Authors, in prep). But this study found that the model with the best reliability (and thus severity that is best matched to the human), may have the potential to introduce additional bias that widens the gaps between student groups, particularly on certain types of questions. While the holistic model (MachineH) was significantly more severe than the human, it was much better at maintaining (or even shrinking) the existing gaps in scores between EL students and non-EL students than the analytic model (MachineA). This would suggest that if we want to build assessments that are both fair and reliable, we also need to examine differential severity across the assessment, and across items, before deciding about which machine model (training on analytic vs. holistic scoring) is the best.

15.8.3 Making the Model Less Biased

If computers are at best reflecting human bias, and at worst amplifying it, what are the implications ethically in terms of how we train and use computer scoring models in the classroom? Researchers at Princeton University suggest that we might need to develop a machine algorithm that can mimic the human ability to "express behavior different from their implicit biases...that can be trained or programmed to avoid the expression of prejudice." (Caliskan et al., 2017). This solution suggests that there may be ways to build AI systems through the lens of equity to "disrupt the status quo." (Perry & Lee, 2019) Computer science researchers have called this idea "Fairness through Awareness" (Dwork et al., 2012), and the body of literature around how this might be done is growing quickly.

But is creating a truly fair algorithm ever possible if the humans who train the algorithm are not representative of the groups against which the algorithm might be biased? According to the Hechinger Report, a nonprofit, independent news organization focused on inequality and innovation in education, "Kids from Black and Latino or Hispanic communities—who are often already on the wrong side of the digital divide—will face greater inequalities if we go too far toward digitizing education without considering how to check the inherent biases of the (mostly white) developers who create AI systems." (Perry & Lee, 2019, para. 2) One solution is for the human scorers who train the computer algorithms to be representative of linguistic, racial, gender, and sexual diversity. This solution is easy to name, but rarely put into practice. Our own human raters were predominately White and monolingual English-speaking. When the Stanford Institute for Human-Centered Artificial Intelligence launched in 2019, the co-directors wrote that "The creators and designers of AI must be broadly representative of humanity. This requires a true diversity of thought — across gender, ethnicity, nationality, culture and age, as well as across disciplines." Yet people were quick to point out that of the 121 original faculty members listed on the Institute's website, over 100 were White, and none of them were Black (Gershgorn, 2019). As we work toward meaningful representation, we can start by being open and transparent in how we report on machine learning in education, and who did the coding that was used to train the computer.

15.9 Conclusion

Assessment developers can learn from the lessons of the past. Consider standardized testing, which promised to democratize the system (e.g. No Child Left Behind [NCLB], 2002). But the designers of these assessment tools "didn't consider how the racism and inequality rife in American society would be baked into the tests if care wasn't taken to make them more fair," (Perry & Lee, 2019, para. 5) and as a result, these tests have become symbols of institutional prejudice (Strauss, 2020), particularly against EL students (Hutchinson & Hadjioannou, 2017). Systemic bias is embedded in the very nature of assessment (Kohn, 2000; Au, 2016; Young & Young, 2019). If we are to disrupt the patterns of inequity propagated by assessment (in lieu of eliminating assessments as we know them), we must intentionally design through the lenses of racial, gender and linguistic equity. As a baseline measure, investigating differential severity/leniency between computers and humans using the tools of Rasch modeling as demonstrated in this study should be a routine part of developing all assessments, and in particular automated scoring models that attempt to mimic human scoring.

As automated scoring goes mainstream, and the EL population in schools grows, researchers are just beginning to think about how to diagnose, monitor, and counteract bias embedded in automated assessment. This study provides evidence from Rasch analyses to inform the nature and the direction of those conversations, so that these automated text scoring technologies can be leveraged to provide opportunities for multilingual students to express competence while learning science.

Appendix: Example of a Facets "Specification File" used in this Project

```
convergence = 0.1; size of largest remaining marginal score residual at
convergence
unexpected = 3.0; size of smallest standardized residual to report
arrange = M; arrange output tables in Measure ascending order
facets = 5; there are 5 facets in this analysis
missing = 9; treat 9 as missing data
Inter-rater = 5; facet 5 is the rater facet
pt-biserial = measure; point-measure correlation
```

```
Mode
?,?B,1,?B,?B,onescale;
?,?B,2,?B,?B,onescale
?,?B,3,?B,?B,threescale
?,?B,4,?B,?B,twoscale
?,?B,5,?B,?B,threescale
?,?B,6,?B,?B,twoscale
?,?B,7,?B,?B,twoscale
?,?B,8,?B,?B,threescale
?,?B,9,?B,?B,threescale
?,?B,10,?B,?B,onescale
?,?B,11,?B,?B,onescale
?,?B,12,?B,?B,threescale
?,?B,13,?B,?B,threescale
?,?B,14,?B,?B,twoscale
?,?B,15,?B,?B,twoscale
?,?B,16,?B,?B,threescale
?,?B,17,?B,?B,threescale
?,?B,18,?B,?B,threescale
?,?B,19,?B,?B,twoscale


*
Rating (or partial credit) scale=onescale, R1, General
0 = lowest score; this is a 2 category (0–1) rating scale (or partial
credit)
1 = highest score
*
Rating (or partial credit) scale=twoscale, R2, General
0 = lowest score; this is a 3 category (0–2) rating scale (or partial
credit)
1 = middle score
2 = highest score
*
Rating (or partial credit) scale=threescale,R3,General
0 = lowest score; this is a 4 category (0–3) rating scale (or partial
credit)
1 = low-middle score
2 = high-middle score
3 = highest score


*
Labels
1,Student
1-932; 932 students in the sample
*
2,Difficulty, D
1, Easy
2, Medium
3, Hard
*
3,Item
```

```
1,G1_0b;01
2,G2_0b;01
3,G3_1a;0123
4,G4_1a;012
5,G5_1a;0123
6,G6_1c;012
7,G7_2a;012
8,G8_2c;0123
9,S1_1c;0123
10,S2_0d;01
11,S3_0d;01
12,S4_2a;0123
13,S5_2c;0123
14,B1_0b;012
15,B2_0d;012
16,B3_1c;0123
17,B4_2d;0123
18,B5_2b;0123
19,B6_1b;012
*
4,EL, D; EL status
1, not EL
2, EL; 2 levels
*
5,Rater
1, MachineH
2, Human
3, MachineA; 3 raters
*
data = April2020LP.csv
```

References

Abedi, J. (2006). Psychometric issues in the ELL assessment and special education eligibility. *Teachers College Record, 108*(11), 2282–2303.

Adamson, A. S., & Smith, A. (2018). Machine learning and health care disparities in dermatology. *JAMA Dermatology, 154*(11), 1247–1248. https://doi.org/10.1001/jamadermatol.2018.2348

Aggarwal, C. C., & Zhai, C. (2012). A survey of text classification algorithms. In *Mining text data* (pp. 163–222). Springer.

Allchin, D. (2013). *Teaching the nature of science. perspectives and resources*. SHiPS Education Press.

Amrhein, V., & Greenland, S. (2018). Remove, rather than redefine, statistical significance. *Nature Human Behaviour, 2*(1), 4–4.

Au, W. (2016). Meritocracy 2.0: High-stakes, standardized testing as a racial project of neoliberal multiculturalism. *Educational Policy, 30*(1), 39–62.

Bradley, C., & Renzulli, L. (2011). The complexity of non-completion: Being pushed or pulled to drop out of high school. *Social Forces, 90*(2), 521–545. Retrieved November 2, 2020, from http://www.jstor.org/stable/41682665

Briggs, D. C., & Wilson, M. (2007). Generalizability in item response modeling. *Journal of Educational Measurement, 44*(2), 131–155.

Buck, Z. E., Lee, H. S., & Flores, J. (2014). I am sure there may be a planet there: Student articulation of uncertainty in argumentation tasks. *International Journal of Science Education, 36*(14), 2391–2420.

Buolamwini, J. (2019, May 7). *AI, Ain't I a woman?* [Posted by Harvard University]. YouTube. https://www.youtube.com/watch?v=HZxV9w2o0FM

Buolamwini, J. (2020, March 31). *Voicing Erasure – A Spoken Word Piece Exploring Bias in Voice Recognition Technology*. The Algorithmic Justice League. https://www.ajl.org/voicing-erasure.

Caliskan, A., Bryson, J. J., & Narayanan, A. (2017). Semantics derived automatically from language corpora contain human-like biases. *Science, 356*(6334), 183–186.

Cavagnetto, A. R. (2010). Argument to foster scientific literacy: A review of argument interventions in K–12 science contexts. *Review of Educational Research, 80*(3), 336–371.

Cronbach, L. J., Rajaratnam, N., & Gleser, G. C. (1963). Theory of generalizability: A liberalization of reliability theory. *British Journal of Statistical Psychology, 16*(2), 137–163.

Crowston, K. (2012). Amazon mechanical turk: A research tool for organizations and information systems scholars. In A. Bhattacherjee & B. Fitzgerald (Eds.), *Shaping the future of ICT research. Methods and approaches. IFIP advances in information and communication technology* (Vol. 389). Springer. https://doi.org/10.1007/978-3-642-35142-6_14137-163

Cutler, D. M., & Lleras-Muney, A. (2006). *Education and health: Evaluating theories and evidence. Cambridge*. National Bureau of Economic Research.

Devine, P. G., Forscher, P. S., Austin, A. J., & Cox, W. T. (2012). Long-term reduction in implicit race bias: A prejudice habit-breaking intervention. *Journal of Experimental Social Psychology, 48*(6), 1267–1278.

Dwork, C., Hardt, M., Pitassi, T., Reingold, O., & Zemel, R. (2012, January). Fairness through awareness. In *proceedings of the 3rd innovations in theoretical computer science conference* (pp. 214–226).

Eckert, P., & Rickford, J. R. (Eds.). (2001). *Style and sociolinguistic variation*. Cambridge University Press.

Erduran, S., Simon, S., & Osborne, J. (2004). TAPping into argumentation: Developments in the application of Toulmin's argument pattern for studying science discourse. *Science Education, 88*(6), 915–933.

Gee, J. P. (2015). Discourse, small d, big D. In *The international encyclopedia of language and social interaction* (pp. 1–5).

Gershgorn, D. (2019). *Stanford's new AI institute is inadvertently showcasing one of tech's biggest problems*. Quartz. https://qz.com/1578617/stanfords-new-diverse-ai-institute-is-overwhelmingly-white-and-male/

Greenwald, A. G., McGhee, D. E., & Schwartz, J. L. (1998). Measuring individual differences in implicit cognition: The implicit association test. *Journal of Personality and Social Psychology, 74*(6), 1464–1480.

Ha, M., & Nehm, R. H. (2016). The impact of misspelled words on automated computer scoring: A case study of scientific explanations. *Journal of Science Education and Technology, 25*(3), 358–374.

Hakuta, K., & Beatty, A. (2000). *Testing English-language learners in US schools*. National Academy Press.

Henard, F., & Roseveare, D. (2012). Fostering quality teaching in higher education: Policies and practices. *An IMHE Guide for Higher Education Institutions, 1*, 7–11.

Hummer, R. A., & Hernandez, E. M. (2013). The effect of educational attainment on adult mortality in the United States. *Population Bulletin, 68*(1), 1–16.

Hutchinson, M., & Hadjioannou, X. (2017). The morphing assessment terrain for English learners in US schools. *English Teaching: Practice & Critique, 16*, 110–126.

Jackson, S. M., Hillard, A. L., & Schneider, T. R. (2014). Using implicit bias training to improve attitudes toward women in STEM. *Social Psychology of Education, 17*(3), 419–438.

Johnson, R. B., & Onwuegbuzie, A. J. (2004). Mixed methods research: A research paradigm whose time has come. *Educational Researcher, 33*(7), 14–26.

Jurka, T. P., Collingwood, L., Boydstun, A. E., Grossman, E., & van Atteveldt, W. (2013). RTextTools: A supervised learning package for text classification. *The R Journal, 5*(1), 6–13.

Kaufman, P., Kwon, J. Y., Klein, S., & Chapman, C. D. (2000). Dropout rates in the United States: 1999. *Education Statistics Quarterly, 2*(4), 37–42.

Kelly, G. J., & Takao, A. (2002). Epistemic levels in argument: An analysis of university oceanography students' use of evidence in writing. *Science Education, 86*(3), 314–342.

Kincheloe, J., & McLaren, P. (1994). *Rethinking critical thinking and qualitative research. Handbook of qualitative research*. Sage Publications.

Koenecke, A., Nam, A., Lake, E., Nudell, J., Quartey, M., Mengesha, Z., et al. (2020). Racial disparities in automated speech recognition. *Proceedings of the National Academy of Sciences, 117*(14), 7684–7689.

Kohn, A. (2000). *The case against standardized testing: Raising the scores, ruining the schools*. Heinemann.

Lave, J., & Wenger, E. (1991). *Situated learning: Legitimate peripheral participation*. Cambridge University Press.

Lee, H. S., Liu, O. L., Pallant, A., Roohr, K. C., Pryputniewicz, S., & Buck, Z. E. (2014). Assessment of uncertainty-infused scientific argumentation. *Journal of Research in Science Teaching, 51*(5), 581–605.

Linacre, J. M. (1989). *Many-facet Rasch measurement*. MESA Press.

Linacre, J. M. (2020). *Facets computer program for many-facet Rasch measurement (version 3.83.2)[computer software]*. Winsteps. Com.

McCullough, L. (2004). Gender, context, and physics assessment. *Journal of International Women's Studies, 5*(4), 20–30.

Monteith, M. J., Devine, P. G., & Zuwerink, J. R. (1993). Self-directed versus other-directed affect as a consequence of prejudice-related discrepancies. *Journal of Personality and Social Psychology, 64*(2), 198–210.

Moschkovich, J. N. (2007). Beyond words to mathematical content: Assessing English learners in the mathematics classroom. *Assessing mathematical proficiency, 53*, 345–352.

Myford, C. M., & Wolfe, E. W. (2003). Detecting and measuring rater effects using many-facet Rasch measurement: Part I. *Journal of Applied Measurement, 4*(4), 386–422.

Myford, C. M., & Wolfe, E. W. (2004). Detecting and measuring rater effects using many-facet Rasch measurement: Part II. *Journal of Applied Measurement, 5*(2), 189–227.

National Research Council. (2007). *Taking science to school: Learning and teaching science in grades K-8*. National Academies Press.

National Center for Educational Statistics [NCES]. (2020, May). *English Language Learners in Public Schools*. Annual Report from the National Center for Educational Statistics. https://nces.ed.gov/programs/coe/indicator_cgf.asp

No Child Left Behind (NCLB). (2002). Act oef 2001, Pub. L. No. 107–110, § 101. *Stat, 1425*, 1059–0889.

Osborne, J. (2010). Arguing to learn in science: The role of collaborative, critical discourse. *Science, 328*(5977), 463–466.

Osborne, J. F., Henderson, J. B., MacPherson, A., Szu, E., Wild, A., & Yao, S. Y. (2016). The development and validation of a learning progression for argumentation in science. *Journal of Research in Science Teaching, 53*(6), 821–846.

Pellegrino, J. W., Chudowsky, N., & Glaser, R. (2001). Knowing what students know: The science and design of educational assessment. National Academy Press.

Perry, A. M., and Lee, N. T. (2019, September 26). *AI is coming to schools, and if we're not careful, so will its biases*. Brookings. https://www.brookings.edu/blog/the-avenue/2019/09/26/ai-is-coming-to-schools-and-if-were-not-careful-so-will-its-biases/

Sandoval, W. A. (2003). Conceptual and epistemic aspects of students' scientific explanations. *The Journal of the Learning Sciences, 12*(1), 5–51.

Shaw, J. M. (2005). Getting things right at the classroom level. *Yearbook of the National Society for the Study of Education 104*(2), 340–357.

Shaw, J. M., Bunch, G. C., & Geaney, E. R. (2010). Analyzing language demands facing English learners on science performance assessments: The SALD framework. *Journal of Research in Science Teaching, 47*(8), 909–928.
Shermis, M. D. (2015). Contrasting state-of-the-art in the machine scoring of short-form constructed responses. *Educational Assessment, 20*(1), 46–65. https://doi.org/10.1080/10627197.2015.997617
Solano-Flores, G., & Nelson-Barber, S. (2001). On the cultural validity of science assessments. *Journal of Research in Science Teaching: The Official Journal of the National Association for Research in Science Teaching, 38*(5), 553–573.
Solano–Flores, G., & Li, M. (2006). The use of generalizability (G) theory in the testing of linguistic minorities. *Educational Measurement: Issues and Practice, 25*(1), 13–22.
Solano-Flores, G. (2008). *Who is given tests in what language by whom, when, and where?* The need for probabilistic views of language in the testing of English language learners. *Educational Researcher, 37*(4), 189–199.
Stage, E. K., Asturias, H., Cheuk, T., Daro, P. A., & Hampton, S. B. (2013). Opportunities and challenges in next generation standards. *Science, 340*(6130), 276–277.
Stage, F. K., & Wells, R. S. (2014). Critical quantitative inquiry in context. *New Directions for Institutional Research, 2013*(158), 1–7.
Strauss, V. (2020, June 21). *It looks like the beginning of the end of America's obsession with student standardized tests.* Washington Post. https://www.washingtonpost.com/education/2020/06/21/it-looks-like-beginning-end-americas-obsession-with-student-standardized-tests/
Symons, C., Palincsar, A. S., & Schleppegrell, M. J. (2017). Fourth-grade emergent bilinguals' uses of functional grammar analysis to talk about text. *Learning and Instruction, 52*, 102–111.
Technavio. (2018, August). *Artificial intelligence market in the US education sector 2018–2022.* Market Research Report.
Tyagi, N. (2020, April 23). *6 Major branches of artificial intelligence (AI).* Analytic Steps. https://www.analyticssteps.com/blogs/6-major-branches-artificial-intelligence-ai
Toulmin, S. (1958). *The uses of argument.* Cambridge University Press.
Urban-Lurain, M., Prevost, L., Haudek, K. C., Henry, E. N., Berry, M., & Merrill, J. E. (2013). Using computerized lexical analysis of student writing to support just-in-time teaching in large enrollment STEM courses. In *Paper presented at the 2013 IEEE frontiers in education conference (FIE)* (pp. 1709–1715). IEEE.
Vygotsky, L. S. (1978). Socio-cultural theory. In *Mind in society, 52–58.* Harvard University Press.
Warren, B., Ballenger, C., Ogonowski, M., Rosebery, A. S., & Hudicourt-Barnes, J. (2001). Rethinking diversity in learning science: The logic of everyday sense-making. *Journal of Research in Science Teaching: The Official Journal of the National Association for Research in Science Teaching, 38*(5), 529–552.
Young, N. D., & Young, D. M. (2019). Understanding and addressing bias in classroom assessment: Promoting fairness through equitable grading practices. In *Acceptance, understanding, and the moral imperative of promoting social justice education in the schoolhouse* (p. 93). Vernon Press.

Zoë Buck Bracey (BSCS Science Learning) conceptualized this aspect of the ArguLex work, produced the analyses, and wrote this chapter. Dr. Buck Bracey is a Senior Science Educator at BSCS Science Learning, and Director of Design for Social and Environmental Justice Outcomes. Her current work focuses on co-design of justice-oriented science curriculum materials with cross-disciplinary teams of educators, curriculum writers, administrators, education researchers, and content specialists.

Molly Stuhlsatz (BSCS Science Learning) was a close thought partner and Rasch methods advisor throughout this work and read and provided feedback on this chapter. Ms. Stuhlsatz is a Senior

Research Scientist at BSCS Science Learning, and Director of People and Culture. She has spent her career leading research and evaluation projects at BSCS and enjoys developing creative solutions to persistent challenges in the assessment of student and teacher outcomes.

Christopher Wilson (BSCS Science Learning) was a thought partner and advisor throughout this work, and provided feedback on the chapter. Dr. Wilson is a Senior Research Scientist and Director of the research division at BSCS Science Learning. His current research focuses on the assessment of teacher and student learning in science education, the impact of lesson analysis based professional development on teacher and student outcomes, the application of automated scoring techniques to the measurement of teacher PCK and student argumentation, and the development and impact of three-dimensional curriculum materials.

Tina Cheuk (CalPoly San Luis Obispo) was instrumental in item and rubric development, coordinated data collection and analysis efforts for this project, and read and provided feedback on drafts of this chapter. Dr. Cheuk is an Assistant Professor in the CalPoly School of Education in San Luis Obispo. Her research interests include disciplinary literacy and language development, curriculum and assessment, and research-pracitce partnerships.

Marisol M. Santiago (Michigan State University) was instrumental in item and rubric development, coordinated analysis at Michigan State University, was part of the team that developed the machine learning algorithms, and read and provided feedback on drafts of this chapter. Dr. Santiago was a Research Associate at the CREATE for STEM Institute at Michigan State University at the time of this project.

Jonathan Osborne (Stanford University) led the team that developed the argumentation learning progression featured in this study. Dr. Osborne is Professor Emeritus in the Stanford Graduate School of Education, and over his career he has researched both policy and pedagogy in the teaching and learning of science.

Kevin Haudek (Michigan State University) led the team that developed the machine learning algorithms featured in this study. Dr. Haudek is Assistant Professor of Teaching in the department of Biochemistry and Molecular Biology at Michigan State University. His research focuses on teaching and learning in undergraduate science courses.

Brian Donovan (BSCS Science Learning) was instrumental in item and rubric development. Dr. Donovan is a Senior Research Scientist at BSCS, and his research employs experimental, quasi-experimental and qualitative methods in formal classroom settings to explore how the learning of biology affects social cognition.

Chapter 16
Using Multi-faceted Rasch Models to Understand Middle School Students' Argumentation Around Scenarios Grounded in Socio-scientific Issues

William Romine, Amy Lannin, Maha K. Kareem, and Nancy Singer

Abstract We describe the multi-faceted Rasch model as a definition of measurement in contexts with multiple independent sources of error. We apply it to validation of open-ended written scenario-based assessments of argumentation around socio-scientific issues which are subject to errors associated with the argumentation competency being assessed, the rater being assigned, and the particular socio-scientific issue given to the student. Through inspection of the hierarchy within each facet and misfit of particular elements, we were able to tease out the strengths and limitations of particular scenarios and raters, and ultimately derive a more general understanding of how students' observed argumentation changes as their ability increases.

16.1 Introduction

The ability to formulate and apply reasoning to evidence-based claims is a central science literacy practice (Lin & Mintzes, 2010). In the context of science education, this practice begs the question: What does it mean to engage in argumentation and

W. Romine (✉)
Department of Biological Sciences, Wright State University, Dayton, OH, USA
e-mail: william.romine@wright.edu

A. Lannin
College of Education and Human Development, University of Missouri, Columbia, MO, USA
e-mail: lannina@missouri.edu

M. K. Kareem
College of Education and Human Development, University of Missouri, Columbia, MO, USA

College of Languages, University of Baghdad, Baghdad, Iraq
e-mail: mkkr3d@mail.missouri.edu

N. Singer
College of Education, University of Missouri—St. Louis, St. Louis, MO, USA
e-mail: singerna@umsl.edu

X. Liu, W. J. Boone (eds.), *Advances in Applications of Rasch Measurement in Science Education*, Contemporary Trends and Issues in Science Education 57,
https://doi.org/10.1007/978-3-031-28776-3_16

how should it be measured? At a basic level, we can think of argumentation as a process which involves making a claim, supporting it with evidence, and making a creative but logic-driven case for how the evidence provided supports the claim (Toulmin, 1958). However, Berland et al. (in Schwarz et al., 2017) argue that argumentation goes deeper than this; they describe the NGSS vision of scientific argumentation (NGSS Lead States, 2013) as an open-ended growth process where students simultaneously become more informed and confident in what they are defending, and also more open to evaluating, critiquing, and revising their ideas. Various forms of this vision have been adopted and contextualized in science education toward measurement of how students understand how to both construct their own arguments and evaluate the arguments of others (Gotwals & Songer, 2010; Osborne et al. 2016; Deane et al., 2019). Gotwals and Songer (2010) advanced the Claim-Evidence-Reasoning (CER) framework which gives emphasis to the use of scientific theory and data toward advancing and reasoning around scientific claims. Under the premise that the various processes behind argument construction and evaluation entail different levels of complexity, Berland and McNeil (2010) proposed a learning progression for argumentation that takes into account the complexity of the instruction and tasks that the students are asked to negotiate, the completeness of the argumentation product (e.g., whether or not a claim is supported with evidence), and the fluidity of the argumentation process (e.g., the extent to which students engage in argumentative discourse without prompting from the instructor). Toulmin's model was utilized more directly by Osborne et al. (2016) to develop and validate a formal quantitative measure of students' ability to not only make valid arguments, but also to compare competing arguments. Similar to Berland and McNeil (2010), Osborne et al. (2016) make the observation that simpler aspects of argumentation such as making a claim tend to be less demanding for students than more complex aspects like identification and scrutiny of evidence, reasoning on how the evidence supports the claim, and evaluating multiple competing arguments. Hence, they use cognitive load theory (Sweller et al., 2011) as a framework for justifying that argumentation can be parametrized in terms of a learning progression where simpler processes such as making a claim must be in place before more complex processes such as identification of evidence and reasoning have context (Osborne et al., 2016). Hence, their progression moves from easier competencies that focus on simply identifying a claim up to the highest levels where students are asked to evaluate multiple competing arguments (Osborne et al., 2016). Most recently, a learning progression in argumentation was explored by Deane et al. (2019), who proposed that negotiation of scenario-based measures of argumentation start with the ability to make a claim and see multiple perspectives (Level 1). This then buttresses the practice of finding evidence and supporting ideas (Level 2), which in turn buttresses written articulation of the argument (Level 3). Students at the highest level are able to use written communication for evaluation and critique (Level 4).

Although the CER framework has been foundational in prior work related to understanding how argumentation changes as students become more literate, scholars have noted that a claim supported by evidence is not necessarily sufficient for students to be fully engaged in scientific argumentation. In addition, many curriculum scholars purposefully avoid the CER framework for argumentation, as it oversimplifies and draws attention away from the ultimate goal of argumentation which centers on growth and progress in one's thinking and ideas as well as learning from and challenging others' ideas (Berland et al., 2017). In light of this limitation, we use the CER framework to provide stepping stones for novice thinkers/writers as they internalize the argumentation practice. However, our framework does not merely ask students to identify claim, evidence, and reasoning. By engaging in authentic writing tasks, the merit and synthesis of the scientific ideas (including appropriate use of quantitative reasoning and discipline-specific vocabulary) help demonstrate students' knowledge of science concepts, the ability to negotiate this information to make an argument, and the ability to identify, analyze and critique multiple perspectives.

Toward the goal of assessing argumentation with authenticity of practice (Berland et al., 2017), understanding how students apply scientific argumentation in the context of socioscientific issues (SSIs) needs further work, in particular due to the complexity of the measurement designs needed to generate authentic and accurate scenario-based measures. The need for authenticity comes to the forefront when we consider the types of data to collect from students. Argumentation is a practice by which scientists communicate, and so it makes sense to derive measures from a context that encourages students to communicate actively either orally or in writing. This need to solicit active engagement in a communication process contra-indicates traditional multiple-choice questions in favor of a rubric-scored open response format based on observations of students' written arguments. An additional layer of complexity is added when we consider that SSIs are context-based (Sadler, 2004). Although it is conceivable that students may have an easier time making arguments around certain SSIs and a more difficult time making arguments around others based on their unique areas of knowledge or interest, we would ideally like to derive measures for argumentation that are transferable across many types of SSIs. In other words, we would like for our measure to indicate *ability to argue around SSI-based scenarios* as opposed to *ability to argue issues related to water pollution* (for example). In this chapter, we illustrate how multi-faceted Rasch analysis can be used to accommodate the complex measurement designs needed to deliver generalizable measures for how students develop written argumentation around SSIs. We first describe the creation and implementation of scenario-based assessments (SBAs) of argumentation, highlighting how we used multi-faceted Rasch modeling to address the methodological challenges associated with use of multiple scenarios and raters in the context of a curricular program evaluation. In doing so, we discuss the practice of using the multi-faceted Rasch model as a standard for construct validity of assessments like this which have multiple facets of variance. We then demonstrate how to utilize multi-faceted Rasch analysis to build theory on how students argue around complex science-related social issues.

16.2 Framework for Measuring Argumentation in the Context of SSIs

Engagement in argumentation around SSIs is both process-based and context-based (Venville & Dawson, 2010). This means that a measure of a student's ability to make arguments related to the negotiation or resolution of SSIs should give students the opportunity to engage actively in communication practices (writing in this case) within a specific SSI context (Zeidler et al., 2019). This extension of the CER framework to SSIs necessitated use of specific scenarios tied to particular SSIs in the construction of SBAs which task students with making an argument around the particular SSI. From a measurement perspective, the greatest level of simplicity would be obtained if all students were given the same SSI and writing task. However, constraining the diverse set of participating teachers in our program to use the same SSI would violate the three-dimensional learning framework put forth in the Next Generation Science Standards (National Research Council, 2014; NGSS Lead States, 2013), which is built on the idea that learning content cannot be separated from the scientific practices in which students are engaging (Krajcik, 2015). Specifically, for our assessments, we wanted to encourage our participating teachers to utilize SSIs that were connected to the content that they were teaching. To help facilitate this, we asked our participating teachers to select from different SBAs which were developed by the program team in cooperation with the participating teachers (Linking Science & Literacy for All Learners, 2018). Some of these SBAs were adapted from previous research (Owens et al., 2021; Kinslow et al., 2019).

Program leaders developed a systematic procedure to guide the development of the SBAs for the project. The first step was to identify an SSI topic (water pollution, genetic modification, vaping, or flooding; for example, one of the SBAs was titled "York's Flooding Problem"). The next step was to specify what we wanted students to know about that topic and potential misconceptions students were likely to have about the topic (Covitt et al., 2017; Womack et al., 2017). For example, with regards to flooding (Linking Science & Literacy for All Learners, 2018), we wanted students to understand the types of land use: in particular differences between developed-permeable surfaces, developed-impermeable surfaces, and natural areas which have not been developed by humans at all. Further, we needed to convey how changes in precipitation and land area coverage are measured from year to year. We also needed the SBA to account for common misconceptions students have about flooding; namely that flooding is solely due to increased precipitation. This led to the third step, which consisted of creating a scenario that showcases the topic and desired content while giving students the opportunity to apply written argumentation. In York's Flooding Problem, students were given a scenario:

> *The city of York has grown a lot over the last 10 years. In that time, the city has built three new schools and many homes for its growing population. York has also expanded roads and built more sidewalks and parking lots. Recently, citizens have noticed that the city floods when it rains heavily. There had never been any flooding problems before 2005. But flooding has occurred more often ever since 2005. Citizens are concerned because flooding can be*

dangerous to people and animals and can cause property damage. A local high school science class is investigating the problem. The students collected data about the city's precipitation over the last 50 years. Measures of precipitation show how much rain and snow the city receives each year.

Once the scenario was built, the next step was identifying or creating data and information to support the scenario with the goal of having 2–3 data representations for students to analyze. In this scenario, the students were given information on the difference between developed permeable surfaces, developed impermeable surfaces, and natural surfaces. They were then presented with two graphs: the first a bar graph showing annual rainfall between 1963 and 2013, the second a line graph showing the percentage of different surfaces each year between 1973 and 2013. After being presented with these representations, the students were given two competing perspectives:

The class invited a meteorologist to look at their data. The meteorologist visited the class and claimed:

The cause of York's flooding problem is the increased precipitation. The data show that rain and snowfall have increased steadily since about 2005. Flooding did not become a problem for the city until 2005. So, increased precipitation has led to local flooding." After the meteorologist visited, a student in the class named Shonda said the following: "I don't think the meteorologist is right. He focused on the precipitation graph and did not consider data from the graph on land area coverage.

Students were then given the task:

Examine the data provided and the differing viewpoints. Write a statement to the class that explains what you determine to be the cause of the flooding. Your statement should present your claim, be supported by the available evidence, and include the following: (1) Reasons why the available evidence supports your claim; (2) Reasons why the other claim is not supported by the available evidence; and (3) Suggestions as to what evidence could be used to support other claims.

We encourage the reader to contact the authors for more information on the SBAs or copies of the SBAs.

Once the writing task was completed by the students, the program team was tasked with assigning each student scores for six competencies (1. making a claim, 2. recognizing evidence, 3. reasoning around the evidence with respect to the claim, 4. integration of content-specific vocabulary, 5. written communication, and 6. a holistic score). Responses were scored with respect to each competency on an ordinal 1–4 scale using a rubric developed by the program team (Table 16.1). In the rubric, we defined a "1" score as the lower benchmark which indicates relative absence of the competency in the written response. Scores of "2" and "3" indicate respective milestones where students are beginning to demonstrate key skills related to application of argumentation such as acknowledging alternative points of view, recognizing the non-factual basis for one's own point of view, integration of multiple information sources, accurate interpretation of data, and use of discipline-specific vocabulary. For example, one student responded:

Table 16.1 Rubric for Scenario-based assessment of argumentation

	4 – Capstone	3 – Milestone	2 – Milestone	1 - Benchmark
Establishing purpose/stating claim	Position/claim is clear and takes into account the complexities of an issue Other viewpoints are explicitly used in the explanation of the position	Position/claim acknowledges or somewhat takes into account complexities of an issue Other viewpoints are acknowledged	Position/claim is stated but is simplistic or unclear Other points of view may be acknowledged	Position/claim is absent or off-topic
Evidence Use of sources to explore issues and analyze evidence	Comprehensively analyzes or synthesizes information taken from sources(s) Evidence and viewpoints are referenced and questioned	Analyzes or synthesizes information taken from sources(s) Evidence and viewpoints are taken as mostly fact, with little questioning	Takes information from sources with minimal or no evaluation Viewpoints are taken as fact, without question	Information is not taken from source(s) Writer does not include or question viewpoints
Domain-specific content and vocabulary	Presents appropriate/ accurate use of science/math/ literacy concepts AND discipline-specific vocabulary	Generally accurate but may have some missing ideas or some misconceptions of concepts **OR** discipline-specific vocabulary	Response based on misconceptions Minimal or unclear use of domain-specific vocabulary	No clear reference to domain-specific content or vocabulary
Reasoning Interpretation and analysis of sources	Effectively demonstrates reasoning using accurate explanation of information Makes appropriate inferences and conclusions based on and referencing data	Demonstrates reasoning in explaining data and information Makes mostly appropriate inferences and conclusions based on that information	Demonstrates some reasoning in attempts to explain data Makes somewhat appropriate inferences and conclusions based on that information but may make some incorrect conclusions	Reasoning is not yet evident Data is not used or may be merely listed
Written communication Development of ideas and clarity of expression	Effectively develops ideas and skillfully communicates meaning to readers with clarity and fluency Is almost error-free	Develops ideas that generally convey meaning to readers The writing has few errors	Is developing ideas. Expressions not consistently clear Writing may include some errors	Ideas not yet developed or clear Writing may have many errors

(continued)

Table 16.1 (continued)

	4 – Capstone	3 – Milestone	2 – Milestone	1 - Benchmark
Holistic - overall	Effectively presents claim, and supports with evidence and logical reasoning to develop ideas using clear written communication	Presents a claim with evidence and reasoning to generally develop ideas in clear written communication	Claim is stated with some information and reasoning that is starting to develop the ideas though not with consistently clear written communication	Claim is not clear or is off topic and does not connect to evidence; written communication needs considerable developing and editing

Note. Adapted from University of Missouri's STEM Literacy Project, Title II, Part A of the Improving Teacher Quality Grant, Missouri Department of Higher Education and based on VALUE Rubrics – Association of American Colleges and Universities

> *It's the rain and ground covering. Because on the charts over the years stuff goes up. And flooding goes up so that means that ground covering goes up because the flooding is moving something around.*

In this response, we observe that the student is able to make a claim and is beginning to see trends on the graphs. However, this student's understanding of how rainfall and surface structure interact and influence the movement of water is incomplete, and so the student has trouble describing how the evidence supports the claim. An example of such an integration can be seen in the following response:

> *I believe Shonda is right. The meteorologist is incorrect. If the meteorologist would have looked at the graph on land area coverage they would have realized that around 2005 the percentage of developed-impermeable surfaces went up, and that the percentage of natural areas actually went down. Natural areas soak up the rain and give the rain somewhere to go but since there is less of it since 2005 the rain has less area to go. The graph on precipitation also points to that I'm right. Rain averages have been this high before, back then there were less impermeable surfaces (roads, buildings, lots) and New York did not flood then.*

The capstone scores were given to responses which indicate a level of proficiency expected of a student with a working knowledge of how to construct and evaluate arguments around the scenario using accurate knowledge and data. The following response illustrates this:

> *The precipitation is not causing the flood. The measures of precipitation is and shows how much rain and how much snow cities receive each year. I don't agree with the person who said precipitation is causing all the flooding because of graph one showing how much precipitation. They said that the flooding had started in 2005. When you look at the graph back in 1936 there was about an average of 53 inches of precipitation. From 1936 the precipitation went down till 1981 and 1983 which both had the same amount of precipitation with 33 inches. From then the precipitation went up again but not by many years since it went back down during the year of 1995 with the precipitation being 31. It went back up through it went up till the year 2003 which was 35 in precipitation. 2 years past and it came to 2005 when flooding began the precipitation when from 47 in 2005 to 67 in 2013. I agree to it by being land. The change of the land started with 75 percent of natural land during 1973 but it went down as the years passed till 2013 with about 45 percent of land. The developed surfaces went up as people built the city. The water would have nowhere to go*

since the natural land isn't there much because of developed-impermeable surfaces being buildings and parking lots which is stopping the water. Developed-permeable would help the water a bit to escape since its opened areas like soccer fields covered with grass or city gardens.

Scoring was done by a panel of 8 raters consisting of teachers and project leaders who participated in a summer scoring event which included training on how to use the rubric. The training included analysis of the SBAs and rubric and aligning the rubric with anchor papers that had been team-scored to represent the four levels on the rubric. After working through all of the scoring documents and processes, raters participated in practice scoring. When raters and facilitators felt prepared, they then started actual scoring. Periodic check-ins and re-calibration sessions were provided after breaks. The rater panel included participants with diverse backgrounds and experiences, including middle school science and language arts teachers, professors, and graduate students.

In training the raters on how to use the rubric to assign scores, we needed to decide whether we wished to treat this considerable diversity as a nuisance to be controlled for or as an asset to be utilized. Should the goal in training a rater be to facilitate perfect agreement with other raters—in essence, treating each member of the scoring team as a scoring machine? Or should the goal in training a rater focus more on consistency as opposed to agreement? A vast majority of studies use the "agreement" model for training raters, defining consistency in terms of perfect agreement which is quantified using an inter-rater reliability index. This model is exemplified by application of pressure to conform to a specific interpretation that is taken to be true and can be difficult for raters when they do not, in concept or in practice, wholeheartedly agree with the rubric score points. For these reasons, an "agreement model" often works best when there are a small number of raters with similar positionality with respect to the data. In the case of the diverse panel in this project, we decided that an "agreement" model, which treats diversity as a nuisance to be controlled, is both unrealistic and undesirable. It did not make sense to expect a science teacher, a language arts teacher, a graduate student, and a professor to necessarily agree on the specific score that should be assigned to a written response. However, any definition of measurement would necessitate that the scores assigned by each rater should nonetheless be consistent (Boone et al., 2016). In other words, although two raters may not agree on the specific score to assign, they should agree on how students' scores are ranked. A lenient rater who tends to give higher scores should do so consistently across the spectrum of responses. Conversely, a difficult rater who tends to give lower scores should be consistently difficult through the duration of scoring. With respect to the Rasch model, this argument can also be framed in terms of the local independence assumption—the scores that raters assign should be independent after accounting for the construct they are rating. By extension, in our panel of independent experts, the differences between raters should be due to understanding of the argumentation construct and how it relates to the rubric, and not due to other factors.

Foregoing the "agreement" model in favor of a "consistency" model for scoring has several advantages. In line with the spirit of argumentation itself, we argue

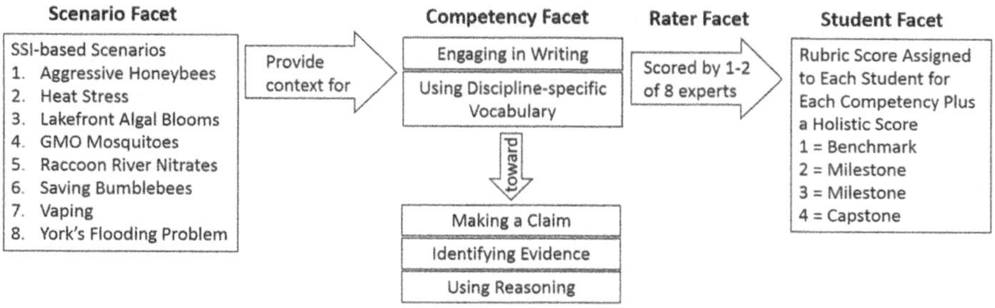

Fig. 16.1 Diagram of the 4-faceted measurement framework to assess argumentation

philosophically that the democratic process of training the raters as a panel of independent experts ensures inclusivity of complementary perspectives, and is likely to generate measures that are robust across diverse perspectives, as opposed to the comparatively autocratic process where the panel is required to agree with a singular perspective. Empirically, a range of rater difficulty allows us to better measure students at the extremes of the scale. For example, if a student is at the "capstone" level on all of the competencies due to very high argumentation ability, it is impossible to give that student a precise location on the scale. One way around this would be to ask a more difficult rater to score that student's written argument. A similar argument could be made for a student who is at the "benchmark" level on all of the competencies, where an easier rater is more likely to assign a score above the minimum. Despite these advantages, moving from the "agreement" model to the "consistency" model presents an additional challenge in that we don't want one student to be unfairly penalized for being assigned a hard rater or another to have an unfair advantage due to being assigned an easy rater. We need a model that yields measures which are independent of which rater(s) were assigned.

In summary, the assessment framework we chose (Fig. 16.1) involves rater-assigned scores on six competencies derived from students' written argumentation on how to negotiate a particular SSI. From a Rasch modeling perspective, this comprises a 4-faceted design where student responses (facet 1) are scored on 6 competencies (facet 2) by one or more raters (facet 3), and that these scores are embedded within one of 8 available SSI-based scenarios (facet 4). Within this design, the maximum amount of information about the facets would be delivered with a fully crossed design where all of the students completed written responses to all 8 scenarios, and all 8 raters scored each response. Although a crossed design is optimal in theory, this would result in considerable student and rater fatigue given the significant effort required for students to complete the writing and raters to score each response. We therefore asked each participating teacher to choose a single SBA to administer to their students which aligned with their instructional goals. In a similar vein, we assigned 1 of the 8 raters at random to score each response with the constraint that participating teachers were not scoring their own students' responses, and then assigned an additional rater to 20% of the responses in order to evaluate the extent of agreement and obtain more reliable estimates of rater severity and consistency. By using a fully nested model for students within SBAs and a partially nested model for students within raters, we were able to reduce dramatically the time demand on our expert raters and potential confounding effects associated with test fatigue. In the next section, we discuss the motivation for using multi-faceted Rasch analysis for multi-faceted measurement designs such as this and discuss the considerable flexibility this approach gives in delivering unbiased measures in cases where it may not be feasible to use a fully crossed design.

16.3 Using Rasch Modeling to Make our Measures Scientific

16.3.1 The Rasch Model Provides Scale Invariance—The Definition of Measurement

In the context of generalized mixed models, the two-faceted Rasch model is often conceived as a mixed effects model with a log-odds (logit) link function where the target of measurement, such as students, is specified as a random factor and the observations of measurement, such as test questions, are specified as a fixed factor (Raudenbush & Bryk, 2002). This conjoint linear measurement system creates a straightforward philosophical criterion for construct validity in its constraint of scale invariance across the continuity of student ability. Further, mapping student ability and observation difficulty on a common scale facilitates qualitative explanations of the types of competencies students at different locations along the scale have and have not mastered and gives further insight into how the competencies are ordered within the construct itself (Wright & Stone, 1979).

Although Rasch models can be specified mathematically as a 1-parameter Item Response Theory (IRT) model, the Rasch and IRT approaches to measurement are antagonistic philosophically and should not be confounded (Andrich, 2004). Unlike IRT, where increasingly complex models are selected to optimize fit with the data, Rasch models hold the philosophical ideal of scale invariance across the domain of measurement *prima facie* (Boone et al., 2013). Through the constraint of scale invariance, the Rasch model constitutes the definition of scientific measurement, and thus facilitates an approach to mental measurement which is not philosophically different from the measurement that takes place in other scientific contexts such as the laboratory (Thurstone, 1928). In this sense, the Rasch approach to measurement is the only approach which can be considered scientific from a measurement perspective (Wright, 1992). Although some scholars criticize this insistence on measurement invariance on the empirical grounds that it cannot be tested (Bergan, 2013), we argue that measurement invariance supersedes empiricism—if measures cannot be held invariant such that they can be compared across contexts, then there is limited scope for an empirical approach in the first place.

16.3.2 Discovery Through Anomaly

We therefore echo Ben Wright's argument (Wright, 1992) that like any measurement system, it is essential that the ordering and distance between the difficulty levels of the various competencies within a construct should not change for students of different ability levels; hence, the probability of a student exceeding a certain response threshold on a competency should be proportional only to the difference between the ability of the student and the difficulty of that threshold (Boone, 2016). This fixed philosophical specification facilitates a hypothetico-deductive approach to understanding the strengths and limitations of a measure and the framework that underlies it in the sense that new information is generated by holding the model as true and rationalizing anomalies that exist in the data. It follows that new information is generated not through fitting increasingly complex models, but instead through the intellectual exercise of rationalizing the anomalies that exist in the measure with respect to a pre-specified model (Popper, 1963). A popular example of this process from astronomy is the discovery of the planet Neptune. In 1845, when astronomers discovered that the orbit of Uranus did not accord with expectations provided by Newton's law of universal gravitation, they were faced with the choice of either changing Newton's equation or rationalizing the anomaly through hypothesizing the existence of a further unknown planet that was perturbing Uranus's orbit (Massey, 2007). Although some astronomers entertained the idea that gravity's inverse square relationship began to break down over large distances (Wallin et al., 2007), the choice to respect the sanctity of Newton's law of universal gravitation led to the mathematical discovery of Neptune (Worrall, 1989). To this day, rationalization of anomaly remains among the most utilized approaches to the prediction and discovery of heavenly bodies. This example is analogous to a researcher's decision on whether to use Rasch or IRT approaches to measurement. Just as Newton's law of universal gravitation was treated as a universal law of gravitational attraction, the Rasch model can be treated as a universal law of measurement; when data violate the law, do we conclude that the law has broken down, or do we use the anomalies we see to make new discoveries about the behavior of our students and assessments?

16.3.3 Multi-faceted Rasch or 2-faceted Rasch?

A majority of Rasch studies use models with two facets of measurement: students (facet 1) and items (facet 2) for example. In our case, a 2-faceted model would be desirable in a case where all students took the same scenario, such that there was no need to generalize across scenarios, and inter-rater agreement was high enough for scores to be combined through averaging or use of a single rater's score. In this case, students and argumentation competencies would be the only facets of variance. In our study, internal consistency of the 66 (of 322 total) double scored SBAs ranged between 0.68 and 0.80 which indicates levels of rater consistency that could be considered reasonable, but nonetheless far from perfect. Cohen's kappa, which treats the scores as nominal and rewards perfect agreement exclusively, ranged between 0.22 and 0.39 which would be considered unacceptable in most scoring applications where an "agreement" model is favored. Previously, we made the case that use of a "consistency" model framework that treats our experts as independent and focuses more on intra-rater consistency than perfect inter-rater agreement will necessarily yield lower levels of agreement. Nonetheless, in cases like this where there is low-to-moderate agreement, treating 'rater' as an additional facet can provide insight into the extent to which the lack of agreement is due to the independent expertise of the raters and the expected uniform disagreement associated with differences in rater severity that follows or the extent to which it is due to non-uniform disagreement associated with intra-rater inconsistency resulting from misunderstanding of the construct measured by the rubric. Given the need to understand and account for rater behavior and the need to generalize students' argumentation measures across multiple SSIs, a four-faceted Rasch modeling framework was necessary.

16.3.4 Our Implementation of Multi-faceted Rasch

Beyond the Rasch model's philosophical attractiveness in its alignment with the hypothetico-deductive approach to knowledge generation, the constraint of scale invariance across facets is mathematically convenient in that it sets up a geometric measurement framework which facilitates the straightforward addition of facets to the model (Linacre & Wright, 2014). Given that use of six competencies, eight raters, and eight scenarios each contributed to variation in the measures, it is conceivable that bias could be introduced by any one of these. We needed a way to quantify this bias as well as to generate measures for students that were independent of the scenarios or raters that generated the measures. With the FACETS software package (Linacre & Wright, 2014), we used a 4-faceted Rasch modeling procedure (facets were "student," "competency," "rater," and "scenario") to generate objective argumentation measures and to evaluate construct validity of the observations within these facets in terms of: (1) the efficacy of the competencies in generating reliable and unidimensional measures and (2) detection of systematic bias with respect to how the specific competencies, scenarios, and raters generate the students' measures of argumentation (Linacre & Wright, 2014).

To explore the extent to which the hierarchy of argumentation competencies form a unidimensional scale, we used principal components analysis (PCA) on standardized residuals with respect to a 2-faceted model which we implemented using WINSTEPS (Linacre, 2006). This model contained just the student and competency facets which were the two primary facets of interest in terms of understanding the argumentation construct as a unidimensional progression. Reduction to a 2-faceted model through averaging across the rater and scenario facets is a recommended procedure for evaluating conformity to the unidimensionality assumption in multi-faceted measurement contexts because the sparsity in the data is too great for PCA on residuals when all facets are considered at once (M. Linacre, Personal Communication, July 8, 2017). If the competencies are measuring argumentation as a single construct, then a first eigenvalue below 2, which indicates randomly distributed residuals, is expected (Linacre & Tennant, 2009).

For all facets, mean squares infit and outfit were used as an indicator of the extent to which responses to each scenario, the raters, and the competency measures fit with the model-expected values derived from a 4-faceted Rasch model. We made the argument previously that having competencies and scenarios with a range of difficulty and raters with a range of severity strengthens the information our assessments can provide about students with a diverse range of argumentation ability. Mean squares fit indices provide a mechanism for evaluating the extent to which competencies, raters, and scenarios provide model-expected consistency with respect to how they differentiate between students with high and low levels of argumentation. Mean squares infit and outfit have expected values of 1 (Wright & Stone, 1979). Values above 1 imply that the responses are less predictable than what the Rasch model would expect, indicating inconsistency in terms of how that competency, rater, or scenario ranks students in terms of their argumentation. Values below 1 indicate that the responses are more consistent than the Rasch model would expect, indicating bias introduced by a construct peripheral to argumentation which artificially inflates the probability of higher scores for students higher on the scale, or conversely, which artificially reduces the probability of higher scores for students lower on the scale (Masters, 1988). In no case will perfect expected fit be achieved for all facet elements, leading to the question of what constitutes an acceptable range for these indices. Currently there is little consensus on this. Wright et al. (1994) suggest that these indices can range between 0.5–1.5 for useful measures, but also recommend stricter ranges for high-stakes decisions. Indeed, the decision on how to use information about fit depends on the goals of the study or program. In the case of this study which was done at a relatively early stage in our program, our goals were more formative than summative. Given the alignment of our tasks and embedded competencies with the CER framework, our primary goal was to diagnose potential measurement issues to consider in the planning of the coming year's efforts. In this sense, we were less interested in strict cutoffs and more interested in looking for anomalies even if they are relatively small. This can be contrasted with the goal of measurement construction where the focus is often on removal of the misfitting or overfitting parts of the assessment.

16.4 Findings

Here we summarize what the 4-faceted Rasch modeling procedure tells us about our assessment and scoring framework and the nature of the argumentation construct itself. The outputs from both WINSTEPS and FACETS are extensive, and it is beyond the scope of this chapter to cover all of the available output. To this end, we highly recommend the respective user manuals for these programs (Linacre, 2006;

Linacre & Wright, 2014). We instead focus on the outputs that we used to help us better understand our assessment framework: namely the dimensionality and ordering of the competencies within the argumentation construct and anomalies with respect to the model indicating how the particular competencies, raters, and scenarios generate measures of argumentation.

We recommend starting by inspecting the Wright map (Fig. 16.2) generated in the analysis which gives insight into the ordering and difficulty range of the observed elements within each facet. Due to the constraint of scale invariance, it follows that

```
+--------------------------------------------------------------+
|Measr|+Student    |-Item                |-Scorer|-Scenario|
|-----+------------+---------------------+-------+---------|
|   9 + *****.     +                     +       +         +
|     | .          |                     |       |         |
|   8 + *.         +                     +       +         +
|     | .          |                     |       |         |
|   7 + **.        +                     +       +         +
|     | *.         |                     |       |         |
|   6 + **.        +                     +       +         +
|     | *.         |                     |       |         |
|   5 + **.        +                     +       +         +
|     | **         |                     |       |         |
|   4 + ***.       +                     +       +         +
|     | *****.     |                     |       |         |
|   3 + **         +                     +       +         +
|     | ***.       |                     |       |         |
|   2 + ***        +                     + 2     +         +
|     | *.         |                     | 4  8  |         |
|   1 + ****.      +                     +       + 3       +
|     | *******    | Holistic  Reasoning | 1  6  | 5  8    |
*   0 * ****.      * Claim     Evidence  * 5     * 2  4  6 *
|     | ********   | Content             | 7     | 7       |
|  -1 + ********   + Writing             +       +         +
|     | ********.  |                     |       | 1       |
|  -2 + *****.     +                     +       +         +
|     | ******     |                     |       |         |
|  -3 + ****.      +                     +       +         +
|     | ****.      |                     |       |         |
|  -4 + .          +                     +       +         +
|     | *.         |                     |       |         |
|  -5 + **.        +                     +       +         +
|     | .          |                     |       |         |
|  -6 + **         +                     + 3     +         +
|-----+------------+---------------------+-------+---------|
|Measr| * = 3      |-Item                |-Scorer|-Scenario|
+--------------------------------------------------------------+
```

Fig. 16.2 Wright map of the student, item, rater, and scenario facets

all of the facets are on the same scale and hence can be compared with each other in order to provide insight into the construct validity of the assessment. In most cases, a Wright map is presented as a person-item map where the item difficulty and student ability are plotted on a common log-odds (logit) scale. Since we have a 4-faceted model, Fig. 16.2 contains student ability (facet 1), item competency difficulty (facet 2), rater severity (facet 3) and scenario difficulty (facet 4). In all of the facets in Fig. 16.2, the location of each facet element aligns with the center of the scale, the location at which achieving 1 = benchmark and 4 = capstone are equally probable.

The Wright map provides important qualitative information about each facet which we will discuss in more detail below. Foremost, we observe that the competencies have a definitive order of difficulty which ranges from communication processes at the easier end to reasoning processes at the more difficult end. Next, when looking at the rater severity measures, we see that four of the raters had a severity rating that is within the range of the difficulty of the competencies. Three of the raters were slightly more severe in scoring and one of the raters was very lenient in scoring. Given this range, it will be important to look at how the raters' scores fit with the model in order to diagnose the extent to which these differences in difficulty are an asset or a hindrance to our goal of independent consistency. Finally, we find that the difficulty range of the scenarios is slightly greater than the difficulty range of the competencies. By inspecting fit or misfit with the multi-faceted Rasch model, we can gain insight into whether this range in difficulty is an asset in measuring argumentation across a range of student ability, or whether the scenarios contain biases peripheral to argumentation which serve to confound students' measures.

16.4.1 Structure of the Argumentation Competencies Along the Scale

The difficulty of the competencies (Table 16.2) ranged from −0.76 and 0.67 logits (reliability = 0.94). The easiest was writing (d = −0.76). Once writing, the students could then integrate content-specific vocabulary (d = −0.39) and then begin engaging in the upper levels of the argumentation construct: making the claim (d = 0.03), recognizing evidence (d = 0.13), and providing reasoning which connects the evidence to the claim (d = 0.67). The holistic score had a difficulty of 0.32. That the holistic score exceeds the measure for recognizing evidence but sits below the

Table 16.2 Difficulty measures and fit indices of the six argumentation competencies

Competency	Measure	SE	Infit	Outfit
Claim	0.03	0.11	1.09	1.28
Evidence	0.13	0.11	1.01	1.03
Content	−0.39	0.11	1.20	1.13
Reasoning	0.67	0.11	1.05	1.06
Writing	−0.76	0.12	1.05	1.01
Holistic	0.32	0.11	0.54	0.50

Table 16.3 Difficulty measures and fit indices for the eight raters

Rater	Measure	SE	Infit	Outfit
1	0.71	0.24	0.84	0.78
2	1.78	0.28	0.85	0.84
3	−5.75	0.59	1.34	1.42
4	1.40	0.17	0.83	0.83
5	0.16	0.07	1.16	1.19
6	0.59	0.16	0.81	0.78
7	−0.60	0.33	1.09	1.12
8	1.72	0.08	0.89	0.91

measure for reasoning indicates that our raters generally considered students who were able to make a claim and recognize evidence as having a satisfactory level of argumentation even if their ability to express their reasoning was still developing. Although this model is informative in terms of understanding how middle school students engage in argumentation, evidence is needed that the data fit with that model. Per the hypothetico-deductive approach to measurement, looking for misfit with the model can help us identify potential issues with the assessment as well as more nuanced patterns in students' thinking.

The student argumentation measures generated by the assessment were reliable (reliability = 0.93) and unidimensional. Since the first eigenvalue from PCA on residuals with respect to a 2-faceted model containing the person and competency facets yielded 1.61 items of variance, we can claim based on simulation work that there is not sufficient evidence to suggest that the scale is multi-dimensional (Linacre & Tennant, 2009). For the individual competencies, mean squares outfit measures ranged between 1.01 and 1.28 and infit measures ranged between 1.01 and 1.20 (Table 16.3). The ability to draw upon discipline-related content, which is one of the easier competencies for students, had an infit of 1.20 and an outfit of 1.13. Although this level of misfit is relatively small with respect to most testing applications, it nonetheless shows some tendency for students to obtain higher scores on this construct despite having an otherwise low level of argumentation ability. With respect to this competency, our previous argument that the scale of competencies is unidimensional can be taken to imply that knowing and drawing on science content is an essential part of argumentation. Nonetheless, the small amount of misfit in this competency indicates some separateness between knowing content and communicating that content within the process of argumentation.

Making a claim was another competency with higher-than-expected misfit. Although the infit value of 1.09 is within reasonable bounds, the outfit value of 1.28 is high enough that it is worthy of consideration and discussion. Since outfit is an unweighted mean squares index, it tends to be more outlier sensitive than infit which is weighted toward the central density of responses (Wilson, 2004). This discrepancy between these two measures indicates that the misfit is being caused by students at the extremes of the scale: there are some low-argumentation students who successfully make a claim, but are nonetheless struggling with the writing and content aspects, and on the other end of the spectrum, there are some high-argumentation students who show the ability to identify evidence and use reasoning but did not elaborate the claim with enough detail to score at the "capstone" level. We reiterate that the magnitude of this misfit is small enough that we don't consider it a validity concern; however, it does indicate that helping students learn to make a well-defined claim is something we should pay more attention to even in students with relatively high levels of argumentation ability.

The holistic score had better-than-expected fit (infit = 0.54, outfit = 0.50) meaning the ratings tended to be unduly biased against students who lack the foundational communication skills which buttress the bottom of the scale. Better-than-expected fit combined with higher difficulty indicates that when assigning a holistic score, the raters uniformly tended to put more emphasis on mastery of the higher-level competencies; hence, these were more impactful on a student's holistic score than the lower-level competencies such as writing, using content, and making a claim. This led to sharper-than-expected discrimination between students of high and lower argumentation ability which is indicated by the higher-than-expected fit with the multi-faceted Rasch model.

16.4.2 *Analysis of Inter-rater and Intra-rater Consistency*

Earlier in the chapter we discussed how treating the scoring panel as a group of independent experts will result in lower levels of perfect agreement, but at the same time provide more robust information across a diverse array of student argumentation ability. However, even in a case like this study where perfect agreement is not emphasized, a certain level of agreement is nonetheless expected by the model. Here we reiterate that the purpose of training our raters was to get everybody on the same page on the meaning of the construct and how to apply the rubric to assign a score for each argumentation competency. How effective was this training in helping our raters retain their independence while also ensuring that their scores were consistent? If observed agreement is higher than model-expected agreement, then that indicates the presence of external pressure for the panel to agree with each other, making the raters non-independent. On the other hand, if observed agreement is lower than model-expected agreement, then that indicates that scoring was more of a "free for all" and that more training is needed to get raters on the same page with respect to how to interpret the rubric and apply it to scoring students' responses. In the case of our raters in this study, the model indicated that 55.6% of the responses were expected to be in exact agreement. This value is greater than the 49.2% of exact agreements observed in the data. Although these values are in proximity to each other, this indicates that some members of our scoring panel may benefit from additional training on how to use the rubric to assign scores, and what low, middle, and high argumentation responses look like.

The deficiency in observed agreement in comparison to expected agreement accompanied with moderate inter-rater reliability suggests that it may be useful to look more closely at rater severity and misfit with the multi-faceted Rasch model toward planning next steps for rater training. Rater severity (reliability = 0.99 which reflects the precision of the severity measures, not inter-rater reliability) had a wide range (−5.75 to 1.78 logits). Mean squares outfit ranged between 0.78 and 1.19 and infit ranged between 0.81 and 1.16 for all of the raters except for the easiest rater which exhibited an outfit value of 1.42 and an infit value of 1.34. Given that this rater (rater 3) was the easier rater (severity = −5.75), the misfit displays a tendency to give uniformly high scores, even to students with relatively low argumentation ability, which makes their ratings less useful for distinguishing between high and low levels of argumentation. Given the high misfit of rater 3 relative to the other raters, it is likely that this rater is largely responsible for the discrepancy between observed and expected agreement discussed previously. This may indicate that additional training on how to recognize and label low and high argumentation responses is needed for this particular rater before upcoming scoring events. In particular, using additional sample student papers (anchor papers) as exemplars for the rater and providing more review during the scoring event may help to illuminate the respective score points.

16.4.3 Analysis of Scenario Difficulty and Bias

Scenario difficulty (reliability = 0.94) ranged between −1.45 and 0.97 logits (Table 16.4). Scenario outfit values ranged between 0.71 and 1.18 and infit values ranged between 0.75 and 1.26 (Table 16.4). Although these levels of misfit or overfit are not necessarily problematic in terms of providing measures with adequate construct validity, it is nonetheless useful from a formative standpoint to inspect more closely the extremes of scenario difficulty and the scenarios with anomalous levels of misfit.

Aggressive Honeybees (d = −1.45) was the easiest scenario. *Raccoon River Nitrates* (d = 0.66) (adapted from Owens et al. 2021) and *Lake Front School Algal Blooms* (d = 0.97) were the most difficult scenarios. The *Aggressive Honeybees* and *Raccoon River Nitrates* scenarios also had the highest levels of misfit. Given that *Aggressive Honeybees* (infit = 1.11, outfit = 1.18) was the easiest

Table 16.4 Difficulty measures and fit indices for the eight scenarios

Scenario	Measure	SE	Infit	Outfit
1: Aggressive honeybees	−1.45	0.13	1.11	1.18
2: Heat stress	0.15	0.13	1.04	1.09
3: Lake front school algal blooms	0.97	0.13	0.90	0.87
4: GMO mosquitoes	−0.23	0.09	0.90	0.93
5: Racoon River nitrates	0.66	0.14	1.26	1.18
6: Saving bumblebees	−0.18	0.11	1.03	1.03
7: Vaping	−0.28	0.38	0.75	0.82
8: York's flooding problem	0.35	0.17	0.76	0.71

scenario, this may indicate that students tended to express higher levels of argumentation on this scenario relatively uniformly regardless of their ability, which attenuates slightly the information it can provide and this is possibly due to the type of task and data that were presented to the students.

The *Aggressive Honeybees* SBA included the background information of the scenario and two graphs. Students were asked to prepare a statement to their local community about the best way to monitor bees:

> Directions: Write a statement with your recommendation for the best approach. Your response should include the claim you are making about the approach you recommend, the evidence provided above that links to your claim, and your justification, or reasons, why the evidence you have cited supports the claim you have made.

In the *Raccoon River Nitrates* SBA (adapted from Owens et al. 2021), students were provided a map, more extensive background information, and a more detailed prompt to address the problem. In doing so, they were asked to address and negotiate the competing arguments made by three different groups:

> Task and Directions: Recommend a Course of Action to Ensure Safe Water
>
> You are one of the local leaders working on a report of recommendations. Based on the information provided, write your report to make a claim for what you propose should be done regarding the likely water crisis. Include the following components:
>
> - An evaluation of arguments from Des Moines Water Works (city utility), Farmer's Drainage Cooperative (farmers), and Iowa Partnership for Clean Water (citizen group).
> - A recommendation for the specific course of action that should be taken to ensure safe water. You could take this in different directions, depending on your perspective. How would an ecologist approach this problem? What about a businessperson or a technological innovator? Where would they look to decide a specific course of action?
> - Evidence to support your recommendation and reasons why the evidence you use supports your recommendation.

The *Lake Front Algal Blooms* SBA included a map and a table. Students were asked to write to the Science Club and address specific perspectives from different people:

> Your task is to consider the data, the map, and both Tom's and Michelle's arguments in light of what you already know about water systems and elevation.
>
> Directions: Write a response addressed to the Science Club at Lake Front School. Your essay should make a claim identifying the two farms you believe are most likely to be the sources of fertilizer pollution and should include the following components:
>
> 1. A clear argument regarding which two farms are most involved.
> 2. Evidence to support that claim regarding the two farms involved.
> 3. Reasons why the evidence you cite does support the claim you've made.
> 4. A critique of Tom's and Michelle's arguments.

Raccoon River Nitrates (infit = 1.26, outfit = 1.18) was among the most difficult scenarios illustrating that the task was more extensive relative to the other SBAs. The higher misfit may indicate that the relatively complex wording in this scenario may have been confusing to the higher-ability students which would lead to both higher difficulty and misfit.

16.5 Theory Building Through Rasch Modeling: Argumentation as a Progression

We have discussed previously the value of Rasch modeling in research programs focusing on the hypothetico-deductive approach which underlies the process of discovery through falsification (Popper, 1963). In the previous section, we discussed that the intellectual burden introduced by rationalizing anomalies in the data with respect to the law of measurement invariance provides a pathway to better understanding the behavior of the scenario-based assessments, our panel of raters, and the competencies being scored. However, in cases where the assessment is built based on theory as is the case with many learning progression-based assessments (Wilson, 2009), the Rasch model can be used as a standard through which the theory underlying the assessment can potentially be falsified or revised. In the case of this study, Berland and McNeil (2010) proposed that argumentation is a progression that involves increasing complexity of both the argumentation product and process. Gotwals and Songer (2010) further specify this progression which starts at making a claim, then identifying evidence for the claim, and then providing reasoning which ties the evidence to the claim. This was built upon by Osborne et al. (2016), which focused on argumentation around scientific explanations, and more recently by Deane et al. (2019) who contextualized a 4-level progression for the practice of argumentation in the context of SBAs. With respect to our assessment, we used this prior research to hypothesize that the argumentation competencies we measured with respect to SSI-based scenarios should also form a progression. Indeed, this is what we see in the Wright map (Fig. 16.2) which appears to align closely with our proposed measurement framework (Fig. 16.1). Specifically, we hypothesized that student's' engagement in the process of writing and use of discipline-specific vocabulary should buttress the process of CER. This process starts with making a claim, which then progresses to finding evidence to support the claim. Once evidence is found, then the student has the opportunity to engage in reasoning around how the evidence supports the claim.

Overall, it appears that the hierarchical progression within the Wright map matches well with our hypothesized framework. At this point, we may be tempted to stop there and conclude that the theory is confirmed. This temptation should be resisted, however, in light of the fact that the difficulty of the competencies can be derived simply from logit-transformed raw scores. Although this has some value in mitigating floor and ceiling effects associated with using non-linear raw scores, Rasch modeling allows us to dig deeper by looking closely at parts of the assessment that misfit with the model (Boone, 2016). In sum, the progression outlined in Fig. 16.2 can only be considered verified to the extent that the data fit the model. The most straightforward type of error to look at is that associated with measurement consistency. The competency difficulty measures have a root mean square error of 0.11 which corresponds to the high item reliability index of 0.94. This in turn corresponds to a separation index of 3.96. Separation is the ratio of the sample standard deviation to the root mean square error of the measures—a value of 3.96

indicates that the difficulty distribution of the measured competencies can be broken up into about 4 measurably different levels (Wright & Masters, 2002). This relatively high precision lends confidence that many of the competencies can be considered distinctly different. For example, the difficulty of using content is over 3 standard errors above the difficulty of the writing competency. The difficulty of making a claim is around 3.8 standard errors above the difficulty of using content. This gives confidence, then, that these baseline competencies in the progression are in fact distinct. When we move from making a claim to identifying evidence, the difficulty estimate increases from 0.03 to 0.13, which is less than 1 standard error. This indicates that it is difficult to predict with confidence that a student is always going to make a clear claim before going about finding evidence. An additional increase of about 5 standard errors is needed before students demonstrate proficient reasoning. In sum, analysis of the precision of the estimates suggests that most of these argumentation competencies are distinct, providing evidence that the easier competencies need to be in place in order to buttress the more difficult competencies. However, there is not sufficient evidence to suggest that making a claim always scaffolds the practice of seeking evidence: the estimates suggest that these could potentially occur in either order or at the same time.

In the discussion above, we made the case that the Wright map provides evidence for a qualitative ordering and the precision of the estimates can be used to indicate the extent to which the observed order in the competencies is real and distinct. But we have not yet gotten into measurement invariance. When a Rasch model is used, we can use the fit indices to quantify the extent to which the progression conforms to the law of measurement invariance. It follows that observed misfit indicates some degree of violation of measurement invariance for that part of the assessment. In terms of the Wright map of the argumentation competencies (Fig. 16.2), high levels of misfit of a competency with the Rasch model would cast doubt on the extent to which the location of that competency reflects its true location which would be obtained by an assessment that conforms adequately to the law of measurement invariance. For the five core competencies in the argumentation progression, the infit indices range between 1.01 and 1.20 and the outfit indices range between 1.01 and 1.28. Since outfit values average to 1 and the holistic score has better-than-expected fit (infit = 0.54, outfit = 0.50), removing the holistic score may provide more realistic measures of fit for the other competencies. With the holistic score removed from the scale, infit values between 0.90 and 1.06 and outfit values between 0.92 and 1.19 are obtained for the other competencies. When only these 5 competencies are considered, making a claim has the highest level of misfit (infit = 1.06, outfit = 1.19) but this level of misfit is small and is in the realm of what may be considered appropriate even for high-stakes diagnostics (Wright, 1994). This level of fit with the 4-faceted Rasch model suggests that these competencies as they were assessed conform adequately to the law of measurement invariance, and hence we can proceed with a degree of confidence that the progression in Fig. 16.2 will likely be reproducible in future research that utilizes invariant measures for these argumentation competencies.

16.6 Conclusions

In this chapter, we described how multi-faceted Rasch modeling can be leveraged to provide relatively context-independent and invariant measures for argumentation. We observe that the competencies tend to progress from establishment of basic communication practices (writing and vocabulary) which buttress less complex declarative processes (stating a claim or purpose), which in turn buttress more complex evidence and reasoning practices. The data also suggest that there is a difficulty range in the scenarios which illustrates the importance of considering the type of SSI in the measurement model in order to generate measures which deliver more information at the extremes of the scale and information with some degree of context-independence. Overall, we found that how students express argumentation is transferable across multiple types of scenarios after the variation in their difficulty is taken into account, opening up the opportunity to develop scenarios that are unique to the disciplinary context of a specific classroom or curricular intervention. In addition, the ability to equate the tasks across multiple scenarios creates opportunities for application of multi-level assessment frameworks which have been shown to be informative in evaluating outcomes associated with SSI-based curricula (Sadler, et al. 2013). Finally, we hope that this chapter has provided additional insight into how to use multi-faceted Rasch modeling to facilitate more useful rater scoring frameworks which go beyond agreement toward a focus on balancing independent expertise with consistency. In sum, we believe that multi-faceted Rasch modeling holds great potential for research programs needing invariant measures for more open-ended constructs which demand more complex measurement frameworks. We look forward to seeing the use of these methods increase as we continue to turn our attention to generating more accurate and authentic measures for difficult-to-measure constructs such as the scientific practices.

Acknowledgements This research was funded by National Science Foundation DRK-12 grant #2010312. The views expressed are those of the authors and do not necessarily reflect the views of the National Science Foundation.

Appendix (Rasch Analysis Codes with Annotations)

Multi-faceted Rasch Analysis in FACETS

```
Title = SBA Cohort 3
; title
MISSING = .
;using a "." to indicate missing data
Facets = 4
;telling how many facets (student, item, rater, scenario)
Inter-rater = 3
;facet 3 is the rater facet
```

```
Delements = LN
;how the data are represented (see manual for details), other options
are N, L, or NL
Non-centered = 1
;First facet (students) is non-centered (i.e. items, raters, scenarios
are centered at 0)
Yardstick = 0,2.
;Yardstick = Horizontal columns, Vertical lines, Low range, High
range, {"Measure" or "End"}
Model=
?,?,?,?,Score
;N2MX,1,8,4,4
;C68MX,1,5,4,4
;N3MX,1,8,4,4
;student identifier, item number, rater number, scenario number, score
on that mix
*
Rating Scale=Score,R4, Keep
;"Score" refers to the 'score on that mix'; R4 refers to a rating scale
with a max of 4. "Keep" means model all ordinal categories even if not
observed.
1 = lowest score
4 = highest score
*
Labels=;this indicates that the section where you are labeling the
facets is beginning.
1, Student;name of first facet, the students
1 = N2MX
2 = C68MX
3 = N3MX
.
.
.
.
320 = J23YX
321 = J24YX
322 = J25YX


*
2, Item ;name of the second facet, the items
1 = Claim
2 = Evidence
3 = Content
4 = Reasoning
5 = Writing
6 = Holistic


*
3, Rater;name of the third facet, the raters. This is the "rater facet".
1 = 1
2 = 2
```

```
3 = 3
4 = 4
5 = 5
6 = 6
7 = 7
8 = 8


*
4, Scenario;name of the fourth facet, the scenarios. This is analogous
to alternate testing forms with parallel items.
1 = 1
2 = 2
3 = 3
4 = 4
5 = 5
6 = 6
7 = 7
8 = 8


*
Query = N;asking the software to not do the query and just run it all the
way through without stopping at each iteration.


*
Missing=.;tells Facets that the "." is the missing data indicator
Data =;tells Facets that you will be pasting the data below.
N2MX,1,8,4,4
C68MX,1,5,4,4
N3MX,1,8,4,4
N4MX,1,8,4,2
W15AX,1,5,1,4
W14AX,1,5,1,4
W13AX,1,5,1,4
V29AX,1,5,1,4
.......
```

Two-Faceted Rasch Analysis in WINSTEPS

```
&INST;indicates beginning of the control file syntax
TITLE = '2-faceted SBA analysis for Cohort 3';title given by the
researcher
NI = 6;six items
ITEM1 = 1;first item begins on first column
XWIDE = 1;each item is one column wide
CODES = 1234;ordinal codes to be modeled. All other codes treated as
missing.
```

```
NCOL = 6;six columns in the data
MODELS = R;default method for dichotomous, rating scale, and partial
credit models
STBIAS=Y;correction for estimation bias
PRCOMP=S;principal components analysis on standardized residuals
GROUPS = 0;each item allowed to have its own unique rating scale
TABLES = 1111111111111111111111;asks for all of the tables
&END;end of the command file syntax
1 = Claim;name of first item
2 = Evidence;name of second item
3 = Content;name of third item
4 = Reasoning;name of fourth item
5 = Writing;name of fifth item
6 = Holistic;name of sixth item
END NAMES;indicates end of the naming syntax. Data are pasted below.
434343
444434
444444
232222
444444
......
```

References

Andrich, D. (2004). Controversy and the Rasch model: A characteristic of incompatible paradigms? *Medical Care, 42*, I7–I16.

Bergan, J. R. (2013). *Rasch versus Birnbaum: New arguments in an old debate*. Assessment Technology.

Berland, L. K., & McNeill, K. L. (2010). A learning progression for scientific argumentation: Understanding student work and designing supportive instructional contexts. *Science Education, 94*(5), 765–793.

Berland, L. K., McNeill, K. L., Pelletier, P., & Krajcik, J. (2017). Engaging in scientific argumentation. In B. Reiser, C. Schwarz, & C. Passmore (Eds.), *Moving beyond knowing science to making sense of the world: Bringing next generation science and engineering practices in our K-12 classrooms*. National Science Teachers Association Press.

Boone, W. J. (2016). *Rasch analysis for instrument development: Why, when, and how? CBE—Life sciences education, 15(4), rm4* (Vol. 15).

Boone, W. J., Staver, J. R., & Yale, M. S. (2013). Rasch analysis in the human sciences. Springer Science & Business Media.

Boone, W. J., Townsend, J. S., & Staver, J. R. (2016). Utilizing multifaceted Rasch measurement through FACETS to evaluate science education data sets composed of judges, respondents, and rating scale items: An exemplar utilizing the elementary science teaching analysis matrix instrument. *Science Education, 100*(2), 221–238.

Covitt, B., Dauer, J., & Anderson, C. (2017). The role of practices in scientific literacy. In C. Schwarz, C. Passmore, & B. Reiser (Eds.), *Helping students make sense of the world using next generation science and engineering practices* (pp. 59–83). NSTA Press.

Deane, P., Song, Y., van Rijn, P., O'Reilly, T., Fowles, M., Bennett, R., et al. (2019). The case for scenario-based assessment of written argumentation. *Reading and Writing, 32*, 1575–1606.

Gotwals, A. W., & Songer, N. B. (2010). Reasoning up and down a food chain: Using an assessment framework to investigate students' middle knowledge. *Science Education, 94*(2), 259–281.

Kinslow, A. T., Sadler, T. D., & Nguyen, H. (2019). Socio-scientific reasoning and environmental literacy in a field-based ecology class. *Environmental Education Research, 25*, 388–410. https://doi.org/10.1080/13504622.2018.1442418

Krajcik, J. (2015). Three-dimensional instruction. *The Science Teacher, 82*(8), 50–52.

Lead States, N. G. S. S. (2013). *Next generation science standards: For states, by states*. National Academies Press.

Lin, S. S., & Mintzes, J. J. (2010). Learning argumentation skills through instruction in socioscientific issues: The effect of ability level. *International Journal of Science and Mathematics Education, 8*(6), 993–1017.

Linacre, J. M. (2006). WINSTEPS Rasch measurement computer program. WINSTEPS.com

Linacre, J. M., & Tennant, A. (2009). More about critical eigenvalue sizes (variances) in standardized residual principal components analysis (PCA). *Rasch Measurement Transactions, 23*(3), 1228.

Linacre, J. M., & Wright, B. D. (2014). Facets. *Computer Program for Many-faceted Rasch Measurement*, 1998. MESA.

Massey, G. J. (2007). A new approach to the logic of discovery. *Theoria, Beograd, 50*(1), 7–27.

Masters, G. N. (1988). Item discrimination: When more is worse. *Journal of Educational Measurement, 25*(1), 15–29.

National Research Council. (2014). *Developing assessments for the next generation science standards*. National Academies Press.

Osborne, J. F., Henderson, J. B., MacPherson, A., Szu, E., Wild, A., & Yao, S. Y. (2016). The development and validation of a learning progression for argumentation in science. *Journal of Research in Science Teaching, 53*(6), 821–846.

Owens, D. C., Sadler, T. D., Petit, D., & Forbes, C. T. (2021). Exploring undergraduates' breadth of socio-scientific reasoning through domains of knowledge. *Research in Science Education, 52*, 1643–1658. https://doi.org/10.1007/s11165-021-10014-w

Popper, K. R. (1963). Science as falsification. *Conjectures and Refutations, 1*(1963), 33–39.

Raudenbush, S. W., & Bryk, A. S. (2002). *Hierarchical linear models: Applications and data analysis methods*. Sage.

Sadler, T. D. (2004). Informal reasoning regarding socioscientific issues: A critical review of research. *Journal of Research in Science Teaching, 41*(5), 513–536.

Sadler, T. D., Romine, W. L., Stuart, P. E., & Merle-Johnson, D. (2013). Game-based curricula in biology classes: Differential effects among varying academic levels. *Journal of Research in Science Teaching, 50*(4), 479–499.

Linking Science & Literacy for All Learners. (2018). *Resources & materials: Multimodal text sets*. University of Missouri. Retrieved April 22, 2022, from https://scienceandliteracy.missouri.edu/resources-materials/

Sweller, J., Chandler, P., & Kalyuga, S. (2011). *Cognitive load theory*. Springer.

Thurstone, L. L. (1928). Attitudes can be measured. *American Journal of Sociology, 33*(4), 529–554.

Venville, G. J., & Dawson, V. M. (2010). The impact of a classroom intervention on grade 10 students' argumentation skills, informal reasoning, and conceptual understanding of science. *Journal of Research in Science Teaching, 47*(8), 952–977.

Wallin, J. F., Dixon, D. S., & Page, G. L. (2007). Testing gravity in the outer solar system: Results from trans-Neptunian objects. *The Astrophysical Journal, 666*(2), 1296–1302.

Wilson, M. (2004). *Constructing measures: An item response modeling approach*. Routledge.

Wilson, M. (2009). Measuring progressions: Assessment structures underlying a learning progression. *Journal of Research in Science Teaching, 46*(6), 716–730.

Womack, A. J., Wulff, E., Sadler, T. D., & Romine, W. (2017, April). *Assessment of next generation science learning*. San Antonio.

Worrall, J. (1989). Structural realism: The best of both worlds? *Dialectica, 43*(1–2), 99–124.
Wright, B. (1992). *IRT in the 1990s: Which models work best? 3PL or Rasch? Ben Wright's opening remarks in his invited debate with Ron Hambleton, session 11.05*, AERA annual meeting 1992.
Wright, B. (1994). Reasonable mean-square fit values. *Rasch Measurement Transactions, 8*, 370.
Wright, B. D., & Stone, M. A. (1979). *Best test design*. MESA Press.
Wright, B. D., Linacre, J. M., Gustafson, J. E., & Martin-Loff, P. (1994). Reasonable mean square fit values. *Rasch Measurement Transactions, 8*(3), 370.
Zeidler, D. L., Herman, B. C., & Sadler, T. D. (2019). New directions in socioscientific issues research. *Disciplinary and Interdisciplinary Science Education Research, 1*(1), 1–9.

William Romine, Ph.D. is an Associate Professor of Biological Sciences at Wright State University. He serves as the director of the *Data Science for Education Laboratory* at Wright State University.

Amy Lannin, Ph.D. is an Associate Professor within the College of Education and Human Development at University of Missouri. She serves as the director of the Campus Writing Program and the Missouri Writing Projects Network.

Maha Kareem is a graduate student within the College of Education and Human Development at University of Missouri. She currently undertakes efforts with research and teaching, specializing in work with international students.

Nancy Singer, Ph.D. is an Associate Professor within the College of Education at University of Missouri—St. Louis. She serves as the Associate Dean of the College of Education.

Chapter 17
Using Linear Logistic Rasch Models to Examine Cognitive Complexity and Linguistic Cohesion in Science Items

Ye Yuan and George Engelhard Jr

Abstract The purpose of this chapter is to describe the use of Rasch measurement theory to examine the impact of cognitive complexity and linguistic cohesion on science items. This chapter describes a linear logistic Rasch model (LLRM), and conducts a case study to illustrate how to apply the LLRM. In order to provide continuity with previous research on the Rasch model, this chapter describes the connections between methods used to explore item covariates including the Many Facet Rasch Model, hierarchical regression models, and LLRM. The case study uses high school biology assessment data to illustrate the connections between these three approaches.

Rasch measurement theory is widely used to guide measurement practices in the human sciences (Bond et al., 2021; Boone & Staver, 2020; Engelhard & Wang, 2021). Rasch measurement theory has also been used extensively in science education (Liu & Boone, 2006). As pointed out by Liu (2020), the science education community has recognized the key role of Rasch measurement theory in order to develop instruments that support the validity of interpretation and use of scores in science education.

There have been numerous advances in Rasch measurement theory, and there are a wide array of Rasch models available to the science education community (de Boeck & Wilson, 2004). Previous models that have been used to investigate the item type effects, especially the rater and task characteristics, include the Many Facet Rasch Model and the hierarchical regression model. This chapter explores an alternative method based on generalized linear logistic models. The recognition that the Rasch model can be extended by using generalized linear mixed models (GLMMs) is one of the major advances in measurement. A major advantage of this conceptualization of Rasch measurement models is that there are widely available packages in R (R Development Core Team, 2004) that can be used to estimate

Y. Yuan (✉) · G. Engelhard Jr
The University of Georgia, Athens, GA, USA
e-mail: ye.yuan@uga.edu; gengelh@uga.edu

X. Liu, W. J. Boone (eds.), *Advances in Applications of Rasch Measurement in Science Education*, Contemporary Trends and Issues in Science Education 57,
https://doi.org/10.1007/978-3-031-28776-3_17

these models. A second advantage is that a variety of substantive questions can be systematically addressed from this perspective. The extensions to the Rasch model based on generalized linear mixed models offer an opportunity for researchers in science education to systematically explore item covariates. Items covariates can add to our understanding of the item characteristics that impact item difficulty.

This chapter provides an application of one important extension called Linear Logistic Rasch models (LLRMs) using GLMMs. Item covariates can be included in LLRMs that allow researchers in science education to explore item components that may affect student performance on science items. The purpose of this chapter is to describe the use of Rasch measurement theory to examine the impact of cognitive complexity and linguistic cohesion on science items based on GLMMs. The specific questions considered in the present chapter are:

1. What are the Linear Logistic Rasch models (LLRMs)?
2. What components of items may influence student responses on science items?

In addition to addressing these questions, this chapter also provides a case study of a biology assessment in order to illustrate the use of LLRMs in science items. The case study also discusses the Many Facet Rasch Model and hierarchical regression model in exploring item covariates. In the following sections, the chapter first provides a conceptual and mathematical explanation of LLRMs. Then, the chapter considers potential item components that may influence item locations on the construct being measured, focusing on the science items. Next, the case study is presented based on a detailed analysis of a biology assessment with several covariates, including cognitive complexity, linguistic cohesion, and content domain. In addition, the chapter includes a science teacher's refection and discussion.

17.1 What Are Linear Logistic Rasch Models?

Rasch measurement theory has been used extensively to guide the development and appropriate use of scales (Bond et al., 2021; Engelhard & Wind, 2022; Wilson, 2005). The parameters of Rasch models can be estimated with specialized computer software including Winsteps (Linacre, 2019a) and Facets (Linacre, 2019b). In this study, we use R packages to estimate linear logistic Rasch models (LLRMs) based on generalized linear logistic mixed models. LLRMs can be estimated using the *glmr* function in R (De Boeck et al. (2011). We use the *eirm* package (Bulut et al., 2021) that serves as a wrapper around the *glmr* function in R (Bates et al., 2015). The *eirm* package provides estimates for various extensions of Rasch models including the LLRMs. The term "*eirm*" is an abbreviation for explanatory item response models.

LLRMs are an important and under-utilized member of the Rasch family of models. Versions of this model have been called the linear logistic test model (Fischer, 1973), unidimensional linear logistic Rasch model (Fischer, 1997), and multicomponent models (Embretson, 1997; Embretson & McCollam, 2000). LLRMs have been used in the formative assessment and instructional research.

For example, Fischer (1973) illustrated an analysis in the context of mathematics with related cognitive components (Fischer, 1973). Other examples of applications of LLRMs in science education include elementary mechanics (Spada & May, 1982) and statistical word problems (Holling, et al., 2008). Essentially, LLRMs add item covariates that can be used to predict item difficulty. The basic idea is that there are underlying components, such as cognitive demand and linguistic cohesion, that can be used to understand item difficulty. These broader components may be used to improve teaching and learning without a focus on individual items. These underlying components can also be used to generate new items that have expected levels of item difficulty.

The simple logistic model developed by Rasch (1960/1980) for responding correctly to by person n to item i ($X_{ni} = 1$) can be written as follows

$$P(X_{ni}=1|\theta_n)=\frac{\exp(\theta_n-\delta_i)}{1+\exp(\theta_n-\delta_i)}=\frac{\exp(\eta_{ni})}{1+exp(\eta_{ni})} \tag{17.1}$$

where

$$\eta_{ni}=(\theta_n-\delta_i) \tag{17.2}$$

and

θ_n is the location of person n on the latent variable, and
δ_i is the location of item i on the latent variable.

Rasch models can be extended using generalized linear mixed models. Generalized linear mixed models are statistical models that include three components: (1) a random component, (2) a systematic component, and (3) a linking component. Azen and Walker (2011) provide detailed descriptions of these three components. The motivation for using generalized linear mixed models is that researchers can apply linear statistical models to situations where the dependent variable is categorical. These models can be used to transform these categorical dependent variables using a linking component (e.g., logistic function) to provide a framework for using well-known regression and ANOVA models. Tutz (2011), Faraway (2016), and Pinheiro and Bates (2006) provide helpful introductions to generalized linear mixed models. These models have been used successfully to estimate several extended Rasch models (De Boeck et al., 2011).

The linear logistic Rasch model examined in this chapter includes item covariates, and it can be written as an extension of Eq. 17.2:

$$\eta_{ni}=\theta_n-\sum_{k=1}^{K}\beta_k X_{ik} \tag{17.3}$$

η_{ni} is monotonically related to the person n's probability of answering item i correctly in a Rasch model,

θ_n is the location of person n on the latent variable with a normal distribution $N(0, \sigma^2)$,
X_{ik} is a design matrix identifying item component $k(k = 1, \ldots, k)$,
β_k is the estimated regression weight for an item component that is similar to the item difficulty in the Rasch model (Bulut, Gorgun, & Yildirim-Erbasli, 2021).

De Boeck and Wilson (2004) refer to measurement models without covariates as descriptive models, and measurement models with covariates as explanatory models. LLRMs are explanatory models with item covariates.

17.2 What Components of Items May Influence Student Responses on Science Items?

This section discusses several item components that can serve as covariates in exploring item characteristics that may impact the difficulty of a test item for a student. First, we discuss cognitive complexity. Next, we introduce a discussion of the potential role of linguistic cohesion in affecting item difficulty. Finally, the role of content domain is discussed.

17.2.1 Cognitive Complexity

Research on the role of cognitive complexity in test development has a long history. An important component of this line of research is the exploration of item characteristics that can influence item difficulty. Early approaches to cognitive complexity included item and test development strategies based on Bloom's taxonomy for the cognitive domain with items classified by both content and cognitive complexity in a table of specifications (Bloom et al., 1956). Anderson and Krathwohl provided a revision of Bloom's taxonomy (2001). The development of taxonomies, tables of item specifications, and test blueprints to guide item development has continued to be important in modern assessments (Lane et al., 2016; Webb, 1999).

One of the widely used approaches to categorize items for cognitive complexity is Webb's Depth of Knowledge (DOK; Webb, 1999). The items examined in this study were classified by the test developers based on DOK. There are four levels of cognitive demand included in DOK:

Level 1: Items involve recall and reproduction, such as remembering facts or defining a procedure.
Level 2: Items involve skills and concepts used to answer questions.
Level 3: Items involve strategic thinking with planning, justification, and complex reasoning.
Level 4: Items reflect extended thinking that requires going beyond standard learning by asking how learning be used in real world contexts.

Level 1 represents student recall of information and the capability to perform a clearly defined simple science process or procedure. The verbs in the item stems include recall, recognize, use, and calculate. Level 2 represents skills and concepts in science that involve going beyond recall and is more complex than Level 1. Illustrative verbs for Level 2 items are classify, organize, estimate, make observations, and collect and display data. Level 3 items involve strategic thinking, such as reasoning, planning, and using evidence. Level 4 items have high cognitive demands and are very complex. Typically, Level 4 items require an extended period of time, and include a scientific investigation.

In science education, Krell et al. (2021) examined the influence of cognitive demand on a scientific reasoning test. They concluded that higher cognitive demand (more text complexity), and the presence of a visual display increased item difficulty. Other studies related to predictors of item difficulty in science have found that the energy concept can be a significant predictor of item difficulty (Park & Liu, 2021). Misconceptions in biology is another area that has been found to be related to item difficulty (Galvin et al., 2015).

The biology assessment analyzed in this study consists of items classified as Types 2 and 3. Two items are included in the Appendix A. They are examples of Type 2 and Type 3 items examined in this chapter. Item 1 is classified as Type 2 (skills and concepts). A correct response to Item 1 requires the student to understand that cellular respiration takes place in the cytoplasm and mitochondria. Item 3 (strategic thinking) requires the student to understand that hypertonic solutions cause cells to shrink because water moves out of the cell.

17.2.2 Linguistic Cohesion

In addition to the cognitive demand required to answer an item correctly, linguistic cohesion is another important item component. Test items in content areas include verbal components that may influence student responses. For example, items that are too short or too long may influence student responses This variation may be construct irrelevant, and it has the potential to interfere with students being able to demonstrate their knowledge and skills in the content domain. Linguistic cohesion is defined by aspects of the text that play a role in aiding the reader to connect ideas in the text. Coh-Metrix Indices are used to identify aspects of language, discourse, and cognition of texts that are aligned with comprehension (Graesser et al., 2011; McNamara et al., 2005). Linguistic cohesion is distinct from other indices of readability and textual characteristics (Flesch, 1948; Stenner et al., 1988).

Text characteristics are important item components. In this study, cohesion is defined by textual characteristics that support a coherent mental representation of the questions posed by the items. This study explores the following text characteristics: Word count, Narrativity, Syntactic simplicity, Word concreteness, Referential cohesion, Deep cohesion, Verb cohesion, and Connectivity. These indices of linguistic cohesion are described in Appendix B.

The two sample items included in Appendix A offer examples of how text characteristics such as word count may influence item difficulty. Items with lower cognitive demands tend to be shorter (e.g., Item 1), while items with higher cognitive demands tend to have more words (e.g., Item 3). Cognitive demand and linguistic cohesion are both important item components that can be examined with LLRMs.

17.2.3 Content Domain

The National Academies of Sciences, Engineering and Medicine (2012) has identified several disciplinary core ideas in science: Physical Sciences, Life Sciences, Earth and Space Sciences, and Engineering, Technology, and Applications of Science. Measurement of these core ideas provides an essential component in the improvement of teaching and learning.

Content areas are also important item components. Content areas provide the context for defining student knowledge, skills and abilities that may influence the difficulty of an item. Five content areas are included in this study related to life sciences in the area of biology: Cells, Cellular Genetics and Heredity, Classification & Phylogeny, Ecology and Theory of Evolution. These content domains provide an important context for understanding the influences of both cognitive demand and linguistic cohesion on items. These content areas are related to the core ideas in life sciences from the National Academies of Sciences, Engineering and Medicine (2012). These content domains match the state standards for the southeastern state that guided the development of the biology assessment.

There are five content domains that align with state standards in this southeastern state. The state standards focus on developing student proficiency in biology. The overall performance standards for each content domain are as follows:

1. Cells: Obtain, evaluate, and communicate information to analyze the nature of the relationships between structures and functions in living cells.
2. Cellular genetics and heredity: Obtain, evaluate, and communicate information to analyze how genetic information is expressed in cells, and to analyze how biological traits are passed on to successive generations
3. Classification and phylogeny: Obtain, evaluate, and communicate information to illustrate the organization of interacting systems within single-celled and multi-celled organisms.
4. Ecology: Obtain, evaluate, and communicate information to assess the interdependence of all organisms on one another and their environment
5. Theory of evaluation: Obtain, evaluate, and communicate information to assess the theory of evolution.

These content domains are based on standards designed to continue the student investigations of the life sciences that began in grades K-8. The overall goal is to provide students with the necessary skills to be proficient in biology by focusing on the identification of patterns, processes, and relationships of living organisms.

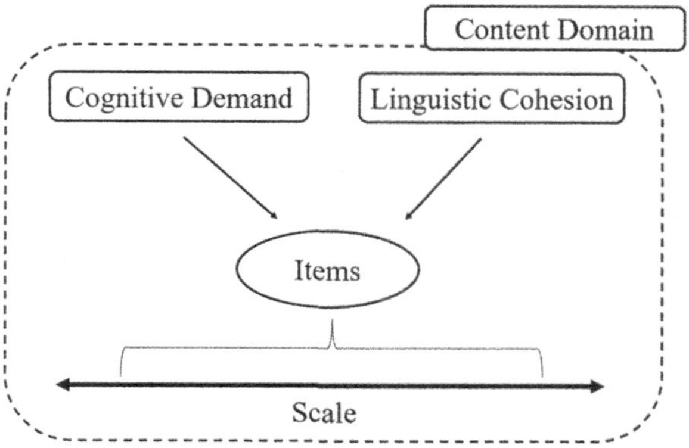

Fig. 17.1 Item Components

In summary, there are a variety of item components that can explored that may play a role in determining student responses. Fig. 17.1 provides a framework to guide our conceptualization of the three major item components. Cognitive complexity and linguistic cohesion directly influence items with the content domain providing an important context for examining their effects on items. The item components play an important role in determining item locations on the underlying latent variable. We have represented this connection with a large bracket to suggest this mapping of items onto the underlying continuum that defines the construct. The construct in this chapter is proficiency in high school biology. Of course, additional factors related to item locations may include the characteristics of students (e.g., opportunity to learn the content).

17.3 Case Study

In order to illustrate the ideas presented in this chapter, we provide a detailed analysis of a biology assessment.

17.3.1 Analyses

We analyzed biology assessment data with three different methods. First, we used the Many Facets Rasch Model that is familiar to many researchers in science education (Liu, 2020). Next, we estimated a hierarchical regression model that has been used by researchers in cognitive psychology to explore the impact of various item components (Embretson, 1997). Finally, generalized linear logistic mixed models were used to show the connections between these three different methods. Our goal is to show a logical link across these methods by moving from a familiar approach based on a Many Facets Rasch Model through hierarchical regression to

LLRMs estimated with generalized linear logistic mixed models. Generalized linear logistic mixed models offer a systematic and coherent approach for examining item covariates in Rasch measurement theory.

17.3.2 Participants

Students enrolled in biology classes (high-school level) from a school district in a southeastern state are included in this case study ($N = 583$). The biology assessment was designed to serve as a formative assessment to aid students and teachers in preparation for a statewide assessment in biology. Additional information regarding student characteristics is not available because this is a secondary data analysis.

17.3.3 Instrument

A 25-item biology assessment is examined in this study. The items were developed to reflect five content domains: Cells (Items 1, 2, 3, 25), Cellular Genetics and Heredity (Items 5, 6, 8, 9, 10, 12, 23), Classification & Phylogeny (Items 13, 14, 15, 16, 17), Ecology (Items 7, 18, 20, 21, 22, 24) and Theory of Evolution (Items 4, 11, 19). These content domains were selected because they matched the test blueprint for the statewide assessment in biology. The 25 items were classified by the test developer into levels of cognitive demand based on Webb's Depth of Knowledge taxonomy (Webb, 1999): Level 2 (skills and concepts) and Level 3 (strategic thinking). Additional cognitive levels were not included in this assessment by the test developer. Table 17.1 shows the classification of the items by content domain and cognitive demand. Two sample items are included in Appendix A. These two items are from the content domain of Cells. Item 1 was classified as requiring skills

Table 17.1 Classification of biology items by content domain and cognitive demand

Content domain	Cognitive demand	Items
1. Cells	Level 2	1, 2
	Level 3	3, 25
2. Cellular Genetics & Heredity	Level 2	5, 6, 8, 10, 12, 23
	Level 3	9
3. Classification & Phylogeny	Level 2	13, 14, 15, 16, 17
4. Ecology	Level 2	7, 18, 20, 21
	Level 3	22, 24
5. Theory of Evolution	Level 2	4, 19
	Level 3	11

Note. Cognitive demand is based on Depth-of-Knowledge Categories widely used in statewide assessments (Webb, 1999). Specifically, the Biology Assessment includes two levels: Level 2 = skills and concepts and Level 3 = strategic thinking

and concepts (Level 2), while Item 3 was classified as requiring strategic thinking (Level 3).

The items were also examined to identify levels of linguistic cohesion. We selected several indices reflecting linguistic cohesion to include as item covariates: word count, narrativity, syntactic simplicity, word concreteness, referential cohesion, deep cohesion, verb cohesion, and connectivity. The description of these indices is included in the Appendix B (McNamara et al., 2005). These indices were selected because they have been successfully used in earlier psychometric research by Wang & Engelhard (2019). The Coh-Metrix system was used to analyze the 25 biology assesslet items. The *z* scores for the indices were obtained using Coh-Metrix 3 online system (http://cohmetrix.com; McNamara et al., 2005).

17.3.4 Results

Many Facets Rasch Model

Table 17.2 provides the summary statistics obtained for the biology data based on the Facets computer program (Linacre, 2019b). Based on the Infit and Outfit MSE statistics, there is evidence for good model-data fit. The Infit and Outfit MSEs for persons and items are close to the expected mean values of 1.00. The reliability of separation for items is .99, while the reliability of person separation is .61. The variance explained by the Rasch measures is approximately 19%. Variance values around 20% are generally considered as an indication of unidimensionality. Table 17.2 also shows person and item statistics including the mean (*M*) and the standard deviation (*SD*) of person and item measures.

Table 17.2 Summary statistics for rasch model based on many facets rasch model

Measures	Person	Item
M	−.53	.00
SD	.69	.79
N	583	25
Infit MSE		
M	.99	1.00
SD	.22	.07
Outfit MSE		
M	1.01	1.01
SD	.27	.17
Reliability of separation	.61	.99
χ^2 statistic	1329.7*	2029.2*
Degrees of freedom	582	24

Note. MSE is the mean square error
$^{*}p < .05$

Table 17.3 describes in detail the item level statistics for the biology items. Overall, the items fit very well. We categorized the items into four fit categories based on Engelhard and Wang (2021):

Mean Square Error (MSE)	Interpretation	Fit Category
$.50 \leq MSE < 1.50$	Productive for measurement	A
$MSE < .50$	Less productive for measurement, but not distorting of measures	B
$1.50 \leq MSE < 2.00$	Unproductive for measurement, but not distorting of measures	C
$2.00 \leq MSE$	Unproductive for measurement, and distorting of measures	D

Mean square errors (MSEs) between 0.50 and 1.50 are placed in the A category, MSEs less than 0.50 are placed in the B category, MSEs fall between 1.50 and 2.00 (including 1.50) are placed in category C, and MSE values larger than 2.00 (including 2.00) are placed in Category D. Items with fit values in Categories A and B are considered productive for measurement, and items with fit values belonging to Categories C and D are potentially unproductive and possibly distorting for measurement. None of the biology items were flagged in the C or D categories.

The Wright map for these items is shown in Fig. 17.2. The first column in the Wright map is the logit scale, the second column is the distribution of the students on the scale, and the final column indicates the locations of the items on the scale. The data do not exhibit any ceiling or floor effects for persons. The difficulties of the biology assessment items range from −1.60 logits for Item 6 (easiest item) to 1.90 logits for Item 24 (hardest item). This variability in item difficulties is explored further using both hierarchical regression and linear logistic Rasch models in the next two sections. In essence, the item difficulties become the dependent variable, and the covariates are added to the model to aid in understanding why items range in their locations on the underlying construct.

The syntax for the Many Facets Rasch Model is given in Appendix C.

Hierarchical Regression Models

Once the item locations are estimated, then a variety of covariates can be used to explore components of item difficulty. The item difficulties obtained based on the Rasch model are viewed as the values of the dependent variable with covariates introduced into the hierarchical regression models. This approach has been used in cognitive psychology (Embretson, 2010).

Table 17.4 provides a summary of the hierarchical regression models estimated in this study. As indicated earlier, the items were analyzed to obtain item covariates related to linguistic cohesion. Specifically, we identified the following aspects of linguistic cohesion: Word count, Narrativity, Syntactic simplicity, Word

Table 17.3 Summary of Rasch calibration of items based on Many Facets Rasch Model

Item	Content	Cognitive demand	Item (δ)	Item (SE)	Infit (MSE)	Infit (Z)	Outfit (MSE)	Outfit (Z)	Pt Bis	Infit category	Outfit category
1	1	2	0.05	0.09	0.98	−0.58	0.95	−1.14	0.21	A	A
2	1	2	0.13	0.09	1.12	3.61	1.18	3.90	−0.01	A	A
3	1	3	0.51	0.10	1.04	0.91	1.05	0.84	0.11	A	A
4	5	2	0.41	0.09	1.05	1.16	1.09	1.58	0.10	A	A
5	2	2	0.34	0.09	1.04	1.16	1.08	1.53	0.10	A	A
6	2	2	−1.60	0.10	0.99	−0.27	0.97	−0.53	0.18	A	A
7	4	2	0.05	0.09	0.98	−0.59	1.00	0.00	0.20	A	A
8	2	2	−0.50	0.09	0.96	−1.77	0.94	−1.81	0.25	A	A
9	2	3	−1.39	0.09	0.97	−0.83	0.97	−0.47	0.20	A	A
10	2	2	−0.48	0.09	0.99	−0.58	0.97	−0.94	0.20	A	A
11	5	3	1.37	0.12	1.15	1.86	1.56	4.56	−0.14	A	B
12	2	2	0.84	0.10	1.02	0.46	1.05	0.68	0.12	A	A
13	3	2	0.49	0.10	1.06	1.48	1.18	3.04	0.07	A	A
14	3	2	0.77	0.10	1.14	2.54	1.24	3.18	−0.05	A	A
15	3	2	−0.64	0.09	0.96	−1.79	0.95	−1.43	0.24	A	A
16	3	2	−0.67	0.09	0.94	−2.42	0.92	−2.49	0.27	A	A
17	3	2	0.24	0.09	0.92	−2.26	0.89	−2.3	0.30	A	A
18	4	2	0.57	0.10	1.1	2.23	1.17	2.67	0.01	A	A
19	5	2	−0.74	0.09	0.93	−2.84	0.90	−2.99	0.29	A	A
20	4	2	−0.55	0.09	0.91	−3.67	0.89	−3.46	0.32	A	A
21	4	2	−0.41	0.09	1.00	0.03	1.00	−0.01	0.18	A	A
22	4	3	−0.05	0.09	0.99	−0.32	0.97	−0.69	0.19	A	A
23	2	2	−1.03	0.09	0.92	−2.55	0.9	−2.53	0.28	A	A
24	4	3	1.90	0.14	0.94	−0.53	0.72	−2.08	0.26	A	A

(continued)

Table 17.3 (continued)

Item	Content	Cognitive demand	Item (δ)	Item (SE)	Infit (MSE)	Infit (Z)	Outfit (MSE)	Outfit (Z)	Pt Bis	Infit category	Outfit category
25	1	3	0.40	0.09	0.89	−2.89	0.83	−3.31	0.35	A	A
Mean			0.00	0.10	1.00	−0.34	1.01	−0.17	0.17		
SD			0.82	0.01	0.07	1.90	0.16	2.33	0.12		

Note. Items fit Categories A and B are considered productive for measurement, and items with fit values belonging to Categories C and D are potentially unproductive and possibly distorting for measurement (Engelhard & Wang, 2021). The content codes are as follows: 1 = Cells, 2 = Cellular Genetics & Heredity, 3 = Classification & Phylogeny, 4 = Ecology, and 5 = Theory of Evolution. The level of cognitive demand is coded as follows: Level 2 = skills and concepts and Level 3 = strategic thinking

Fig. 17.2 Wright m0061p

```
+------------------------------+
|Logit|+Students     |-Items    |
|-----+-------------+----------|
|   3 + High         + Hard     |
|     |              |          |
|     |              |          |
|     |              |          |
|     |              |          |
|     | .            |          |
|     |              |          |
|   2 +              +          |
|     | .            | 24       |
|     |              |          |
|     | .            |          |
|     |              | 11       |
|     | .            |          |
|     |              |          |
|   1 + .            +          |
|     | *.           | 12       |
|     | *.           | 14       |
|     |              | 3  18    |
|     | ****.        | 4  13 25 |
|     | ***.         | 5  17    |
|     | ***.         | 2        |
*   0 *              * 1  7  22 *
|     | ******.      |          |
|     | ******.      |          |
|     | *********.   | 8  10 21 |
|     |              | 15 20    |
|     | *********    | 16 19    |
|     | *******.     |          |
|  -1 +              + 23       |
|     | ******.      |          |
|     | ***.         |          |
|     |              | 9        |
|     | ****.        | 6        |
|     |              |          |
|     | **           |          |
|  -2 +              +          |
|     |              |          |
|     | .            |          |
|     |              |          |
|     |              |          |
|     |              |          |
|     |              |          |
|  -3 + Low          + Easy     |
|-----+-------------+----------|
|Logit| * = 8        |-Items    |
+------------------------------+
```

concreteness, Referential cohesion, Deep cohesion, Verb cohesion, and Connectivity. Next, we include cognitive demand in the model (Level 2 = skills and concepts and Level 3 = strategic thinking). Finally, we explore whether or not content domains play a role in explaining item locations on the latent variable.

The Pearson correlations between item difficulty and the covariates are shown in Table 17.4. None of the correlations are statistically significant. The regression coefficients identified three statistically significant covariates: word concreteness and two content domains (Cells and Ecology). Next, a stepwise regression analysis was conducted with a backward deletion process The detailed results of these stepwise analyses are also shown in Table 17.4. The full regression model accounts for 72.5% of the variance in item difficulty, while the stepwise model accounts for 68.4% of the variance. The adjusted variances are 37.4% and 47.3% for the two models respectively. The stepwise model is a better fit based on the adjusted r squared. Based on the stepwise model, word count, word concreteness, deep cohesion, verb cohesion and cognitive demand are major covariates retained in the model. The effects of two content domains (Cells and Ecology) are also retained in the model.

Table 17.4 Summary of hierarchical regression models with item locations as dependent variables

Item covariates	r	Regression coefficients	SE	Stepwise coefficients	SE
Linguistic cohesion					
C1. Word count	.232	−.683	.433	−.616	.333
C2. Narrativity	.115	.197	.283		
C3. Syntactic simplicity	.241	.211	.216		
C4. Word concreteness	−.309	−.598*	.244	−.588*	.184
C5. Referential cohesion	−.135	.159	.237		
C6. Deep cohesion	.014	.288	.282	.362	.234
C7. Verb cohesion	−.007	.314	.243	.322	.221
C8. Connectivity	−.061	−.051	.166		
Cognitive demand	.321	.395	.283	.534*	.229
Content domain					
1. Cells		1.573*	.610	1.482*	.496
2. Cellular genetics and heredity		−.427	.370	−.318	.284
3. Classification & Phylogeny		.171	.452	.413	.312
4. Ecology		.979*	.375	.798*	.275
5. Theory of evolution		.597	.496	.423	.436
R-squared		.725		.684	
Adjusted R-squared		.374		.473	

Note. All the item covariates are based on z-scores. r = Pearson correlation between item covariates and item difficulty. Cognitive demand is based on Depth-of-knowledge with Level 2 = skills and concepts and Level 3 = strategic thinking
*$p < .05$

The statistically significant stepwise regression coefficients shown in Table 17.4 suggest items with higher cognitive demand tend to be harder for students, while items with higher levels of word concreteness are easier for these students. The content domains of cells and ecology also seem to be harder than the other content domains.

Linear Logistic Rasch Models (LLRMs)

The first two analyses were based on the Many Facets Rasch Model and the hierarchical regression models respectively. The Many Facets Rasch Model includes both persons and items, but does not provide a systematic way to add both categorical and metric covariates. In the hierarchical regression approach, person characteristics are ignored because the items are the focus of the analyses. LLRMs includes both persons and items in the model, and also provide a systematic way to examine categorical and metric covariates within the framework of the model.

First of all, it is important to recognize that item estimates obtained under both the Many Facets Rasch Model and LLRMs are comparable. The correlation between the item locations estimated under each model is very high ($r\,(23) = .999, p < .01$). The details of the separate calibrations are shown in Table 17.5. The differences between the re-scaled or standardized item locations are small with the mean difference essentially zero.

In order to estimate the parameters for linear logistic Rasch models (LLRM), we used the *eirm* package (Bulut et al., 2021). As pointed out earlier, the *eirm* package is a wrapper around a R package called lme4 (Bates, et al., 2015). We estimated the seven models that are shown in Table 17.6. The eirm syntax has a simple structure where "Response" is the dichotomous response to an item (dependent variable), "-1" indicates that an intercept term is not included, "(1|x)" indicates a random effect, and "x" are independent variables modelled as fixed effects. The information criteria for each model are reported in Table 17.6. The specific information criteria are Akaike's information criteria (AIC), Bayesian information criteria (BIC), log likelihood, and Deviance for several models. For information criteria, smaller values indicate better model-data fit and differences between models of 10 points or less are considered negligible.

The first set of models included: 1. Baseline ("Response ~ (1|Students)") and 2. Rasch model ("Response ~ −1 + Items + (1|Students)"). As expected, the Rasch model fits the data better than the baseline model with smaller information criteria across all of the indices.

Next, we estimated LLRMs that represent three models for examining the effects of two item components: cognitive demand (DOK) and content domains. This set of models included: 3. DOK ("Response ~ −1 + DOK + (1|Items) + (1|Students)"), 4. content domains ("Response ~ −1 + Content + (1|Items) + (1|Students)"), and 5. DOK and Content combined ("Response ~ −1 + Content + DOK + (1|Items) + (1| Students)"). All of these models lead to better fit than the baseline model. BIC

Table 17.5 Comparison of Many Facets Rasch Model and LLRM item locations (δ)

Item	Facets item (δ)	LLRM item (δ)	Facets standardized	LLRM standardized	Difference
1	0.05	0.52	0.06	0.06	0.00
2	0.13	0.59	0.16	0.15	0.01
3	0.51	0.97	0.62	0.63	−0.01
4	0.41	0.86	0.50	0.49	0.01
5	0.34	0.80	0.41	0.42	−0.01
6	−1.60	−1.07	−1.95	−1.96	0.01
7	0.05	0.52	0.06	0.06	0.00
8	−0.50	0.00	−0.61	−0.60	−0.01
9	−1.39	−0.86	−1.70	−1.69	−0.01
10	−0.48	0.02	−0.59	−0.58	−0.01
11	1.37	1.79	1.67	1.67	0.00
12	0.84	1.28	1.02	1.03	−0.01
13	0.49	0.94	0.60	0.59	0.01
14	0.77	1.21	0.94	0.94	0.00
15	−0.64	−0.14	−0.78	−0.78	0.00
16	−0.67	−0.17	−0.82	−0.82	0.00
17	0.24	0.70	0.29	0.29	0.00
18	0.57	1.02	0.70	0.70	0.00
19	−0.74	−0.24	−0.90	−0.91	0.01
20	−0.55	−0.06	−0.67	−0.68	0.01
21	−0.41	0.08	−0.50	−0.50	0.00
22	−0.05	0.42	−0.06	−0.07	0.01
23	−1.03	−0.52	−1.26	−1.26	0.00
24	1.90	2.31	2.32	2.34	−0.02
25	0.40	0.85	0.49	0.48	0.01
Mean	0.00	0.47	0.00	0.00	0.00
SD	0.82	0.79	1.00	1.00	0.01

indicated that these three models fit better than the Model 2 (simple logistic Rasch model).

Table 17.6 presents two additional models based on LLRMs. These are the Full Model (Model 6) that includes all of the covariates ("Response ~ −1 + C1 + C2 + C3 +C4 + C5 +C6 +C7 +C8 + DOK + Content + (1|Items) + (1|Students)"). The Reduced Model (Model 7) includes the covariates identified in the stepwise regression model: word count, word concreteness, deep cohesion, verb cohesion and cognitive demand ("Response ~ −1 + Content + DOK + C1 + C4 + C6 + C7 + (1| Items) + (1|Students)"). All of the LLRMs in this final set included the five content domains.

The summary of η coefficients for Models 6 and 7 are shown in Table 17.7. The Full Model includes all of the covariates. The Reduced Model shows the η coefficients, and these indicate that lower word count and word concreteness tend to make

Table 17.6 Model comparisons for Linear Logistic Rasch Models

<table>
<tr><th>Models</th><th>eirm syntax</th><th>AIC</th><th>BIC</th><th>LogLik</th><th>Deviance</th><th>Residual DF</th></tr>
<tr><td>1. Baseline</td><td>“Response ~ (1|students)”</td><td>19453.1</td><td>19468.3</td><td>−9724.6</td><td>19449.1</td><td>14,573</td></tr>
<tr><td>2. Rasch</td><td>“Response ~ −1 + items + (1|students)”</td><td>17856.6</td><td>18053.9</td><td>−8902.3</td><td>17804.6</td><td>14,549</td></tr>
<tr><td>3. LLRM (DOK)</td><td>“Response ~ −1 + DOK + (1|items) + (1|students)”</td><td>17943.0</td><td>17965.8</td><td>−8968.5</td><td>17937.0</td><td>14,572</td></tr>
<tr><td>4. LLRM (content)</td><td>“Response ~ −1 + content + (1|items) + (1|students)”</td><td>17944.6</td><td>17997.7</td><td>−8965.3</td><td>17930.6</td><td>14,568</td></tr>
<tr><td>5. LLRM (content, DOK)</td><td>“Response ~ −1 + content + DOK + (1|items) + (1|students)”</td><td>17944.8</td><td>18005.5</td><td>−8964.4</td><td>17928.8</td><td>14,567</td></tr>
<tr><td>6. LLRM (full model)</td><td>“Response ~ −1 + C1 + C2 + C3 + C4 + C5 + C6 + C7 + C8 + DOK + content + (1|items) + (1|students)”</td><td>17943.8</td><td>18065.2</td><td>−8955.9</td><td>17911.8</td><td>14,559</td></tr>
<tr><td>7. LLRM (reduced model)</td><td>“Response ~ −1 + content + DOK + C1 + C4 + C6 + C7 + (1|items) + (1|students)”</td><td>17939.3</td><td>18030.4</td><td>−8957.7</td><td>17915.3</td><td>14,563</td></tr>
</table>

Note. Number of observations = 14,575 (583 students, 25 items)

Table 17.7 Summary of item covariates based on Linear Logistic Rasch Models

	Full model		Reduced model	
Item covariates	η	SE	η	SE
Linguistic cohesion				
C1.Word count	.658*	.281	.596*	.252
C2. Narrativity	−.186	.184		
C3. Syntactic simplicity	−.206	.140		
C4. Word concreteness	.583*	.158	.572*	.139
C5. Referential cohesion	−.160	.154		
C6. Deep cohesion	−.280	.183	−.352*	.177
C7. Verb cohesion	−.309	.158	−.317	.165
C8. Connectivity	.049	.107		
Cognitive demand	−.379*	.184	−516*	.174
Content domains				
1. Cells	−1.564*	.404	−1.476*	.362
2. Cellular genetics and heredity	.428	.246	.316	.261
3. Classification & Phylogeny	.175	.300	−.411	.251
4. Ecology	−.976*	.249	−.794*	.235
5. Theory of evolution	−.598	.328	−.425	.375

Note. Cognitive demand has two levels: Level 2 (skills and concepts) and Level 3 (strategic thinking)
*$p < .05$

items easier (positive coefficients). Also, the Reduced Model indicates that deep cohesion and cognitive demand, as well as the content domains of Cells and Ecology, tend to be associated with harder items (negative coefficients).

The relationships between cognitive demand and content domains were examined graphically. The expected values are plotted in Fig. 17.3 (Panel A), while the expected values based on the reduced model including covariates is shown in Fig. 17.3 (Panel B). The data suggest that higher levels of cognitive demand are more difficult across content areas, and that this relationship is even stronger when additional covariates that control for linguistic complexity are added into the model.

17.4 Teacher Reflections

In addition to our empirical analyses, we shared our results with an experienced biology teacher. She also had experience with the construction of standardized tests in science. She examined the 25 items and suggested reasons why these items varied in difficulty.

The teacher made several general observations. First of all, she suggested that an important item component is whether or not an item includes any mathematics. All else being equal, she indicated that biology items involving calculations are harder. Next, she stressed the importance of vocabulary. Items with a more specialized

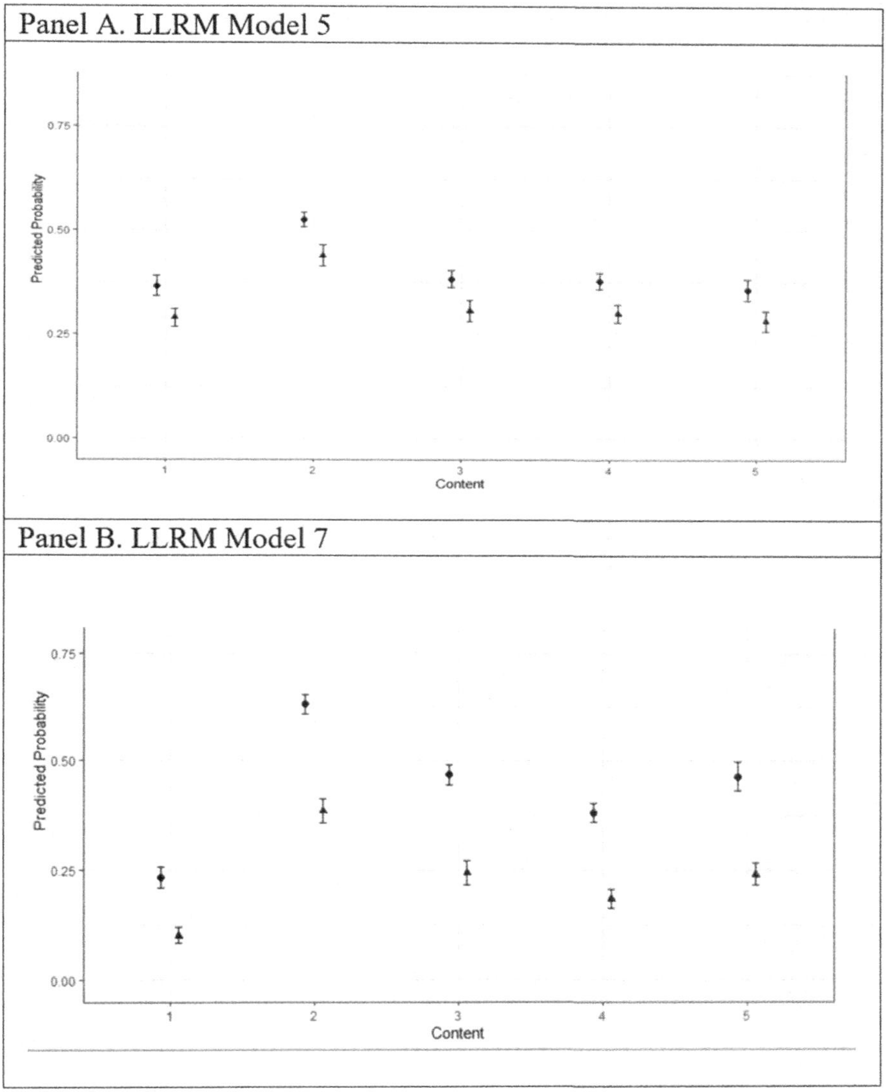

Fig. 17.3 Expected effects of cognitive demand within different content areas
Note. Triangles = DOK Level 3 (strategic thinking) and circles = DOK Level 2 (skills and concepts)

vocabulary tended to be more difficult based on her experience as a biology teacher. One essential message stressed by this teacher was that item difficulty is closely related to the background characteristics of the students including their educational experiences related to learning goals, instruction and practice.

She also commented specifically on each item. Her detailed comments on the four sample items included Appendix A are as follows:

Item 1 was judged to be of moderate difficulty.
Identifying the structures in both plant and animal cells narrows the answer options to A and D; understanding how they work together means the student needs to understand the function of ribosomes and mitochondria. D is the correct answer. Teachers and texts don't usually emphasize that the mitochondria work with the cytoplasm, which will also make this question more challenging for students.
Item 3 was judged to be of moderate difficulty.
Many students should be familiar with a semipermeable membrane experiment and the concept of diffusion from high concentration to low concentration. This question is an application of that knowledge. Answer option A accounts for students who are confused the direction of diffusion; B accounts for students who do not understand diffusion; and C accounts for students who miss the concept.
Item 6 was judged to be of low difficulty.
Most students should have a basic understanding of DNA making RNA and RNA providing the template for protein synthesis. The answer options are broad overviews and not detailed explanations.
Item 11 was judged to be of high difficulty.
Students need to know the Hardy-Weinberg equation and understand the variables in the content of a given percentage of homozygous individuals. Then they need to apply it to solve for q. Students commonly struggle with math as applied to biology!

17.5 Discussion

Developing an understanding of science is of critical importance for members of modern societies. Understanding issues, such as climate change, require a basic understanding of science. In educational settings, the development of science assessments is essential for improving student learning in science. An important component of high-quality assessment is the exploration of item characteristics that can be used to predict item difficulty.

This study focused on the use of Rasch measurement theory to examine science assessments. The linear logistic Rasch model (LLRM) is described in detail as an approach for examining the influence of cognitive complexity and linguistic cohesion on science items. Linear logistic Rasch models (LLRM) are an important and under-utilized member of the Rasch family of models. Essentially, the LLRM adds covariates that can be used to predict item locations. To answer the research questions of (1) what are the LLRMs, and (2) what components of items impact student responses in science assessment, the chapter provides an informative description of the LLRM and conducts a detailed case analysis of a biology assessment in order to illustrate how to apply the LLRM.

Detailed descriptions of three models (Many Facets Rasch Model, hierarchical regression models, and linear logistic Rasch models) were used to illustrate approaches to examine science items. The analyses suggested that several item covariates influence item locations on the scale. This information can provide diagnostic information to both students and teachers by identifying items features that may influence the assessment of student proficiency in science.

Three sets of item covariates were included in this study: cognitive demand, linguistic cohesion and content domain. Cognitive demand was defined by Webb's depth of knowledge framework (1999), linguistic cohesion was defined using several Coh-Metrix indices (Graesser et al., 2011), and the content domain was high school biology (Cells, Cellular Genetics and Heredity, Classification & Phylogeny, Ecology, and Theory of Evolution).

The results of the case study suggest that lower word count and word concreteness tend to make items easier. Deep cohesion and cognitive demand, as well as the content domains of Cells and Ecology, tend to be associated with harder items. Future research should explore whether or not these item covariates also influence science items in other content areas.

There are several limitations of this study that should be addressed in future research. First of all, the biology assessments used in the case study consisted of a small number of items. It would be interesting to replicate this study with a larger number of items. Another limitation is that linguistic cohesion was defined using the Coh-Metrix framework. There are other indicators of linguistic complexity that should be included in future research. Cognitive demand was measured by Webb's depth of knowledge taxonomy, and it is important to consider other frameworks for examining cognitive complexity (Embretson, 2010). For example, it is important to consider why some items are harder than others from the perspectives of science teachers and item writing experts.

There are several important implications of this study for research on item components in science assessment. The explicit use of cognitive models in item development is a keystone for improving science assessments. Measurement models, such as LLRMs, offer a rich perspective on item development with implications for improving learning and teaching in science. If research identifies key components for solving science problems, then analyses can predict the difficulty of science items that can be used to create item banks with calibrated items measuring science content in various areas.

In summary, the main goal of this study was to describe linear logistic Rasch models (LLRMs) for understanding item components in the context of science education. We demonstrated the use of the LLRM with a high school assessment of student proficiency in Biology. Linear logistic Rasch models represent an underutilized methodology that has promise to assist science educators in understanding item components that may influence the measurement of proficiency in science.

Appendices

Appendix A

Sample Items from Cell Domain (Levels 2 and 3)

Level 2 (Skills and concepts)

1. Which structures, found in both plant and animal cells, work together to help the cell maintain homeostasis and how?

   A. cytoplasm and ribosomes by storing energy

   B. chloroplasts and cytoplasm by releasing energy

   C. chloroplasts and mitochondria by storing energy

   D. cytoplasm and mitochondria by releasing energy

Level 3 (Strategic thinking)

3. A group of students used vinegar to remove the shell of a chicken egg to expose its semipermeable membrane so that osmosis could take place. The egg contained a 15% concentration of solutes, including sugars. The students then placed the egg in a beaker containing a 30% sugar solution. The students left the egg in the solution for 24 hours to determine the effects of osmosis on the cell.

   What would be the **MOST LIKELY** response of the egg to being placed in this solution, and why would this occur?

   A. The egg would swell because water would flow into the egg in order to reach equilibrium.

   B. The egg would shrink because water would move out of the egg in order to reach equilibrium.

   C. The egg would remain the same size because the net movement of water into and out of the egg would be equal.

   D. The egg would dissolve because the high concentration of the sugar solution would cause the membrane to rupture.

Sample Items from Cell Domain (Levels 2 and 3) (Cont.)

Level 2 (Skills and concepts)

6. Which sequence of molecules correctly identifies how genetic information is expressed in the cell?

   A. DNA → protein → RNA

   B. DNA → RNA → protein

   C. RNA → DNA → protein

   D. RNA → protein → DNA

Level 3 (Strategic thinking)

11. Suppose that in humans, there are only two alleles for eye color and that eye color is inherited in a pattern of simple dominance. Blue eyes are caused by a recessive trait. In the United States, approximately 17% of the population has blue eyes.

    What is the expected allele frequency for blue eyes?

    A. 3%

    B. 35%

    C. 41%

    D. 59%

Appendix B Description of the Coh-Metrix Text Features

Text feature	Description
1.Word count	This is the total number of words in the item stem.
2. Narrativity	Narrative text tells a story, with characters, events, places, and things that are familiar to the reader. Nonnarrative texts on less familiar topics lie at the opposite end of the continuum.
3. Syntactic simplicity	This component reflects the degree to which the sentences in the text contain fewer words and use simpler, familiar syntactic structures, which are less challenging to process. At the opposite end of the continuum are texts that contain sentences with more words and use complex, unfamiliar syntactic structures.

(continued)

Text feature	Description
4. Word concreteness	Texts that contain content words that are concrete, meaningful, and evoke mental images are easier to process and understand. Abstract words on the other end are more difficult to process.
5. Referential cohesion	A text with high referential cohesion contains words and ideas that overlap across sentences and the entire text, forming explicit threads that connect the text for the reader.
6. Deep cohesion	This component reflects the degree to which the text contains causal and intentional connectives when there are causal and logical relationships within the text. These connectives help the reader to form a more coherent and deeper understanding of the causal events, processes, and actions in the text. If the text is high in deep cohesion, then those relationships and global cohesion are more explicit.
7. Verb cohesion	This component reflects the degree to which there are overlapping verbs in the text. When there are repeated verbs, the text likely includes a more coherent event structure that will facilitate and enhance situation model understanding.
8. Connectivity	This component reflects the degree to which the text contains explicit adversative, additive, and comparative connectives to express relations in the text.

Appendix C

Many Facets Rasch Model Syntax

```
title = HS Biology
facets = 2
Gstats=yes
barchart=no
dvalue=2,1-25
Model = ?,?
*
Labels=
1,Students
1-583
*
2,Items
1-25
*
data =
1,0,1,0,1,1,1,1,1,1,0,0,0,0,0,1,1,0,0,1,1,1,1,0,1,1
2,1,0,1,1,1,0,0,0,1,1,0,1,0,0,1,1,0,0,1,1,1,1,0,1,1
3,1,1,0,1,0,0,1,1,1,0,0,0,0,0,0,1,0,0,0,0,1,0,0,0,0
4,1,0,0,1,0,0,0,1,1,0,0,0,0,0,1,1,0,0,1,1,0,0,0,0,0
5,1,0,0,1,0,0,1,1,0,1,0,0,1,1,1,1,1,1,1,1,0,1,0,0,0
6,0,0,0,1,0,0,0,1,0,0,0,1,0,0,0,0,0,0,0,0,0,1,0,0,0
7,1,0,1,0,0,1,1,1,1,1,0,0,0,0,1,1,1,1,1,1,1,1,0,0,0
8,0,1,0,1,0,0,0,0,1,0,0,0,0,0,1,0,0,1,0,0,1,0,0,0,0
9,0,1,0,0,1,0,0,0,0,1,0,0,0,0,0,1,0,1,0,1,1,1,1,0,1
10,0,0,0,0,0,0,1,1,1,0,0,0,0,0,1,1,0,1,1,1,0,1,1,0,0
...
```

Appendix D

R syntax for Hierarchical Regression Models

```
rm(list=ls())
library(eirm)
library(dplyr)
library(MASS)
library(olsrr)
library(Hmisc)

x <- read.csv(DATA.csv", header = TRUE)
head(x)

Item_Difficulty <- x[,2]
x$C1 <- (x$C1-mean(x$C1))/sd(x$C1)
x$C2 <- (x$C2-mean(x$C2))/sd(x$C2)
x$C3 <- (x$C3-mean(x$C3))/sd(x$C3)
x$C4 <- ((x$C4-mean(x$C4))/sd(x$C4))
x$C5 <- (x$C5-mean(x$C5))/sd(x$C5)
x$C6 <- (x$C6-mean(x$C6))/sd(x$C6)
x$C7 <- (x$C7-mean(x$C7))/sd(x$C7)
x$C8 <- (x$C8-mean(x$C8))/sd(x$C8)
x$DOK <- recode(x$DOK,"2"=0,"3"=1)
x$DOK <- (x$DOK-mean(x$DOK))/sd(x$DOK)
x$Content <- as.character(x$Content)

r <- cor(as.matrix(x[,c(2,4:13)]))
round(r,3)

Full.model <- lm(Item_Difficulty~-1+C1+C2+C3+C4+C5+C6+C7+C8+DOK+Content, data = x)
summary(Full.model)
Full <- predict(Full.model)
plot(Full,Item_Difficulty,xlim=c(-2,2),ylim=c(-2,2))

step.model <- step(Full.model, direction = "backward", trace = FALSE)
summary(step.model)
Covariates <- predict(step.model)
plot(Covariates,Item_Difficulty,xlim=c(-2,2),ylim=c(-2,2))
ols_step_best_subset(step.model)
fitplot <- ols_step_best_subset(step.model)
plot(fitplot)
```

Appendix E

R syntax for Linear Logistic Rasch Models

```
rm(list=ls())
library(eirm)
library(dplyr)
x <- read.csv("DATA.csv", header = TRUE)
x$Items <- as.character(x$Items)
x$Students <- as.character(x$Students)
x$SciBio <- as.character(x$SciBio)
x$Content <- as.character(x$Domain)
x$DOK <- recode(x$DOK,"2"=0,"3"=1)
x$C1 <- (x$C1-mean(x$C1))/sd(x$C1)
x$C2 <- (x$C2-mean(x$C2))/sd(x$C2)
x$C3 <- (x$C3-mean(x$C3))/sd(x$C3)
x$C4 <- (x$C4-mean(x$C4))/sd(x$C4)
x$C5 <- (x$C5-mean(x$C5))/sd(x$C5)
x$C6 <- (x$C6-mean(x$C6))/sd(x$C6)
x$C7 <- (x$C7-mean(x$C7))/sd(x$C7)
x$C8 <- (x$C8-mean(x$C8))/sd(x$C8)
x$DOK <- (x$DOK-mean(x$DOK))/sd(x$DOK)
head(x)
cor(x[,8:15])
# Model 1 Baseline
modBase <- eirm( "Response ~ (1|Students)", data = x)
summary(modBase$model)
# Model 2 Rasch model
modRasch<- eirm( "Response ~ -1 + Items + (1|Students)", data = x)
summary(modRasch$model)
# Model 3 Cognitive domain (DOK)
modDOK <- eirm( "Response ~ -1 + DOK + (1|Items) + (1|Students)", data = x)
summary(modDOK$model)
# Model 4 Content
modCon <- eirm( "Response ~ -1 + Content + (1|Items) + (1|Students)", data = x)
summary(modCon$model)
# Model 5 Cognitive demand and content
modBoth <- eirm( "Response ~ -1 + Content + DOK + (1|Items) + (1|Students)", data = x)
summary(modBoth$model)
# Model 6 Full model
modFull <- eirm( "Response ~ -1 + C1 + C2 + C3 +C4 + C5 +C6 +C7 +C8 + DOK + Content + (1|Items) + (1|Students)",
data = x)
summary(modFull$model)
# Model 7 Reduced Model
modReduced <- eirm( "Response ~ -1 + Content + DOK + C1 + C4 + C6 + C7  + (1|Items) + (1|Students)", data = x)
summary(modReduced$model)
```

References

Azen, R., & Walker, C. M. (2011). *Categorical data analysis for the behavioral and social sciences*. Routledge.

Bloom, B. S., et al. (1956). *Taxonomy of educational objectives, handbook 1: Cognitive domain.* Longman.

Bates, D., Mächler, M., Bolker, B., & Walker, S. (2015). Fitting linear mixed-effects models using lme4. *Journal of Statistical Software, 67*(1), 1–48.

Bond, T. G., Yan, Z., & Heene, M. (2021). *Applying the Rasch model: Fundamental measurement in the human sciences*. Routledge.

Boone, W. J., & Staver, J. R. (2020). *Advances in Rasch analyses in the human sciences* (p. 13). Springer.

Bulut, O., Gorgun, G., & Yildirim-Erbasli, S. N. (2021). Estimating explanatory extensions of dichotomous and Polytomous Rasch models: The eirm package in R. *Psych, 3*(3), 308–321.

De Boeck, P., Bakker, M., Zwitser, R., Nivard, M., Hofman, A., Tuerlinckx, F., & Partchev, I. (2011). The estimation of item response models with the lmer function from the lme4 package in R. *Journal of Statistical Software, 39*, 1–28.

De Boeck, P., & Wilson, M. (2004). *Explanatory item response models: A generalized linear and nonlinear approach*. Springer.

Embretson, S. E. (1997). Multicomponent latent trait models. In W. van der Linden & R. Hambleton (Eds.), *Handbook of modern item response theory* (pp. 305–322). Springer.

Embretson, S. E. (Ed.). (2010). *Measuring psychological constructs: Advances in model-based approaches*. American Psychological Association.

Embretson, S. E., & McCollam, K. M. S. (2000). A multicomponent Rasch model for measuring covert processes: Application to life span ability changes. In M. Wilson & G. Engelhard (Eds.), *Objective measurement: Theory into practice*. Ablex Publishing Company.

Engelhard, G., & Wang, J. (2021). *Rasch models for solving measurement problems: Invariant measurement in the social sciences*. Sage.

Engelhard, G., & Wind, S. A. (2022). A history of Rasch measurement theory. In B. Clauser & M. Bunch (Eds.), *History of educational measurement in the United States* (pp. 346–360). Routledge.

Faraway, J. J. (2016). *Extending the linear model with R: Generalized linear, mixed effects and nonparametric regression models*. Chapman and Hall/CRC.

Fischer, G. H. (1973). The linear logistic test model as an instrument in educational research. *Acta Psychologica, 37*(6), 359–374.

Fischer, G. H. (1997). Unidimensional linear logistic Rasch models. In W. van der Linden & R. Hambleton (Eds.), *Handbook of modern item response theory* (pp. 225–243). Springer.

Flesch, R. (1948). A new readability yardstick. *Journal of Applied Psychology, 32*, 221–233.

Galvin, E., Simmie, M. G., & O'Grady, A. (2015). Identification of misconceptions in the teaching of biology: A pedagogical cycle of recognition, reduction and removal. *Higher Education of Social Science, 8*(2), 1–8.

Graesser, A. C., McNamara, D. S., & Kulikowich, J. M. (2011). Coh-Metrix: Providing multilevel analyses of text characteristics. *Educational Researcher, 40*(5), 223–234.

Holling, H., Blank, H., Kuchenbacker, K., & Kuhn, J. T. (2008). Rule-based item design of statistical word problems: A review and first implementation. *Psychology Science, 50*(3), 363.

Krell, M., Khan, S., & van Driel, J. (2021). Analyzing cognitive demands of a scientific reasoning test using the linear logistic test model (LLTM). *Education Sciences, 11*(9), 472.

Lane, S., Raymond, M. R., & Haladyna, T. M. (Eds.). (2016). *Handbook of test development* (pp. 3–18). Routledge.

Linacre, J. M. (2019a). *A User's guide to WINSTEPS® Rasch-model computer programs: Program manual 4.4. 6*. MESA Press.

Linacre, J. M. (2019b). *Facets computer program for many-facet Rasch measurement, version 3.81.2*. Winsteps.com

Liu, X. (2020). *Using and developing measurement instruments in science education: A Rasch modeling approach* (2nd ed.). Information Age Publishing, Inc.

Liu, X., & Boone, W. J. (2006). *Applications of Rasch measurement in science education*. JAM Press.

National Academies of Sciences, Engineering, and Medicine. (2012). *A framework for K-12 science education: Practices, crosscutting concepts, and core ideas*. The National Academies Press. https://doi.org/10.17226/13165

McNamara, D. S., Louwerse, M. M., Cai, Z., & Graesser, A. (2005). *Coh-Metrix Version 1.4*. Retrieved from http://cohmetrix.com

Park, M., & Liu, X. (2021). An investigation of item difficulties in energy aspects across biology, chemistry, environmental science, and physics. *Springer Research in Science Education, 51-* (Supplement 1), S43–S60.

Pinheiro, J., & Bates, D. (2006). *Mixed-effects models in S and S-PLUS*. Springer.

Rasch, G. (1960/1980). *Probabilistic models for some intelligence and attainment tests*. Danish Institute for Educational Research (Expanded edition, Chicago: University of Chicago Press, 1980).

R Development Core Team. (2004). *R: A Language and Environment for Statistical Computing*. R Foundation for Statistical Computing, Vienna, Austria. ISBN 3-900051-00-3, URL http://www.R-project.org

Spada, H., & May, R. (1982). The linear logistic test model and its application in educational research. *The improvement of measurement in education and psychology* (pp. 67–84).

Stenner, A. J., Horabin, I., Smith, D. R., & Smith, R. (1988). *The Lexile framework*. Metametrics, Inc.

Tutz, G. (2011). *Regression for categorical data*. Cambridge University Press.

Wang, J., & Engelhard, G. (2019). Exploring the impersonal judgments and personal preferences of raters in rater-mediated assessments with unfolding models. *Educational and Psychological Measurement, 79*(4), 773–795.

Webb, N. (1999). *Alignment of science and mathematics standards and assessments in four states* (Research monograph No. 18). CCSSO.

Wilson, M. (2005). *Constructing measures: An item response modeling approach* (2nd ed.). Erlbaum.

Ye Yuan is a Ph.D. student in the Quantitative Methodology, Department of Educational Psychology at the University of Georgia. She received her M.A. degree in Department of Teaching and Learning from Ohio State University. Her research focuses on educational and social science measurement using IRT models, especially Rasch models, explanatory IRT models and unfolding (ideal point) models. She has applied various measurement models to study attitudes, writing assessment (including formative assessment, instructional feedback, and rater accuracy), performance assessment, and certification testing.

George Engelhard is a Professor of Educational Measurement and Policy at the University of Georgia. Professor Engelhard received his Ph.D. in 1985 from The University of Chicago (MESA Program – measurement, evaluation, and statistical analysis) where he had the opportunity to study with Professors Ben Bloom and Ben Wright. He is a fellow of the American Educational Research Association. He recently published a book with Jue Wang (The University of Miami) entitled: *Rasch models for solving measurement problems: Invariant measurement in the social sciences*.

Chapter 18
Development and Application of a Questionnaire on Teachers' Knowledge of Argument as an Epistemic Tool

Gavin W. Fulmer, William E. Hansen, Jihyun Hwang, Chenchen Ding, Andrea Malek Ash, Brian Hand, and Jee Kyung Suh

Abstract We report on the development and pilot testing of a new instrument for teachers' knowledge of argument as an epistemic tool. Our construct-based approach involved domain analysis, item writing, expert reviews, piloting, and applications of the Rasch rating scale model. We began with an initial set of 20 items that we piloted with a sample of 158 pre-service and in-service teachers. After analyses, we retained 14 items as originally written or with minor revisions. The instrument is applicable for researchers working on argument-driven inquiry science as a means to study potential growth in teachers' views of argumentation.

18.1 Background

Argumentation is widely encouraged in the field as a method for students to raise knowledge claims and challenge their own and others' knowledge claims (Cavagnetto, 2010; Henderson et al., 2018; NRC, 2012). This process of generating and validating knowledge is therefore inherently epistemic; argument thus serves as an important *epistemic tool*: a device, implement, or heuristic that serves students to create or validate knowledge (Boon & Van Baalen, 2019) such as by simplification, efficiency, or leveraging available concepts or models (Mathayas et al., 2019). For students to develop and use argument to generate and validate knowledge requires

G. W. Fulmer (✉) · C. Ding · A. M. Ash · B. Hand
University of Iowa, Iowa City, IA, USA
e-mail: gavin-fulmer@uiowa.edu

W. E. Hansen
Iowa Wesleyan University, Mount Pleasant, IA, USA

J. Hwang
Korea National University of Education, Cheongju, South Korea

J. K. Suh
University of Alabama, Tuscaloosa, AL, USA

X. Liu, W. J. Boone (eds.), *Advances in Applications of Rasch Measurement in Science Education*, Contemporary Trends and Issues in Science Education 57,
https://doi.org/10.1007/978-3-031-28776-3_18

teachers to develop the respective pedagogical content knowledge of argument as an epistemic tool. Yet, so far there is relatively little rigorous measurement work from which to draw to help us in gauging teachers' *knowledge of argument as an epistemic tool* so that we could measure and study possible changes. This chapter reports on our work to develop and implement an instrument for this purpose. We focus on teachers' knowledge of argument as an epistemic tool.

As we discuss below, this work was undertaken during an NSF-supported research project, but we attempted to broaden the focus so it would be useful beyond any specific argumentation framework, teaching strategy, or approach. The goal was thus to allow for application in the particular research project of our work, while allowing broad applicability to other scholars' operationalization of argumentation.

Argument is believed to be fundamental and essential to science education as it is how students develop communication skills, critical thinking, metacognitive awareness and confidence in scientific literacy (Cavagnetto, 2010). Despite the push for argument-based activities and teaching in science education, it can be purported that argument is beneficial to any academic field in which language plays an important role. María Pilar Jiménez-Aleixandre (2007) states that classrooms are conceived as communities of intentional learning environments and that an emphasis on social interaction, which is an essential component of both cognitive development and deep learning.

Utilizing argumentation as a method of inquiry for science learning has been shown to be particularly effective in primary and secondary classrooms. One such approach is the Science Writing Heuristic (SWH), a student-centered approach to science learning first described by Brian Hand and Carolyn Keyes in 1998, and since implemented in classrooms worldwide and studied for its effects on critical thinking skills, argumentation, and writing (Burke et al., 2005). Schoerning et al. (2015) undertook a randomized-controlled trial on the SWH approach and a didactic teaching approach with grade 3 through 5 students from schools in the Midwestern United States. They found that the argumentation and the negotiation allowed by the SWH approach allowed students to develop authentic language skills with scientific terminology, stronger writing capabilities, and behavior that indicated a drive to learn science beyond the classroom. The control class did not exhibit the same gains (Schorning et al., 2015). Another study by Hand et al. (2012) utilized a similar, randomized comparison of SWH and traditional didactic teaching methods in high school chemistry. The results showed that students in the SWH classes not only learned chemistry content but also had significantly higher gains on summative writing assessments than the students of the traditional classroom (Hand et al., 2012). These and other studies have demonstrated effectiveness of this approach (e.g., Hand et al., 2012).

The SWH approach emphasizes an atmosphere of continuous inquiry and encourages student discussion and argumentation in every level of their learning (Hand et al., 2012). This approach takes traditional science teaching and gives the students the reins to their learning and the teacher becomes a facilitator in structuring an environment that encourages discussion and argumentation to enhance student learning and evolve previously established ideas. In the SWH classroom students generate complex questions that require investigation, develop the relevant investigations to gather data to form evidence, and finally read from multiple sources to compare their claims to others. Regardless of the activity or topic the students always reflect on how their ideas have changed and write out their ideas and incorporate the ideas of their peers and other sources.

18.2 Validity Framework

This study is based on unified validity theory, which holds that most educational and psychological measures of import are constructs (Kane, 2001; Messick, 1995) that cannot be measured directly but must be inferred from responses to tasks, prompts, or questions. While there is no single coefficient or statistic to demonstrate validity, we may assert a case for validity using evidence from instrument development, application, and interpretation (Kane, 2001). Here we address two validity aspects: *Content* and *Structure*. Content refers to the contents of the items: how well they match and represent the proposed domain. We address the content validity by performing a domain analysis, reinforced by review from knowledgeable experts. Structure refers to the appropriateness of the scoring process and the interpretation of response patterns with respect to the domain. For our purpose, this means examining whether the responses demonstrate the items are functioning well across the range of the construct, and that the pattern of responses is consistent. We address the structure validity by estimating Rasch measurement models to test whether the responses are consistent with this strict model of good measurement.

18.3 Instrument Development

We followed a construct-driven design process informed by our theoretical framework on validity theory, consistent with and in parallel to our process as reported in Fulmer et al. (2021). The process included the following phases that we revisited as needed: domain analysis, item writing, expert review, pilot testing, Rasch modeling, and item selection and revision. We summarize these steps here for this particular instrument.

18.3.1 Domain Analysis

Our process began with a domain analysis to explore the key concepts underlying teachers' knowledge of argumentation as an epistemic tool. The domain analysis drew on literature on argumentation—such as its purposes, structures, and students' engagement in it— followed by interviews with expert teachers and

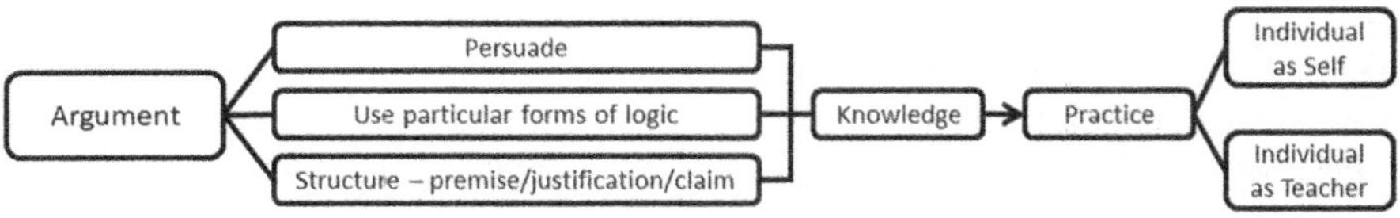

Fig. 18.1 Visualizing the domain analysis for argument as an epistemic tool

teacher-educators who had experience with argument-based inquiry. Figure 18.1 shows a figural representation of the key concepts emerging from the analysis.

Research on argumentation's role in science education has explored its connections to language use as well as the roles of reasoning and evidence (Erduran, Ozdem, & Park, 2015). While some scholars argue that argumentation's importance may be overstated (Allchin & Zemplén, 2020), others point to the value of understanding the special role that argumentation plays in scientific advances and science literacy (Henderson et al., 2018).

Our domain analysis revealed three sub-domains for an argument's purpose as an epistemic tool: that an argument is used to negotiate or *persuade* others of the validity of their beliefs (Toulmin, 2003) with either explicit or implicit dialectical process (Nielsen, 2013); that arguments use particular *forms of logic* (whether abductive, deductive, or inductive reasoning); and that arguments may have *special structures* when applying this logic (e.g., Claims-Evidence-Reasoning or Premise-Justification-Claim).

Arguments Serve to Persuade Persuasion is a set of beliefs that form the foundation of an argument. Invoking persuasion in the classroom is a way to have students voice their ideas regarding a specific subject. The conception and duration of argumentation is often based upon the participants' (in this case students) ability to be persuasive and convince others of the validity of their beliefs (Toulmin, 2003). These processes are fundamentally dialectical (Nielsen, 2013) in that they are not simply logical inferences connecting some evidence to a claim, but rather require certain logical, rhetorical, or epistemic moves to account for and respond to competing alternatives (Rex et al., 2010). This coheres with the view that scientific arguments are as much rhetorical as they are logical (Taber, 2017). Teachers supporting students in argumentation should, therefore, understand this role and how it may be distinct from the lay interpretation of argument as a *disputation* rather than a *deliberation* (e.g., Allchin & Zemplén, 2020).

Argument Uses Particular Forms of Logic Arguments in science follow one or more "logic styles": deductive, inductive, and abductive reasoning. Inductive and deductive reasoning received relatively greater research attention in earlier eras of science education research, and while they are sometimes contrasted they are often used in tandem when developing and practicing argument: inductive reasoning is used to determine the strength of an argument and deductive reasoning is focused on the validity of an argument (Heit & Rotello, 2010). Abductive reasoning has been increasingly visible within the field in recent years for its prevalence in science itself

(Osborne et al., 2018), and results in more probabilistic and contextualized arguments that create promising hypotheses that could then be tested or generalized (Oh, 2017). Yet the various forms of reasoning and their relative purposes in argumentation are not always clearly understood by teachers (e.g., Yoon et al., 2012).

Arguments Have Structures Previous research on argumentation in science education has explored the different roles that can help students in recognizing, creating, and critiquing arguments, which can be characterized as argument structures (Sampson et al., 2013) or schemes (Konstantinidou & Macagno, 2013). The particular structure of an argument that is encouraged can vary according to the approach that is proposed: premise-justification-claim (e.g., Konstantinidou & Macagno, 2013), question-claim-evidence (e.g., Choi et al., 2010; Keys et al., 1999), or claim-evidence-reasoning (e.g., Krajick et al., 2008). All of these schemes provide opportunities for students to create and critique arguments and provide some guidance on how to indicate a clear relationship between an intended assertion and the evidence available for it. These structures can also provide a common basis for students to identify gaps in the argument. Therefore, teachers with an understanding of argument will recognize the potential uses of these structures and the effectiveness for students to communicate using a comparable argument scheme in class.

The three sub-domains mentioned above—that arguments serve to persuade, using particular forms of logic, and possessing some structure—are interrelated concepts that are existent for all arguments. That is, they are not posited to be distinguishable sub-domains that warrant separate assessments. Instead, it appears that having the knowledge of argument as an epistemic tool would likely draw on all three sub-domains to some extent.

18.3.2 Characteristic Statements & Initial Item Writing

We combine the above three notions about argument to then unpack how we could characterize teachers' knowledge of argumentation. The ability to persuade plays a key role in argumentation, however one has the ability to be persuasive without necessarily possessing a sound argument. As we found in our domain analysis, teachers' work with students needs to go beyond persuasion to consider the underlying logic of argument and the varying structures that an argument could have. Considering the logic and structure of arguments allows teachers to move beyond the presence of an argument structure to support quality of argumentation.

Drawing on our experience with Rasch measurement, we aimed to distinguish participating teachers along a range of the trait. We focused on ways to distinguish teachers of varying level. Starting from the domain analysis, we created long-form, descriptive passages to help us characterize teachers of high, moderate, and low knowledge of argument as an epistemic tool. We then shared a draft of the characteristic statements with outside experts: experienced teachers, regional education professional development leaders, and the project's external advisory board. We integrated this feedback into the characteristic statements, using all sources to help underscore the knowledge or views that could help distinguish teachers across these levels.

We used the long-form distinctions outlined in the characteristic statements to draft an initial set of item stems that teachers may endorse. Again using notions from Rasch measurement, we intended that some items would be "easy", that is, agreeable to most teachers even from lower levels of the trait. Other items were intended to be "moderate" (endorsable by teachers at a moderate level and above) or "hard" (endorsable by teachers at a high level). We then shared the initial items with our experienced teachers, teacher-educators, and the advisory board for further review and to clarify wording. This resulted in a pool of 20 Likert-type items for piloting.

18.4 Piloting and Item Analyses, Selection, and Revision

18.4.1 Data Context and Sources

Data for pilot testing and analyses were collected from 158 teachers from across Iowa, including in-service elementary teachers from multiple districts and pre-service elementary and secondary teachers from the University of Iowa. This selection process allowed for recruiting respondents who had a wide range of prior experience with the SWH approach. The in-service teachers came from districts and schools where there had been partial coverage of a previous SWH project, so some but not all of the in-service teachers had prior experience with the SWH approach and related concepts. The pre-service teachers were in various stages of the teacher education program, and so had varied amounts of exposure to SWH or related ideas in their coursework and practicum experiences. We viewed this range of prior exposure as an important consideration in our pilot population, because we wanted to be sure that any instrument emerging from this work would be informative for teachers whether they had previously experienced the SWH approach or no prior knowledge of it. Respondents were guaranteed anonymity during recruitment for this pilot study, so we were unable to ascertain what proportions of the in-service teachers and pre-service teachers had prior direct exposure to the SWH approach. We were unable to obtain data on the respondents' identified teaching level and area (e.g., elementary, secondary science), but based on the sampling procedure we expect that the data oversamples from the elementary level. The respondents identified as 68.8% in-service teachers; 87.6% self-identified as women (NB. a gender imbalance is common among elementary teachers and teacher candidates).

18.4.2 Analysis Method

Our analyses of the pilot data seek to evaluate the structural aspect of validity, by determining whether the 5-point scaling structure works well and whether the pattern of responses to the items matches the intended item difficulty. We used Rasch measurement modeling to achieve this because its purpose is "to maximize the homogeneity of the trait and to allow greater reduction of redundancy at no sacrifice of measurement information by decreasing items and/or scoring levels to yield a more valid and simple measure" (Granger, 2008, p. 1122). We present the 20 statements in Table 18.1 on a 5-point Likert-type scale. Though all of the items use the same 5-point scale, it is important to note that the numerical difference of 1 point from responses 1 to 2 ("Strongly Disagree" to "Disagree") cannot be assumed to be equal to a 1-point difference from responses 3 to 4 ("Neither Agree nor Disagree" to "Agree"). Thus, we tested a Rasch rating scale model, which is well-suited to the present purpose given that allows for a common scale for the entire instrument (Linacre, 2000) unless there is strong evidence to prefer a partial-credit model. Given our sample size is sufficient for polytomous models (Linacre, 1994) but not overly

Table 18.1 Initial Argument Questionnaire Items for Piloting

Item ID	Statement
AQ1	Students have to know the structure of an argument: question, claims, evidence
AQ2	A good template can help students easily master an argument structure
AQ3	Students should persuade others by providing claims based on evidence
AQ4	We argue to solve one problem at a time
AQ5	Answering a question means the argument has been completed
AQ6	Better arguments have coherent links among question, claim, evidence
AQ7	Argumentation, or "Arguing to learn", should be a cycle to resolve questions
AQ8	Argument structures can be used in a cycle
AQ9	The focus of building arguments is not only on finding correct answers, but also a process to generate knowledge
AQ10	Specific forms of reasoning are needed to construct arguments in science
AQ11	Scientific argument has a unique structure different from argument in other subjects
AQ12	Students can argue with an idea, not just another person
AQ13	Students' ability to identify weakness of arguments has a role in learning of science
AQ14	New evidence might create new questions or claims that need to be argued
AQ15	Any statement can be interpreted as a claim requiring evidence
AQ16	Argument is valuable when all the participants contribute to it
AQ17	Students generate their own knowledge by arguing with other people
AQ18	Knowledge is tentative and developed by other's feedback and revision in response to one's argument
AQ19	Connectedness among questions, claims, and evidence improves students' learning of science
AQ20	The function of argumentation is not only on finding disciplinary acceptable answers, but also a process to generate knowledge

large, the rating scale model is also less likely to have unstable estimates than a partial credit model.

The rating-scale model (RSM) can be formulated as follows: the probability of teacher n student selecting rating level k on item i can be estimated based on the following equation:

$$\ln\left[\frac{P_{n_i(x=k)}}{P_{n_i(x=k-1)}}\right] = \theta_n - (\delta_i + \tau_k)$$

where θ_n is the nth student's latent ability, δ_i is the item difficulty of item i, and τ_k is the kth threshold location of the rating scale. This statistical model does not include other parameters for guessing or attenuated item discrimination—as such, the Rasch measurement model upholds a strict interpretation of "good measurement" based on how test-takers respond to items. For our analyses, we fit our data to RSM using the TAM package (Robitzsch, Kiefer, & Wu, 2021) for the R statistical environment (Ihaka & Gentleman, 1996). The scale is fixed so that the average θ of all teachers is zero. We also wanted to control for potential differential item functioning (DIF), which occurs "when equally able test takers differ in their probabilities of answering a test item correctly as a function of group membership" (Joint Committee, 2014, p. 51). Prior research has considered the potential effects of self-reported gender in how individuals respond to items despite having the same underlying level of the measured trait. To make this possible at the scale level and the item level, we used the model formula approach in the TAM package. As we show in the R script in the Appendix, the basic RSM is written as the formula ~item+step, and for estimating DIF we use a faceted approach with the formula ~item+step+{facet} + item*{facet}.

Lastly, since 20 items may be too burdensome in practical settings, especially if this instrument were used in conjunction with other measures (as is the case in this study context; Fulmer et al., 2021), we sought to identify the most informative items to retain or revise for future use and drop any items that were not productive for measurement. We aimed to select items that sample from the domain appropriately, that cover a broad range of item difficulty, and that function well according to the strict definition of "good measurement" consistent with the Rasch measurement model. We consider an item to be informative for measurement when the responses are not redundant with other items, nor too unpredictable. This decision is based on the mean-square fit statistic: a value near 1 is considered "just right," and the window for acceptable mean-square fit statistics is between 0.6 and 1.4 when developing rating scales (Boone, Staver, & Yale, 2013; Wright, 1996). There are two mean-square fit statistics to review: infit is "inlier-sensitive" or information-weighted, which prioritizes respondents near the center of the distribution; outfit is outlier-sensitive and prioritizes respondents near the upper and lower ends of the distribution (Bond & Fox, 2007; Boone, Staver, & Yale, 2013). Items with mean-squares below 0.6 are too redundant; items with mean-squares above 1.4 are too unpredictable. We could also examine t-values for infit and outfit, but these can be sensitive to sample size so may be misleading with samples of 150 or greater.

18.4.3 *Findings*

Initial Item Estimation and Fit

Before conducting our analyses we reviewed the response patterns. Table 18.2 shows the frequencies of response options selected by participants. Some response options are rarely or never selected for given items. With this pattern, a partial credit model would be difficult to estimate, but across the entire set of items all response levels have sufficient numbers to estimate a rating scale that would be common use for all items (Linacre, 2000).

We therefore estimated a Rasch rating scale model, with item difficulty estimates and fit statistics shown in Table 18.3. We discuss fit first, then move on to difficulty and other issues. Findings showed there was overall acceptable fit of the items to the Rasch model: most items have mean-square fit statistics in the acceptable range. Just three items had mean-square fit statistics above 1.4, indicating unpredictable responses outside the acceptable range, and none of the items had fit statistics below the lower cutoff of 0.6. Threshold fit statistics are high, though, indicating that responses are not as predictable as desirable under the Rasch model. Despite this, the initial items showed a reliability coefficient of .672. This is somewhat low but acceptable for rating-scale instruments for research purposes, and we consider

Table 18.2 Initial item piloting response frequencies

	Response				
Item	1	2	3	4	5
AQ1	62	99	15	9	0
AQ2	24	106	39	15	1
AQ3	0	2	10	86	87
AQ4	15	61	38	64	6
AQ5	5	32	25	77	46
AQ6	0	1	19	94	69
AQ7	0	7	22	118	37
AQ8	0	1	41	110	31
AQ9	0	3	13	87	80
AQ10	38	101	33	11	1
AQ11	2	34	40	83	23
AQ12	0	1	7	49	125
AQ13	0	4	21	93	64
AQ14	0	0	6	58	118
AQ15	1	15	39	82	44
AQ16	0	9	14	81	78
AQ17	0	11	21	92	58
AQ18	0	2	20	106	54
AQ19	0	0	11	81	90
AQ20	0	3	15	92	72

Table 18.3 RSM item and threshold estimates with item fit statistics

Item	δ_i	MNSQ outfit	MNSQ infit
AQ01R	−1.756	1.087	1.043
AQ02R	−0.970	0.991	0.971
AQ03	−2.365	0.890	0.898
AQ04R	−0.041	**1.564**	**1.480**
AQ05R	1.005	**1.916**	**1.895**
AQ06	−2.011	0.760	0.737
AQ07	−1.450	0.709	0.689
AQ08	−1.312	0.584	0.538
AQ09	−2.193	0.921	0.918
AQ10R	−1.226	1.009	0.957
AQ11	−0.606	**1.592**	1.397
AQ12	−3.158	0.991	1.109
AQ13	−1.843	0.939	0.901
AQ14	−3.073	0.775	0.870
AQ15	−1.146	1.283	1.185
AQ16	−1.994	1.202	1.247
AQ17	−1.609	0.961	0.995
AQ18	−1.788	0.803	0.764
AQ19	−2.480	0.770	0.778
AQ20	−2.062	1.001	0.936
Threshold	#xD835#xDF0F$_k$	MNSQ outfit	MNSQ infit
Step1	−2.179	**3.233**	**3.209**
Step2	−0.113	**3.793**	**3.804**
Step3	−0.110	**3.538**	**3.550**
Step4	2.402	**3.772**	**3.804**

Note. Boldface indicates mean-square fit values exceeding 1.4. Item numbers ending in R are reverse-coded

this a promising start for a construct as complex as we are studying. We found a person separation reliability of .852, which is also in the acceptable range.

Regarding the items' difficulty, we see that the levels are low, with an average of −1.604, and a range from −3.158 to 1.005. Here, negative values indicate that an item is easy to agree with or endorse. Figure 18.1 shows a Wright map with the respective thresholds for the items. As can be seen in the figure and tables, the initial set of items includes many that are very easy to endorse, and while they are not redundant from the perspective of item fit statistics, there are several items located near each other on the relatively lower end of the logit scale. Additionally, looking at the thresholds, or steps, we see that difficulty estimates for two of the thresholds are too near each other and may be almost indistinguishable: −.113 for step 2 (the threshold from "2″ to "3″ on the original scale) and − .110 for step 3 (the threshold from "3″ to "4″ on the original scale). This is concomitant with high mean-square fit statistics for all of the steps—having steps that are too near each other will inhibit the capacity to distinguish individuals in a reliable way as intended under the Rasch approach (Table 18.3 and Fig. 18.2).

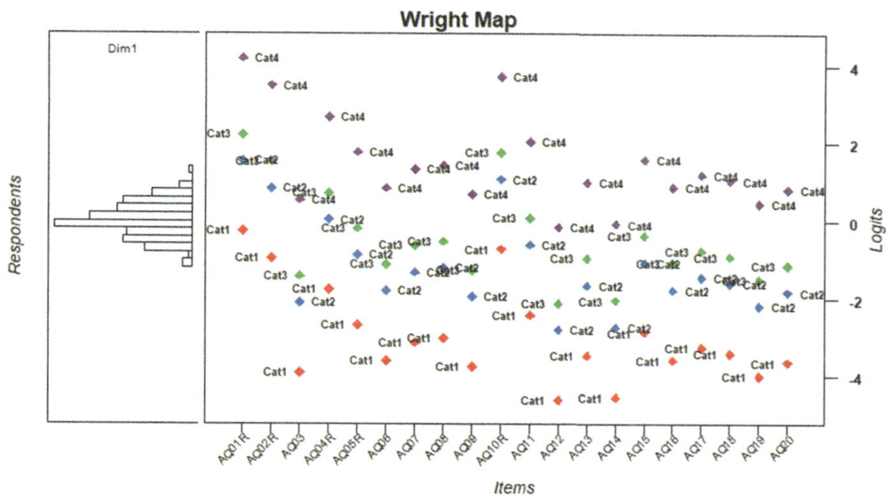

Fig. 18.2 Wright map for original 20-item questionnaire & responses
Note. The left side is a distribution of persons' estimated level of the trait. The right side indicates the thresholds of the respective items

18.4.4 Differential Item Functioning

We also examined issues of Differential Item Functioning (DIF) according to two potential sources of variation: self-reported gender, and in-service/pre-service status. We fit faceted Rasch models that included the same rating-scale model and both a facet and an item*facet interaction (i.e., item + step + gender + item*gender). We found that there was a modest DIF effect for gender. Figure 18.3 show this graphically. As the figure shows, most items are relatively close to the reference line, which indicates zero DIF, along the entire range of difficulty, with a few exceptions. Numerically, the difference in item difficulty due to gender averaged .105 logits (ranging from .001 to .276, with 4 having significant differences at the $p < .05$ level). For example, item AQ14, "New evidence might create new questions or claims that need to be argued", is easier for self-identified women to endorse.

18.4.5 Item Selections and Revisions

As mentioned, for the purpose of the larger project we needed to reduce the number of items to lessen the overall response burden for our future uses of the instrument. That also included a need to have room for potential growth in teachers' knowledge of argument as an epistemic tool, meaning that we anticipated needing to include some items that were relatively easier to endorse but also to identify and retain items that were somewhat harder to endorse so that we would be able to estimate and

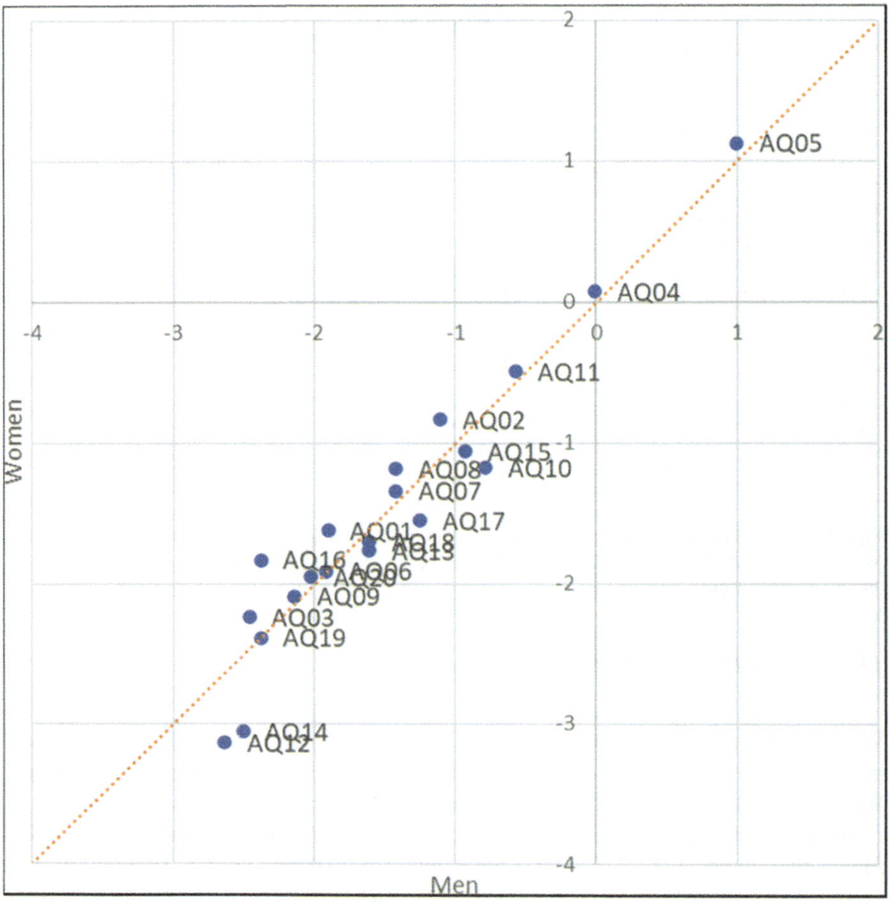

Fig. 18.3 Plot of item estimates to test for differential item functioning (DIF) due to gender
Note. The central line is a reference line for isometry: items near this line have no DIF

potentially differentiate teachers at the upper end of the scale. To select items to retain for a shorter instrument, we examined the raw response patterns, the infit and outfit mean-square fit statistics, the items' difficulty estimates, findings from the domain analysis, and input from expert reviewers.

As Table 18.3 shows, three items had fit statistics that were too high, indicating that responses were more unpredictable than we usually accept for good measurement. At the same time, reviewing Table 18.2 showed us that some items had few or no participants endorse one or more of the "extreme" ends of the scale (i.e., "Strongly Disagree" or "Strongly Agree"). So, based on the pilot data, we wanted to retain some items where it appeared the full response scale was being used, such as items AQ04 and AQ05. At the same time, some items had similar content and similar level of difficulty per the Rasch item estimates. In feedback from our expert reviewers, they pointed out that some of these low-difficulty items address ideas that are exceedingly easy to endorse, to the extent that they are essentially apparent *prima facie* to be socially desirable. So, some items that had acceptable fit to the Rasch model were also argued by our expert reviewers to be less informative—particularly items with very low mean-square fit statistics, which are indicative of items that could be redundant. This focused on six items that were either very easy to endorse, had low mean-square fit statistics, or both, ordered from lowest to highest difficulty:

- 14. New evidence might create new questions or claims that need to be argued. [$\delta = -3.073$]
- 19. Connectedness among questions, claims, and evidence improves students' learning of science. [$\delta = -2.480$].
- 3. Students should persuade others by providing claims based on evidence. [$\delta = -2.365$]
- 9. The focus of building arguments is not only on finding correct answers, but also a process to generate knowledge. [$\delta = -2.193$]
- 20. The function of argumentation is not only on finding disciplinary acceptable answers, but also a process to generate knowledge. [$\delta = -2.062$]
- 16. Argument is valuable when all the participants contribute to it. [$\delta = -1.994$]

Expert reviewers pointed out that some of these ideas were addressed with more informative items, such as items 15 (Any statement can be interpreted as a claim requiring evidence; $\delta = -1.146$) and 17 (Students generate their own knowledge by arguing with other people; $\delta = -1.609$).

Three items had mean-square fit statistics that were too high, indicating that the response might be too unpredictable for useful measurement; AQ04, AQ05, and AQ11. We noted that these items had relatively higher item difficulty statistics compared to the others—that is, teachers found them more difficult to endorse (or, for the negatively worded items, harder to disagree). We considered deleting these items, but discussions with our expert reviewers underscored that the fact that these items' relative difficulty may contribute to their low fit but also make them useful for potential longitudinal applications of the questionnaire. That is, if we hope to use the questionnaire to gauge future growth on the construct, we would need items of higher difficulty to be sensitive to teachers' higher ability later in the larger study. For this reason, we decided to retain these items despite the concerns about fit. This left us with 14 items, shown in Table 18.4, that draws on a mix of all three sub-domains of argument structure, argument logic, and argument as persuasion. We examined all the items and made minor adjustments to the language to increase clarity for 5 of them (items AQ01, AQ02, AQ04, AQ07, & AQ08, marked with an asterisk in Table 18.4). We have adopted these items for application with our summer teacher professional development workshops, and aim to report further on the functionality and interpretation of these instruments as the project proceeds to completion.

Table 18.4 Selected and Revised items of the Questionnaire on Teachers' Knowledge of Argument as an Epistemic Tool

ID [1]	Revised Statement	Sub-Domain[2]
AQ01R*	Students have to memorize the structure of an argument (Question- > Claim- > Evidence) before they engage in argumentation	S
AQ02R*	A structured template for arguments can help students easily understand argumentation (Arguing to Learn)	S
AQ04R*	We argue to generate one correct answer at a time	P
AQ05R	Answering a question means the argument has been completed	P
AQ06	Better arguments have coherent links among question, claim, evidence	S
AQ07*	Argumentation can be used in an iterative cycle to resolve a question	S
AQ08R*	Argument structures should be used in the designated order: Question- > Claim- > Evidence	S
AQ10	Specific forms of reasoning are needed to construct arguments in science	L
AQ11	Scientific argument has a unique structure different from argument in other subjects	L
AQ12	Students can argue with an idea, not just another person	L
AQ13	Students' ability to identify weakness of arguments has a role in their learning of science	L
AQ15	Any statement can be interpreted as a claim requiring evidence	S
AQ17	Students generate their own knowledge by arguing with other people	L
AQ18	Knowledge is tentative and developed with other's feedback and revision in response to one's argument	P

Note. * Indicates this item was revised after piloting. [1] R indicates items that should be reversed before scoring. [2] S = Structure; L = Logic; P = Persuasion

18.5 Discussion

We aimed to develop and refine an instrument for assessing teachers' knowledge of argument as an epistemic tool, as such an instrument would allow researchers to gauge the level of and potential changes in teachers' knowledge of argument as an epistemic tool. Our construct-based approach involved analyzing the domain to understand how teachers' varying level of knowledge could be characterized, drafting items for piloting, then analyzing and selecting the items to identify those most appropriate for use. In this sense, the current chapter's reporting on a specific instrument fits within the broader push in the science education community toward more explicit attention to the epistemic basis of science knowledge and science learning under the NGSS era.

Our findings show that the initial instrument has moderate separation reliability, suitable for rating scales for research purposes. We consider this an important first start for this line of research. Prior research has shown mixed results of attempts to measure pedagogical content knowledge—it is a very complex body of knowledge, tied to the specific topic, population, and context (Gess-Newsome, 2015; Suh & Park, 2017). It has often been studied with in-depth qualitative work (e.g., Mazibe,

Coetzee, & Gaigher, 2020; Park & Oliver, 2008). For this reason, while there is certainly room for improvement we also argue that an instrument of this form has potential for impact in the field.

Because one of our main purposes with this overarching NSF-funded project is to study and model teachers' growth over time on the identified knowledge of epistemic tools (Fulmer et al., 2021; Lammert, Suh, Hand, & Fulmer, 2022), we intentionally included items that were relatively harder, with fewer teachers in our sample endorsing the "strongly agree" levels, because we anticipated that this could allow us to study potential increases in the trait. That decision draws on the unique perspective of Rasch measurement, which differs from classical test theory precisely in recognizing that items of varying difficulty are essential to distinguish respondents across a range of the trait.

Our findings show that there was modest differential item functioning, but in these cases it did not drastically alter the relative difficulty of the items overall. This is promising because it is conceptually possible for pedagogical knowledge and pedagogical content knowledge to differ dramatically between pre-service and in-service teachers—such as due to the roles of reflection-on-action (Park & Oliver, 2008)—or among teachers who identify as women vs as men. Yet for the purpose of our argumentation questionnaire, this was not a major issue.

One area for concern and future work is the aforementioned quality of the rating scale steps. We found that two steps were too near in estimated difficulty to allow for good separation, and fit statistics for the steps are high enough that responses are less predictable than expected under the Rasch measurement model. This could be a limitation of the scale structure itself, it could be indicative of differences in how respondents interpret the notion of "argument", or an interaction thereof. For instance, if some respondents have negative connotations for the word "argument"then they may not respond in the same ways as respondents who do not have this stance—but as yet we do not have a known, observable variable that we could use to estimate the potential differential effects of this. Given the complexity of the constructs of pedagogical knowledge and pedagogical content knowledge, and the intersections of these constructs with other social and cultural factors, we see a need for more work to improve our measurement capability.

18.5.1 Limitations and Opportunities for Further Research

While we argue that our application of a construct-driven instrument development approach provides a strong basis for claims for the potential validity of the proposed instrument, we acknowledge there are areas for further improvement or follow-up study. In particular, as we note above, we report here on just two validity aspects identified by Messick (1995): Content and Structure. There are four additional aspects of validity, namely Substance, External Criteria, Consequence, and Generalizability. For an instrument of the present type, we may focus less on the more broad-focused issues of consequence and generalizability because they are, for the time being, relatively more low stakes. For example, there is little reason at this stage of development to anticipate there would be value implications of score interpretation that could create opportunities for bias. Yet on the aspects of substance and external criteria, there is a need for further research to help instantiate the instrument as used.

To address the need for further evidence of validity with respect to substance and criteria, we suggest two opportunities for further work. For substance, which addresses the extent to which a response matches what participants have "in their heads''so to speak, one option is to engage in clinical interviews with teachers who respond to the instrument, and use these interviews to understand how the interpretations indicated by the instrument response may match or depart from what teachers describe in their own words. For external criteria, which addresses the extent to which the interpretations match an established behavioral outcome or alternative measure of interest, we have to think more expansively. Unlike some standardized assessment settings (e.g., comparing SAT scores to future academic performance), there exists neither a widely-adopted external criterion for the particular topic of teachers' knowledge of argument as an epistemic tool nor a generalized criterion on teachers' pedagogical content knowledge. Even so, it would be possible to compare instrument responses with an alternative behavioral outcome, such as by triangulating teachers' responses on the instrument with observations of classroom practice by trained outside observers.

On both fronts we are fortunate to have funding from the current NSF award to allow us to gather this data—data collection using interviews with participating teachers and observations of their classrooms can allow us to study these aspects in additional depth. However, that data collection with our project-supported teachers is ongoing, and furthermore is a different sample from the piloting sample reported here. We anticipate being able to describe evidence of substance and external criteria in a future publication.

An additional line of work that is emerging from this instrument development is the extent to which the proposed instrument is amenable to a variety of interpretations and applications of the notion of "argument." The professional development model for the research team uses the Science Writing Heuristic (SWH) approach to argument-driven inquiry. Yet during the instrument developing phases such as domain analysis and expert review, we intentionally incorporated literature and expert input both related to SWH and from other scholars. It is our goal that the tool would therefore be applicable to scholars unaffiliated with the SWH approach. Still, it is an empirical question as to how well the instrument is applicable, and how well it measures possible changes in teachers' knowledge over time. Addressing this will require disseminating the instrument and its scoring with a wide audience, and soliciting participating projects that could allow for comparing findings across settings and across different professional development models.

Acknowledgement This material is based upon work supported by the National Science Foundation under Grant No. 1812576. Any opinions, findings, and conclusions or recommendations expressed in this material are those of the authors and do not necessarily reflect the views of the National Science Foundation.

Appendix

```
################################################################################
### This is a script for Rasch analysis in the R environment
### developed by Gavin Fulmer with input from Jihyun Hwang and Andrea Ash.
### Please cite this chapter if you adapt this script.
################################################################################

################################################################################
###  Preparations
################################################################################

### Load required packages for use in the R environment
library(readr)
library(TAM)
library(tidyr)
library(pastecs)
library(WrightMap)
library(RColorBrewer)

### set a working directory
### this will be a default for finding data files and saving outputs if used
setwd("C:/ ... /Documents")

### read in the data set
data1 <- read.csv("Arg data clean inverse cut.csv", na.strings = c("NA","NaN",""))

### preview the data structure
str(data1)

### sub-set the items from other variables
item_data<-data1[,23:42]
```

```
### create facets for later use with DIF
gender<-data1$Female                    #gender factor
inserve<-data1$Inserv             #inservice status factor
facet.2 <- as.data.frame(gender)
facet.3 <- as.data.frame(inserve)


##################################################################################
### Estimating the Rasch model without facets
##################################################################################

### define the rating scale formula
formRSM <- ~item+step

### estimate the model and print a summary
mod1 <- TAM::tam.mml.mfr( resp=item_data, formulaA = formRSM,
        verbose = FALSE )
summary(mod1.gen)

### Obtain the fit indices and print a summary
fit1 <- tam.fit(mod1)
summary(fit1)

### Prepare a figure for the Wright map
ncat <- 4          # number of threshold parameters
I <- ncol(mod1$resp)   # number of items
itemlevelcolors <- matrix(rep( RColorBrewer::brewer.pal(ncat, "Set1"), I),
             byrow = TRUE, ncol = ncat) # set a color pattern
```

```
### Create the Wright map
IRT.WrightMap(mod1 , prob.lvl=.625 ,
        label.items.srt = 45, thr.sym.col.fg = itemlevelcolors,
        thr.sym.col.bg = itemlevelcolors ,
        label.items = colnames( mod1$resp) )


##############################################################################
### Estimating the Rasch model with facets for DIF
##############################################################################

###################
### DIF for gender
###################

### create the rating scale formula with
###     main effect and DIF effect for gender
DIF.gen.form2 <- ~item+step+gender+item*gender

### estimate the model and print a summary
mod1.dif.gen <- tam.mml.mfr( resp= item_data, facets= facet.2 ,
          formulaA = DIF.gen.form2, verbose = "FALSE")
summary(mod1.dif.gen)

### save the estimated parameters
write.csv(mod1.dif.gen$xsi.facets,"mod1 DIF gen results.csv")

### estimate fit indices and print a summary

fit2.dif.gen<-tam.fit(mod1.dif.gen)
summary(fit2.dif.gen)
```

References

Allchin, D., & Zemplén, G. Á. (2020). Finding the place of argumentation in science education: Epistemics and Whole Science. *Science Education, 104*(5), 907–933.

Bond, T. & Fox, C. (2007). *Applying the Rasch model: Fundamental measurement in the human sciences (2nd)*. LEA.

Boon, M., & Van Baalen, S. (2019). Epistemology for interdisciplinary research: Shifting philosophical paradigms of science. *European Journal for Philosophy of Science, 9*, 16.

Boone, W. J., Staver, J. R., & Yale, M. S. (2013). *Rasch analysis in the human sciences*. Springer.

Burke, K. A., Hand, B., Poock, J., & Greenbowe, T. (2005). Using the Science Writing Heuristic. *Journal of College Science Teaching., 35*(1), 36–58.

Cavagnetto, A. R. (2010). Argument to foster scientific literacy: a review of argument interventions in science contexts. *Review of Educational Research, 80*(3), 336–371.

Choi, A., Notebaert, A., Diaz, J., & Hand, B. (2010). Examining arguments generated by year 5, 7, and 10 students in science classrooms. *Research in Science Education, 40*(2), 149–169.

Erduran, S., Ozdem, Y., & Park, J. Y. (2015). Research trends on argumentation in science education: a journal content analysis from 1998–2014. *International Journal of STEM Education, 2*(1), 1–12.

Fulmer, G. W., Hwang, J., Ding, C., Hand, B., Suh, J. K., & Hansen, W. (2021). Development of a questionnaire on teachers' knowledge of language as an epistemic tool. *Journal of Research in Science Teaching, 58*(4), 459–490.

Gess-Newsome, J. (2015). *A model of teacher professional knowledge and skill including PCK: results of the thinking from the PCK summit. Re-examining pedagogical content knowledge in science education* (pp. 28–42). Routledge Press.

Granger, C. (2008). Rasch analysis is important to understand and use for measurement. *Rasch Measurement Transactions, 21*(3), 1122–1123.

Hand, B., Nam, J., & Choi, A. (2012). Argument-Based General Chemistry Laboratory Investigations for Pre-Service Science Teachers. *Educación Química, 23*, 96–100.

Heit, E., & Rotello, C. M. (2010). Relations between inductive reasoning and deductive reasoning. *Journal of Experimental Psychology: Learning, Memory, and Cognition, 36*(3), 805–812.

Henderson, J. B., McNeill, K. L., González-Howard, M., Close, K., & Evans, M. (2018). Key challenges and future directions for educational research on scientific argumentation. *Journal of Research in Science Teaching, 55*(1), 5–18.

Ihaka, R., & Gentleman, R. (1996). R: A language for data analysis and graphics. *Journal of Computational and Graphical Statistics, 5*(3), 299–314.

Jimenez-Aleixandre, M. P. & Erduran, S. (2007). Argumentation in science education: an overview. Argumentation in science education: perspectives from classroom-based research. Dordrecht, Netherlands: Springer. 3–35.

Kane, M. T. (2001). Current concerns in validity theory. *Journal of Educational Measurement, 38*(4), 319–342.

Keys, C. W., Hand, B., Prain, V., & Collins, S. (1999). Using the Science Writing Heuristic as a tool for learning from laboratory investigations in secondary science. *Journal of Research in Science Teaching, 36*(10), 1065–1084.

Konstantinidou, A., & Macagno, F. (2013). Understanding Students' Reasoning: Argumentation Schemes as an Interpretation Method in Science Education. *Science & Education, 22*, 1069–1087. https://doi.org/10.1007/s11191-012-9564-3

Krajcik, J., McNeill, K. L., & Reiser, B. J. (2008). Learning-goals-driven design model: Developing curriculum materials that align with national standards and incorporate project-based pedagogy. *Science Education, 92*(1), 1–32.

Lammert, C., Suh, J. K., Hand, B., & Fulmer, G. (2022). Is epistemic orientation the chicken or the egg in professional development for knowledge generation approaches? *Teaching and Teacher Education, 116*, 103747.

Linacre, J. M. (1994). Sample Size and Item Calibration Stability. *Rasch Measurement Transactions, 7*(4), 328. https://www.rasch.org/rmt/rmt74m.htm

Linacre, J. M. (2000). Partial Credit Models (PCM) and Rating Scale Models (RSM). *Rasch Measurement Transactions, 14*(3), 768. https://www.rasch.org/rmt/rmt143k.htm

Mathayas, N., Brown, D. E., Wallon, R. C., & Lindgren, R. (2019). Representational gesturing as an epistemic tool for the development of mechanistic explanatory models. *Science Education, 103*(4), 1047–1079.

Mazibe, E. N., Coetzee, C., & Gaigher, E. (2020). A comparison between reported and enacted pedagogical content knowledge (PCK) about graphs of motion. *Research in Science Education, 50*(3), 941–964.

Messick, S. (1995). Validity of psychological assessment: Validation of inferences from Persons' responses and performances as scientific inquiry into score meaning. *American Psychologist, 50*(9), 741–749.

National Research Council [NRC]. (2012). *A Framework for K-12 Science Education: Practices, Crosscutting Concepts, and Core Ideas*. The National Academies Press.

Nielsen, J. A. (2013). Dialectical features of students' argumentation: A critical review of argumentation studies in science education. *Research in Science Education, 43*(1), 371–393.

Oh, P. S. (2017). The roles and importance of critical evidence (CE) and critical resource models (CRMs) in abductive reasoning for earth scientific problem solving. *Journal of Science Education, 41*(3), 426–446.

Osborne, J., Rafanelli, S., & Kind, P. (2018). Toward a more coherent model for science education than the crosscutting concepts of the next generation science standards: The affordances of styles of reasoning. *Journal of Research in Science Teaching, 55*(7), 962–981.

Park, S., & Oliver, J. S. (2008). Revisiting the conceptualisation of pedagogical content knowledge (PCK): PCK as a conceptual tool to understand teachers as professionals. *Research in Science Education, 38*(3), 261–284.

Rex, L. A., Thomas, E. E., & Engel, S. (2010). Applying Toulmin: teaching and logical reasoning in argumentative writing. *The English Journal, 99*(6), 56–62.

Robitzsch, A., Kiefer, T., & Wu, M. (2021). Package 'TAM'. https://cran.r-project.org/web/packages/TAM/TAM.pdf

Sampson, V., Enderle, P., & Grooms, J. (2013). Argumentation in science education. *The Science Teacher, 80*(5), 30.

Schoerning, E., Hand, B., Shelley, M., & Therrien, W. (2015). Language, Access, and Power in the Elementary Science Classroom. *Science Education, 99*(2), 238–259.

Suh, J. K., & Park, S. (2017). Exploring the relationship between pedagogical content knowledge (PCK) and sustainability of an innovative science teaching approach. *Teaching and Teacher Education, 64*(1), 246–259.

Taber, K. S. (2017). Reflecting the nature of science in science education. In K. S. Taber & B. Akpan (Eds.), *Science Education* (pp. 21–37). Sense Publishers.

Toulmin, S. (2003). *The uses of argument* (updated ed.). Cambridge University Press. pp. 12–46.

Wright, B. D. (1996). Comparing Rasch measurement and factor analysis. *Structural Equation Modeling: A Multidisciplinary Journal, 3*(1), 3–24.

Yoon, H. G., Joung, Y. J., & Kim, M. (2012). The challenges of science inquiry teaching for pre-service teachers in elementary classrooms: Difficulties on and under the scene. *Research in Science Education, 42*(3), 589–608.

Gavin W. Fulmer is an Associate Professor of Science Education at the University of Iowa. His research addresses the development, applications, and implications of assessment in science education.

William E. Hansen is an Assistant Professor of Biology and Science Education at Iowa Wesleyan University. His research includes collaboratively developing instruments to evaluate immersive environments in education.

Jihyun Hwang is an Assistant Professor of mathematics education at South Korea National University of Education. His research topics are epistemic actions, epistemological belief in mathematics, and attitude toward mathematics in relation to mathematics achievement.

Chenchen Ding is a doctoral student in Science Education at the University of Iowa. Her research interests are in the development of assessments and learning environments in science education.

Andrea Malek Ash is a Research Consultant at Gallup where she specializes in quantitative methodology. Her research interests include measurement, evaluation, and the role of student agency in learning.

Brian Hand is a Professor of Science Education at the University of Iowa. His research addresses the roles of language in science learning, including the Science Writing Heuristic.

Jee Kyung Suh is an Assistant Professor of Science Education at the University of Alabama. Her research examines teachers' epistemic orientation and its role in their classroom learning contexts.

Epilogue

Xiufeng Liu and William J. Boone

Xiufeng Liu is a Professor of Science Education at University at Buffalo, State University of New York, and a Fellow of the American Association for the Advancement of Science (AAAS). Liu conducts research on applications of Rasch measurement in STEM education such as student long-term development of cross-cutting concepts on matter and energy.

William Boone is a Professor of Educational Psychology (Program in Learning Sciences and Human Development) at Miami University, Oxford, Ohio, USA. Boone received his Ph.D from the University of Chicago's Measurement, Evaluation and Statistical Analysis Program (Education) with Benjamin Wright as his thesis director. His research involves applications of Rasch measurement in the fields of science teacher education, biology education, physics education, chemistry education, and geology education.

Over the last 15+ years since the publication of our previous volume (Liu & Boone, 2006) many advances have taken place in applications of Rasch measurement within the field of science education. As we compared the chapters of our previously edited volume to this volume we see continued interest in applying Rasch measurement within the field of science education by scholars in the USA and Germany. We were pleased to now see researchers in China, Canada, and Philippines using Rasch measurement to develop instruments and to study learning progression (see Chaps 9, 11 and 14 of this book). Also, we know firsthand that Rasch measurement has been actively applied in a few other countries, particularly Australia, England, and Eastern European countries (please see Liu, 2020), although we don't have chapters from authors of those countries. The international scope of

X. Liu
Department of Learning & Instruction, Graduate School of Education, University at Buffalo, Buffalo, NY, USA
e-mail: xliu5@buffalo.edu

W. J. Boone
Department of Educational Psychology, Program in Learning Sciences and Human Development, Miami University, Oxford, OH, USA

X. Liu, W. J. Boone (eds.), *Advances in Applications of Rasch Measurement in Science Education*, Contemporary Trends and Issues in Science Education 57,
https://doi.org/10.1007/978-3-031-28776-3

applications of Rasch measurement documented in this volume indicates that Rasch measurement is being utilized throughout the world and there is promise for its continued expansion of use.

Not only has the geographical regions of Rasch applications in science education expanded, but also the domains of applications have broadened. In the 2006 volume (Liu & Boone, 2006), applications generally concerned introducing basic concepts of Rasch measurement and applying Rasch measurement to develop items and to validate measures of assessing science conceptual understanding and affective variables. In this volume, in addition to measurement of conceptual understanding of chemistry (Chap. 14), we see applications of Rasch models applied to the topics of measurement of scientific argumentation (Chaps. 7 and 16), scientific reasoning (Chap. 9), three-dimensional learning, knowledge-in-use and cross-cutting concepts of the Next Generation Science Standards (Chaps. 10, 11 and 13), medical education learning experiences (Chap. 12), machine-scoring bias (Chap. 15), formative assessment (Chap. 17), and teacher knowledge of argument (Chap. 18). We also see advances in Rasch analysis techniques and technology such as R (Chap. 3), Bayesian estimation (Chap. 4), comparison between joint maximum likelihood (JML) and marginal maximum likelihood (MML) estimations on model-data-fit (Chap. 8); we further see advances in enhancement to Rasch models by Cognitive Diagnostic Models (Chap. 5) and Latent Class Analysis (Chap. 6). In addition to continued uses of such common Rasch models as the dichotomous Rasch model, Rating Scale Rasch model, Partial-Credit Rasch model, and multidimensional Rasch models, in this volume, we also see applications of Many-Facet Rasch model (Chaps. 15 and 16) and Linear Logistic Rasch Model (Chap. 17). Chapters 1 and 2 touch upon the fact that there are still confusions, inconsistencies, and even inappropriate uses of Rasch measurement in science education. Achieving rigor in the application of Rasch measurement in science education remains an important goal to strive toward.

As readers review the applications of Rasch measurement in the 18 chapters of this book, we hope that readers will better appreciate and comprehend the diverse Rasch techniques that can be used in analyses (science education as well as other fields of course). For us, we hope a foundational takeaway for readers will be that when one conducts a Rasch analysis one is carrying out highly rigorous measurement. And that this measurement is theory driven. In each and every chapter of this book, the chapter authors demonstrate how to improve the measurement they conduct. With a specific science education data set not all of the Rasch techniques of this book will be applicable. But by using this book's chapters readers can continue to improve the quality of measurement taking place in their own science education research, and of course readers can influence other science educators. Ben Wright in his classes at the University of Chicago would often talk about high quality measurement devices developed by scientists, and he was appalled that there was not the same commitment to rigorous measurement in social science research. This book adds to the resources which are available to science educators (as well as others conducting research) so that (1) high quality instrumentation can be developed, and (2) the measurement functioning of those instruments can be monitored and clearly documented. It seems to us that science educators should find it easy to embrace Rasch measurement when developing and using measurement devices.

There are of course added broad hopes that we have as the result of putting together this book. Our second wish (dream) is that in not-so-distant future it will be commonplace that when a science education test data set or a science education survey data set is analyzed that Rasch theory and techniques are used to prepare/evaluate measures and that Rasch item and/or person measures are used for statistical analyses (e.g., comparing group differences).

We would like to close with one more hope of ours. It is of great importance to explain Rasch techniques succinctly and clearly in tables, figures, and words (be those words spoken or written). We encourage readers to look through the text, tables, and figures provided by chapter authors of this book. We know that each author has thought long and hard about such critical components of their chapter. With any technique it is absolutely critical to explain what one is doing and why. No article using Rasch is perfect; no Rasch chapter is the final word, but chapters such as these in this book can be used by all science educators to guide the explanation of how and why Rasch was utilized in a study.

This book is not the final word or the final lesson for anyone learning and appreciating how to use Rasch methods and Rasch theory. *Advances in Applications of Rasch Measurement in Science Education* provides a snapshot of how Rasch measurement has been applied in science education research, and an additional road map for those utilizing Rasch to guide the development, and use, of rigorous theory driven science education instrumentation

References

Liu, X. (2020). *Using and developing measurement instruments in science education: A Rasch Modeling approach* (2nd ed.). Information Age Publishing.

Liu, X., & Boone, W. J. (Eds.). (2006). *Applications of Rasch measurement in science education*. JAM Press.

Name Index

X. Liu, W. J. Boone (eds.), *Advances in Applications of Rasch Measurement in Science Education*, Contemporary Trends and Issues in Science Education 57,
https://doi.org/10.1007/978-3-031-28776-3

U

V

W

X

Y

Subject Index

X. Liu, W. J. Boone (eds.), *Advances in Applications of Rasch Measurement in Science Education*, Contemporary Trends and Issues in Science Education 57,
https://doi.org/10.1007/978-3-031-28776-3

D

T